Encyclopaedia

of

ORDINARY DIFFERENTIAL EQUATION

Encyclopaedia
of

ORDINARY DIFFERENTIAL EQUATION

Volume 1

Dr. Rakesh Kumar Pandey

ANMOL PUBLICATIONS PVT. LTD.
NEW DELHI - 110 002 (INDIA)

ANMOL PUBLICATIONS PVT. LTD.
Regd. Office: 4360/4, Ansari Road, Daryaganj,
New Delhi-110 002 (India)
Ph.: 23278000, 23261597
Branch Office: No. 1015, Ist Main Road, BSK IIIrd Stage
IIIrd Phase, IIIrd Block,
Bangalore-560 085 (India)
Tel.: 080-41723429
Visit us at: www.anmolpublications.com

Encyclopaedia of Ordinary Differential Equation

First Edition, 2009
ISBN 978-81-261-4106-7 (Set)

PRINTED IN INDIA

Printed at Mehra Offset Press, Delhi.

Contents

Preface

The Encyclopaedia of Ordinary Differential Equation is intended both as an introductory text and as a reference book for those interested in studying Ordinary Differential Equation. This encyclopaedia gives an up-to-date account of the theories for the *Equations and their applications*. All the chapters are beautifully illuminated towards the approach found in these volumes. The Encyclopaedia starts with Introduction to First Order Equation making it suitable for those with little background in mathematics. Both the volumes are good size covering all the essential topics is in adequate depth. It covers topics similar to other books targeted at the same audience.

Written in a clear, precise and readable manner, this encyclopaedia is designed to provide undergraduate mathematics students with a sound and inspiring introduction to the main themes of Ordinary Differential Equations. The encyclopaedia explains the origins of various types of differential equations. Gives in detail suitable method of solution to different situation. This encyclopaedia is designed to cover the basics thoroughly and then move on. Each page of these books is packed with direction, examples, and a world of understanding. This encyclopaedia is an excellent support and resource for the teacher. This is an outstanding resource for anyone interested on knowing more about the art of teaching.

This Encyclopaedia is well written and easy to follow chapters on many different areas. The layout and style are cleverly designed to keep students working through the chapters. As a reminder of useful techniques, these volumes are definitely valuable. The scope of this encyclopaedia is limited to linear differential equations of the first order, linear differential equation of higher order, differential equations of higher degree and special methods of solution of differential equations of second order, keeping in view the requirement of undergraduate students.

Author

List of Symbols

$\pm$	$\not\subset$	$\subset$	$\supset$	$\subseteq$	$\in$	$\notin$	∂
$\neq$	$\equiv$	$\approx$	$\angle$	∇	$\sum$	Π	Ω
ϕ	α	β	χ	ε	γ	η	λ
μ	ν	$\perp$	π	θ	ρ	σ	τ
$\Re$	ω	ξ	ψ	ζ	Υ	$\exists$	$\int$
$\iint$	$\iiint$	$\oint$	$\times$	$\sqrt{\ }$	$\therefore$	$\because$	Δ
$\geq$	$\leq$	$\Rightarrow$	$\Downarrow$	$\Leftrightarrow$	∞	$\propto$	$\mathbb{R}$
$\mathbb{Z}$	$\mathbb{N}$	$\mathbb{C}$	$\mathbb{Q}$				

Chapter 1

Introduction to First Order Equations

In this introductory chapter we define ordinary differential equations, give examples showing how they are used and show how to find solution of some differential equations of the *first order*.

What is ordinary differential eqution?

A differential equation is one relating a function and its derivatives, and the objective is to find a function satisfying the given equation. An example may help to fix ideas.

Example: Let x be an independent variable on an interval I of the exis of real numbers, and consider the equation

$$\frac{u'}{u} - 2 = 0,$$

Where u is an unspecified function of x aqnd the prime indicates differentiation: $u' = du / dx$. This equaiton can also be written in the alternative form

$$u' = 2u.$$

The function $u = f(x) = C \exp(2x)$ satisfies this differential equation for any choice of the constant c. This can be verifed by forming the derivative $f'(x) = 2C \exp(2x)$ appearing on the left-hand side of the alternative form, and the the combination $2f(x) = 2C \exp(2x)$ appeariang on the right-hand side, and checking that they are indeed equal for each value of x.

The equation is a differential equation. The function $f(x) = C \exp(2x)$ satisfying it will be referred to as a *solution* of the given differential equation. The alaternative form of the equation, equation, in which the derivative is expressed in terms of the funciton, is the *standard* form of the equation.

A more general version of a differential equation is the following. Suppose you're given, for some function $\phi(y,z)$ of two variables, the equation

$$\phi(u(x), u'(x)) = 0,$$

With the objective of finding a function $y = u(y)$ on an interval of the x axis. This is the problem of finding a function $u(x)$ of the varible x (the *independent* variable) such that equation becomes an identity on some interval of the x-axis. In the preceding exampe, the function ϕ took the form

$$\phi(y,z) = \frac{z}{y} - 2$$

and the solution of the differential equation was given as $u = C \exp(2x)$. In the more general case of equation it may not be possible to give a simple formula representive the solution. In fact, the problem arises whether there is *any* function $u(x)$ such that $\phi(u(x), u'(x))$ vanishes identically on an interval of the x axis. This problem, and related problems, will be discussed later in this book. These issues are most convenientily discussed for differential equations written in satandard form and most of the general results of the theory of ordinary differential equations are given for equaitons in this form.

Rewriting the differnetial equation in standard form was a simple matter in example. In the more genral case of equation this requires solving the functional equation $\phi(y,z) = 0$ for z as a function of y, say $z = g(y)$. If this can be achieved then the substitutions $y = u$, $z = u'$ ahieve the standard form

$$u' = g(u).$$

Equation is actually a restrcted version of the standard form; the more genral version is given below.

The goals of this book are not only to establish conditions on the function g so that equations like and possess solution, but also to develop techniques for inding or describing these solutions.

What's "ordinary" about an ordinary differential equation? The qualifier "ordinary" is used to indicate differential equations having one independent variable in contrast to those having two or more independent variables; the latter are called *partial differential equations.* An example of the latter is

$$\frac{\delta u}{\delta s} + \frac{\delta u}{\delta t} = 0$$

How the object is to find a solution $u(s, t)$ depending on the two independent variables s and t. It therefore involves partial derivatives, We will not investigate partial differential equations in any depth in this book. There are, however, instances wherein ordinary differential equations play important roles in the study of partial differential equations, and we will mention some of the latter to motivate our study of the associated ordinary differential equaions.

Since we'll be concerned mostly with ordinary rather than partial differential equations, we'll often drop the qualifier " ordinary differential equation" unless the constrary is explicity stated.

Example: Suppose the function is $\phi(y, z) = y^2 + z^2 - 1$. The differential equaion for $y = u(x)$ is

$$u^2 + u^2 - 1 = 0$$

In this case, it is easy to verify that $u(x) = \sin(x + c)$, for any choice of the constant c, is a solution. The stnadard form of the differential equation is either $y' = \sqrt{1-y^2}$ or $y' = -\sqrt{1-y^2}$.

Example: According to the laws of mechanics, a particle of mass m moving in the gravitational field of the earth has a constant energy

$$\mathrm{E} = (1/2)mv^2 - mMG/r$$

where v is its speed, G is the gravitational constant, M is the mass of the earth, and r is the distance of the particle form the center of the earth; if R is the radius of the earth, $r \geq R$. If the particle is moving raqdially outward, then $v = dr/dt > 0$ where t represents time, and the postion of the particle is governed by the differenetial equation (obtained by solving for v in the energy equation above)

$$\frac{dr}{dt} = \sqrt{\frac{2e}{m} + \frac{2MG}{r}}.$$

The constant value of the energy E can be determined from the initial values of the postion and speed. Here the differential equation for r is in the standard form of equation (1.4) with $f(r) = \sqrt{2E/m + 2MGr^{-1}}$.

In this example we have used t as the independent variable instead of x, since it represents the time. The designation of variables is of course arbitrary, but, some usages are tradtional, and we shall adhere to them. It is also traditional to use a dot to indicate a derivative with respect to time $(\dot{x} \equiv dx/dt)$.

Example: Suppose the speed $\dot{x}$ of a particle on the x axis in controlled so that $x\dot{x} = 1$. Then, solving the functional problem, we find

$$\frac{dx}{dt} = \frac{1}{x}$$

This can be solved by noticing that the original form of the equation implies that

$$\frac{d}{dt}\frac{x^2}{2} = 1.$$

If $x(0)$ is the starting point of the motion when $t = 0$, then intergrating each side of the equation gives

$$x(t)^2 = x(0)^2 + 2t.$$

The solution is then found by taking a square root.

Example: If you have money deposited in an interest- paying account, it will grow with time if you do not make any withdrawals.

How fast it grows depends on the annual interest rate I and the frequency of compounding. If the interest rate is 5% and it is only compounded annually, at the end of one year you will have 1.05 P_0 on deposit, where P_0 is your initial deposit.

At the end of two years you will have $(1.05)^2 P_0$, etc. If it is compounded quarterly, at the end of one quarter your deposit will have increased to $(1 + 0.05/4) P_0$ and after one year to $(1 + 0.05/4)^4 \; P_0 \approx 1.0509 P_0$.

If it is compounded monthly the initial deposit gets increased by a factor fo $(1 + 0.05/12)$ at the end on each month, so that by the end of the year it has increased by the factor

$$(1 + 0.05/12)^{12} \approx 1.0512.$$

Ink general, if an annul interest I (expressed as a decimal: $I = 0.05$ in the examples above) is compounded in N equal intervals, the amount P_1 on deposit after one of these intervals is

$$P_1 = (1 + I/N)P_0.$$

More generally, if $P(t)$ is ghe principal at time t, then after an interval of length $1/N$ the principal will be

$$P = (t + 1/N) = \;(1 + I/N)P(t) + \;I/N)P(t).$$

Suppose your bank compounds *continuously*. This means if your balance at time t is $P(t)$ and Δt is a sufficiently small interval of time, then

$$P(t + \Delta t) \approx P(t) + I\Delta t P(t),$$

the approximation improving as Δt tends to zero; there is an implicit assumptionj in this formula that t is measured in years so that if the time interval is one day, $\Delta t = 1/365$. From this formula one deduces that the balance $P(t)$ on deposit at any time satisfies the differential equation

$$\frac{dP}{dt} = IP.$$

The general, first-order case

There is a more general form of differential equation than that of equation , Suppose $\phi = \phi\ (t, x, y)$ is a prescribed function of three variables and the problem is that of finding a function $x(t)$ such that

$$\phi(t, x(t), \dot{x}(t)) = 0$$

on an interval of the t axis. The standard form of this equation would be

$$\frac{dx}{dt} = f(f, x)$$

where the function f is obtained by solving the equation $\phi\ (t, x, y) = 0$ for y as a function of t and x.

There are rules whereby one can dtermine whether the equation can be solved and put into the standard form. Suppose $\phi\ (t, x, y)$ is continuously differentiable (C^1 for short) in a domain D of the (txy-space. Suppose further that the following two conditions hold:

$$\phi(t_0, x_0, y_0) = 0 \text{ for some } (t_0, x_0, y_0) \in D$$

and

$$\frac{\partial \phi}{\partial y}(t_0, x_0, y_0) \neq 0.$$

Then, according to the implicit-funciton theorem, the equation $\phi\ (t, x, y) = 0$ Possesses a solution $y = f(t, x,)$ in the tx-space. This solutin reduces to y_0 when $(t,x,) = (t_0, x_0)$ Moreover, if N is sufficiently small, it is the *unique* solution reducing to y_0 when $(t, x) = (t_0, x_0)$. This function f is C^1 on N.

We have noted in the contect of equation above that the question arises whether there is in fact *any* solution.

For a general ordinary differential equation into an identity on that interval of the real t axis which turns the equation into an identity on that interval. For the time being we will be concerned with simple problems for which we can express the solutions explicitly in terms of know functions and their integraqls; for such problems the issue of existence is clearly sttled in the affirmative.

By way of example, note that there is one particularly simple class of differential equaions for which the solution is immediate. Suppose the function f in equation does not depend of x: $f = f(t)$. Then the fundamental theorem of calculus shows that the solution is

$$x(t) = \int^t f(s)ds + \text{constant}.$$

Thus to obtain the solution one need only integrate of known function. Such a solution involving the integration of a know function is sometimes referred to as a solution by *quadrature*.

The Initial-value Problem

The solution by quadrature includes an arbitrary constant; this was likewise true of earlier example. We can only have a unique solution of the differential equation if we somehow specity this constant. One way to do this is to require that the function $x(t)$ not only satisfy the differential equation but also satisfy the condition that it take on a prescribed value x_0 at a prescribed point t_0. This lack of uniqueness turns out to be characteristic of more general differential equations as well, and the remedy suggested above to render the solution unique turns out to be appropriate in a variety of naturally occurring circumstances. For this reason we define the *initial-value problem*:

$$\frac{dx}{dt} = f(t,x), x|_{t=t_0} = x_0.$$

Here the function f and the *initial data* t_0, x_0 are prescribed. The initialvalue problem therefore consists of finding that solution of the differential equation taking on the value x_0 when $t = t_0$.

Example: Suppose that the equation is

$$\dot{x} = t,$$

and the initial data are:

$$x = 3 \text{ when } t = 1.$$

The solution of the equation is

$$x = (1/2)t^2 + 5/2.$$

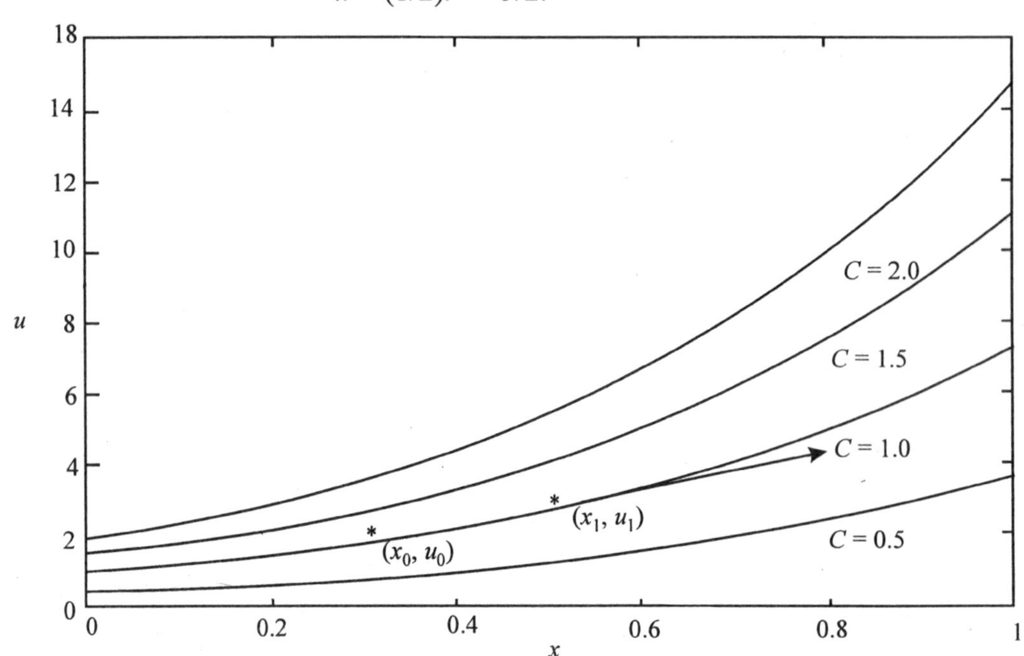

Figure: A family of solution curves for the equation $u' = g(u)$ is shown, in this case with $g(u) = 2u$ as in equation. The solution $u = C\, exp(2x)$ represent curves in the x, u plane. They are shown for several choices of the constant C. If initial data $u(x_0) = u_0$ are specifed, the curve must pass through the point (x_0, u_0): this effects of choice of the constant C. Any choice of a point in the x, u plane–say (x_1, u_1)– produces a slope $g(u)$ which the solution through that point must have. The arrow indicates the direction of a line having that slope.

There is a geometric picture associated with the initial-value problem. The essence of it consists of viewing the solution of a differential equation as a curve in space. The initial condition that $x|_{t=t_0} = x_0$ represents the requirement that that curve pass throught a specified point (t_0, u_0). It can be illustrated by drawing a family of solutions as in Figure and noting that only one of them passes through a specified point –(t_0, u_0) in the notation of the diagram.

A closely related geometric piecture is based on the slopes of these curves. If $x(t)$ is a solution of equation then $\dot{x}(t) = f(t, x, (t))$, and therefore at any point $(t_1, x(t_1)$ of its graph the slope is $f(t_1, x, (t_1))$. Conversely, if (t_1, x_1) is a point in the tx- plane then $f(t_1, x_1)$. is the slope of the tangent line of the graph on any solution passing through the point (t_1, x_1). This leads to the notion of a *direction field.* This is a family of vectors. The vector whose tail lies at the point (t, x) has the components $(h, f(t, x)h)$ for some convenient choice of the constant h, and represents the tangent to any solution curve passing through that point.

DIFFERENTIAL EQUATIONS

Differential equations are the equations containing derivatives or differentials of one or more variables, for example the followings are differential equations:

1. $\dfrac{dy}{dx} = e^x + \sin x$
2. $y'' - 2y' + y = \cos x$
3. $\dfrac{\partial^2 u}{\partial x^2} + \dfrac{\partial^2 u}{\partial y^2} = \dfrac{\partial u}{\partial t}$
4. $2x^2 dx + 2y dy = 0$

An ordinary differential equation (ODE) is an differential equation in which all derivatives of one or more dependent variables with respect to *a* single independent variable.

Clearly, equations (1), (2), and (4) are ODEs, while (3) is not.

In fact, (3) is *a* partial differential equation. *A* partial differential equation (PDE) is *a* differential equation containing at least one partial derivative of some dependent variable. Partial differential equations will be discussed in Unit 21 and Unit 22.

Definition: *A* solution of differential equation is the known function which satisfies the equation.

Definition: The order of differential equations is defined as the order of the highest-order derivative which appears in the equation.

Definition: *A* differential equation along with subsidiary conditions on the unknown function and its derivatives, all given at the same value of the independent variable, constitutes an initial-value

Problem: The subsidiary conditions are initial conditions. If the subsidiary conditions are given at more than one value of the independent variable, the *problem* is *a* boundary-value problem and the conditions are boundary conditions.

For Example: The problem $y'' + 2y' = e^x$; $y(p) = 1$, $y'(\pi) = 2$ is an initial-value problem, because the two subsidiary conditions are both given at $x = \pi$. The problem $y'' + 2y' = e^x$; $y(0) = 1$, $y'(1) = 1$ is *a* boundary-value problem, because the two subsidiary conditions are given at the different values $x = 0$ and $x = 1$.

FIRST-ORDER ODES

An differential equation

$$M(x, y)dx + N(x, y)dy = 0$$

is an exact differential equation if and only if there is *a* function *f* such that $M = \frac{\partial f}{\partial x}$ and $N = \frac{\partial f}{\partial y}$ throughout some region. Thus, we have $\frac{\partial f}{\partial x}dx + \frac{\partial f}{\partial y}dy = 0$ or $d[f(x, y)] = 0$. The solution is $f(x, y) = k$, k is the arbitrary constant.

Although some exact equations are easy to identify and solve, it is general impossible to tell by inspection whether *a* given first-order differential equation is exact or not. The following is *a* simple criterion for exactness.

Theorem: If $M(x, y) = \frac{\partial f}{\partial x}$ and $N(x, y) = \frac{\partial f}{\partial y}$ are continuous, then the differential equation $M(x, y)dx + N(x, y)dy = 0$ is exact if and only if $\frac{\partial M}{\partial y} = \frac{\partial N}{\partial x}$.

Proof: If the differential equation is exact, there exists *a* differential function $f(x, y)$ for which $df = 0$. We have

$$M(x, y) = \frac{\partial f}{\partial x} \text{ and } N(x, y) = \frac{\partial f}{\partial y}$$

as *a* condition of exactness. If, in addition, *M* and *N* are differentiable, it follows that

$$\frac{\partial M}{\partial y} = \frac{\partial^2 f}{\partial y \partial x} = \frac{\partial^2 f}{\partial x \partial y} = \frac{\partial N}{\partial x}$$

whenever the mixed partial derivatives of *f* exist and are continuous.

Hence, whenever $\frac{\partial M}{\partial y}$ and $\frac{\partial N}{\partial x}$ exists, are continuous, and are equal.

To prove the converse of the theorem, we assume that $\frac{\partial M}{\partial y} = \frac{\partial N}{\partial x}$. Then there is *a* function *f* such that $\frac{\partial f}{\partial x} = M$ and $\frac{\partial f}{\partial y} = N$. Let us first integrate $M(x, y)$ with respect to x, holding y fixed.

Introducing the dummy variable t gives us the expression

1. $f(x, y) = \int_{x_0}^{x} M(t, y)dt + c(y)$

Our proof will be completed if we can determine $c(y)$ so that $\frac{\partial f}{\partial y} = N(x, y)$. We have from (1)

$$\frac{\partial f}{\partial y} = \frac{\partial}{\partial y}\int_{x_0}^{x} M(t, y)dt + c'(y) = \int_{x_0}^{x} \frac{\partial M(t, y)}{\partial y} dt + c'(y)$$

$$= \int_{x_0}^{x} \frac{\partial N(t, y)}{\partial t} dt + c'(y) = N(x, y) - N(x_0, y) + c'(y)$$

Thus $\frac{\partial f}{\partial y}$ will equal to $N(x, y)$, as required, if $c(y)$ is determined so that $c'(y) = N(x_0, y)$, that is, if

$$c'(y) = \int_{y_0}^{y} N(x_0, t)\,dt$$

We have shown that if $\frac{\partial M}{\partial y} = \frac{\partial N}{\partial x}$, then

2. $f(x, y) = \int_{x_0}^{x} M(t, y)\,dt + \int_{y_0}^{y} M(x_0, t)\,dt$

This establish the "if" assertion of the theorem, and our proof is completed.

Under the same hypothesis, we can prove the "if" part of the preceding theorem by first integrating $N(x, y)$ with respect to y, holding x fixed.

In this way we find that the function g represented by

3. $g(x, y) = \int_{x_0}^{x} M(t, y_0)\,dt + \int_{y_0}^{y} N(x, t)\,dt$

Once an equation is known to be exact, we can apply the procedure set out in the definition to solve it. Having proved the equation to be exact, we use the definition of exact differential equation to solve it. The following corollary gives two ways of writing down solutions:

Corollary: If the differential equation $M(x, y)dx + N(x, y)dy = 0$ is exact, then for (x_0, y_0) and arbitrary point (x, y), either

$$f(x, y) = \int_{x_0}^{x} M(t, y)\,dt + \int_{y_0}^{y} N(x_0, t)\,dt = k$$

or

$$g(x, y) = \int_{x_0}^{x} M(t, y_0)\,dt + \int_{y_0}^{y} N(x, t)\,dt = c$$

is *a* general solution of the differential equation.

INTEGRATING FACTOR

Sometimes, *a* non-exact differential equation $M(x, y)dx + N(x, y)dy = 0$ can be turned into an exact differential equation by multiplying the whole equation by an appropriate factor, called an integrating factor.

The following observations are often helpful in finding integrating factor:

1. If *a* first-order differential equation contains the combination

$$xdx + ydy = \frac{1}{2}d(x^2 + y^2),$$

try some function $x^2 + y^2$ as *a* multiplier.

2. If a first-order differential equation contains the combination $x\,dy + y\,dx = d(xy)$, try some function x y as *a* multiplier.

3. If a first-order differential equation contains the combination x dy $- y\,dx$, try $\frac{1}{x^2}$ or $\frac{1}{y^2}$ as *a* multiplier. If neither of these works, try $\frac{1}{xy}$ or $\frac{1}{x^2 + y^2}$ or some

function of these expressions, as an integrating factor, remember that $d\tan^{-1}\left(\frac{y}{x}\right) = \frac{xdy - ydx}{xy}$ and $\tan^{-1}\left(\frac{y}{x}\right) = \frac{xdy - ydx}{x^2 + y^2}$.

Also, for the general first-order differential equation $M(x, y)dx + N(x, y)dy = 0$, the following theorem indicates other possibilities of finding integrating factor.

Theorem:

(a) If $\frac{1}{N}\left(\frac{\partial M}{\partial y} - \frac{\partial N}{\partial x}\right)$ is a function of x only, say $f(x)$, then $\exp\left(\int f(x)\,dx\right)$ is an integrating factor of the equation $M(x, y)dx + N(x, y)dy = 0$

(b) If $\frac{1}{M}\left(\frac{\partial M}{\partial y} - \frac{\partial N}{\partial x}\right)$ is *a* function of y only, say $g(y)$, then $\exp\left(\int g(y)\,dy\right)$ is an integrating factor of the equation $M(x, y)dx + N(x, y)dy = 0$

Proof:

(a) By the hypothesis, if $m(x)$ is the integrating factor which depends on variable x only. We see that

$$m(x)M(x, y)dx + m(x)N(x, y)dy = 0$$

is the exact differential. From theorem 13.1, we have the necessary condition

$$\frac{\partial}{\partial y}(\mu M) = \frac{\partial}{\partial x}(\mu N)$$

$$\mu\frac{\partial M}{\partial y} = \mu\frac{\partial N}{\partial x} + N\frac{d\mu}{dx}$$

Eventually, we obtain

$$\frac{d\mu}{dx} = \mu\left(\frac{\frac{\partial M}{\partial y} - \frac{\partial N}{\partial x}}{N}\right) \mu f$$

which has *a* general solution $\mu(x) = \exp\left(\int f(x)\,dx\right)$

The proof for (b) is similar and it is left for exercise.

Example: Show that the following differentials are exact and solve the corresponding differential equations:

1. $(9x^2 + y - 1)dx - (4y - x)dy = 0$

2. $(e^x siny - 2ysinx)dx + (e^x cosy + 2cosx)dy = 0$

Solution:

1. For this ODE,

$$M(x, y) = 9x^2 + y - 1 \Rightarrow \frac{\partial M}{\partial y} = 1$$

$$N(x, y) = -4y + x - 1 \Rightarrow \frac{\partial N}{\partial x} = 1$$

The ODE is exact.

The differential function is

$$\frac{\partial f}{\partial x} = 9x^2 + y - 1 \Rightarrow f(x,y) = \int(9x^2 + y - 1)dx = 3x^3 + xy - x + C_1(y)$$

$$\frac{\partial f}{\partial y} = -4y + x \Rightarrow f(x, y) = -2y^2 + xy + C_2(x)$$

By comparing the above equations, we have

$$f(x, y) = 3x^3 + xy - x - 2y^2$$

Therefore the general solution of the ODE is

$$3x^3 + xy - x - 2y^2 = k$$

2. For this ODE,

$$M(x, y) = e^x \sin y - 2y \sin x \Rightarrow \frac{\partial M}{\partial y} = e^x \cos y - 2\sin x$$

$$N(x, y) = e^x \cos y + 2\cos x \Rightarrow \frac{\partial N}{\partial x} = e^x \cos y - 2\sin x$$

The ODE is exact.

$$\frac{\partial f}{\partial x} = e^x \sin y - 2y \sin x \Rightarrow f(x, y) = e^x \sin y + 2y \cos x + C_2(y)$$

$$\frac{\partial f}{\partial x} = e^x \cos y + 2y \cos x \Rightarrow f(x, y) = e^x \sin y + 2y \cos x + C_2(x)$$

The differential function is

By comparing the above equations, we have

$$f(x, y) = e^x \sin y + 2y \cos x$$

Therefore the general solution of the ODE is

$$e^x \sin y + 2y \cos x = k$$

Example: Find the integration factor of the following ODE and solve the corresponding equation

$$(3x^2y + 2xy + y^3)dx + (x^2 + y^2)dy = 0$$

Solution: For this ODE,

$$M(x, y) = 3x^2y + 2xy + y^3$$

$$\Rightarrow \frac{\partial M}{\partial y} = 3x^2 + 2x + 3y^2 \text{ and}$$

$$\frac{\partial M}{\partial x} = 6xy + 2y$$

$$N(x, y) = x^2 + y^2 \Rightarrow \frac{\partial N}{\partial x} = 2x \text{ and } \frac{\partial N}{\partial y} = 2y$$

We can see that $\frac{1}{N}\left(\frac{\partial M}{\partial y} - \frac{\partial N}{\partial x}\right) = 3$ depends on variable x only. Therefore the integration factor is given by $\exp\left(\int 3\,dx\right) = e^{3x}$ and the ODE is reduced into

$$e^{3x}(3x^2y + 2xy + y^3)dx + e^{3x}(x^2 + y^2)dy = 0$$

The differential function is $f(x, y)$

$$\frac{\partial f}{\partial x} = e^{3x}(3x^2y + 2xy + y^3)$$

$$\frac{\partial f}{\partial y} = e^{3x}(x^2 + y^2) \Rightarrow f(x, y) = \left(x^2y + \frac{y^3}{3}\right) + C(x)$$

$$\Rightarrow \frac{\partial f}{\partial x} = e^{3x}(2xy) + 3e^{3x}\left(x^2y + \frac{y^3}{3}\right) + C'(x)$$

$$\Rightarrow \frac{\partial f}{\partial x} = e^{3x}(2xy + 3x^2y + y^3) + C'(x)$$

By comparing the above equations, we have

$$f(x, y) = e^{3x}(2xy + 3x^2y + y^3)$$

Therefore the general solution of the ODE is

$$e^{3x}(2xy + 3x^2y + y^3) = \text{k}$$

SEPARABLE FIRST-ORDER DIFFERENTIAL EQUATIONS

A separable differential equation is *a* first-order ordinary equation that is algebraically reducible to *a* standard differential form in which each of the non-zero terms contains exactly one variable solutions to this kind of equation is usually quite straightforward. For example $f(x)dx + g(y)dy = 0$, the solution will be, of course

$$\int f(x)\,dx + \int g(x)\,dy = c$$

One way of discovering whether or not *a* given equation is separable is to collect coefficients on the two differentials and see if the result can be put in the form

$$f(x)G(y)dx = F(x)g(y)dy$$

Another way is to solve for *a* derivative and compare the result with

$$\frac{dy}{dx} = M(x)\,N(y)$$

A general solution of the form

$$f(x)G(y)dx = F(x)g(y)dy$$

can be found by first dividing by the product $F(x)G(y)$ to separate the variables and then integrating

$$\int \frac{f(x)}{F(x)}dx = \int \frac{g(y)}{G(x)}dy + c$$

Similarly, the second form $\frac{dy}{dx} = M(x)N(y)$ can be solved by first multiplying by dx, then dividing by $N(y)$ and subsequently integrating

$$\int \frac{dy}{N(y)} = \int M(x)dx + c$$

The process of solving *a* separable equation will often involve division by one or more expressions. In such cases the results are valid where the divisors are not equal to z ero but may or may not be meaningful for values of the variables for which the division is undefined. Such values require special consideration and may lead to singular solutions!

Example: Solve the following differential equations by separation of variables:

1. $\frac{dy}{dx} = \frac{x^2}{y}$

2. $\dfrac{dy}{dx} = \dfrac{x^2}{y(1+x^3)}$

Solution:

1. The ODE $\dfrac{dy}{dx} = \dfrac{x^2}{y}$ becomes $ydy = x^2dx$ and

$$\int yd\,y = \int x^2 dx \Rightarrow \frac{y^2}{2} = \frac{x^3}{3} + C \Rightarrow y = \sqrt{\frac{2x^3}{3} + 2C}$$

2. The ODE $\dfrac{dy}{dx} = \dfrac{x^2}{y(1+x^3)}$ becomes $\dfrac{dy}{dx} = M(x)\,N(y)$ and

$$\int yd\,y = \int \frac{x^2}{1+x} dx \Rightarrow \frac{y^2}{2} = \frac{1}{3}\ln(1+x^3) + C \Rightarrow y = \sqrt{\frac{2}{3}\ln(1+x^3) + C'}$$

Higher-order Equations

The example given above are ofr firstorder differential equations, i.e. those in which only the first derivative of the dependent variable occurs.

Example: A particle of mass m moving on the x-axis under the influence of a force $f(x)$ depending on its position on the axis is governed by *Newton's Second Law of Motion*:

$$m\ddot{x} = f(x),$$

where $\ddot{x} \equiv d^2x/dt^2$ is shorthand notation for the second time derivative.

This is an example of a *second-order* differential equation since the highest-order derivative occurring in it is the second. In applications differential equaions of arbitrarily high order can occur. For positive integer n, the general n^{th}- order differential equation has the form

$$\phi(t, x, dx/dt, \ldots, d^n x/dt^n) = 0.$$

It is put into standard form by solving this equation for the highest derivative:

$$\frac{d^n x}{dt^n} = f\left(t, x, \frac{dx}{dt}, \ldots, \frac{d^{n-1}x}{dt^{n-1}}\right).$$

It is traditional to begin discussions of the theory of ordinary differential equations with those of the first order. This introduces the student to many features of the theory in a relatively uncluttered context, so we shall adhere to this sensible tradition in the remainder of this chapter. In the examples above we have assumed that the number system is the *real* number system, and accordingly that all functions are real-valued. We shall continue to assume this unless there is an explicit statement otherwise.

Linear Equations

A function of one variable $f(x)$ is linear if, for arbitrary number α, x,

$$f(\alpha x) = \alpha f(x).$$

It is easy to see from this definition that any linear function has the form $f(x) = kx$ for some constant k form some constant k and that any such function is linear. A function $f(t, x)$ of two variables is then linear in the second (x) variable if and only $f(t, x) = k(t)x$, where k can be an arbitrary function of t. A linear differential equation is one which, when written in the standard form, has a linear function $f(t, x)$ on the right-hand side:

$$\frac{dx}{dt} = k(t)x.$$

The Linear, Homogeneous Equation

The equation will be referred to as *linear, homogeneous* equation, for reasons that will soon become clear. It's obvious by substitution that the function $x(t) \equiv 0$ is always a solution of the linear, homogeneous equation. However, if at some point $t = t_0$ $x(t_0) \neq 0$, then one can, for values of t near t_0, divide through in equation by x. Then

$$\frac{1}{x}\frac{dx}{dt} = \frac{d}{dt}\text{In}\, x = k(t).$$

From this we easily infer, by integrating from t_0 to t and solving for $x(t)$, that the solution is

$$x(t) = x(t_0)\exp\left\{\int_{t_0}^{t} k(s)ds\right\}.$$

This formula gives this general solution to the linear, homogeneous equation of first order provided that k is a continuous function of t. It agrees with the previous observation that $x \equiv 0$. is a solution: if $x(t_0) = 0$, this formula implies that $x(t) \equiv 0$. On the other hand, if $x(t_0) \neq 0$, this formula implies that $x(t) \neq 0$ for any value of t.

Many so-called *rate* problem take the form of equation (1.16) where $k(t) = k$ is constant, For these the solution is immediate:

$$x(t) = x(t_0)e^{k(t-t_0)}$$

Example: Let $x(t)$ represet the number of individuals in an isoclated population. We ignore the fact that x should be an integer and assume it can viewed as a smooth function of time, a plausible assumption if the population is large. The population could be aq numan, insect or some other biological group.

That it is isolated means we can ignore migration, inwards or outwards. The simplest modle of population changbe then assumes that there is a constant birth rate b and a constant death rate d.

Then the increase in x attributable to births during a short time-interval Δt is approximately $x(t)\ b\Delta t$, and the decrease in x, attributable to deaths, is $x(t)d\Delta t$. The not change is therefore $x(t+\Delta t)-x(t)\approx(b-d)x(t)\Delta t$. As in the derivation of equation above we infer that the population is governed by the equation

$$\dot{x}=kx$$

where $k = b - d$ and reflects the difference between the birth and death rate the solution is

$$x(t)=x_0e^{k(t-t_0)}$$

. If births exceed deaths, $k > 0$ and the population increases exponentially; if deaths exceed briths, $k < 0$ and the population decreases exponentially.

In the previous example, if $k < 0$ so that the population decays, it never decays to zero: the exponential function is always positive. The population would approach zero as $t\to\infty$, but would never reach it in finite time Likewise, if $k > 0$, the population would become arbitrarily large as $t\to\infty$ but would not become infinite in finite time

Example: The initial - value problem

$$\frac{dx}{dt}=-tx, x\,|_{t=0}=1$$

has the solution $x = exp\{-t^2/2\}$ according to the forumla.

The Inhomogeneous Equation

The equation

$$\frac{dx}{dt}=k(t)x+a(t),$$

where a is specified function of t, differs from the linear equation by the addition to the right-hand side of a term independent of x. The resulting equation, called the *linear inhomogeneous* equation, may be solved in closed form as follows. Multiply each aside by

$$K(t)=\exp\left\{-\int_{t_0}^{t}k(s)ds\right\},$$

where t_0 is an appropriate initial point, An easy calculation shows that equation may be rewritten

$$\frac{d}{dt}(Kx) = Ka,$$

so, integrating above from t_0 to t and then multiplying each side by

$$K(t)^{-1} = \exp\left\{+\int_{t_0}^{t} k(s)ds\right\},$$

we find

$$x(t) = \exp\left\{\int_{t_0}^{t} k(s)ds\right\}$$

$$\left[x(t_0) + \int_{t_0}^{t} \exp\left\{-\int_{t_0}^{s} k(u)du\right\} a(s)ds\right],$$

where we have unsed the facat that $K(t_0) = 1$. This can be rewritten as

$$x(t) = \exp\left\{\int_{t_0}^{t} k(s)ds\right\}$$

$$x(t_0) + \int_{t_0}^{t} \exp\left\{-\int_{s}^{t} k(u)du\right\} a(s)ds,$$

This somewhat cumbersome formula provides the complete solution to the linear, inhomogeneous differential equation of the first order. The exact representation of the solution of a differential equation in literal terms, reducing its solution to quadratures as in equation , is rare. We' ll stroke of good luck.

Example: The initial-value problem

$$\dot{x} = tx + t, x(0) = 1$$

has, according to the fomula, the solution

$$x(t) = \exp(t^2/2) + \exp(t^2/2)[-\exp(-t^2/2) + 1] = 2\exp(t^2/2)-1.$$

Other solvable Equations

There are certain other classes of first-order equations that can be refuced to quadratures. These differential equations represent rather special cases but they do arise in a variety of circumstaqnces and the methods described below are then useful.

Separable Equations

One of these classes of differential equations, called *separable,* has the form

$$y' = f(x)g(y).$$

It can be reduced to quadratures in the form

$$\int_{y_0}^{y} \frac{du}{g(u)} = \int_{x0}^{x} f(v)dv$$

where x_0 and y_0 are appropriate constants; if the equation is provided with the initial data $y(x_0) = y_0$, the initial-value problem is solved by the formulae in an interval of the x-axis containing x_0 provided $g(y_0) \neq 0$.

The linear, homogeneous equation falls into this category, but so do many others. The quadratures appearing in equation do not complete the solution: one still needs to solve the remaining equation for for y as a function of x.

This can always be done in principle: Problem of the following problem set addresses this issue. However, obtaining an explicit expression for the solution, in terms of known functions, may not be possible.

The differential equation

$$\dot{x} = \sqrt{(\alpha^2 - x^2)(\beta^2 - x^2)},$$

for appropriate values of the constants α, β,arises in mechanics in the study of the motion of a rigid body. It is separable, but the quadrature

$$\int^{x} \frac{dx}{\sqrt{(\alpha^2 - x^2)(\beta^2 - x^2)}}$$

is an *elliptic integral* and the solutions are expressed in *elliptic functions,* which represent a new class of functions, effectively invented to deal with this and related examples.

The following exampe, on the other hand, *can* be integrated in elementary functions.

Example: The *logistic* equation

$$\dot{x} = kx(1 - x/N).$$

describes population growth limited by competition for resources. The constant k reflects the difference between the birth and death rates ($k > 0$) for a growing population) and N is the "carrying capacity" of the environment. This equation is nonlinear, but it is separable. If $x(0) = x_0$ expresses the initial population, then

$$\int_{x_0}^{x} \frac{dx}{(1 - u/N)} = kt.$$

Example: Imagine an object dropped from a great height with initial velocity $v = 0$.

Measure distance and velocity downward toward the surface of the earth. Then the equation governing the velocity of the object may be written

$$m\dot{v} = mg - Cv^2$$

where m is the mass of the object, g is acceleration of gravity (regarded as a constant) and the second term reflects the slowing effect of air resistance. C is a positive constant depending on the nature of the object. This equation is separable.

Homogeneous Equations

A class of differential equations in the standard form that can be reduced to separable form is that for which the right-hand side is *homogenous of degree zero.* A function $f(x, y)$ is said to be *homogeneous of degree k* if

$$f(ax, ay) = a^k f(x, y) \text{ for arbitrary values of } a.$$

For example, the function $f(x) = kx$ for constant k is homogneous of degree one (and that's why the linear equation is alos referred to as homogeneous0. A function that is homogeneous of degree zero has the form $f(x, y) = g(y/x)$, as can be verified by choosing $a = 1/x$ in the definition of homogeneity. The differential equation $y' = g(y/x)$ can then be put into separable form by the substitution $y = xu$: $u' = x^{-1}(g(u)-u)$.

Example: Consider the initial-value problem

$$y' = y^2/x^2, \; y(1) = 2.$$

The substitution $y = xu$ leads to the equation $u' = (u^2 - u)/x$ or

$$\int_2^u \frac{dv}{v^2 - v} = \text{In}\, x, \quad \int_2^u \frac{dv}{v^2 - v} = \text{In}\, x,$$

where we have made use of the initial data, which imply that $u = 2$ when $x = 1$. The integration is elementary and leads to the formula $u = (1 - x/2)^{-1}$ and therefore to

$$y = x\,(1 - x/2)^{-1}.$$

Exact Equatoins

Both linear and separable equations discussed above are examples of a more general category, that of *exact* equations. The idea of *exactness* hgas its origin in the calculus of two (or more) variables. Recall that a differentiable function $x(x, y)$ of two variables has a differential

$$du = M(x, y)\, dx + N(x, y)\, dy$$

where $M = \partial u/\partial x$ and $N = \partial u/\partial y$. It follows from this that the functions M and N satisfy the relation

$$\frac{\partial M}{\partial y} = \frac{\partial N}{\partial x}$$

throughout the domain D provided the function u has continuous second partial-derivatives there. Such a function is said to be of class C^2 in that region, or , more briefly, is said to be C^2 there. The reason that equation holds is that mixed second partial- derivatives fo u are then equal: $\partial^2 u\partial/\partial x\ \partial y = \partial u^2/\partial y\partial x$ .Moreover, the converse is also true if the domain D is simply connected (a domain is simply connected, but that of an annulus is not). In other words, if equation holds throughout such a region, then there exist a C^2 functions u for which equation holds. An equation

$$Mdx + Ndy = 0$$

for which the condition holds is called an *exact* equation. We supose in the following discussion that the functions under consideration are sufficiently differentiable in a simply connected domain D.

Consider the differential equation

$$\frac{dy}{dx} = f(x,t);$$

This is the general differential equation is standard form, although nojw we have written x for the independent variable and y for the dependent variable. Rewrite equation (1.32) in the form

$$f(x, y)\ dx = dy = 0.$$

If there should be a function u such that

$$du = f(x, y)\ dx = dy,$$

then equation requires that u be constant. Solving the equation $u(x, y) = C$ for y as a function of x should then provide the solution of the differential equation. However, this result is of little value because the condition , which is necessary for u to exist, implies that f is independent of y in D, that i s, that is equation is of the form $y' = f(x)$. This is in deed solvable as in the trivial case of equation above, but this is not interesting.

Observe, however, that the differential equation, in either of the forms, is unaffected by multiplying each side of the equaqtion with a monzero funcation p (x, y). Doing so transform equaqtion for the existence of a function u satisfying equation is then

$$f\frac{\partial p}{\partial y} + \frac{\partial p}{\partial x} + p\frac{\partial f}{\partial y} = 0$$

If such a function p can be found, the solution can be found by then finding the function f and setting $u(x, y) = C$ for an appropriate constant C. The following example shows that the method of solving the linear, inhomogenous equation is an example of the present method.

Example: Suppose the equation is linear; $f(x, y) = a(x)y + b(x)$. Then equation takes the special form

$$(a(x)y+b(x))\frac{\partial p}{\partial y}+\frac{\partial p}{\partial x}+a(x)p=0.$$

Observe that we can seek a solution of this partial differential equation in the form $p = p(x)$, independent of y, since that eliminates the first term and reduces the equation to the differential equation $dp/dx + a(x)p = 0$. It gives us precisely the factor K given in equation, if the changes of notation are noted.

The equation is a partial-differential equation. Since partial-differential equaions are typically more difficult to solve than are ordinary-differential equaions, it may appear that very little progress has been made. There are, however, cases – usually involving educated guesses – for which one can indeed solve equation explicitly for p, then find the function u, and finally solve the differential equation by the method outlined in this section.

More generally, if the form $Mdx + Ndy$ is not exact but $p(M\,dx + Ndy)$ is, we call p an integrating factor. It is a solution of the partial-differential equation

$$\frac{\partial(pM)}{\partial y}+\frac{\partial(pN)}{\partial x}$$

Bernoulli and Riccati Equations

Some examples of nonlinear, first-order differential equations of special interest are those associated with the names Bernoulli and Riccati. Bernoulli's equation may be written

$$y'+a(x)y=b(x)y^n,$$

where a and b are given functions. The substitution $u = y^{1-n}$ converts this to the linear, inhomogneous equation

$$u' + (1-n)\,a(x)u = (1-n)b\,(x),$$

which can be solved ' by quadratures' with the aid of equation above. A Riccati equation is one of the form

$$y' + a(x)y + b\,(x)y^2 = c(x).$$

It cannot in general be reduced to quadratures like the Bernoulli equation, but has the following special property. Suppose a solution $y = Y(x)$ is know. Then any *other* solution may be found by quadratures. To see this, let $y = Y + \eta$; the equation for η is

$$\eta'+(a(x)+2Y(x))\eta=-b(x)\eta^2.$$

This is a Bernoulli equation with $n = 2$ so the substitution $\eta = y^{-1}$ reduces it to a linear equation which can be solved by quadratures.

There are other "named" equations, associated with the researches of particular mathematicians, but we pass on now to more general issues.

Inequalities, Uniqueness and Existence

In general the initial-value problem,

$$\frac{dy}{dx} = f(x, y), y\,|_{x=x_0} = .y_0,$$

can neither be solved explicity in terms of known elementary functions nor reduced to quadratures. The question arises: can it be solved at all? In other words, are ther well-defined functions f for which there is no solution to the initial-value problem? In fact, this problem possesses a unique solution under quite general, rather mild conditions on the function $f(x, y)$ appearing on the right-haqnd side.

By *uniqueness* is meant that only one solution can satisfy the initial condition $x(t_0) = x_0$ or, put another way, two distinct solutions cannt intersect at the point (t_0, x_0). The uniqueness of the solution to the basic initial-value problem can be inferred in a more elementary manner than its existence, and it is useful to address this now. It is based on a simple inequality, and we next turn to this.

Gronwall's Lemma

There is a very simple result, based on the same reasoning that gave us the formula for the solution of the general linear problem, which is "unreasonably" effective in the analysis of differential equations. It is the following:

Lemma: *(Gronwall's Lemma) Let u be a continuous function on the interval* $a \le x \le b$ *and let* α *and* β *be constants with* $\beta \ge 0$. *If*

$$u(x) \le \alpha + \beta \int_a^x u(s)ds$$

on this interval, then

$$u(x) \le \alpha \exp(\beta(x-a))$$

there.

Proof: Define

$$R(x) = \int_a^x u(s)ds$$

Then

$$\frac{d}{dx}(R\exp(-\beta x)) \le \alpha\exp(-\beta x).$$

Next, integrating from a to $x > a$ and noting that $R(a) = 0$, we find

$$\exp(-\beta x)R(x) \le -\frac{\alpha}{\beta}(\exp(-\beta x) - \exp(-\beta a)),$$

so

$$R(x) \le -\frac{\alpha}{\beta}(\exp\beta(x-a)-1).$$

Substituting this expression for the integral in equation then gives the conclusion.

This version of Gronwall's lemma holds when the current point x is greater than a. There is a correspoinding version when $x < a$:

Corollary: *Suppose, under the same conditons on u, α, β that c < x < a and*

$$u(x) \le \alpha + \beta\int_x^a u(s)ds.$$

Then

$$u(x) \le \alpha\exp(\beta(a-x))$$

there.

The proof may be obtained by reducing this ot the form given in the Lemma by means of the substitution $x \to 2a - x$.

The Lipschitz Condition

The differential equation is determined by the properties of the function on the right-hand side of equation, so we need to know something about this function before we can say anything about the solution. Most functions $f(x, y)$ encountered in practice are at least continuous, and usually have continuous derivatives (often infinitely many). A convenient assumptioin, which is general enough for most situations, is that f satisfies a *Lipschitz condition* in some region of the xy- phone. We define this in the case of a function $f(y)$ that does not depend on the independent variable x:

Definition: *The function f defined for y in the interval I satisfies a Lepschitz conditon there if there is a positive constant L such that*

$$| f(y_1) = f(y_2) | \le L | y_1 = y_2 |$$

for all y_1, y_2 *in I.*

It's easy to see from this definition that the function f is continuous in y but not necessarily differentiable. For example, the function $f(y) = |y|$ satisfies a Lipschitz condition everywhere but fails to be differentiable at $y = 0$.

In the general version of the initial-value problem, as in equation below, the function f appearing on the right-hand side of the equation depends on x as well as y. The generalization of the Lipschitz condition that is needed for the uniqueness and existence theorems is placed on the y dependence only, and will be referred to a " a Lipschitz condition with respect to y."

Definition: *The function f satisfies a Lipschitz condition with respect to y in a region D of the xy-plane if there is a positive constant L such that*

$$| f(x, y_1) = f(x, y_2) | \leq L | y_1 = y_2 |$$

for each pair (x, y_1), (x, y_2) in D.

If we assume that f is a continuous function on D *and* satisfies the condition, we have then made an assumption stronger than continuity but weaker than differentiability.

There is a simple, sufficient condition for a function f to satisfy a Lipschitz condition with respect to y in a region D of the xy-plane. One of the requirements in this condition is that the region be *convex*. A convex region C is one with property that if a pair of point p_1 and p_2 belongs to C, so also does the entire line segment joing them. The region inside a circle or a rectable is convex, whereas that inside an annulus is not.

Theorem: *Suppose f and its first partial derivatives are continuous in the closed, bounded and convex region D of the xy plane. Then f satisfies a Lipschitz condition with respect to y there.*

Proof: We recall from analysis the theorem that a continuous function on a closed, bounded set has a maximum and a minimum there. Consequently, there is some number L such that

$$\left|\frac{\partial f}{\partial y}\right| \leq L$$

on D. By the mean-value theorem (with fixed x),

$$f(x, y_1) - f(x, y_2) = \frac{\partial f}{\partial y}(x, y^*)(y_1 - y_2),$$

where y^* lies between y_1 and y_2. Taking absolute values and applying the preceding inequality, we immediately obtain the required result.

The assumpations in this theorem are somewhat stronger than necessary for the result. For example, if the domain is unbounded but the partial derivative in nevertheless known to be bounded, the conclusion holds. Furthermore, the convexity condition is stronger than is needed: it would suffice if the vertical lines form (x, y_1) to (x, y_2) lie in D for each fixed x. This formulation is nevertheless convenient since it is easy to verify in a great many cases of interest.

Theorem: *Suppose* $y = \varphi(x)$ *is a solution of initial-value problem* (1.39) *on an interval* $c < x < b$, *with graph* $(x, \varphi(x))$ *remaining in a regioni D of the xy-plane in which f is continuous and satisfies a Lipschitz condition with respect to y. Then* φ *is only such solution.*

Proof: Since $\varphi(x)$ is a solution it satisfies the relation $\varphi'(x) = f(x, \varphi, (x))$, and furthermore $\varphi(x_0) = y_0$.

Integrating the equation for φ from x_0 to x and taking the initial condition into account gives the equation

$$\varphi(x) = y_0 + \int_{x_0}^{x} f(s,\varphi(s))ds$$

Suppose now there is a second solution ψ having the same initial condition. It satisfies the same equation so, subtracting, we obtain

$$\varphi(x) - \psi(x) = \int_{x_0}^{x} \{f(s,\varphi(s)) - f(s,\psi(s))\}ds.$$

Taking absolute values gives

$$|\varphi(x) - \psi(x)| \leq \int_{x_0}^{x} |f(s,\varphi(s)) - f(s,\psi(s))|\, ds.$$

where we have assumed for definiteness that $x > x_0$.

The integrand can be estimated with the aid of the Lipschitz condition:

$$|f(s,\varphi(s)) - f(s,\psi(s))| \leq L|\varphi(s) - \psi(s)|.$$

Then, defining $u = |\varphi - \psi|$, we see from the preceding two inequalities that

$$u(x) \leq L\int_{x_0}^{x} u(s)ds$$

for $x_0 < b$. Referring to Gronwall's lemma with $\beta = L$ but $\alpha = 0$, we infer that $u \leq 0$ on this entire interval. But in view of its definition $u \geq 0$ at each point. Therefore $u \equiv 0$ on $[x_0, b]$, i.e. $\psi = \varphi$ there. A similar analysis on the interval $c < x < x_0$ using the corollary to Gronwall's lemma shows that this holds on that interval as well.

The Lipschitz condition with respect to y is sufficient for uniqueness but not necessary.

For example, the following, somewhat less stringent, condition suffices as well. Consider the initial -value problem and suppose that, in place of the inequality, we haqve the inequality

$$| f(x, y_1) - f(x, y_2) | \leq \phi(| y_1 - y_2 |),$$

where ϕ is defined on an interval [0, *a*] of the real axis $(a > 0)$, is continuous and increasing there, satisfies the two conditions

$$\phi(0) = 0 \text{ and } \int_x^a \frac{du}{\phi(u)} \to \infty \text{ as } x \to 0+.$$

For example, in the case of the Lipschitz condition, $\phi(u) = Lu$. The proof of uniqueness under the conditions is considered in the next problem set.

The Existence Theorem

The uniqueness theorem assest that there cannot be more one solution of the initial-value problem but does not assert that there is one. In fact, under the same conditions as those of the uniqueness theorem, there does indeed exist a solution.

Theorem: *suppose the function f of equation is continuous and satisfies a Lipschitz condition with respect to y in a domain D of the xy-plane containing the point* (x_0, y_0). *Then there exist an interval a < x < b containing* x_0 *on which a solution* φ *of exist; the graph* $(x, \varphi(x))$ *lies in D on this interval.*

The existence theorem is local: it asserts the existence of a solution curve only in a neighborhood of the point (x_0, y_0) where the initial data are assigned. Thus the interval (*c*, *b*) might be very small.

The situation is actually better than it appears in that interval of existence of the solution can often be extended in a very natural way.

We are not through with first-order equationjs. W'll need some other properties of them from time to time as we explore other subjects, but we'll develop these properties as we need them.

SOME ELEMENTS OF THE THEORY

A *differential equation* is an equation involving an unknown function and one or more of its derivatives. The equation is an *ordinary differential equation* (ODE) if the unknown function depends on only one independent variable. Some examples of ODEs follow:

$$\frac{du}{dt} = F(t)G(u), \qquad \text{the growth equation}$$

$$\frac{d^2\theta}{dt^2} + \frac{g}{l}\sin(\theta) = F(t), \qquad \text{the pendulum equation;}$$

$$\frac{d^2y}{dt^2}+\varepsilon(y^2+1)\frac{dy}{dt}+y=0,$$ the van der Pol equation;

$$L\frac{d^2Q}{dt^2}+R\frac{dQ}{dt}+\frac{Q}{C}=E(t)$$ the LCR oscillator equation;

$$\frac{dp}{dt}=-2a(t)p+\frac{b(t)^2}{u(t)}p^2-u(t),$$ a Ricati oscillator equation;

In the equation t is the independent variable; the dependent variables are u θ ,and p, respectively.

In what follows we will frequently use the notation $\dot{y}$ to represent dy/dt, $\ddot{y}$ to represent d^2y/dt^2, $y^{(3)}$ to represent d^3y/dt^3, and, in general $y^{(n)}$, to represent d^ny/dt^n.

The *order* of a differential equation is the order of the highest derivative appearing in the equation.

A *solution* of a general differential equation of the nth order,

$$f(t,y,\dot{y},\cdots,y^{(n)})=0$$

is a real-valued function $y(t)$ defined over some interval I having the following properties:1) $y(t)$ and its first derivatives exist for all t in I, so $y(t)$ and its first $n-1$ derivatives must be continuous in I, and 2) $y(t)$ satisfies the differential equation for all t in I.

Example: Two differential equations and their solutions.

1. The function,

$$y(t)=\frac{3}{4}\left(1+t^2\right)+\frac{5}{1+t^2},$$

is a solution to the differential equation

$$\frac{1}{t}\dot{y}+\frac{2}{1+t^2}y=3.$$

2. The function

$$y(t)=c_1e^{-2t}+c_2e^{-3t}+\frac{1}{2}e^{2t},-\infty<t<\infty,$$

where c_1 and c_2 are arbitrary constants, is a solution to the differential equation $\ddot{y}+5\dot{y}+6y=10e^{2t}$

In this case, $y(t)$ is also referred to as a *general* solution because all solutions to the differential equation can be represented in this form for appropriate choices of the constants $c_1$and c_2.

The function $y(t) = \frac{1}{2}e^{2t}$ is a *particular* solution because it contains no arbitrary constants.

With a differential equation, we can associate *initial conditions* or *boundary conditions*, auxiliary conditions on the unknown function and its derivatives.

If these conditions are specified at a single value of the independent variable, they are referred to as initial conditions and the combination of the differential equation and an appropriate number of initial conditions is called an *initial value problem* (IVP).

If these conditions are specified at more than one value of the independent variable, they are referred to as boundary conditions and the combination of the differential equation and the boundary conditions is called a *boundary value problem* (BVP).

Example: Tow examples of IVP*s*.

1. The logistic equation,

$$\dot{p} = ap - bp^2,$$

with initial condition $P(t_0) = p^0$; $p^0 = 10$ for the solution is

$$p(t) = \frac{10a}{10b + (a - 10b)e^{-a}(t - t_0)}.$$

2. The mass-spring system equation,

$$\ddot{x} + (a/m)\dot{x} + (k/m)x = g + (F(t)/m),$$

with the initial conditions $x(0) = x_0, \dot{x}(0) = v_0$; for $m = 10$, $k = 140$, $a = 90$, $F(t) = 5\sin(t)$,$x_0 = 0$, $v_0 = -1$ the solution is

$$x(t) = \frac{1}{500}\left(-90e^{2t} + 99e^{-7t} + 13\sin t - 9\cos t\right).$$

Example: tow examples of BVP*s*.

1. The differential equation,

$$\ddot{y} + 9y = \sin t,$$

with the boundary conditions $y(0) = 1$, $\dot{y}(2\pi) = -1$ the solution $y(t) = \frac{1}{8}\sin t + \cos 3t + \quad \sin 3t$.

2. The differential equation,

$$\ddot{y} + \pi^2 y = 0$$

with boundary conditions $y(0) = 2, y(1) = -2$; the solution is $y(t) = \cos \pi t + c \sin \pi t$, for c an arbitrary constant.

An *n*th-order differential equation is said to be *linear* if it can be written in the form

$$a_n(t)y^{(n)} + a_{n-1}(t)y^{(n-1)} + \cdots + a_1\dot{y} + a_0(t)y = h(t).$$

A *nonlinear* differential equation is simply one that is not linear.

Equation is linear when $G(u)$ is a linear function of; otherwise, it is nonlinear.

Differential equations arising from first principle models are generally nonlinear. Nonlinear equations do not usually yield to analytical approaches and computational methods are called for.

Linear equations constitute a highly important class of differential equations in physics and engineering and are used in idealized models of such phenomena as mechanical vibrations, electrical circuits, planetary motions, etc.

An important property of linear equations is that of *superposition*: To illustrate the superposition principle, consider the following IVP:

$$\ddot{y} = g(t), \qquad y(0) = a, \dot{y} \qquad (0) = \beta.$$

The data for this problem are $\{g(t), \alpha, \beta\}$. If $y_1(t)$ is a solution with data $\{g_1, \alpha_1, \beta_1\}$, and $y_2(t)$ is a solution with data $\{g_2, \alpha_2, \beta_2\}$, then the principle states that $c_1y_1(t) + c_2y_2(t)$ is a solution for the data $\{c_1g_1 +, c_2g_2, c_1\alpha_1 + c_2\alpha_2, c_2\beta_2\}$.

This idea extends readily to th order differential equations.

In practice, superposition permits us to decompose a problem with complicated data into simpler parts, to solve each problem separately, and then to combine these solutions to find the solution to the original problem.

Example: An illustration of superposition.

To solve the IVP,

$$\ddot{y} + 6\dot{y} + 9y = 4e - 3t$$

$$y(0) = 1, \dot{y}(0) = 0,$$

we solve the following two problems:

$$\ddot{y} + 6\dot{y} + 9y = 0$$

$$y(0) = 1, \ \dot{y}(0) = 0$$

a homogeneous equation with the original initial conditions, and

$$\ddot{y} + 6\dot{y} + 9y = 4^{e-3t}$$

$$y\ (0) = 0, \ \dot{y}(0) = 0,$$

an inhomogeneous equation with zero initial conditions. The solution to the first problem is

$$y_1(t) = \frac{1}{4}e^{-3t} + \frac{3}{4}te^{-3t},$$

and the second is

$$y^2 = (y) = 2t^2\ e^{-3t}$$

So, the solution to the original problem is the sum

$$y(t) = y1\ (t) + y2\ (t) = \frac{1}{4}e^{-3t} + \frac{3}{4}te^{-3t} + 2t^2e^{-3t}$$

A solution, $y(t)$, of a differential equation is said to be *stable* if any other solution whose initial data is sufficiently close to that of $y(t)$ remains in a "tube" enclosing $y(t)$; if the solution is not stable, it is said to be *unstable*.

If the diameter of the tube approaches zero as t becomes large, then $y(t)$ is said to be *asymptotically stable*.

In elementary treatments of differential equations it is assumed that the initial value problem has a unique solution that exists throughout the interval of interest and which can be obtained by analytical techniques.

However, many of the differential equations encountered in practice cannot be solve explicitly, so we are led to methods for obtaining approximations to solutions. Such solutions are usually called *numerical* solutions. Matters are also complicated by the fact that solutions can fail to exist over the desired interval of interest. Even more troublesome are problems with more than one solution.

Example: same examples of difficulties in solving IVPs.

1. The differential equation,

$$\dot{y} = e^{t^2},$$

does not have a solution that can be expressed in terms of elementary functions.

2. The IVP,

$$\dot{y} = 1 + y2 , y (0) = 0,$$

has the solution $y(t) = \tan t$ which exists on the interval $0 \leq t \leq 1$ but does not exist on the interval $0 \leq t \leq$ p/2.

3. The IVP,

$$\dot{y} = \sqrt{|1-y^2|}, y(0) = 1$$

does not have a unique solution. In fact, it is not difficult to show that:

(a) $y(t) = 1$ is a solution on any interval containing $t = 0$;

(b) $y(t) = \cosh(t)$ is a solution on any interval $0 \leq t \leq b$ for any $b > 0$;

(c) $y(t) = \cos(t)$ is a solution on $-\pi \leq t \leq 0$ and this is the largest such interval on which $\cos(t)$ is a solution.

Chapter 2

Differentiable Functions

This chapter is devoted to developing some tools from Branch space valued function theory which will be needed in the following chapters. We first define the concept of a Banch space and introduce a number of examples of such which will be used later. We then discuss the notion of differentiabililty of Banach–space valued functions and state an infinite dimensional version of Taylor's theorem.

As we shall see, a crucial result in the implicity function theorem in Banch spaces, a version of this important result, suitable for our purposes is started and proved. As a consequence we derive the Inverse Function theorem in Banach spaces and close thie chapter with an extension theorem for functions defined on proper subsets. of the domain space (the Durgundji extension theorem).

In this chapter we shall mainly be concerned with results for not necessarity linear functions; results about linear operators which are needed in these notes will be quoted as needed.

Banach Spaces

Let E be real (or complex) vector space which is equipped with a norm $\|\cdot\|$, i.e. a function $\|\cdot\|: E \to R_+$ having the properties:

1. $\|\cdot\| \geq 0$, for every $u \in E$,
2. $\|u\| = 0$ is equivalent to $u = 0 \in E$,
3. $\|\lambda u\| = |\lambda| \|u\|$, for every scalar λ and every $u \in E$,
4. $\|u+v\| \geq \|u\| + \|v\|$, for all *u, v,* $u, v, \in E$ (*triangle inequality*).

A norm $\|\cdot\|$ defines a metric $d: E \times E \to \mathbb{R}_+$ by $d(u,v) = \|u-v\|$ and $(E, \|\cdot\|)$ or simply E (if it is understood which norm is being used) is called a Banach space if the metric space (*E*, *d*), *d* defined as above, is complete (i.e. all Cauchy sequences have limits in *E*).

If E is a real (or complex) vector space which is equipped with an inner product, i.e., a mapping

$$\langle \cdot,\cdot \rangle : E \times E \to \mathbb{R}(\text{or } \mathbb{C} \text{ (the complex number)})$$

satisfying

1. $\langle u,v \rangle = \overline{\langle v,u \rangle}, u,v \in E$
2. $\langle u+v,w \rangle = \langle u,w \rangle + \langle v,w \rangle, u,v,w \in E$
3. $\langle \lambda u,v \rangle = \lambda \langle u,v \rangle, \lambda \in \mathbb{C}, u,v, \in E$
4. $\langle u,u \rangle \geq 0, u \in E$, and $\langle u,u \rangle = 0$ if and only if $u = 0$, then E is a normed space with the norm defined by

$$\| u \| \sqrt{\langle u,u \rangle}, u \in E.$$

If E is complete with respect to this norm, then E is called a Hilbert space.

An inner product is special case of what is known as a conjugate linear form, i.e. a mapping $b : E \times E \to \mathbb{C}$ having the properties (i)–(iv) above (with $\langle \cdot . \cdot \rangle$ replaced by $b(\cdot,\cdot)$); in case E is a real vector space, then b is called a bilinear form.

The following collection of spaces are examples of Banach spaces. They will frequently be employed in the applications presented later. The verification that the spaces defined are Banach spaces may be found in the standard literature on analysis.

Spaces of Continuouos Functions

Let Ω be an open subset of $\mathbb{R}^n$, define

$$C^0(\Omega, \mathbb{R}^m) = \{ f : W \to \mathbb{R}^m \text{ such that } f \text{ is continuous on } \Omega \}.$$

Let

$$\| f \|_0 = \sup_{x \in \Omega} | f(x) |,$$

where $| \cdot |$ is a norm in $\mathbb{R}^m$.

Since the uniform limit of a sequence of continuous functions is again continuous, it follows that the space

$$E = \{ f \in C^0(\Omega, \mathbb{R}^m) : \| f \|_0 < \infty \}$$

is a banach space.

If Ω is as above and Ω' is an open set with $\Omega \subset \Omega'$, we let $C^0(\Omega, \mathbb{R}^m) = \{$ the restriction to Ω of $f \in C^0(\Omega', \mathbb{R}^m)$ $\}$.

If Ω is bounded and $f \in C^0(\Omega, \mathbb{R}^m)$, then $\| f \|_0 < +\infty$. Hence $f \in C^0(\Omega', \mathbb{R}^m)$ is a Banach space.

Spaces of Differentiable Functions

Let Ω be an open subset of $\mathbb{R}^n$. Let $\beta = (i_1, \ldots, i_n)$ be a multindex, i.e. $i_k \in \mathbb{Z}$ (the nonnegative integers), $1 \le k \le n$. We let $|\beta| = \sum_{k=1}^{n} i_k$. Let $f : \Omega \to \mathbb{R}^m$, then the partial derivative of f of order β, $D^\beta f(x)$, is givenk by

$$D^\beta (x) = \frac{\partial^\beta f(x)}{\partial^i x_1 \ldots \partial_{in} x_n},$$

where $x = (x_1 \ldots x_n)$ Define $C^j(\Omega, \mathbb{R}^m) = \{f : \Omega \to \mathbb{R}^m$ such that $D^\beta f$ is continuous for all $\beta, |\beta| \le j\}$

Let

$$\| f \| = \sum_{j=0}^{j} \max_{|\beta| \le k} \| f \|_j < +\infty\}$$

Then, using further convergence results for families of differentiable functions it follows that the space

$$E = \{f \in C^j(\Omega, \mathbb{R}^m) = \| f \|_j < +\infty\}$$

is a Banach space.

The space $C^j(\Omega, \mathbb{R}^m)$ is defined in a manner to the space $C^0(\Omega, \mathbb{R}^m)$ and if Ω is bounded $C^j(\Omega, \mathbb{R}^m)$ is a banach space.

Holderspaces

Let Ω be an open setup $\mathbb{R}^n$ A function $f : \Omega \to \mathbb{R}^m$ is called Holdercontinuous with exponent $\alpha, 0, \alpha \le 1$, at a point $x \in \Omega$, if

$$\sup_{x \ne y} \frac{| f(x) - f(y) |}{| x - y |^\alpha} < \infty,$$

and Holder continuous with exponent $\alpha, 0, \alpha \leq 1$, on Ω if it is Holder continuous with the same exponent α at every $x \in \Omega$. For such f we define

$$H_{\Omega}^{\alpha}(f) = \sup_{\substack{x \neq y \\ x,y \in \Omega}} \frac{|f(x) - f(y)|}{|x - y|^{\alpha}}.$$

If $f \in C^{j}(\Omega, \mathbb{R}^{m})$ and $D^{\beta} f, |\beta| = j$, is Holder continuous with exponent α on Ω, we say $f \in C^{j}(\Omega, \mathbb{R}^{m})$. Let

$$\| f \|_{j,\alpha} = \| f \| \, j \, | \max_{|\beta| = j} H_{\Omega}^{\alpha}(D^{\beta} f),$$

then the space

$$E = \{ f \in C^{j,\alpha}(\Omega, \mathbb{R}^{m}) : \| f \|_{j,\alpha} < \infty \}$$

is a Banach space.

As above, one may define the space $C^{j,\alpha}(\overline{\Omega}, \mathbb{R}^{m})$. And again, if Ω is bounded, $C^{j,\alpha}(\overline{\Omega}, \mathbb{R}^{m})$ is a Banach space.

We shall also employ the following convention

$$C^{j,0}(\Omega, \mathbb{R}^{m}) = C^{j}(\Omega, \mathbb{R}^{m})$$

and

$$C^{j,0}(\overline{\Omega}, \mathbb{R}^{m}) = C^{j}(\overline{\Omega}, \mathbb{R}^{m})$$

Functions with compact support

Let Ω be an open subset of $\mathbb{R}^{n}$. A function $f : \Omega \to \mathbb{R}^{m}$ is said to have compact support in Ω if the set supp f = closure $\{x \in \Omega : f(x) \neq 0\} = \overline{\{x \in \Omega : f(x) \neq 0\}}$ is compact.

We let

$C_{0}^{j,\alpha}(\overline{\Omega}, \mathbb{R}^{m}) = \{ f \in C^{j,\alpha}(\Omega, \mathbb{R}^{m}) : =$ supp f is compact subset of $\Omega\}$ and define $C_{0}^{j,\alpha}(\overline{\Omega}, \mathbb{R}^{m})$ similarly.

Then, again, if Ω is bounded, the space $C_{0}^{j,\alpha}(\overline{\Omega}, \mathbb{R}^{m})$ is a Banach space and $C_{0}^{j,\alpha}(\overline{\Omega}, \mathbb{R}^{m}) = \{ f \in C^{j,\alpha}(\Omega, \mathbb{R}^{m}) : f(x) = 0, x \in \partial\Omega \}$.

L^p Spaces

Let Ω be a Lebesgue measurable subset of $\mathbb{R}^n$ and let $f:\Omega \to \mathbb{R}^m$ be a measurable function. Let, for $1 \le p < \infty$,

$$\left(\int_\Omega | f(x) |^p \, dx\right)^{1/p},$$

and for $p = \infty$, let

$$\| f \|_{L\infty} \text{ essup}_{x\in\Omega} | f(x) |,$$

where essup denotes the essential supremum.

For $1 \le p \le \infty$, let

$$L^p(\Omega, \mathbb{R}^m) = \{ f : \| f \|_{Lp} < +\infty \}.$$

Then $L^p(\Omega, \mathbb{R}^m)$ is a Banach space for $1 \le p \le \infty$. The space $L^2(\Omega, \mathbb{R}^m)$ is a Hilbert space with inner product defined by

$$\langle f, g \rangle = \int_\Omega | f(x) | . g(x) dx,$$

where $f(x).g(x)$ is the inner product of $f(x)$ and $g(x)$ in the Hilbert space (Euclidean space) $\mathbb{R}^n$.

Weak derivatives

Let Ω be an open subset of $\mathbb{R}^n$. A function $f:\Omega \to \mathbb{R}^m$ is said to belong to class $L^p_{loc}(\Omega, \mathbb{R}^m)$, if for every compact subset $\Omega' \subset \Omega . f \in L^p(\Omega', \mathbb{R}^m)$. Let $\beta = (\beta_1 \ldots, \beta_n)$ be a multindex. Then a locally integrable fucntion $v \in L^1_{loc}(\Omega, \mathbb{R}^m)$ is called the βth weak derivative of f if it satisfies

$$\int_\Omega v\phi dx = (-1)^{|\beta|} \int_\Omega f D^\beta \phi dx, \text{ for all } \phi \in C_0^\infty(\Omega).$$

We write $v = D^B f$ and note that, up to a set of measure zero, v is uniquely determined. The concept of weak derivative extends the classical concept of derivative and has many similar properties (see e.g. [18]).

Sobolev spaces

We say that $f \in W^k(\Omega, \mathbb{R}^m)$, if f has weak derivatives up to order k, aqnd set $W^{k,p}(\Omega, \mathbb{R}^m) = \{ f \in W^k(\Omega, \mathbb{R}^m) : D^\beta f \in L^p(\Omega, \mathbb{R}^m), | \beta | \le k \}$.

Then the vector space $W^{k,p}(\Omega,\mathbb{R}^m)$ equipped with the norm

$$\| f \| W^{k,p} = \left(\int_\Omega \sum_{|\beta|\le k} | D^\beta f |^p \, dx \right)^{1/p}$$

is a Banach space. The space $C_0^k(\Omega,\mathbb{R}^m)$ is a subspace of $W^{k,p}(\Omega,\mathbb{R}^m)$, it's closure in $W_0^k(\Omega,\mathbb{R}^m)$, is a Banach subspace which, in general, is a proper subspace. For $p = 2$, the spaces $W^{k,2}(\Omega,\mathbb{R}^m)$ and $W_0^{k,2}(\Omega,\mathbb{R}^m)$ are Hilbert spaces with inner product $\langle f,g \rangle$ given by

$$\langle f,g \rangle = \int_\Omega \sum_{|\alpha|\le k} | D^\alpha f . D^\alpha g dx.$$

These spaces play a special role in the linear theory of partial differential equations, and in case Ω satisfies sufficient regularity conditions, they may be identified with the following spaces.

Consider the space $C^k(\bar{\Omega},\mathbb{R}^m)$, as as normal space using the $\|\cdot\| W^{k,p}$ norm. It's completion is denoted by $H^{k,p}(\Omega,\mathbb{R}^m)$. If $p = 2$ it is a Hilbert space with inner product. $H_0^{k,p}(\Omega,\mathbb{R}^m)$ is the completion of $C_0^\infty(\Omega,\mathbb{R}^m)$ in $H^{k,p}(\Omega,\mathbb{R}^m)$.

Spaces in linear operators

Let E and X be normed linear spaces with norms $\|\cdot\|_E$ and $\|\cdot\|_X$, respectively.

Let $£(E;X) = \{f : E \to X$ such that f in linear and continuous$\}$.

For $f \in £(E;X)$, let

$$\| f \|_£ \equiv \sup_{\|x\|_E \le 1} \| f(x) \| x.$$

Then $\|\cdot\|_£$ is a norm for $£(E;X)$. This space is a banach space, whenever X is

Let $E_1,\ldots,E_n$ and X be $n + 1$ normed linear space, let

$£(E_1,\ldots E_n;X) = \{f : E_1 \times \ldots \times E_n \to X$ such that f is multilinear (i.e. f is linear in each variable separately) and continuous$\}$. Let

$$\| f \| = \sup\{\| f(x_1,\ldots,x_n) \| x : \| x_1 \|_{E_1} \le 1,\ldots,\| x_n \| E_n; \le 1\},$$

then $£(E_1,\ldots E_n;X)$ is a normed linear space. It again is a Banach space, whenever X is.

If E and X are normed spaces, one may define the spaces

$$\pounds_1(E;X) = \pounds(E;X)$$
$$f \in \pounds_2(E;X) = \pounds(E;\pounds_1 E;X))$$
$$\vdots \quad \vdots \quad \vdots$$
$$\pounds_n(E;X) = \pounds(E;\pounds_{n-1} E;X)), \; n \geq 2.$$

We leave it as an exercise to show that the spaces $\pounds(E,_{:}..E;X)$ (E repeated n times) and $\pounds_n(E;X)$ may be identified (i.e. there exists an isomorphism between these spaces which is norm preserving.

Differentiability, Taylor's Theorem

Gateaux and Frechet Differentiability

Let E and X be Banach spaces and let U be an open subset of E. Let

$$f : U \to X$$

be a function. Let $x_0 \in U$, then f is said to be Gateaux differentiable (g-differentiable) at x_0 in direction h, if

$$\lim_{t \to 0} \frac{1}{t} \{f(x_0 + th) - f(x_0)\}$$

exists. It said to b e Frechet differentiable (f-differentiable) at x_0, if there exist $T \in E;X)$ such that

$$f(x_0 + h) - f(x_0) = \dot{T}(h) + o(\| h \|)$$

for $\| h \|$ small, here o($\| h \|$) means that

$$\lim_{\|h\| \to 0} \frac{o(\| h \|)}{\| h \|} = 0.$$

(We shall use the symbol $\| \cdot \|$ to denote both the norm of E and the norm of X, since it will be clear from the context in which space we are working.) We note that Frechet differentiability is a more restrictive concept.

It follows from this definition that the Frechet-derivative of f at x_0, if it exists, is unique. We shall use the following symbols interchangeably for the Frechet-derivative of f at x_0; $Df(x_0)$, $f\,'(x_0)$, $df(x_0)$, where the latter is usually used in case $X = \mathbb{R}$. We say that f is of class C^1 in a neighborhood of x_0 if f is Frechet differenetiable there and if the mapping

$$Df; x \to \pounds(E;X)$$

is Frechet-differentiable at $x_0 \in U$, we say that f is twice Frechet -differentiable and we denote the second F- derivative by $D2f(x_0)$, or $f''(x_0)$, or $d2f(x_0)$.

Thus $D^2 f(x_0) \in £_2(E;X)$ .In an analogous way one difines higher order differentiability. If $h \in E$, we shall write

$$D^n f(x_0)(h,\ldots,h) \text{ as } D^n f(x_0)h^n,$$

If $f:\mathbb{R}^n \to \mathbb{R}^n$ is Frechet differentiable at x_0, then $Df(x_0)$ is given by the Jacobian matrix

$$Df(x_0) = \left(\left. \frac{\partial f_i}{\partial x_i} \right|_{x=x_0} \right),$$

and if $f:\mathbb{R}^n \to \mathbb{R}$ is Frechet differentiable, then $Df(x_0)$ is represented by the gradient vector $\nabla f(x_0)$, i.e.

$$Df(x_0)(h) = \nabla f(x_0).h,$$

where "." is the dot product in $\mathbb{R}^n$, and the second derivative $D^2 f(x_0)$ is given by the Hessian matrix $\left. \left(\frac{\partial^2 f}{\partial x_i \partial x_j} \right) h \right|_{x=x_0}$,*i.e.*

$$D^2 f(x_0)h^2 = h^T \left(\frac{\partial^2 f}{\partial x_i \partial x_j} \right) h,\Big|$$

where h^T is the transpose of the vector h.

Tayolor's Formula

1. Theorem *Let* $f : E \to X$ *and all of its Frechet-derivatives* $Df,\ldots,D^m f$ *be continuous on an open set U. Let x and x+h be such that the line segment connecting these points lies in U. Then*

$$f(x+h) - f(x) = \sum_{k=1}^{m-1} \frac{1}{k!} D^k f(x)h^k + \frac{1}{m!} D^m f(z)h^m,$$

where z is a point on the line segment connecting x to x + h. The remâinder $\frac{1}{m!} D^m f(z)h^m$ *is also given by*

$$\frac{1}{(m=1)!}\int_a^b (1-s)^{m-1} D^m f(x_0+sh)h^m ds.$$

We shall not give a proof of this result here, since the proof is similar to the one for functions $f:\mathbb{R}^n \to \mathbb{R}^m$.

Euler- Lagrange Equations

In this example we shall discuss a fundamental problem of variational calculus to illustrate the concepts of differentiation just introduced; specifically we shall derive the so called Euler-Lagrange differential equations.

The equations derived give necessary conditions for the existence of minima (or maxima) of certain functionals.

Let $g:[a,b]\times\mathbb{R}^2 \to \mathbb{R}$ be twice continuously differentiable. Let $E = C_0^2[a,b]$ and let $T: E \to \mathbb{R}$ be given by

$$T(u) = \int_a^b g(t,u(t),u'(t))dt.$$

It then follows from elementary properties of the integral, that T is of class C^1.

Let $u_0 \in E$ be such there exists an open neighborbhood U of u_0 such that

$$T(u_0) \leq T(u)$$

for all $u \in U$ (u_0 is called an extermal of T). Since T is of class C^1 we obtain that for $u \in U$

$$T(u) = T(u_0)DT(u_0)(u-u_0)+a(\| u-u_0 \|).$$

Hence for fixed $v \in E$ and $\varepsilon \in \mathbb{R}$ small,

$$T(u_0+\varepsilon v) = T(u_0)+DT(u_0)(\varepsilon v)+o(|\,\varepsilon\,|\|\,v\,\|).$$

It follows from (13) that

$$0 \leq DT(u_0)(\varepsilon v)o(|\,\varepsilon\,|\|\,v\,\|).$$

and hence. dividing by $\|\,\varepsilon v\,\|$,

$$0 \leq DT(u_0)\left(\pm\frac{v}{\|\,v\,\|}\right)+\frac{|\,o\,|\,\varepsilon v\,\|}{\|\,\varepsilon v\,\|},$$

where $\|\cdot\|$ is the norm in E.

It therefore follows, letting $\varepsilon \to 0$, that for every $v \in E, DT(u_0)(v) = 0$, to derive the Euler- Lagrange equation, we must compute $DT(u_0)$.For arbitrary $h \in E$ we have

$$T(u_0+h) = \int_a^b g(t,u_0(t)+h(t),u_0'(t)+h'(t))dt$$

$$= \int_a^b g(t,u_0(t),u_0'(t)dt$$

$$+\int_a^b \frac{\partial g}{\partial p}(t,u_0(t),u_0'(t))h(t)dt$$

$$+\int_a^b \frac{\partial g}{\partial q}(t,u_0(t),u_0(t))h'(t)dt + o(\| h \|),$$

where p and q denote generic second, respectively, third variables in g.Thus

$$DT(u_0)(h) = \int_a^b \frac{\partial g}{\partial p}(t,u_0(t),u_0'(t))h(t)dt$$

$$+\int_a^b \frac{\partial g}{\partial q}(t,u_0(t),u_0'(t))h'(t)dt.$$

For notation's sake we shall now drop the arguments in g and its partial derivatives. We compute

$$\int_a^b \frac{\partial g}{\partial q} h' dt = \left[\frac{\partial g}{\partial q} h\right]_a^b - \int_a^b h \frac{h}{dt}\left(\frac{\partial g}{\partial q}\right) dt,$$

and since $h \in E$, it follows that

$$DT(u_0)(h) = \int_a^b \left[\frac{\partial g}{\partial p} - \frac{d}{dt}\frac{\partial g}{\partial q}\right] hdt.$$

Since $DT(u_0)(h) = 0$ for all $h \in E$, it follows that

$$\frac{\partial g}{\partial p}(t,u_0(t),u_0'(t)) - \frac{d}{dt}\frac{\partial g}{\partial p}(t,u_0(t),u_0'(t)) = 0,$$

for $a \le t \le b$ (this fact is often referred to as the fundamental lemma of the calculus of variations).

Equaton is called the Euler-Lagrange equationi. If g is twice continuously differentiable becomes

$$\frac{\partial g}{\partial p} - \frac{\partial^2 g}{\partial t \partial q} - \frac{\partial^2 g}{\partial p \partial q} u_0' - \frac{\partial^2 g}{\partial p^2} u_0'' = 0,$$

where it again is understood that all partial derivatives are to be evaluated at $(t, u_0(t), u_0'(t))$.

We hence conclude that an extremal $u_0 \in E$ must solve the nonlinear differential equation.

Some Special Mappings

Throughout our text we shall have occasion to study equations defined by mappings which enjoy special kinds of properties.

We shall briefly review some such properties and refer the reader for more details discussions to standard texts on analysis and functional analysis.

Completely continuous mappings

Let E and X be Banach spaces and let Ω be an open subset, of E, let

$$f : \Omega \to X$$

be a mapping. Then f is called compact, whenever $f(\Omega')$ is compact is X). We call f completely continuous whenever f is compact and continuous. We note that if f is linear and compact, then f is completely continuous.

Lemma: *Let W be an open set in E and let $f : \Omega \to X$ be completely continuous, let f be F-differentiable at a point $x_0 \in \Omega$. Then the linear mapping $T = Df(x_0)$ is compact, hence completely continuous.*

Proof: Since T is linear it suffices to show that $T(\{x : \| x \| \leq 1\})$ is precompact in X. (We again shall use the symbol $\| \cdot \|$ to denote both the norm in E and in X.) If this were not the case, there exists $\varepsilon > 0$ and a sequence $\{xn\}_{n=1}^{\infty} \subset E, \| x_n \| \leq 1, n = 1, 2, 3, \ldots$ such that

$$\| Tx_n - Tx_m \| \geq \varepsilon, n \neq m.$$

Choose $o < \delta < 1$ such that

$$\| f(x_0 + h) - f(x_0) - Th \| < \frac{\varepsilon}{3} \| h \|,$$

for $h \in E, \| h \| \leq \delta$. Then for $n \neq m$

$$\| f(x_0 + \delta x_n) - f(x_0 + \delta x_m) \| \geq \delta \| Tx_n - Tx_m \|$$

$$- \| f(x_0 + \delta x_n) - f(x_0) - \delta Tx_n \| - \|$$

$$f(x_0+\delta x_m)-f(x_0)-\delta Tx_m \|$$

$$\geq \delta\varepsilon - \frac{\delta\varepsilon}{3} - \frac{\delta\varepsilon}{3} = \frac{\delta\varepsilon}{3}.$$

Hence the sequence $\{f(x_0+\delta x_n)\}_{n=1}^{\infty}$ has no convergent subsequence. On the other hand, for δ > 0, small the set $\{f(x_0+\delta x_n)\}_{n=1}^{\infty} \subset \Omega$, and is bounded, implying by the complete continuity of *f* that $\{f(x_0+\delta x_n)\}_{n=1}^{\infty}$ is precompact. We have hence arrived at a contradiction.

Proper mappings

Let $M \subset E, Y \subset X$ and let $f : M \to Y$ be continuous, then *f* is called a proper mapping if for every compact subset *K* of Y, $f^{-1}(K)$ is compact in *M*. (Here we consider *M* and *Y* as metric spaces with metrics induced by the norms of *E* and *X*, respectively.)

Lemma: *Let $h : E \to X$ be completely continous and let $g : E \to X$ be proper, then f = g – h is a proper mapping, provided that f is coercive, i.e.*

$$\| f(x) \| \to \infty \text{ as} \| x \| \to \infty.$$

Proof: Let *K* be a compact subset of *X* and let *N* - $f^{-1}(K)$. Let $\{x_n\}_{n=1}^{\infty}$ be a sequence in *N*. Then exists $\{y_n\}_{n=1}^{\infty} \subset K$ such that

$$y_n = g(x_n) - h(x_n).$$

Since *K* is compact, the sequence $\{y_n\}_{n=1}^{\infty}$ has a convegent subsequence, and since *f* is coercive the sequence $\{x_n\}_{n=1}^{\infty}$ must be bounded, further, because *h* is completely continuous, the sequence $\{h(x_n)\}_{n=1}^{\infty}$ must have a convergent subsequence. It follows that the sequence $\{g(x_n)\}_{n=1}^{\infty}$ has a convergent subsequence. Relabeling, if necessary, we may assume that all three sequences $\{y_n\}_{n=1}^{\infty}, \{g(x_n)\}_{n=1}^{\infty}$ and $\{h(x_n)\}_{n=1}^{\infty}$ are convergent. Since

$$g(x_n) = y_n + h(x_n)$$

and *g* is proper, it follows that $\{x_n\}_{n=1}^{\infty}$ converges also, say $x_n \to x$; hence *N* is precompact. That *N* is alos closed follows from the fact that *g* and *h* are continuous.

Corollary: Let $h : E \to E$ *be a completely continuous mapping, and let f = id –h be coercive, then f is proper (here id is the identity mapping).*

Proof: We note that id: $E \to E$ is a proper mapping.

In finite dimensional spaces the concepts of coercivity and properness are equivalent, i.e. we have:

Lemma: *Let $f:\mathbb{R}^n \to \mathbb{R}^m$ be continuous, then f is proper if and only if f is coercive.*

Contraction Mappings, the Banach Fixed Point Theorem

Let *M* be a subset of a Banach space E. A function $f: M \to E$ is called a contraction mapping if there exists a constant $k, 0 \le k < 1$ such that

$$\| f(x) - f(y) \| \le k \| x - y \|,$$

Theorem: Let *M be a closed subset of E and $f: M \to M$ be a contraction mapping, then f has a unique fixed point in M; i.e. there exists a unique $x \in M$ such that*

$$f(x) = x.$$

Proof: If $x, y \in M$ both satisfy (20), then

$$\| x - y \| - f(x) - f(y) \| \le k \| x - y \|,$$

hence, since k < 1, *x* must equal y, establishing uniqueness of a fixed point.

To proof existence, we define a sequence $\{x_n\}_{n=1}^{\infty} \subset M$ inductively as follows: Choose $x_0 \in M$ and let

$$xn = f(x_{n-1}),\ n \ge 1.$$

implies that for any $j \ge 1$

$$\| f(x_j) - f(x_j - 1) \| \le k^j \| x_1 - x_0 \|$$

and hence if $m > n$

$$x_m - x_n = x_m - x_{m-1} + x_{m-1} - \ldots + x_{n+1-x_n}$$

$$= f(x_{m-1}) - f(x_{m-2}) + \ldots + f(x_n) - f(x_{n-1})$$

and therefore

$$\| x_m - x_n \| \le \| x_1 - x_0 \| (k^n + \ldots + k^{m-1}) = \frac{k^n - k^m}{1-k} \| x_1 - x_0 \|.$$

It follows from that $\{x_n\}_{n=0}^{\infty}$ is a Cauchy sequence in *E,* hence

$$\lim_{n \to \infty} x_n = x$$

exists and since *M* is closed, $x \in M$.

Remark: We note that the above theorem, Theorem 6, alos holds if E is a complete metric space with metric d.This easily seen by replacing $||x-y||$ by $d(x, y)$ in the proof.

In the following example we provide an elementary approach to the existence and uniqueness of a solution of a nonlinear boundary value problem. The approach is based on the L^P theory of certain linear differentail operators subject to boundary constraints.

Let $T > 0$ be given and let

$$f:[0,T]\times\mathbb{R}\times\mathbb{R}\rightarrow\mathbb{R}$$

be a mappinkg stisfying Caratheodory condition; i.e. $f(t,u,u')$ is continuous in $(u.u')$ for almost all t and measurable in t for fixed (u, u').

We consider Dirichlet problem, i.e. the problem of finding a function u satisfying the following differential equation subject to boundary conditions

$$\begin{cases} u''=f(t,u,u')0<t<T, \\ y=0,t\in\{0.T\}. \end{cases}$$

In what is to follow, we shall employ the notation that $|\cdot|$ stands for absolute value in $\mathbb{R}$ and $||\cdot||_2$ the norm in $L^2(0,T)$.

We have the following results:

Theorem: Let *f satisfy*

$$|f(t,u,v)-f(t,\tilde{u},\tilde{v})|\leq a|u-\tilde{u}|+b|v-\tilde{v}|,$$

$$\forall u,\tilde{u},v,\tilde{v}\in\mathbb{R},0<t<T,$$

where a,b are nonegative constants such that

$$\frac{a}{\lambda_1}+\frac{b}{\sqrt{\lambda_1}}<1,$$

and λ_1 *is principal eigenvalue of* $-u''$ *subject to the Dirichlet boudary conditions* $u(0) = 0 = u(T)$ *(i.e. the smallest number* λ *such that the problem*

$$\begin{cases} -u''=\lambda u,0<t<T, \\ u=0,t\in\{o,T\} \end{cases}$$

has a nontrivial solution). Then problem has a unique solution $u \in C_0^1([0,T])$, *with* u' *absolutely continuous and the equation being satisfied almost everywhere.*

Proof: Results from elementary differential equations tell us that λ_1 is the first postive number λ such that the problem has a nontrivial solution, i.e. $\lambda_1=\frac{r^2}{T^2}$

To prove the theorem, let us, for $v \in L^1(0,T)$, put

$$Av = f(.,w,w'),$$

where

$$w(t) = \frac{t}{T}\int_0^T \int_0^T v(s)dsdr$$

$$+\int_0^t \int_0^t v(s)dsdr,$$

which, in turn may be rewritten as

$$w(t) = \int_0^T G(t,s)v(s)ds,$$

where
$$G(t,s) = -\frac{1}{T}\begin{cases} (T-t)s, \text{if } 0 \leq s \leq t \\ t(T-s)s, \text{if } t \leq s \leq T. \end{cases}$$

It follows that the operator A is a mappinkg of $L^1(t,T)$ to any $L^q(o,T)$ $q \geq 1$. On the other hand we have that the imbedding

$$L^q(o,T) \mapsto L^1(0,T), q \geq 1,$$

$$u \in L^q(0,T) \mapsto u \in L^1(0,T),$$

is a continuous mapping, since

$$\| u \|_{L1} \leq T\frac{9}{q-1} \| u \|_{Lq.}$$

We hence may consider

$$A : L^q(0,T) \to L^q(0,T),$$

for any $q \geq 1$. In carring out the computations in the case q =2, the following inequalities will be used; their proofs may be obtained using Fourier series methods, and will be left as an exercise. We have for $w(t) = \int_0^T G(t,s)v(s)ds$ that

$$\| w \| L \leq \frac{1}{\lambda_1} \| v \|_{L^2},$$

from which easily follows, view an integration by parts, that

$$\| w' \| L \leq \frac{1}{\lambda_1} \| v \|_{L^2},$$

Using these facts in the computatons one obtains that A is a contraction mapping. On the other hand, if $v \in L^2(0,T)$ is a fixed point u of A. Then

$$u(t) = \int_0^T G(t,s)v(s)ds$$

is in $C_0^1(0,T)$ and $u'' \in L^2(0,T)$.

Remark: It is clear from the proof that in the above the real line $\mathbb{R}$ may be replaced by $\mathbb{R}^m$ thus obtaining a result for systems of boundary value problems

Remark: In case $T = \pi$, $\lambda_1 = 1$ and condition becomes

$$a + b < 1,$$

whereas a classical result of Picard requires

$$a\frac{\pi^2}{8} + b\frac{\pi}{8} < 1,$$

Remark: Theorem may be somewhat extended using a result of Opial which says that for $u \in C_0[0,T]$, with u' absolutely continuous, we have that

$$\int_0^T |u(x)||u'(x)|\,dx \le \frac{T}{4}\int_0^T |u'(x)|^2\,dx.$$

The derivation of such a statement is left as an exercise.

The Implicit Function Theorem

Let us now assume we have Banach spaces E,X,A and let

$$f : U \times V \to X,$$

(where U is open in E, V is open in A) be a continuous mapping statisfying the following condition:

1. For each l ∈ V the map $f(., l)$: $U \to X$ is frechet-differentiable on U with Frechet derivative

   $$D_u f(u, l)$$

   and the mapping $(u, \lambda) \mapsto D_u f(u,\lambda)$ is a continuous mapping from $U \times V$ to £(E, X).

Theorem: (Implicit Function Theorem): *Let f satisfy* (31) *and let there exist* $(u_0, \lambda_0) \in U \times V$ *such that* $D_u f(u_0, \lambda_0)$ *is a linear homeomorphism of E onto X* (i.e. $D_u f(u_0, \lambda_0) \in$ £(E, X) *and* $[D_u f(u_0, \lambda_0)]^{-1} \in$ £(X, E)). *Then there exist* $\delta > 0$ *and* $r > 0$ *and unique mapping u:* $B_\delta(\lambda_0) = \{\lambda / \|\lambda - \lambda\| \swarrow \delta\} \to E$ *such that*

$$f(u(\lambda), \lambda) = f(u_0, \lambda_0),$$

and $\|u(\lambda) - u_0\| \swarrow r, u(\lambda_0) = u_0$.

Proof: Let us consider the equation

$$f(u, \lambda) = f(u_0, \lambda_0)$$

which is equivlent to

$$[D_u f(u_0, \lambda_0)]^{-1} (f(u, \lambda) - f(u_0, \lambda_0) \underline{\underline{def}}\ G(u, \lambda).$$

The mapping G has the following properties:

1. $G(u_0, l_0) = u_0,$
2. G and $D_u G$ are continuous in (u, λ),
3. $D_u G(u_0, \lambda_0) = 0$

Hence

$$\|G(u_1, \lambda) - G(u_2, \lambda)\| \leq (\sup_{0\leq t\leq 1} \| D_u G(u_1 + t(u_2 - u_1), \lambda) \|) \| u_1 - u_2 \|$$

$$\leq \frac{1}{2} \| u_1 - u_2 \|$$

provided $\| u_1 - u_0 \| \leq r, \| u_2 - u_0 \| \leq r,$ *where r is small enough, Now*

$$| G(u,\lambda) - u_0 \| = \| G(u,\lambda) - \dot{G}(u_0,\lambda_0)$$

$$\| \leq \| G(u,\lambda) - G(u_0,\lambda) \|$$

$$+ \| G(u_0,\lambda) - G(u_0,\lambda_0) \| \leq \frac{1}{2} \| u - u_0 \|$$

$$+ \| G(u_0,\lambda) - G(u_0,\lambda_0) \|$$

$\leq \frac{1}{2} + \frac{1}{2} r$, provided $\|\lambda - \lambda_0\| \swarrow \delta$ is small enough so that $G(u_0,\lambda) - G(u_0,\lambda_0) \| \leq \frac{1}{2} r$

Let $B_\delta(\lambda_0) = \{\lambda : \| \lambda - \lambda_0 \| \leq \delta\}$ and define $M = \{u : B_\delta(\lambda_0) \to E$ such that u is continuous, $u(\lambda_0) = u_0, \| u(\lambda) - u_0 \|_0 \leq r,$ and $\| u \|_0 = \sup_{\lambda \in B_\delta(\lambda_0)} \| u(\lambda) \| < +\infty\}$.

Then M is a closed subset of a Banach space and defines an equation

$$y(\lambda) = G(u(\lambda), \lambda)$$

in M.

Define g by (here we think of u as an element of M)

$$g(u)(\lambda) = G(u(\lambda), \lambda).$$

then $g : M \to M$ and it follows by (36) that

$$\| g(u) \rightarrow g(v) \|_0 \leq \frac{1}{2} \| u - v \|_0,$$

hence g has a unique fixed point by the contraction mapping principle.

Remark: If in the implicit function theorem f is k times continuously differentiable, then mapping $\lambda \mapsto u(\lambda)$ inherits this property.

Example: *As an example let us consider the nonlinear boundary value problem*

$$u'' + \lambda e^u = 0, 0 < t < \pi, u(0) = 0 = u(\pi)$$

This is a one space-dimensional mathmatical model from the theory of combustion and u represents a dimensionless temperature. We shall show, by an application of Theorem that for $\lambda \in \mathbb{R}$, in a neighborhood of 0, has unique solution of small norm in $C^2([0,\pi], \mathbb{R})$.

To this end we define

$$E = C_0^2([0,\pi], \mathbb{R})$$

$$X = C^0[0,\pi]$$

$$A = \mathbb{R},$$

these spaces being equipped with their usual norms. Let

$$f : E \times A \rightarrow X$$

be given by

$$f(u,\lambda) = u'' + \lambda e^u.$$

Then f is continuous and $f(0,0) = 0$. (When $\lambda = 0$ (no heat generation) the unique solution is $u \equiv 0$.) . Furthermore, for $u_0 \in E, D_u f(u_0, \lambda)$ is given by 9the reader should carry out the verfication)

$$D_u f(u_0, \lambda) = v'' + \lambda e^{u_0(x)} v,$$

and hence the mapping

$$f(u,\lambda) \mapsto D_u f(u,\lambda)$$

is continuous. Let us consider the linear mapping

$$T = D_u f(0,0) + E \rightarrow X.$$

We must show that this mapping is a linear homeomorphism. To see this we note that for every $h \in X$, the unique solution of

$$v'' = h(t), 0 < t < \pi, v(0) = 0 = v(\pi).$$

is given by $$v(t) = \int_0^{\pi} G(t,s)h(s)ds,$$

where

$$G(x,s) = \begin{cases} -\frac{1}{\pi}(\pi - t)s, 0 \leq s \leq t \\ -\frac{1}{\pi}(\pi - s), t \leq s \leq \pi. \end{cases}$$

From the representation we may conclude that there exists a constant c such that

$$\| v \|_2 = \| T^{-1}h \|_2 \leq c \| h \|_0,$$

i.e. T^{-1} is one to one and continuous. Hence all conditions of the implicit function theorem are satisfied and we may conclude that for each λ, λ sufficiently small, has a unique small solution $u \in C^2([0,\pi],\mathbb{R})$, furthermore the map $\lambda \to u(\lambda)$ is continuous from a neighborhood of $0 \in \mathbb{R}$ to $C^2([0,\pi],\mathbb{R})$.

We later shall show that thsi 'solution branch' $(\lambda, u(\lambda))$ may be globallyil continued. To this end we note here that the set is bounded above. We observe that if $\lambda > 0$ is such that has a solution, then the corresponding solution u must be postive, $u(x) > 0$, $0 > x < \pi$. Hence

$$0 = u'' + \text{l}e^u > u'' + \text{l}\, u.$$

Let $v(t) = \sin t$, then v satisfies

$$v'' + v = 0, 0 < t < \pi, v(0) = 0 = v(\pi)$$

we obtain

$$\int_0^{\pi} (u''v - v''u)dt + (\lambda - 1)\int_0^{\pi} uvdt,$$

and hence, integrating by parts,

$$0 > (\lambda - 1)\int_0^{\pi} uvdx,$$

implying that $\lambda < 1$.

INVERSE FUNCTION THEOREMS

We next proceed to the study of the inverse of a given mapping and provide two inverse function theorems. Since that first result is proved in exactly the same way as its finite dimensioanal snalogure (it is an immediate consequence of the implicit function theorem) we shall not prove it here.

Theorem: Let E and X be Branch spaces and let U be an open neighborhood of $a \in E$. Let $f: U \rightarrow X$ be a C^1 mapping with $Df(a)$ a linear homeomorphism of E onto X. Then there exist open sets U' and v, $a \in U'$, $f(a) \in V$ and a uniquely determined function g such that:

1. $V = f(U')$,
2. *f is one to one on* U',
3. $g: V \rightarrow U'$, $g(V) = U', (V(u)) = u$, *fro every* $u \in U'$,
4. *g is a* C^1*function of* V *and* $Dg(f(a)) = [Df(a)]^{-1}$.

Example: *Consider the forced nonlinear oscillator (periodic boudary value problem)*

$$u'' + \lambda u + u^2 = g, u(0) = u(2\pi), u'(0) = u'(2\pi)$$

where g is a continous 2π*– periodic function and* $l \in \mathbb{R}$, *is a parameter,*

Let $E = C^2([0, 2\pi], \mathbb{R}) \cap \{u : u(0) = u(2\pi), u'(0) = u'(2\pi)\}$, and $C^0([0, 2\pi], \mathbb{R})$, *where both spaces are equipped with the noms discussed earlier, Then for certain values of* l, *has a unique solution for all forcing terms g of small norm.*

Let

$$f : E \rightarrow X$$

be given by

$$f(u) = u'' + \lambda u + u^2.$$

Then $Df(u)$ is defined by

$$(Df(u))(v) = v'' + lv + 2uv.$$

and hence the mapping

$$u \mapsto Df(u)$$

is a continuous mapping of E to £$(E;X)$, i.e. f is a C^1 mapping. It follows from elementary differential equations theory (see eg. [4]) that the problem

$$v'' + \lambda v = h,$$

has a unique 2π–periodic solution for every 2p- periodic h as long as $\lambda \neq n^2$, $n = 1,2....$, and that $\|v\|_2 \swarrow C \|h\|_0$ for every 2π–periodic h as long as $\lambda \neq n^2$, has a unique solution $u \in E$ of small norm for every $g \in X$ of small norm.

We note that the above example is prototypical for forced nonlinear oscillators. Virtually the same arguments can be applied (the reader might carry out the necessary calculations) to conclude that the forced pendulum equation

$$u'' + \lambda \sin u = g$$

has a 2π-. periodice response of small norm for very 2π - periodic forcing terms g of small norm, as long as $\lambda \neq n^2$, $n = 1,2,\ldots$.

In many physicall situations (see the example below) it is of interest to know the number of solutions of the equation describing this situation. The following result describles a class of problems where the precise number of solutions (for every given forcing terms) may be obtained by simply knowing the number of solutions for some fixed forcing term.

Let M and Y metric spaces (e.g. subsets fo Banach spaces with metric induced by the norms).

Theorem: Let $f : M \to Y$ be continuous, proper and locally invertible (e.g. Theorem 15 is applicable at each point). For $y \in Y$, let

$$N(y) = \text{cardinal number of } \{f^{-1}(y)\} = \#\{f^{-1}(y)\}$$

Then the mapping

$$N \to N(y)$$

is finite, Since $\{y\}$ is compact $\{f^{-1}(y)\}$

Proof: We first show that for each $y \in Y$, $N(y)$ is finite. Since $\{y\}$ is compact also, because f is proper mapping. Since f is locally invertible, there exists, for each $u \in \{f^{-1}(y)\}$ a neighborhood O_u such that

$$O_u \cap (\{f^{-1}(y)\} \setminus \{u\}) = 0,$$

and thus $\{f-^1(y)\}$ is a discrete and compact set, hence finite.

We next show that N is continuous mapping to the nonnegative integers, which will imply that N is constant–valued. Let $y \in Y$ and let $\{f^{-1}(y)\} = \{u_1,\ldots,u_n\}$. We choose disjoint open neighborhoods O_i of u_i $1 \swarrow i \swarrow n$ and let $I = \bigcap_{i=1}^{n} f(O_i)$, Then there exist open sets V_i, $u_i \in V_i$ such that f is a homeomorphisim from V_i to I. We next claim that there exists a neighborhood $W \cdot \subset I$ of y such that N is constant on W. For it not, there will exist a squence $\{y_m\}$, with $y_m \to y$, such that $m \to \infty, N(y_m) > N(y)$ (note that for any $v \in I$, v has a preimage in each V_i Which implies $N(v) \mp N(y), v \in I!$).

Hence there exists a sequence $\{\xi_m\}, \xi_m \notin \bigcap_{i=1}^{n} V_i$, such that $f(\xi_m) = y_m$. Since $f^{-1}(\{y_n\} \cup \{y\})$ is compact, the sequence (ξ_n) will have a convergent subsequence, say $\xi_{nj} \to \xi$. And since f is continuous, $f(\xi) = y$. Hence $\xi = u_i$, for some i, a contradiction to $\xi \notin \bigcap_{i=1}^{n} V_i$.

18.Corollary *Assume Y is connected, then N(Y)* is constant.

Examples illustrating this result will be given later in the text.

The Dugundji Extension Theorem

In the course of developing the Brouwer and Leray–Schauder degree and in proving some of the classical fixed point theorems we need to extend mappings defined on proper subsets of a Banach space to the whole space in a suitable manner. The result which guarantees the existence of extensions having the desired properties is the Dugundji extension theorem which will be established in this section.

In proving the theorem we need a result from general topology which we state here for convenience. We first give some terminology.

Let M be a metric space and let $\{O_\lambda\}_{\lambda\in A,}$ where A is an index set, be an open cover of M. Then $\{O_\lambda\}_{\lambda\in A,}$ is called locally finite if every point $u \in M$ has a neighborhood U such that U intersects at most finitely many elements of $\{O_\lambda\}_{\lambda\in A,}$.

Lemma: Let M be metric space. Then every open cover of M has a locally finite refinement.

Theorem: Let E and X be Banach spaces and let $f: C \to K$ be a contiuous mapping, where C is closed in E and K is convex in X. Then there exists a continuous mapping

$$\tilde{f}: E \to K$$

such that

$$\tilde{f}(u): f(u), u \in C.$$

Proof: For each $u \in C \setminus C$ let

$$r_u = \frac{1}{3} dist(u, C)$$

and

$$B_u = \{v \in E : \| v - u < r_u\}.$$

Then

$$diam B_u \leq dist(B_u, C).$$

The collection $\{B_u\} u \in E \setminus C$ is an open cover of the metric space E/C and hence has a locally finite refinement $\{O\lambda\}_{\lambda\in A,}$ i.e.

1. $U_{\lambda\in A} O\lambda \subset E/C,$

2. For each $\lambda \in A$ there exists B_u such that $O\lambda \subset B_u$
3. $\{O_\lambda\}_{\lambda\in A}$ is locally finite.

Define

$$q : E \setminus C \to (0,\infty)$$

by

$$q:(u) = \sum_{\lambda\in A} dist(u, E \setminus O\lambda).$$

The sum in the right hand side contains only finitely many terms, since $\{O_\lambda\}_{\lambda\in A}$ is locally finte. This also implies that q is a continuous function.

Define

$$\rho_\lambda(u) = \frac{dist(u, E \setminus O_\lambda)}{q(u)}, l \in A, u \in E \setminus C.$$

It follows for $\lambda \in A$ and $u \in E \setminus C$ that

$$0 \le \rho_\lambda(u) \le 1, \sum_{\lambda\in A} \rho_\lambda(u) = 1$$

For each $\lambda \in A$ choose $u\lambda \in C$ such that

$$\text{dist}(u_\lambda, O_\lambda) \le 2\text{dist}(C, O_\lambda)$$

and define

$$\tilde{f}(u) = \begin{cases} f(u), u \in C \\ \sum_{\lambda\in A} \rho_\lambda(u) f(u_\lambda), u \notin C. \end{cases}$$

Then $\tilde{f}$ has the following properties:

1. $\tilde{f}$ is defined on E and is an extension of f.
2. $\tilde{f}$ is continuous on the interior of C.
3. $\tilde{f}$ is continuous of $E \setminus C$.

These properties follow immediately from the definition of $\tilde{f}$. To show that $\tilde{f}$ is continuous of E it suffices therefore to show that $\tilde{f}$ is continuous of ∂C. Let $u \in \partial C$, then since f is continuous we may, for given $\varepsilon > 0$, find $0 < \delta = \delta(u, \varepsilon)$ such that

$$\| f(u) - f(v) \| \le \varepsilon, if \| u - v \| \le \delta, v \in C.$$

Now for $v \in E \setminus C$

$$\| \tilde{f}(u) - \tilde{f}(v) \| = \| f(u) - \sum_{\lambda \in A} \rho_\lambda(v) f(u_\lambda) \| \leq$$

$$\sum_{\lambda \in A} \rho_\lambda(v) \| f(u) - f(u_\lambda) \|$$

If $\rho_\lambda(v) \neq 0$, $\lambda \in A$, then dist $(v, E \backslash O_\lambda) > 0$, i.e. $v\ O_\lambda$. Hence $\| v - u_\lambda \| \leq \| v - w \| + \| w - u_\lambda \|$ for any $w \in O_\lambda$. Since $\|v - w\| \leq$ diamO_λ we may take the infimum for $w \in O_\lambda$ and obtain

$$\| v - u_\lambda \| \leq \text{diam} O_\lambda + \text{dist}(u_\lambda, O_\lambda).$$

Now $O_\lambda \subset B_{u1}$ for some $u_1 \in E \setminus C$. Hence, since

$$\text{diam} O_\lambda \leq \text{diam} B_{u1} C) \leq \text{dist}(C, O_\lambda),$$

we get

$$\| v - u_\lambda \| \leq 3\text{diam}(C, O_\lambda) \leq 3(v - u)$$

Thus for λ such that $\rho_\lambda(v) \neq 0$ we get $\| u - u_\lambda \| \leq \| v - u \| + \| v - u_\lambda \| \leq 4 \| u - v \|$. Therefore if $\| u - v \| \leq \delta/4$, then $\| u - u_\lambda \| \leq \delta$, and $\| f(u) - f(u_\lambda) \| \leq \varepsilon$, and therefore $\| \tilde{f}(u) - \tilde{f}(v) \leq \varepsilon \sum_{\lambda \in A} \rho_\lambda(v) = \varepsilon$.

Corollary: Let *E*, *X* be banach spaces and let $f : C \to X$ be continuous, where *C* is closedd in *E*. Then *f* has a continuous extension $\tilde{f}$ to *E* such that

$$\tilde{f}(E) \subset \infty f(C),$$

where $\infty f(C)$ is the convex hull of *f*(*C*).

Corollary: Let *K* be a closed convex subset of a Banach space. *E*. Then there exists a continuous mapping $f : E \to K$ such that *f*(*E*) = *K* and *f*(*u*) = *u*. *u* ∈ *K*, *i.e*. *K* is a continuous retract of *E*.

$$u' = Au \mid g(t, u)$$

and show how to construct Lyapunov functionals to test the stability of the trivial solution of this system.

The type of Lyapunov functional we shall be looking for are of the form

$$v(x) = x^T Bx,$$

where B is a constant $N \times N$ matrix, i.e. we are looking for v as a quadratic form. If u is a solution of then

$$\frac{dv^*}{dt} = \frac{dv(t,u(t))}{dt}$$

$$= u^T(A^T B + BA)u + g^T(t,u)Bu + u^T Bg(t,u).$$

Hence, given, A if B can be found so that $C = A^T B + BA$ has certain definniteness properties, then the results of the trivial solution. To proceed along these lines we need some linear algebra results.

Proposition: Let A be a constant $N \times N$ matrix having the property that for any eigenvalue λ of A, $= \lambda$ is not an eigenvalue. Then for any $N \times N$ matrix C. there exists a unique $N \times N$ matrix B such that $C = A^T B + BA \cdot$

Proof: On the space of $N \times N$ matrices define the bounded linear operator L by

$$L(B) = A^T B + BA \cdot$$

Then L may be viewed as a bounded linear operator of $\mathbb{R}^{N\times N}$ to itself, hence it will be a bijection provided it does not have 0 as an eigenvalue, Once we show the latter, the result follows. Thus let μ be an eigenvalue of L, i.e., there exists a nonzero matrix B such that

$$L(B) = A^T B +, \mu B$$

Hence

$$A^T B + B(A - \mu I) = 0$$

From this follows

$$B(A - \mu I)^n = (-A^T)^n B,$$

for any integer $n \geq 1$, hence for any polynomial p

$$Bp(A - \mu I)^n = p(-A^T)B.$$

Since, on the other hand if F and G are two matrices with no common eigenvalues, there exists a polynomial p such that $p(F) = I, p(G) = 0$, implies that $A - \mu I$ and $_-A^T$ must have a common eigenvalue,. From which follows taht μ is the sum of two eigenvalues of A, which, by hypothesis connot equal 0.

This proposition has the following corollary.

Corollary: Let A be a constant $N \times N$ matrix. Then for any $N \times N$ matrix C, there exists $\mu > 0$ and a unique $N \times N$ matrix B such that $2\mu B + C = A^T B + BA$.

Proof: Let

$$S = \{\lambda \in \mathbb{C} : \lambda = \lambda_1 + \lambda_2\},$$

where λ_1 and λ_2 are eigenvalues of A. Since S is a finite set, there exists $r_0 > 0$. such that $\lambda(\neq 0) \in S$ implies that $|\lambda| > r_0$. Choose $0 < \mu \leq r_0$ and consider the matrix $A_1 = A - \mu I$. We may now apply Proposition 21 to the matrix A_1 and find for a given matrix C, a unique matrix B such that $C = A_1^T B + BA_1$, i.e. $2\mu BC = A^T B + BA$.

Corollary: Let A be a constant $N \times N$ matrix haivng the property that all eigenvalues λ of A have negative real parts. Then for any negative definite $N \times N$ matrix B such that $C = A^T B + BA.$

Proof: Let C be a negative definite matrix and let B be given by Proposition, which may be applied since all eigenvalues of A have negative real part. Let $v(x) = x^T Bx$, and let u be a solution of $u' = Au, u(0) = x_0 \neq 0$ Then

$$\frac{dv^*}{dt} = \frac{dv(u(t))}{dt} = u^T(A^T B + BA)u$$

$$= u^T Cu \leq -\mu |u|^2,$$

since C is negative definite. Since $\lim_{t\to\infty} u(t) = 0$ (all eigenvalues of A have negative real part!), it follows that $\lim_{t\to\infty} v(u(t)) = 0$. We also have

$$v(u(t)) \leq v(x_0) - \int_0^t \mu |u(s)|^2 \, ds,$$

from which follows that $v(x_0) > 0$. Hence B is positive definite.

The next corollary follows from stability theory equations and what has just been discussed.

Corollary: A necessary and sufficient condition that an $N \times N$ matrix A have all of its eigenvalues with negative real part is that there exists a unique positive definite matrix B such that

$$A^T B + BA = -I.$$

We next consider the nonlinear problem with

$$g(t, x) = o(|x|),$$

uniformly with respect to $t \in [t_0, \infty)$, and show that for certain types of matrices A the trivial solution of the perturbed system has the same stability property as that of the unperturbed problem. The class of matrices we shall consider is the following.

Definition: We call an $N \times N$ matrix A critical if all its eigenvalues have non-positive real part and there exists at least one eigenvalue with zero real part. We call it noncritical otherwise.

Theorem: Assume A is a noncritical $N \times N$ matrix and let g satisfy.

Then the stability behavior of the trivial solution of is uniformly asymptotically stable if all eigenvalues of A have negative real part and it is unstable if A has an eigenvalue with positive real part.

Proof: Assume all eigenvalues of A have negative real part. By the above exists a unique positive definite matrix B such that

$$A^T B + BA = -I.$$

Let $v(x) = x^T Bx.$ Then v is positive definite and if u is a solution it satisfies

$$\frac{dv*}{dt} = \frac{dv(u(t))}{dt}$$

$$= -[u(t)|^2 | g^T(t,u)Bu + u^T Bg(t,u).$$

Now

$$| gT(t,u)bu + u^T Bg(t,u) | \leq 2 | g(t,u) || B || u |.$$

Choose $r > 0$ such that $| x | \leq r$ implies

$$| g(t,u) | \leq \frac{1}{4} | B |^{-1} | u |,$$

then

$$\frac{dv*}{dt} \leq -\frac{1}{2} | u(t) |^2,$$

as long as $| u(t) | \leq r.$

Let A have an eigenvalue with positive real part. Then there exists $\mu > 0$ such that $2\mu < | \lambda_i + \lambda_j |,$ for all eigenvalues λ_i, λ_j of A, and a unique matrix B such that

$$A^T B + BA = 2\mu B = I,$$

as follows from Corollary 22. We note that B connot be positive definite nor positive semidefinite for otherwise we must have, letting $v(x) = x^T Bx,$

$$\frac{dv*}{dt} = 2\mu v*(t) - | u(t) |^2,$$

or

$$e^{-2\mu t} v^*(t) - v^*(0) = -\int_0^t e^{-2\mu s} |u(s)^2 ds$$

i.e.
$$0 \le v^*(0) - \int_0^t e^{-2\mu s} |u(s)|^2 ds$$

for any solution, *u*, contradicting the fact that solution *u* exist for which

$$\int_0^t e^{-2\mu s} |u(s)|^2 ds$$

becomes unbounded as $t \to \infty$. Hence there exists $x_0 \neq 0$, of arbitrarily small norm, so that $v(x_0) < 0$. Let *u* be a solution. If the trivial solution were stable, then $|u(t)| \le r$ for some $r > 0$. Again letting $v(x) = x^T Bx$ we obtain

$$\frac{dv^*}{dt} = 2\mu v^*(t) - |u(t)|^2 + g^T(t,u) \quad Bu + u^T Bg(t,u)$$

We can choose *r* so small that

$$2|g(t,u)||B||u| \le \frac{1}{2}|u|^2, |u| \le r,$$

hence

$$e^{-2\mu t} v^*(t) - v^*(0) \le -\frac{1}{2}\int_0^t e^{-2\mu s} |u(s)|^2 ds,$$

i.e.
$$v^*(t) \le e^{-2\mu t} v^*(0) \to -\infty,$$

contradicting that $v^*(t)$ is bounded for bounded *u*. Hence *u* connot stay bounded, and we have instability.

It it is the case that the matrix *A* is a critical matrix, the trivial solution of the linear system may still be stable or it may be unstable. In either case, one may construct examples, where the trivial solution of the perturbed problem has either the same or opposite stability behavior as the unperturbed system.

Chapter 3

Calculus of Several Variables

PARTIAL DERIVATIVES

Consider *a* function f of three variables x, y and z: $f = f(x, y, z)$

If y and z are held constant and only x is allowed to vary, the partial derivative with respect to x is denoted by $\frac{\partial f}{\partial x}$ or f_x and is defined as the limit

$$\frac{\partial f(x,y,z)}{\partial x} = \lim_{\Delta x \to 0} \frac{f(x+\Delta x,y,z)-f(x,y,z)}{\partial x}$$

Similarly we can define the functions $\frac{\partial f}{\partial f}$ and $\frac{\partial f}{\partial z}$. In all cases two of the three variables explicitly appearing in the definition of f are held at constant, and f is differentiated with respect to the third variable.

Partial derivatives of higher order, of *a* function $f(x, y, z)$ are calculated by successive differentiation. Thus we write, for example

$$f_{yx} = \frac{\partial^2 f}{\partial y \partial x} = \frac{\partial f}{\partial y}\left(\frac{\partial f}{\partial x}\right)$$

and so forth. In this connection, we review the important fact that crossed partial derivative are equals

$$\frac{\partial^2 f}{\partial y \partial x} = \frac{\partial^2 f}{\partial x \partial y} \text{ or } f_{yx} = f_{xy}$$

if the derivatives involved are continuous.

Example: Find the partial derivatives $\frac{\partial^2 f}{\partial x^2}, \frac{\partial^2 f}{\partial y \partial x}, \frac{\partial^2 f}{\partial y^2}$ of the following functions

1. $f(x,y) = \sqrt{x^2+y^2}$
2. $f(x, y) = e^x \sin y$

Solution:

1. $\dfrac{\partial f}{\partial x} = \dfrac{x}{\sqrt{x^2+y^2}}, \dfrac{\partial^2 f}{\partial x^2} = \dfrac{y^2}{(x^2+y^2)^{3/2}}, \dfrac{y}{\sqrt{x^2+y^2}}, \dfrac{\partial^2 f}{\partial y^2} = \dfrac{x^2}{(x^2+y^2)^{3/2}}$

and $\dfrac{\partial^2 f}{\partial y \partial x} = \dfrac{\partial}{\partial y}\left(\dfrac{x}{\sqrt{x^2+y^2}}\right) = \dfrac{-xy}{(x^2+y^2)^{3/2}}$

2. $\dfrac{\partial f}{\partial x} = e^x \sin y, \quad \dfrac{\partial^2 f}{\partial x^2} = e^x \sin y, \quad \dfrac{\partial f}{\partial y} = e^x \cos y, \quad \dfrac{\partial^2 f}{\partial y^2} = -e^x \sin y$

and $\dfrac{\partial^2 f}{\partial y \partial x} = \dfrac{\partial}{\partial y}(e^x \sin y) = e^x \cos y$

The total differential of f is defined by the equation

$$df = \frac{\partial f}{\partial x}dx + \frac{\partial f}{\partial y}dy + \frac{\partial f}{\partial z}dz$$

whether or not x, y and z are independent of each other, provided only that the partial derivatives involved are continuous.

Proof: To understand the above formula in an informal way, we can consider the difference of the functional values at two adjacent points $P(x, y, z)$ and $Q(x + \Delta x, y + \Delta y, z + \Delta z)$, namely

$$\begin{aligned}
\Delta f &= f(x+\Delta x, y+\Delta y, z+\Delta z) - f(x,y,z) \\
&= [f(x+\Delta x, y+\Delta y, z+\Delta z) - f(x, y, + \Delta y, z+\Delta z)] \\
&\quad + [f(x+y+\Delta y, z+\Delta z) - f(x,y,z,+\Delta z)] + [f(x,y,z,+\Delta z) - f(x,y,z)] \\
&= \left[\left(\frac{f(x+\Delta x, y+\Delta y, z+\Delta z) - f(x, y+\Delta y, z+\Delta z)}{\Delta x}\right)\Delta x\right] \\
&\quad + \left[\left(\frac{f(x,y,+\Delta y, z+\Delta z) - f(x,y,z+\Delta z)}{\Delta y}\right)\Delta y\right] \\
&\quad + \left[\left(\frac{f(x,y,z+\Delta z) - f(x,y,z)}{\Delta z}\right)\Delta z\right]
\end{aligned}$$

In the limiting case as $\Delta x \to 0$, $\Delta y \to 0$ and $\Delta z \to 0$, we have $\Delta x \cong dx$, $\Delta y \cong dy$, $\Delta z \cong dz$ and$\Delta f \cong df$. The above equation can be reduced into

$$df = \left[\lim_{\Delta x \to 0}\left(\frac{f(x, +\Delta x, y + \Delta y, z + \Delta z) - f(x, y + \Delta y, z + \Delta z)}{\Delta x}\right)\right]dx$$

$$+\left[\lim_{\Delta x \to 0}\left(\frac{f(x, y + \Delta y, z + \Delta z) - f(x, y, z + \Delta z)}{\Delta y}\right)\right]dy$$

$$+\left[\lim_{\Delta x \to 0}\left(\frac{f(x, y, z + \Delta z) - f(x, y, z)}{\Delta z}\right)\right]dz$$

$$\therefore \qquad df = \frac{\partial f}{\partial x}dx + \frac{\partial f}{\partial y}dy + \frac{\partial y}{\partial z}dz$$

Remarks:

1. If x, y and z are all functions of *a* single variables, say t, then f may also be considered as truly *a* function of the one independent variable t, then

$$\frac{df}{dt} = \frac{\partial f}{\partial x}\frac{dx}{dt} + \frac{\partial f}{\partial y}\frac{dy}{dt} + \frac{\partial y}{\partial z}\frac{dz}{dt}$$

2. If we suppose that y and z are functions of x, then f is the function of the one independent variable x, we obtain

$$\frac{df}{dx} = \frac{\partial f}{\partial x} + \frac{\partial f}{\partial y}\frac{dy}{dt} + \frac{\partial f}{\partial z}\frac{dz}{dx}$$

3. If we suppose that x and y are independent but that z is *a* function of x and y, then $\frac{dy}{dx} = 0$ and we have

$$\left(\frac{\partial f}{\partial x}\right)_y = \frac{\partial f}{\partial x} + \frac{\partial f}{\partial z}\frac{\partial z}{\partial x}$$

The subscript indicates the variable y held constant.

N.B. If z is *a* function of x and y which is defined by the equation $f(x, y, z) = 0$, the partial derivative of f with respect to either x or y must vanish

$$\left(\frac{\partial f}{\partial x}\right)_y = \frac{\partial f}{\partial x} + \frac{\partial f}{\partial z}\frac{\partial z}{\partial x} = 0$$

This gives

$$\frac{\partial z}{\partial x} = -\frac{\frac{\partial f}{\partial x}}{\frac{\partial f}{\partial z}}$$

Similarly

$$\frac{\partial f}{\partial z} = \frac{\frac{\partial f}{\partial y}}{\frac{\partial f}{\partial z}}$$

Example:

1. Find the total differential of the function $f(x, y, z) = e^z\cos(x + y)$.
2. Given *a* function $x^2 + y^2 + z^2 = 1$, find the partial derivatives $\frac{\partial z}{\partial x}$ and $\frac{\partial z}{\partial y}$.
3. The three variables *x, y* and *z* are related by an equation in the form $F(x, y, z) = 0$

Show that $\left(\frac{\partial z}{\partial y}\right)\left(\frac{\partial y}{\partial x}\right)\left(\frac{\partial x}{\partial z}\right) = -1$.

Solution:

1. $\frac{\partial f}{\partial x} = -e^z\sin(x+y),\quad \frac{\partial f}{\partial y} = -e^z\sin(x+y),\quad \frac{\partial f}{\partial z} = e^z\cos(x+y)$

$$df = \frac{\partial f}{\partial x}dx + \frac{\partial f}{\partial y}dy + \frac{\partial f}{\partial z}dz$$

$$= -e^x\sin(x+y)dx - e^x\sin(x+y)dy + e^z\cos(x+y)dz$$

2. Differentiating both sides w.r.t. *x*, we obtain

$$2x + 2z\frac{\partial z}{\partial x} = 0 \quad\Rightarrow\quad \frac{\partial z}{\partial x} = -\frac{x}{z}$$

Similarly, if we differentiate both sides w.r.t. *y*, we have

$$2y + 2z\frac{\partial z}{\partial x} = 0 \quad\Rightarrow\quad \frac{\partial z}{\partial x} = -\frac{x}{z}$$

3. The implicit function $f(x, y, z) = 0$ can be used to define another function $z = f(x, y)$.

We get

$$\frac{\partial z}{\partial y} = -\frac{\dfrac{\partial F}{\partial y}}{\dfrac{\partial F}{\partial z}}$$

Similarly we have $\dfrac{\partial z}{\partial y} = -\dfrac{\dfrac{\partial F}{\partial y}}{\dfrac{\partial F}{\partial z}}$ and $\dfrac{\partial x}{\partial y} = -\dfrac{\dfrac{\partial z}{\partial z}}{\dfrac{\partial F}{\partial x}}$. The product of the above equations gives the required result.

TAYLOR SERIES IN SEVERAL VARIABLES

Functions of two or more variables often can be expanded in power series which generalize the familiar one-dimensional expansions. The more general situation may be illustrated here by a consideration of the two-variable case. For this purpose, let us consider *a* function $f(x, y)$ which can be expanded in the form

$$f(x+h, y+k) = f(x, y) + \left(\frac{\partial f}{\partial x}h + \frac{\partial f}{\partial y}k\right) +$$

$$+\frac{1}{2!}\left(\frac{\partial^2 f}{\partial x^2}h^2 + 2\frac{\partial^2 f}{\partial x \partial y}hk + \frac{\partial^2 f}{\partial y^2}k^2\right) + \ldots$$

$$= \sum_{p=0}^{N-1}\frac{1}{p!}\left(\frac{\partial f}{\partial x}h + \frac{\partial f}{\partial y}k\right)^p f(x, y) + R_N$$

where R_N is the "remainder after N terms" which is given by

$$R_N = \frac{1}{N!}\left(\frac{\partial}{\partial x}h + \frac{\partial}{\partial y}k\right)^N f(x+\tau h, y+\tau k) \qquad (0 < \tau < 1)$$

for some t between 0 and 1. The expansion Equation is known as the Taylor series of the function $f(x, y)$.

Proof: For this purpose, we begin by defining *a* function F(t), such that

1. $F(t) = f(x + ht, y + kt)$

   where x, y, h and k are temporarily to be held fixed and, in any case, are to be independent of t.

   Then, is $F(t)$ has *a* continuous Nth derivative in some interval about $t = 0$, we may write

2. $F(t) = \sum_{p-0}^{N-1} \frac{F^{(n)}(0)}{n!} t^n + \frac{F^{(N)}(\tau)}{N!} t^N$ for some value of t between 0 and t. Now, since

$$\frac{d}{dt} F(t) = h \frac{\partial f(x+ht, y+kt)}{\partial x} + k \frac{\partial f(x+ht, y+kt)}{\partial y}$$

$$= \left(h \frac{\partial}{\partial x} + k \frac{\partial}{\partial y} \right) f(x+ht, y+kt)$$

there follows also

$$\frac{d^n}{dt^n} F(t) = \left(h \frac{\partial}{\partial y} + k \frac{\partial}{\partial y} \right)^n f(x+ht, y+k\dot{t}) \qquad (n = 0, 1, \cdots)$$

Hence we have the results

3. $F^{(n)}(0) = \left(h \frac{\partial}{\partial y} + k \frac{\partial}{\partial y} \right)^n f(x, y)$

and

4. $F^{(n)}(\tau) = \left(h \frac{\partial}{\partial y} + k \frac{\partial}{\partial y} \right)^N f(x+h\tau, y+k\tau)$

If we introduce Equations (2), (3), and (4) into (1), and specialize the results by taking $t = 1$, we thus obtain the form

$$f(x+h, y+k) = \sum_{p-0}^{N-1} \frac{1}{p!} \left(h \frac{\partial}{\partial y} + k \frac{\partial}{\partial y} \right)^p f(x, y) + R_N$$

where R_N is the "remainder after N terms" which is given by

$$R_N = \frac{1}{N!} \left(\frac{\partial}{\partial x} h + \frac{\partial}{\partial y} k \right)^p f(x+\tau h, y+tk) \qquad (0 < \tau < 1)$$

for some τ between 0 and 1.

When $f(x, y)$ is sufficiently well behaved, the remainder R_N tends to zero for sufficiently small values of the increments, y ielding *a* power series of the form.

Although the form given is perhaps the most compact one, *a* form which more closely resembles the most familiar one-dimensional form can be obtained from Equation by first replacing (x, y) by (x_0, y_0) and then replacing h and k by $x - x_0$ and $y - y_0$, respectively. We obtain

$$f(x,y) = f(x_0,y_0) + \left[(x-x_0)\frac{\partial}{\partial x}f(x_0,y_0) + (y-y_0)\frac{\partial}{\partial y}f(x_0,y_0\right] +$$

$$\frac{1}{2!}\left[(x-x_0)^2\frac{\partial^2}{\partial x^2}f(x_0,y_0) + 2(x-x_0)(y-y_0)\frac{\partial^2}{\partial x\partial y}f(x_0,y_0) + (y-y_0)^2\frac{\partial^2}{\partial y^2}f(x_0,y_0)\right] + .$$

The above series expansion is known as the Taylor series about *a* point (x_0, y_0).

Example: Find the Taylor series of the function $f(x, y) = e^x$ siny up to second-order about *a* point (0, 0).

Solution: Using the results in Example 1.1b, we have

$$\frac{\partial f}{\partial x} = e^x \sin y, \frac{\partial^2 f}{\partial x^2} = e^x \sin y, \frac{\partial f}{\partial y} = e^x \cos y, \frac{\partial^2 f}{\partial y^2} = -e^x \sin y,$$

and $\dfrac{\partial^2 f}{\partial y\partial x} = \dfrac{\partial}{\partial y}(e^x \sin y) = e^x \cos y$

Hence $\dfrac{\partial f}{\partial x}(0,0) = 0$, $\dfrac{\partial^2 f}{\partial x^2}(0,0) = 0$, $\dfrac{\partial f}{\partial y}(0,0) = 1$, $\dfrac{\partial^2 f}{\partial y^2}(0,0) = 0$ and $\dfrac{\partial^2 f}{\partial y\partial x}(0,0) = 1$.

Then by Taylor expansion, we obtain

$$f(x, y) = f(0,0) + \left(x\frac{\partial}{\partial x}f(0,0) + y\frac{\partial}{\partial y}f(0,0)\right)$$

$$+\frac{1}{2}\left(x^2\frac{\partial}{\partial x^2}f(0,0) + 2xy\frac{\partial^2}{\partial x\partial y}f(0,0) + y^2\frac{\partial}{\partial y^2}f(0,0)\right) + \cdots$$

$$\therefore \qquad e^x \sin y = y + xy + \cdots$$

MAXIMA AND MINIMA

The Taylor series are particular helpful in studying maxima and minima of functions of several variables. We again restrict attention here to the two-dimensional case.

$$f(x+h, y+k) = f(x, y) + \left(\frac{\partial f}{\partial x}h + \frac{\partial f}{\partial y}k\right) + R_2$$

where $R_2 = \dfrac{1}{2}\left(h^2\dfrac{\partial^2}{\partial x^2}h + 2hk\dfrac{\partial^2}{\partial x\partial y} + k^2\dfrac{\partial^2}{\partial y^2}\right)f(x+\tau h, y+\tau k) \qquad (0 < \tau < 1)$

If we write $\Delta f = f(x + h, y + k) - f(x, y)$ for the increment of f, corresponding to the increments h and k in x and y, respectively, we say that f has *a* relative minimum at $P(x_0, y_0) \Delta f(x_0, y_0) \geq 0$ if for all sufficiently small permissible increments h and k, and that f has *a* relative maximum at $P(x_0, y_0)$ if for all such increments in x and y.

If the point P is an interior point of *a* region in which f, $\frac{\partial f}{\partial y}$ and $\frac{\partial f}{\partial z}$ exits, the necessary condition that f assume *a* relative maximum or *a* relative minimum at $P(x_0, y_0)$ is that

$$f_x = f_y = 0 \text{ at } P(x_0, y_0)$$

and the extrema can be classified according to the sign of the following expression:

$$\Delta f(x_0, y_0) = \frac{1}{2}\left(h^2 \frac{\partial^2}{\partial x^2} + 2hk\frac{\partial^2}{\partial x \partial y} + k^2 \frac{\partial^2}{\partial y^2}\right) f(x_0, y_0)$$

For this purpose, it is convenient to define an expression

$$\delta = f_{xx} f_{yy} - f_{xy}^2 \text{ at } P(x_0, y_0)$$

1. If $d < 0$ at *a* point $P(x_0, y_0)$, then Df is positive for some increments in x and y and negative for others, and the point P is said to be *a* saddle point or minimax.
2. If $d > 0$ at *a* point $P(x_0, y_0)$, then f_{xx} and f_{yy} must be either both positive or both negative at that point. Since $\Delta f(x_0, y_0)$ is of constant sign in either case, when h and k are sufficiently small, then at that point the function f has
   (a) *a* relative maxima if $f_{xx} < 0, f_{xx} f_{yy} > f_{xy}^2$ at P,
   (b) *a* relative minima if $f_{xx} < 0, f_{xx} f_{yy} > f_{xy}^2$ at P.

Example: Locate and identify all relative maxima and minima and saddle points (if such exist) of the function $f(x, y) = x^2y^2 - x^2 - y^2$.

Solution:

$$\frac{\partial f}{\partial x} = 2xy^2 - 2x, \quad \frac{\partial^2 f}{\partial x^2} = 2y^2 - 2, \quad \frac{\partial f}{\partial y} = 2x^2y - 2y, \quad \frac{\partial^2 f}{\partial y^2} = 2x^2 - 2$$

and $\frac{\partial^2 f}{\partial y \partial x} = \frac{\partial}{\partial y}(2xy^2 - 2x) = 4xy$

Put $\frac{\partial f}{\partial x} = 2xy^2 - 2x = 0 \Rightarrow x = 0 \text{ or } y = \pm 1$

and $\frac{\partial f}{\partial y} = 2x^2 y - 2y = 0 \Rightarrow y = 0 \text{ or } x = \pm 1$

The extremum points are (0, 0), (1, 1), (1, –1), (–1, 1) and (–1, –1)

$$\delta(x, y) = \frac{\partial^2 f}{\partial x^2}\frac{\partial^2 f}{\partial y^2} - \left(\frac{\partial^2 f}{\partial y \partial x}\right)^2 = (2y^2 - 2)(2x^2 - 2) - 16x^2y^2$$

$$= 4 - 4x^2 - 4y^2 - 12x^2y^2$$

Since

For the point (0, 0), we have

$$\delta(0, 0) = 4 \text{ and } \frac{\partial^2 f}{\partial x^2}(0,0) = -2 < 0$$

For the other points:

Hence (0, 0) is *a* maximum point while (1, 1), (1, –1), (–1, 1) and (–1, –1) are saddle points

Problem: Find the partial derivatives $\frac{\partial^2 f}{\partial x^2}, \frac{\partial^2 f}{\partial y \partial x}, \frac{\partial^2 f}{\partial y^2}$ of the following functions

(a) $f(x,y) = \frac{xy}{\sqrt{x^2 + y^2}}$

(b) $f(x, y) = \sin(x - y)$

(c) $f(x, y) = \sinh x \cosh y$

Problem: The relationship between the pressure *P*, the volume *V* and the temperature *T* of *a* one mole of CO_2 gas is given by the *V* an der Waals equation of state,

$$P = \frac{RT}{V - b} - \frac{a}{V^2}$$

where $R = 8.31$ J/mol·K is the universal gas constant, *a* and b are proportionality constants.

1. Evaluate and.
2. The critical point for one mole of CO_2 gas is given by setting both of the above derivatives equal to *z* ero.
   (a) Express the constants *a* and b in terms of the temperature T_c and pressure P_c at critical point.
   (b) Experiment has shown that T_c = 304 K and P_c = 0.7510^7 *Pa*, estimate the numerical values for *a* and *b*.

Answer: $a = 0.364$ J·m³/mol³ and $b = 4.27 \times 10^{-5}$ m³/mol)

Problem: Show that the wave motion $\Psi(x, z) = A\sin(x - vt)$ satisfying the wave equation

$$\frac{\partial^2 \psi}{\partial t^2} = v^2 \frac{\partial^2 \psi}{\partial x^2}$$

Problem: The velocity field of *a* fluid flow is given by the following equations,

$$\begin{cases} v_x = a\cos\omega t \\ v_y = a\sin\omega t \\ v_z = x^2 + y^2 \end{cases}$$

Find the acceleration of *a* fluid element at $x = p$, $y = q$, and $t = t_0$.

Problem: Find the Taylor series of the function $f(x, y)$ up to second-order about *a* point $P(x_0, y_0)$.

(a) $f(x, y) = \ln(x^2, y^2)$, $P(0, 0)$

(b) $f(x, y) = \cosh(xy)$, $P(1, 1)$

Problem: (Double Pendulum) The figure below shows *a* pendulum of length L_2 and mass m_2 capable of swinging freely from the end of another pendulum of length L_1 and mass m_1, attached to the point *O*.

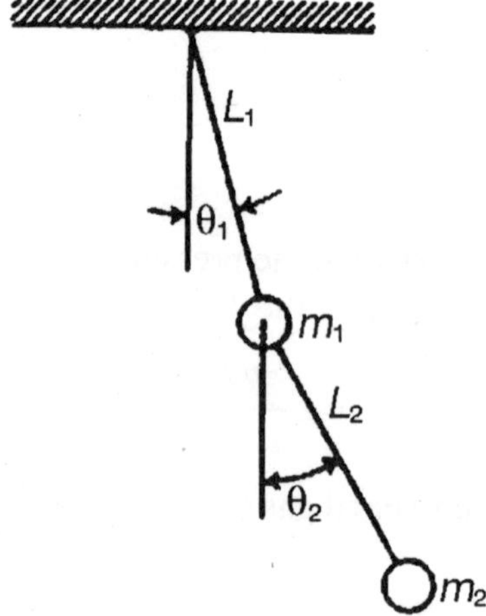

The total potential energy of the double pendulum is given by

$$V(\theta_1, \theta_2) = (m_1 + m_2)gL_1(1 - \cos q_1) + m_2gL_2(1 - \cos\theta_2)$$

Obtain the terms of degree two or less in the Taylor expansion.

Problem: Locate and identify all relative maxima and minima and saddle points (if such exist) of the following functions

(a) $f(x, y) = 2x - 2y - x^2 - y^2$

(b) $f(x, y) = x^2 - xy + y^2$

DOUBLE INTEGRAL OVER RECTANGLES

Let us consider *a* function f of two variables x and y. Suppose that $f(x,y) \geq 0$ and f is defined on the rectangle $R = [a,b] \times [c,d]$.

The graph of f is *a* surface with equation $z = f(x, y)$. Let S be the solid that lies above R and under the graph of f, that is

$$S = \{(x,y,z) \mid 0 \leq z \leq f(x,y), (x,y) \in R\}$$

If we partition R into mn subrectangles R_{ij} and choose in (x_{ij}^*, y_{ij}^*) R_{ij}, then the volume of each box is $V_{ij} = f(x_{ij}^*, y_{ij}^*)\Delta A_{ij}$. We can get the approximation to the total volume by

$$V \approx \sum_{i=1}^{m}\sum_{j=1}^{n} V_{ij} = \sum_{i=1}^{m}\sum_{j=1}^{n} f(x_{ij}^*, y_{ij}^*)\Delta A_{ij}$$

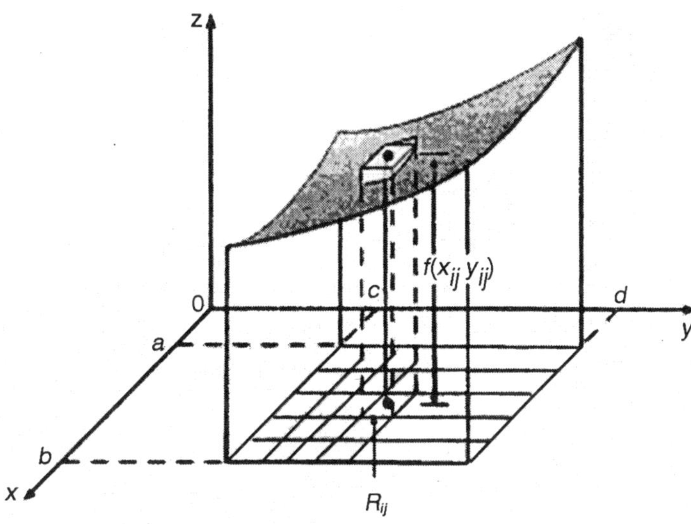

Fig. Double Integral over Rectangles

If this sum approaches *a* unique limit as m and n become infinite and the size of each box approaches 0, then the volume V is defined as the double integral of the function f over the region R and it is defined as that limit:

$$\iint_R f(x,y)\,dA = \lim \sum_{i=1}^{m}\sum_{j=1}^{n} f(x_{ij}^*, y_{ij}^*)\Delta A_{ij}$$

The following theorem gives *a* practical method for evaluating *a* double integral by expressing it as an iterated integral (in either order).

Fubini's Theorem

Theorem: If f is continuous on the rectangle $R = \{(x, y) \mid a \le x \le b, a \le y \le d\}$, then

$$\iint_R f(x,y)\,dA = \int_a^b \left[\int_c^d f(x,y)\,dy \right] dx = \int_c^d \left[\int_a^d f(x,y)\,dy \right] dy$$

Proof: The proof of Fubini's theorem in general case is too difficult to include in this section, but we can at least give an intuitive indication why it is true for the case where $f(x,y) \ge 0$.

Recall that if f is positive, then we can interpret the double integral $\iint_R f(x,y)\,dA$ as the volume *V* of the solid of *S* that lies above *R* and under the surface $z = f(x, y)$. But we have where *A*(*x*) is the area of *a* cross-section of *S* in the plane through *x* perpendicular to the *x*-axis.

$$V = \int_a^b A(x)\,dx$$

From Figure we can see that *A*(*x*) is the area under the curve *C* whose equation $z = f(x, y)$, where x is held constant and $c \le y \le d$.

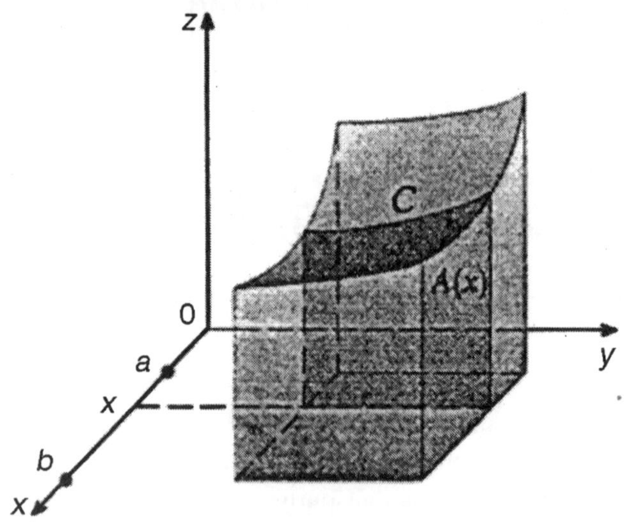

Fig. Area Under the Curve

Therefore

$$A(x) = \int_c^d f(x,y)\,dy$$

and we have

$$\iint_R f(x,y)\,dA = V = \int_a^b A(x,y)\,dy = \int_a^b \int_c^d f(x,y)\,dy\,dx$$

A similar argument, using cross-sections perpendicular to the y -axis as in Figure.

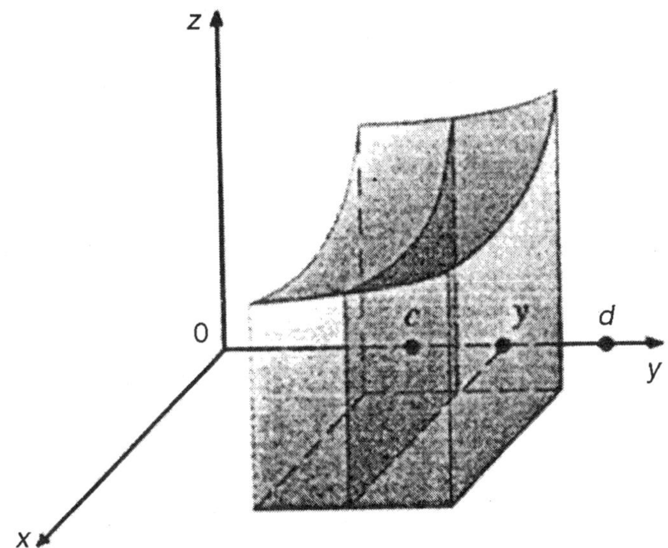

Fig. Cross-Sections Perpendicular to the y-axis

Example: Find the following double integrals:

1. $\iint_R x\,y\,dx\,dy \; R = [0,1]\times[0,1]$

2. $\iint_R \cos(x+y)\,dx\,dy \; R = [0,\pi]\times[0,\pi]$

Solution: By Fubini's theorem

1. $\iint_R x\,y\,dx\,dy = \int_0^1 \int_0^1 xy\,dx\,dy = \int_0^1 \left[\frac{x^2 y}{2}\right]_0^1 dy = \int_0^1 \frac{y}{2} dy = \frac{1}{4}$

2. $\iint_R \cos(x+y)\,dx\,dy = \int_0^x \int_0^x \cos(x+y)\,dx\,dy = \int_0^x [\sin(x+y)]_0^x\,dy$

$$= \int_0^x (\sin(x+y) - \sin y)\,dy\,[\cos y - \cos(\pi + y)]_0^x = 0$$

DOUBLE INTEGRAL OVER GENERAL REGIONS

For double integrals, we want to be able to integrate not just over rectangles but also over regions D of more general shape, such as the one illustrated in Figure below. We define a new function F with domain R as in Figure by

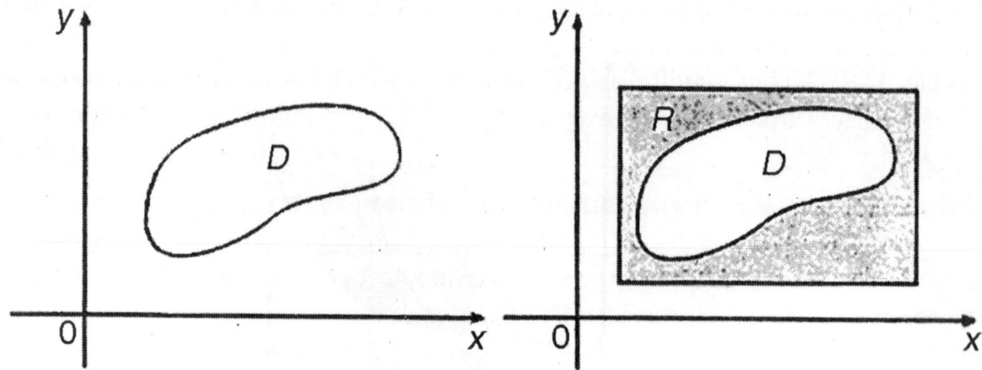

Fig. Double Integral over General Regions

$$F = \begin{cases} f(x,y) & \text{if } (x,y) \text{ is in } D \\ 0 & \text{if } (x,\ y) \text{ is in } R \text{ but not in } D \end{cases}$$

So

$$\iint_D f(x,y)\,dA = \iint_R F(x,y)\,dA$$

A plane region D is said to be of type I if it lies between the graphs of two continuous functions of x that is

$$D = \{x,y) \mid a \le x \le b, g_1(x) \le y \le g_2(x)\}$$

where g_1 and g_2 are continuous on $[a, b]$. Some examples of type I regions are shown in Figure below

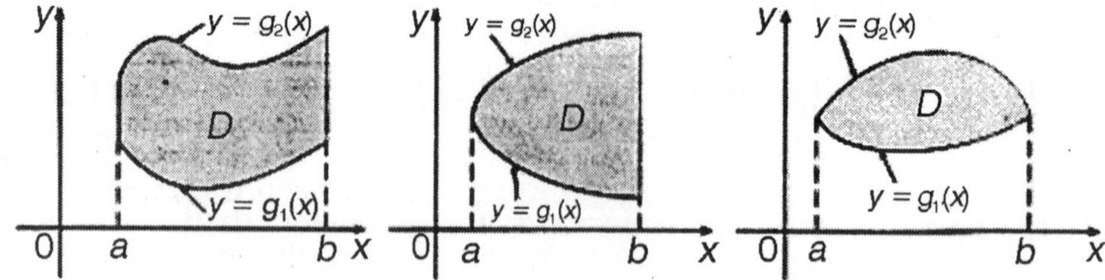

Fig. Plane Region of Type I

In order to evaluate $\iint_D f(x,y)\,dA$ where D is a region in type I we choose a rectangle that contains D and we let $F(x, y)$ be the function given by above. Then by

Fubin's Theorem, we have

$$\iint_D f(x,y)\,dA = \iint_R F(x,y)\,dA = \int_a^b \int_c^d F(x,y)\,dy\,dx$$

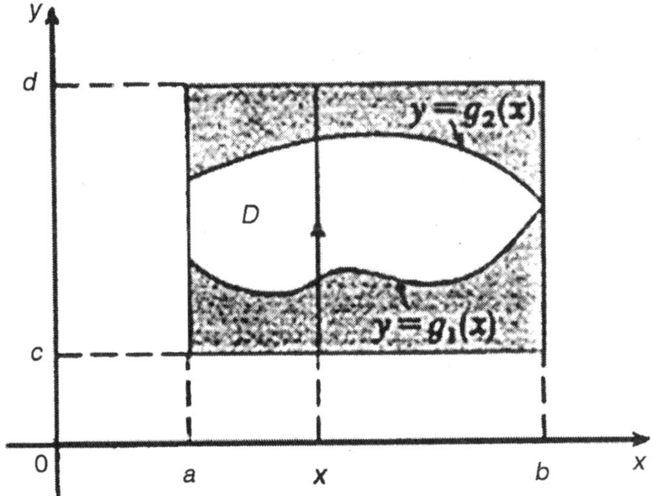

Fig. Plane Rectangle Region

Observe that $F(x, y) = 0$ if $y < g_1(x)$ or $y < g_2(x)$ since then (x, y) lies outside D. Therefore

$$\int_c^d F(x,y)\,dx = \int_{g_1(x)}^{g_2(x)} F(x,y)\,dy = \int_{g_1(x)}^{g_2(x)} f(x,y)\,dy$$

because $F(x, y) = f(x, y)$ when $g_1(x) \leq y \leq g_2(x)$

So if f is continuous on *a* type I region

$$\iint_D f(x,y)\,dA = \int_a^b \left[\int_{g_1(x)}^{g_2(x)} f(x,y)\,dy \right] dx$$

We also consider plane regions of type II, which can be expressed as

$$D = \{x,y) \mid c \leq y \leq d, h_1(y) \leq x \leq h_2(y)\}$$

where h_1 and h_2 are continuous. Two such regions are illustrated below

Using the same methods that were used above we can show that

$$\iint_D f(x,y)\,dA = \int_c^d \left[\int_{h_1(x)}^{h_2(x)} f(x,y)\,dx \right] dy$$

where D is *a* type II region.

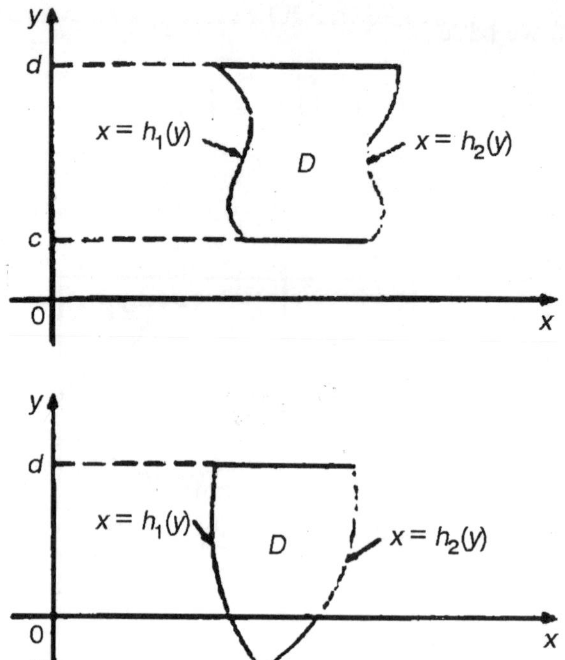

Fig. Plane Region of Type II

Example: Let $R = \left\{ (x, y) \middle| \frac{x^2}{a^2} + \frac{y^2}{b^2} \le 1 \right\}$ be *a* region bounded by an ellipse. Find the double integral $\iint_R x^2 dA$.

Solution:

$$\iint_R x^2 dA = \int_{-b}^{a} \left[\int_{-a\sqrt{1-\frac{y^2}{b^2}}}^{a\sqrt{1-\frac{y^2}{b^2}}} x^2 dx \right] dy = \int_{-b}^{a} \left[\frac{x^3}{3} \right]_{a\sqrt{1-\frac{y^2}{b^2}}}^{a\sqrt{1-\frac{y^2}{b^2}}} dy = \frac{\pi}{4} a^3 b$$

PROPERTIES OF DOUBLE INTEGRALS

Given two continuous functions $f(x, y)$, $g(x, y)$ and *a* real constant *c*, we have the following properties of double integrals:

1. $\iint_D [f(x,y) + g(x,y)] dA = \iint_D [f(x,y) dA + \iint_D [g(x,y) dA$

Proof:

$$\iint\limits_D [f(x,y)+g(x,y)]\,dA = \lim \sum_{i=1}^{m}\sum_{j=1}^{n}\left[f\left(x_{ij}^*, y_{ij}^*\right)+g\left(x_{ij}^*, y_{ij}^*\right)\right]\Delta A_{ij}$$

$$= \lim \sum_{i=1}^{m}\sum_{j=1}^{n} f\left(x_{ij}^*, y_{ij}^*\right)\Delta A_{ij} + \lim \sum_{i=1}^{m}\sum_{j=1}^{n} g\left(x_{ij}^*, y_{ij}^*\right)\Delta A_{ij}$$

$$= \iint\limits_D f(x,y)\,dA + \iint\limits_D g(x,y)\,dA$$

2. $\iint\limits_D c\,f(x,y)\,dA = c\iint\limits_D [f(x,y)\,dA$

Proof:

$$\iint\limits_D c\,f(x,y)\,dA = \lim \sum_{i=1}^{m}\sum_{j=1}^{n} c\,f\left(x_{ij}^*, y_{ij}^*\right)\Delta A_{ij} = c\lim \sum_{i=1}^{m}\sum_{j=1}^{n} f\left(x_{ij}^*, y_{ij}^*\right)\Delta A_{ij}$$

$$= c\iint\limits_D [f(x,y)\,dA$$

3. If $D = D_1 \cup D_2$, where D_1 and D_2 do not overlap except perhaps on their boundaries, then

$$\iint\limits_D f(x,y)\,dA = \iint\limits_{D_1} f(x,y)\,dA + \iint\limits_{D_2} f(x,y)\,dA$$

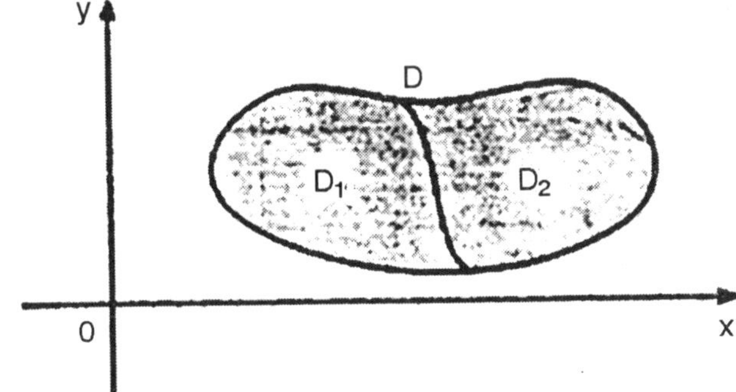

Fig. D_1 and D_2 Do not Overlap Except Perhaps on their Boundaries.

DOUBLE INTEGRALS IN POLAR COORDINATES

The polar coordinates (r, q) of *a* point are related to the rectangular coordinates by the equations

$$r^2 = x^2 + y^2 \quad x = r\cos\theta \quad y = r\sin\theta$$

a polar rectangle $R = \{(r,\theta) \mid a \le r \le b, \alpha \le \theta \le \beta\}$

If f is continuos on polar rectangle R given by $0 \le a \le r, \le b$, $\alpha \le \theta \le \beta$ where $0 \le \beta - \alpha \le 2\pi$ then

$$\iint_R f(x,y)\,dA = \int_\alpha^\beta \int_a^b f(r\cos\theta, r\sin\theta)\,r\,dr\,d\theta \;(aA = rdrd\theta)$$

Proof: In order to compute $\iint_R f(x,y)\,dA$, where R is *a* polar rectangle, we start with a partition of [*a*, *b*] into m subintervals:

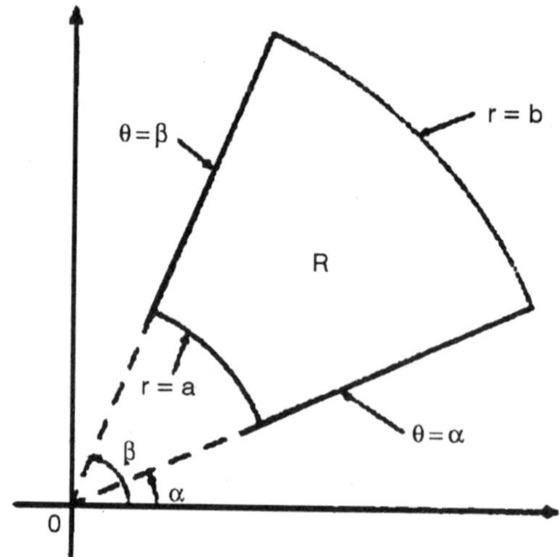

Fig. Polar Rectangle

$$a = r_0 < r_1 < r_2 < \dots < r_{i-1} < r_i < \dots < r_m = b$$

and *a* partition of [α, β] into *n* subintervals:

$$\alpha = \theta_0 < \theta_1 < \theta_2 < \dots < \theta_{j-1} < \theta_i < \dots < \theta_n = \beta$$

The "centre" of the polar subrectangle R_{ij} has polar coordinates

$$r_i^* = \frac{1}{2}(r_i + r_{i-1}) \text{ and } \theta_j^* = \frac{1}{2}(\theta_i + \theta_{j-1})$$

we find that the area of R_{ij} is

$$\Delta A_{ij} = \frac{1}{2}r_i^2 \Delta\theta_j - \frac{1}{2}r_{i-1}^2 \Delta\theta_j = \frac{1}{2}r_i^2 - r_{i-1}^2)\Delta\theta_j$$

$$= \frac{1}{2}(r_1 + r_{i-1})\Delta\theta_j = r_i^* \Delta r_i \Delta\theta_j$$

where $\Delta r_i = r_i - r_{i-1}$ and $\Delta\theta_j = \theta_j - \theta_{j-1}$.

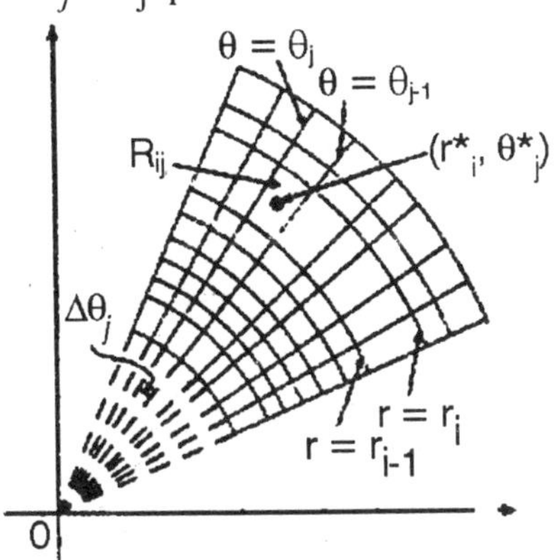

Fig. Polar Subrectangle

Therefore the double integral $\iint\limits_R f(x,y)\,dA$ can be approximated by

$$\iint\limits_R f(x,y)\,dA \approx \sum_{i=1}^{m}\sum_{j=1}^{n} f(x_{ij}^*, y_{ij}^*)\Delta A_{ij} = \sum_{i=1}^{m}\sum_{j=1}^{n} f(r_i^* \cos\theta_j^*, r_i^* \sin\theta_j^*) r_i^* \Delta r_i \Delta\theta_j$$

In the limiting case, we have

$$\iint\limits_R f(x,y)\,dA = \lim \sum_{i=1}^{m}\sum_{j=1}^{n} f(r_i^* \cos\theta_j^*, r_i^* \sin\theta_j^*) r_i^* \Delta r_i \Delta\theta_j$$

$$= \int_\alpha^\beta \int_a^b f(r\cos\theta, r\sin\theta)\, r\, dr\, d\theta$$

Example: Let $R = \{(x,y) \mid x^2 + y^2 \le 4\}$ be a region bounded by *a* circle. Find the double integral $\iint\limits_R \sqrt{x^2+y^2}\,dx\,dy$.

Solution: By using the polar coordinates $\begin{cases} x = 2\cos\theta \\ y = 2\sin\theta \end{cases}$, we have

$$\iint\limits_R \sqrt{x^2+y^2}\,dx\,dy = \int_0^{2x}\left[\int_0^2 r(r\,dr)\right]d\theta = \int_0^{2x}\left[\frac{r^3}{3}\right]_0^2 d\theta = \frac{16x}{3}$$

SURFACE AREA

Let S be *a* surface with equation $z = f(x, y)$ we assume that $f(x, y) \geq 0$ and the domain D of f is *a* rectangle. We consider *a* partition P into small rectangles R_{ij} with area $\Delta A_{ij} = \Delta x_i \Delta y_j$. If (x_i, y_i) is the corner R_{ij} closest to the origin, let be *a* point on S directly above it. The tangent plane to S at P_{ij} is an approximation to S near P_{ij}. So the area ΔT_{ij} of the part of this tangent plane that lies directly above R_{ij} is an approximation to the area ΔS_{ij} of the part of S that lies directly above R_{ij}. Therefore, we can define the surface area of S to be

$$A(S) = \lim \sum_{i=1}^{m} \sum_{j=1}^{n} \Delta T_{ij}$$

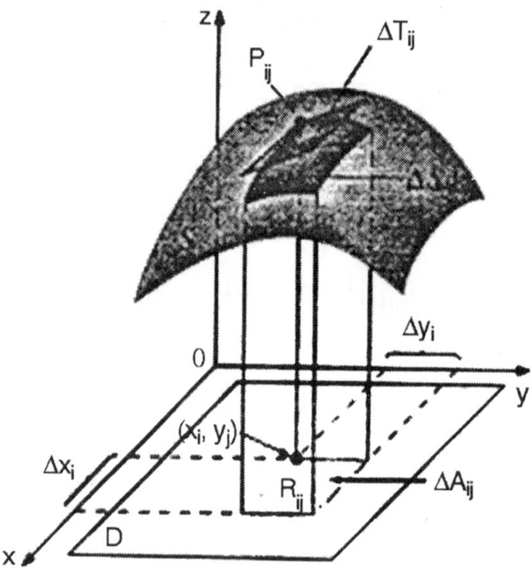

Fig. Surface Area

To find *a* practical formula that is more convenient than Equation for computational purpose, we let *a* and b be the vectors that start at P_{ij} and lie along the sides of the parallelogram with area

$\Delta A_{ij} = \Delta x_i \Delta y_j$. Then $\Delta T_{ij} = | a \times b |$. Remember that $f_x(x_i, y_j)$ and $f_y(x_i, y_j)$ are the slopes of the tangent lines through P_{ij} in the directions of *a* and *b*. Therefore

$$a = \Delta x_i i + f_x(x_i, y_j)\Delta x_i k \text{ and } b = \Delta y_j j + f_y(x_i, y_j)\Delta y_i k$$

and

$$a \times b = \begin{vmatrix} i & j & k \\ \Delta x_i & 0 & f_x(x_i, y_i)\Delta x_i \\ 0 & \Delta y_j & f_y(x_i, y_i)\Delta y_i \end{vmatrix}$$

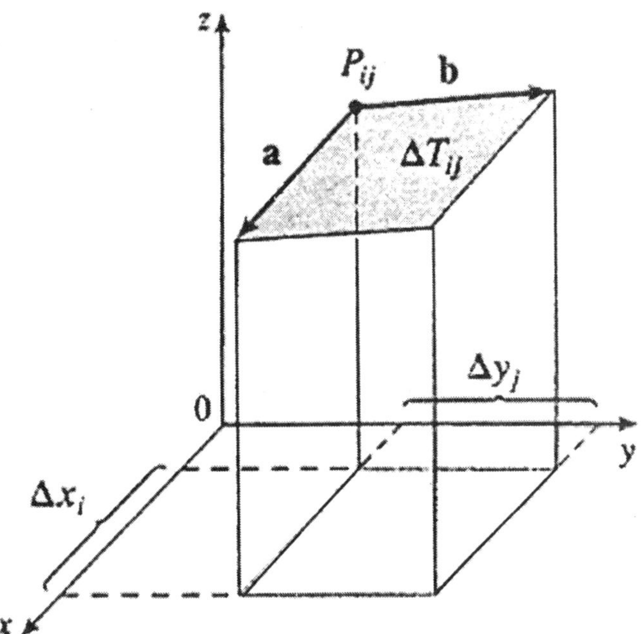

Fig. The Slopes of the Tangent Lines

$$= -f_x(x_i, y_i)\Delta x_i \Delta y_i\, i - f_y(x_i, y_i)\Delta x_i \Delta y_i\, j + \Delta x_i \Delta y_j k$$

$$= [-f_x(x_i, y_i)i - f_y(x_i, y_i)j + k]\Delta A_{ij}$$

Thus

$$\Delta T_{ij} = | a \times | = \sqrt{1 + \left(\frac{\partial z}{\partial x}\right)^2 + \left(\frac{\partial z}{\partial y}\right)^2}\ \Delta A_{ij}$$

From the definition, we have

$$A(S) = \lim \sum_{i=1}^{m} \sum_{j=1}^{n} \Delta T_{ij}\ \iint_D \sqrt{1 + \left(\frac{\partial z}{\partial x}\right)^2 + \left(\frac{\partial z}{\partial y}\right)^2}\ dA$$

Hence, we have the following theorem

Theorem: The area of the surface with equation $z = f(x, y)$ where (x, y) is in D where and are continuous, is

$$A(S) = \iint_D \sqrt{1 + \left(\frac{\partial z}{\partial x}\right)^2 + \left(\frac{\partial z}{\partial y}\right)^2}\ dA$$

Example: Show that the surface area of a sphere with radius R is $A = 4\pi R^2$.

Solution: The region bounded by the sphere is given by

$$E = \{(x, y, z) \mid x^2 + y^2 + z^2 \le R^2\}.$$

Therefore the surface of one octant is described by the equation

$$z(x, y) = \sqrt{R^2 - x^2 - y^2}$$

where $0 < x < R$ and $0 < y < R$.

Also we can obtain $\dfrac{\partial z}{\partial x} = \dfrac{-x}{\sqrt{R^2 - x^2 - y^2}}$ and $\dfrac{\partial z}{\partial y} = \dfrac{-y}{\sqrt{R^2 - x^2 - y^2}}$

$$\therefore \quad \sqrt{1 + \left(\frac{\partial z}{\partial x}\right)^2 + \left(\frac{\partial z}{\partial y}\right)^2} = \sqrt{1 + \frac{x^2}{\sqrt{R^2 - x^2 - y^2}} + \frac{y^2}{\sqrt{R^2 - x^2 - y^2}}}$$

$$= \frac{R}{\sqrt{R^2 - x^2 - y^2}}$$

Hence the surface of the sphere is

$$A = 8 \iint \frac{R}{\sqrt{R^2 - x^2 - y^2}} dxdy = 8 \int_0^{\frac{x}{2}} \left[\int_0^R \frac{R}{\sqrt{R^2 - r^2}} (rdr) \right] d\theta$$

$$= 8 \int_0^{\frac{x}{2}} R^2 \, d\theta = 4\pi R^2$$

TRIPLE INTEGRAL OVER RECTANGLES

Let us first deal with the simplest case where f is defined on a rectangular box:

$$B = \{(x, y, z) \mid a \le x \le b, c \le y \le d, r \le z \le s\}$$

The first step is to partition the interval $[a, b]$ $[c, d]$, and $[r, s]$ as follows:

$$a = x_0 < x_1 < x_2 < \dots < x_{i-1} < x_i < \dots < x_i = b$$

$$c = y_0 < y_1 < y_2 < \dots < x_{j-1} < y_j < \dots < y_m = d$$

$$r = z_0 < z_1 < z_2 < \dots < x_{k-1} < z_k < \dots < z_n = s$$

The planes through these partition points parallel to the coordinate planes divide the box B into *lmn* sub-boxes

$$B_{ijk} = [x_{i-1}, x_i] \times [y_{j-1}, y_j] \times [z_{k-1}, z_k]$$

which is shown in Figure

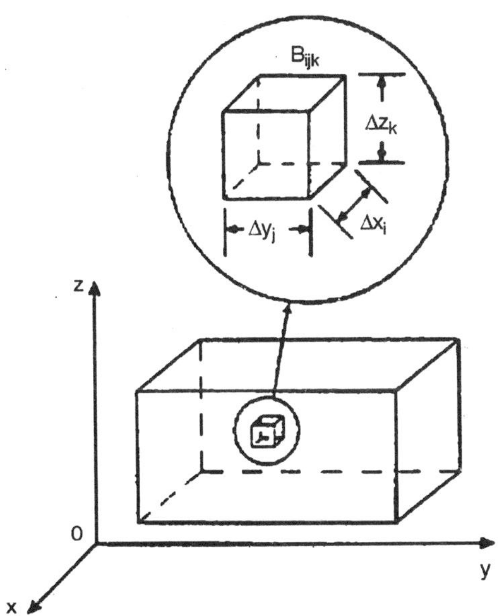

Fig. The Planes Through Partition Points Parallel to the Coordinate Planes

The volume of B_{ijk} is $\Delta V_{ijk} = \Delta x_i \Delta y_j \Delta z_k$ where $\Delta x_i = x_i - x_{i-1}$, $\Delta y_j = y_j - y_{j-1}$, and $\Delta z_k = z_k - z_{k-1}$. Now we can choose $(x^*_{ijk}, y^*_{ijk}, z^*_{ijk})$ in B_{ijk} and define the triple integral of the function f over the region B as the limit:

$$\iiint_B f(x,y,z)\,dV = \lim \sum_{i=1}^{l} \sum_{j=1}^{m} \sum_{k=1}^{n} f(x^*_{ijk}, y^*_{ijk}, z^*_{ijk}) \Delta V_{ijk}$$

If f is integrable over the rectangular box $B = [a,b] \times [c,d] \times [r,s]$, then we will have the Fubini's theorem in three dimensions:

$$\iiint_B f(x,y,z)\,dV = \int_r^s \int_c^d \int_a^b f(x,y,z)\,dx\,dy\,dz$$

If we integrate with respect to y, then z, and then x, we have

$$\iiint_B f(x,y,z)\,dV = \int_a^b \int_r^s \int_c^d f(x,y,z)\,dy\,dz\,dx$$

Example: Let $E = [0,1] \times [0,1] \times [0,1]$, find the following triple integrals

1. $\iiint_E x\,y\,z\,dx\,dy\,dz$

2. $\iiint\limits_E x\,y\,z\sin(x^2+y^2+z^2)\,dx\,dy\,dz$

Solution:

1. $\iiint\limits_E x\,y\,z\,dx\,dy\,dz = \int_0^1\int_0^1\left[\int_0^1 x\,dx\right]yz\,dy\,dz = \int_0^1\left[\int_0^1 \frac{y}{2}dy\right]z\,dz$

$$= \int_0^1 \frac{z}{4}dz = \frac{1}{8}$$

2. $\iiint\limits_E x\,y\,z\sin(x^2+y^2+z^2)\,dx\,dy\,dz$

$$= \int_0^1\int_0^1\left[\int_0^1 x\sin(x^2+y^2+z^2)\,dx\right]yzdy\,dz$$

$$= \int_0^1\int_0^1\left[-\frac{1}{2}\cos(x^2+y^2+z^2)\right]_0^1 yzdy\,dz$$

$$= \int_0^1\left[\int_0^1 \frac{1}{2}\left[\cos(y^2+z^2)-\cos(1+y^2+z^2)\right]y\,dy\right]zdz$$

$$= \int_0^1 \frac{1}{4}\left[\sin(y^2+z^2)-\sin(1+y^2+z^2\right]_0^1 zdz$$

$$= \int_0^1 \frac{1}{4}\left[2\sin(1+z^2)-\sin(2+z^2)-\sin(z^2)\right]zdz$$

$$= \frac{1}{8}\left[-2\cos(1+z^2)+\cos(2+z^2)+\cos(z^2)\right]_0^1$$

$$= \frac{1}{8}(\cos 3-3\cos 2+3\cos 1-1)$$

TRIPLE INTEGRAL OVER GENERAL REGIONS

Now we defined the triple integral over *a* general bounded region E in three dimensional space (a solid) by the same procedure that we used for double integrals. *A* solid region *E* is said to be type I, that is

$$E = \{(x,y,z)\mid(x,y)\in D, \phi_1(x,y)\le z\le\phi_2(x,y)\}$$

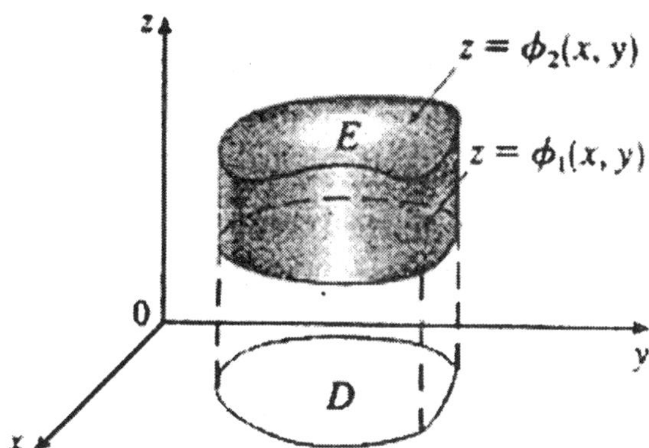

Fig. Triple Integral over *a* General bounded Region

where D is the projection of E onto the x y-plane then

$$\iiint_E f(x,y,z)\,dV = \iint_D \left[\int_{\phi_1(x,y)}^{\phi_2(x,y)} f(x,y,z)\,dz \right] dA$$

A solid region E is changed into type II where

$$E = \{(x,y,z) \mid (x,y) \in D, \phi_1(x,y) \leq z \leq \phi_2(x,y)\}$$

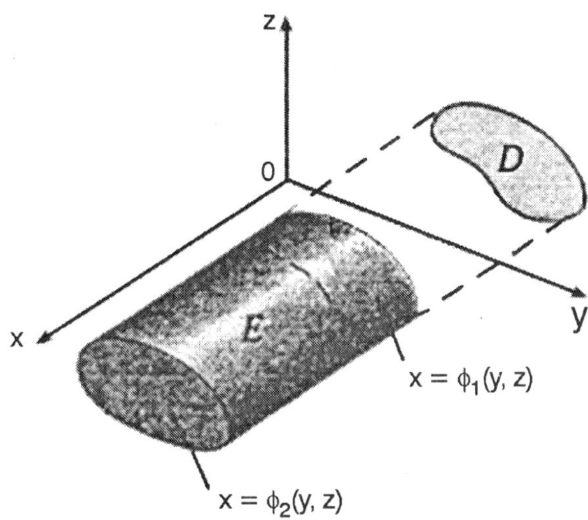

Fig. *A* Solid Region

we have

$$\iiint_E f(x,y,z)\,dV = \iint_D \left[\int_{\phi_1(y,z)}^{\phi_2(y,z)} f(x,y,z)\,dz \right] dA$$

a type III region is of the form

$$E = \{(x,y,z) \mid (x,y) \in D, \phi_1(x,y) \leq z \leq \phi_2(x,y)\}$$

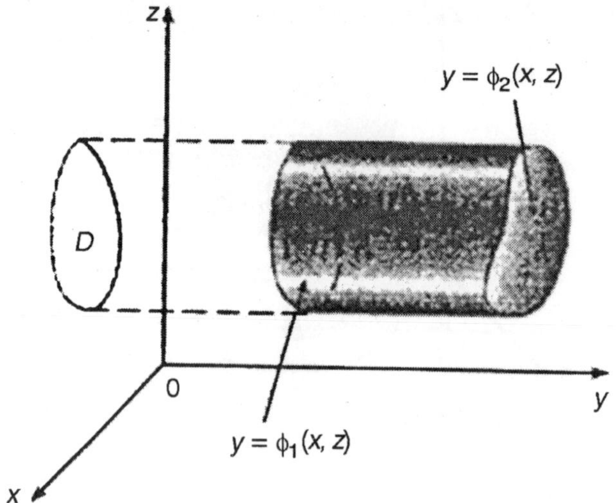

Fig. *A* Solid Region

For this type of region we have

$$\iiint_E f(x,y,z)\,dV = \iint_D \left[\int_{\phi_1(x,z)}^{\phi_2(x,z)} f(x,y,z)\,dz \right] dA$$

Example: Let $E = \{(x,y,z) \mid x^2 + y^2 + z^2 \le 1, x \ge 0, y \ge 0, z \ge 0\}$, find the triple integral $\iiint_E x\,yz\,dV$

Solution:

$$\iiint_E x\,yz\,dV = \int_0^1 x\,dx \int_0^{\sqrt{1-x^2}} y dy \int_0^{\sqrt{1-x^2-y^2}} z dz$$

$$= \frac{1}{2}\int_0^1 x\,dx \int_0^{\sqrt{1-x^2}} y(1-x^2-y^2)\,dy$$

$$= \frac{1}{2}\int_0^1 x\,dx \int_0^{\sqrt{1-x^2}} y(1-x^2-y^2)\,dy$$

CYLINDRICAL COORDINATES

We define the cylindrical coordinates (r, θ, z) of *a* point and we have the following relationships between rectangular coordinates and cylindrical coordinates:

$$\begin{cases} x = r\cos\theta \\ y = r\sin\theta \\ z = z \end{cases} \quad E = \{(r,\theta,z) \mid a \le r \le b, \alpha \le \theta \le \beta, c \le z \le d\}$$

If f is continuos on E given by $0 \le a \le r \le b, \alpha \le \theta \le \beta$, where $0 \le \beta - \alpha \le 2\pi$ and $0 \le c \le z \le d$ then

$$\iiint_E f(x,y,z)\,dV = \int_\alpha^\beta \int_z^b \int_c^d f(r\cos\theta, r\sin\theta, z)\,r\,dr\,d\theta\,dz$$

Proof: In order to compute $\iiint_E f(x,y,z)\,dV$, where R is *a* polar rectangle, we start with *a* partition of [α, β] into l subintervals, *a* partition of into m subintervals and *a* partition of into n subintervals.

$$a = r_0 < r_1 < r_2 < \ldots < r_{i-1} < r_i < \ldots < r_i = b$$
$$\alpha = \theta_0 < \theta_1 < \theta_2 < \ldots < \theta_{j-1} < \theta_j < \ldots < \theta_m = \beta$$
$$c = z_0 < z_1 < z_2 < \ldots < z_{k-1} < z_k < \ldots < z_n = d$$

The "centre" of the polar subrectangle V_{ijk} has polar coordinates

$$r_i^* = \frac{1}{2}(r_i + r_{i-1}),\ \theta_j^* = \frac{1}{2}(\theta_j + \theta_{j-1}),\ \text{and } z_k^* = \frac{1}{2}(z_k + z_{k-1})$$

we find that the volume of V_{ijk} is $V_{ijk} = (r_i^* \Delta\theta_j)\Delta r_i \Delta z_k = r_i^* \Delta r_i \Delta\theta_j \Delta z_k$, ,

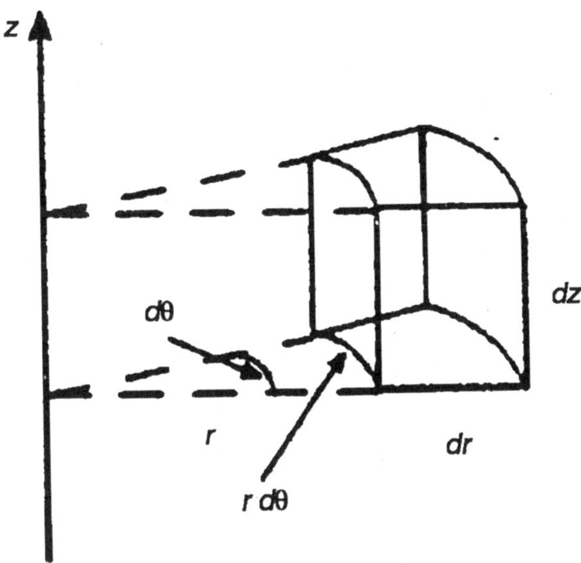

Fig. The "Centre" of the Polar Subrectangle

where $\Delta r_i = r_i + r_{i-1}$, $\Delta q_j = q_j - q_{j-1}$, and $\Delta z_k = z_k + z_{k-1}$.

Therefore the double integral $\iiint_E f(x,y,z)\,dV$ can be approximated by

$$\iiint_E f(x,y,z)\,dV = \sum_{i=1}^{l}\sum_{j=1}^{m}\sum_{k=1}^{n} f(x_{ijk}^*, y_{ijk}^*, z_{ijk}^*)\Delta V_{ijk}$$

$$= \sum_{i=1}^{l}\sum_{j=1}^{m}\sum_{k=1}^{n} f(x_i^* \cos\theta_j^*, r_i^* \sin\theta_j^*, z_k^*)\Delta r_i \Delta\theta_j \Delta z_k$$

In the limiting case, we have

$$\iiint_E f(x,y,z)\,dV = \lim \sum_{i=1}^{l}\sum_{j=1}^{m}\sum_{k=1}^{n} f(x_i^* \cos\theta_j^*, r_i^* \sin\theta_j^*, z_k^*) r_i^* \Delta r_i \Delta\theta_j \Delta z_k$$

$$= \int_c^d \int_\alpha^\beta \int_a^b f(r\cos\theta, r\sin\theta, z)\, r\, dr\, d\theta\, dz$$

Example: Let $E = \{(x,y,z) \mid x^2 + y^2 \le 4, 0, \le z \le 1\}$, evaluate the triple integral

$$\iiint_E (x^2y^2 + 2y^4z)\,dV$$

Solution: By using the cylindrical coordinates $\begin{cases} x = r\cos\theta \\ y = r\sin\theta \\ z = z \end{cases}$

The triple integral becomes

$$\iiint_E (x^2y^2 + 2y^4z)\,dV = \int_0^{2x}\int_0^1\left[\int_0^1 (x^2y^2 + 2y^4z)\,dz\right] rdrd\theta$$

$$.= \int_0^{2x}\int_0^2 \left[x^2y^2z + y^4z^2\right]_0^1 rdrd\theta$$

$$= \int_0^{2x}\left[\int_0^2 \left[x^2y^2z + y^4\right] rdr\right] d\theta$$

$$= \int_0^{2x}\left[\int_0^2 y^2\left(x^2 + y^2\right) rdr\right] d\theta$$

$$= \int_0^{2x}\left[\int_0^2 r^2 \sin^2\theta \left(r^2\right) rdr\right] d\theta$$

$$= \int_0^{2x} \sin^2\theta\, d\theta \int_0^2 r^5\, dr = \frac{32}{3}\pi$$

SPHERICAL COORDINATES

We define the spherical coordinates (ρ, θ, ϕ) of *a* point and we have the following relationships between rectangular coordinates and spherical coordinates:

$$\begin{cases} x = \rho \sin f\theta \\ y = r\sin\theta \\ z = z \end{cases} \quad E = \{(r,\theta,z) \mid a \le r \le b, \alpha \le \theta \le \beta, c \le z \le d\}$$

Fig. Relationships between Rectangular Coordinates and Spherical Coordinates

If *f* is continuos on *E* given by $0 \le a \le \rho \le b$, $\alpha \le \theta \le \beta$ where $0 \le \beta - \alpha \le 2\pi$ and $c \le \theta \le d$ where $0 \le d - c \le 2\pi$. We consider *a* spherical partition *P* of E into smaller spherical wedges E_{ijk} by means of sphere $\rho = \rho_i$, half-plane $\theta = \theta_j$ and half-cone $\theta = \theta_k$. Then the volume of E_{ijk} is approximated by *a* rectangular box with dimensions $\theta\rho_i$, $\rho_i\Delta\phi_k$ (arc of *a* circle with radius ρ_i, angle $\Delta\phi_k$), and $\rho_i \sin \phi_k \Delta\theta_j$ (arc of *a* circle with radius $\rho_i \sin \phi_k$, angle $\Delta\theta_j$). So an approximation to the volume of E_{ijk} is given by

$$\Delta V_{ijk} = \rho_i^2 \sin\phi_k \Delta\rho_i \Delta\theta_j \Delta\phi_k$$

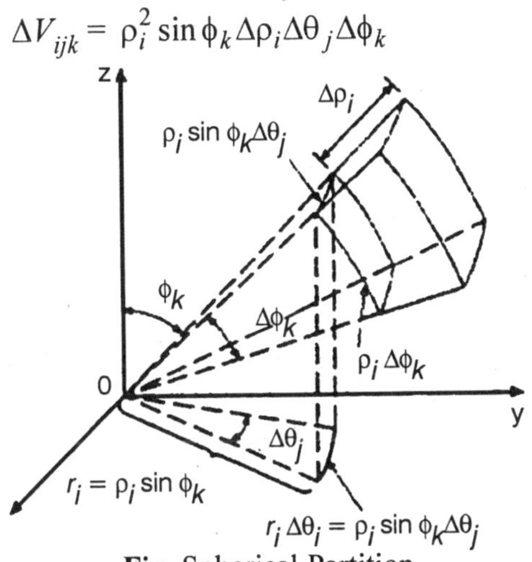

Fig. Spherical Partition

Now we can choose $(x_{ijk}^*, y_{ijk}^*, z_{ijk}^*)$ in E_{ijk} and the triple integral of the function f over the region E as the limit:

$$\iiint_E f(x,y,z)\,dV = \lim \sum_{i=1}^{l}\sum_{j=1}^{m}\sum_{k=1}^{n} f(x_{ijk}^*, y_{ijk}^*, z_{ijk}^*)\Delta V_{ijk}$$

$$= \lim \sum_{i=1}^{l}\sum_{j=1}^{m}\sum_{k=1}^{n} f(\rho_i \sin\phi_k \cos\theta_j, \rho_i \sin\phi_k \sin\theta_j, \rho_i \cos\phi_k)$$

$$\rho_i^2 \sin\phi_k \Delta\rho_i \Delta\theta_j \Delta\phi_k$$

$$= \iiint_E f(\rho\sin\phi\cos\theta, \rho\sin\phi\sin\theta, \rho\cos\phi)\,\rho^2 \sin\phi\, d\rho\, d\theta\, d\phi$$

or

$$\iiint_E f(x,y,z)\,dV = \int_c^d\int_\alpha^\beta\int_a^b f(\rho\sin\phi\cos\theta, \rho\sin\phi\sin\theta, \rho\cos\phi)\rho^2 \sin\phi\, d\rho\, d\theta\, d\phi$$

This formula can be extended to include more general spherical regions.

Example: Let $E = \{(x,y,z) \mid x^2+y^2+z^2 \le 4\}$, evaluate the triple integral

$$\iiint_E \sqrt{x^2+y^2}\exp\left[-(x^2+y^2+z^2)\right]dV.$$

Solution: By using the spherical coordinates $\begin{cases} x = \rho\sin\theta\cos\phi \\ y = \rho\sin\theta\sin\phi \\ z = \rho\cos\theta \end{cases}$, the triple integral becomes

$$\iiint_E \sqrt{x^2+y^2}\exp\left(\sqrt{x^2+y^2+z^2}\right)dV$$

$$= \int_0^{2x}\int_0^{x}\int_0^{2} \rho\sin\theta\exp(-\rho)(\rho^2 \sin\theta d\,\rho d\,\rho d\theta d\phi)$$

$$= \int_0^{2x} d\phi \int_0^{x} \sin^2\theta d\theta \int_0^2 \rho^3 \exp(-\rho)\,d\rho - 2\pi^2(3-19e^{-2})$$

TRANSFORMATION OF COORDINATES

We consider *a* change of variables that is given by *a* transformation *T* from the uv-plane to the *xy*-plane:

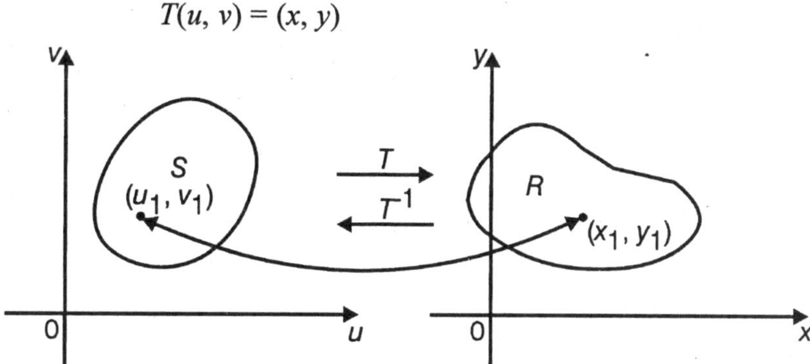

Fig. Transformation of Coordinates

where x and y are related to u and v by the equations

$$x = g(u, v) \quad y = h(u, v)$$

or, as we sometimes write,

$$x = x\,(u, v) \quad y = y\,(u, v)$$

We usually assume that T is *a* C^1 transformation, which means that g and h have continuous first-order partial derivatives.

Next we partition *a* region S in the *uv*-plane into rectangle S_{ij} and call their images in the *xy*-plane R_{ij}.

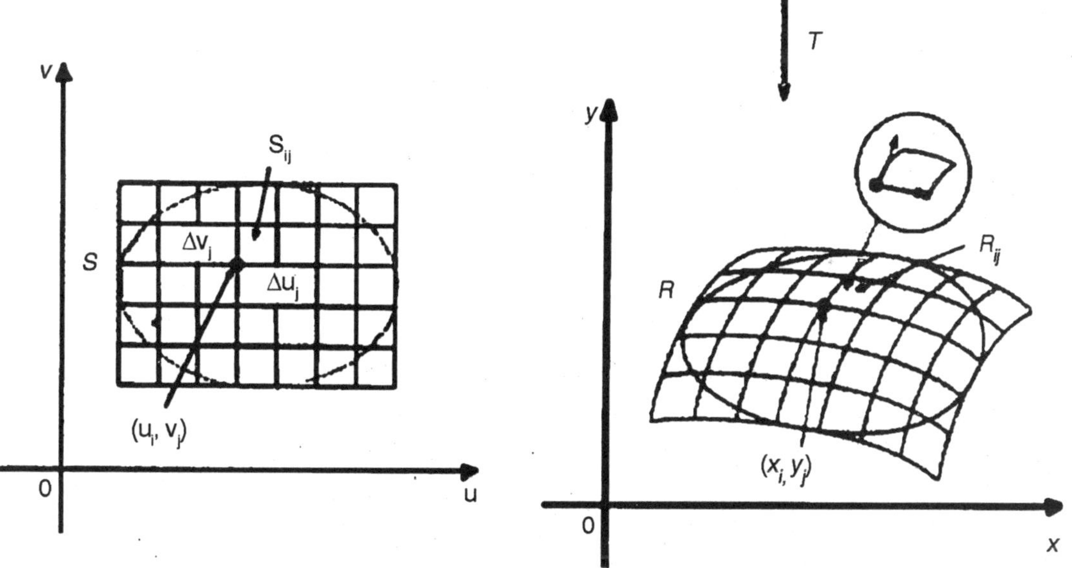

Fig. Transformation of Coordinates with first-order Partial Derivatives

Applying the approximation to each R_{ij}, we approximate the double integral of f over R as follows:

$$\iint_R f(x,y)\,dA \approx \sum_{i=1}^{m}\sum_{i=1}^{n} f(g(u_i,v_j),h(u_i,v_j))\left|\frac{\partial(x,y)}{\partial(u,v)}\right|\Delta u_i \Delta v_j$$

$$= \iint_S f(g(u,v),h(u,v)\left|\frac{\partial(x,y)}{\partial(u,v)}\right| du\,dv$$

where the Jacobian is determined by

$$\frac{\partial(x,y)}{\partial(u,v)} = \begin{vmatrix} \frac{\partial x}{\partial u} & \frac{\partial x}{\partial v} \\ \frac{\partial y}{\partial u} & \frac{\partial y}{\partial v} \end{vmatrix} = \frac{\partial x}{\partial u}\frac{\partial x}{\partial v} - \frac{\partial y}{\partial u}\frac{\partial y}{\partial v}$$

and is evaluated at (u_i, v_j).

As an example, let us show that the formula for integration in polar coordinates is just *a* special case. The transformation from the $r\theta$-plane to the xy-plane is given by

$$r^2 = x^2 + y^2 \quad x = \text{rcos}\theta \quad y = \text{rsin}\theta$$

The Jacobian of the transformation is given by

$$\frac{\partial(x,y)}{\partial(r,\theta)} = \begin{vmatrix} \frac{\partial x}{\partial r} & \frac{\partial x}{\partial \theta} \\ \frac{\partial y}{\partial r} & \frac{\partial y}{\partial \theta} \end{vmatrix} = \frac{\partial x}{\partial r}\frac{\partial y}{\partial \theta} - \frac{\partial x}{\partial \theta}\frac{\partial y}{\partial r}$$

$$= \cos\theta(r\cos\theta) - (-\sin\theta)\sin\theta = r$$

Hence the transformation of integrals can be recovered as

$$\iint_R f(x,y)\,dA = \iint_S f(r\cos\theta, r\sin\theta)\,r\,dr\,r\,d\theta$$

Example: Let R be *a* region bounded by four straight lines $x + y = 1$, $x + y = -1$, $x - y = 1$ and $x - y = -1$.

1. Find the Jocobian of the following transformation

$$\begin{cases} u = x + y \\ v = x - y \end{cases}$$

2. Hence, evaluate the double integral $\iint_R (x^2 + 2xy)\,dx\,dy$

Solution:

1. The inverse of the transformation is given by

$$\begin{cases} x = \dfrac{1}{2}(u+v) \\ y = \dfrac{1}{2}(u-v) \end{cases}$$

Hence, the Jocobian of the transformation is

$$\frac{\partial(x,y)}{\partial(u,v)} = \begin{vmatrix} \dfrac{\partial x}{\partial v} & \dfrac{\partial x}{\partial v} \\ \dfrac{\partial y}{\partial u} & \dfrac{\partial y}{\partial v} \end{vmatrix} = \frac{\partial x}{\partial u}\frac{\partial y}{\partial v} - \frac{\partial x}{\partial v}\frac{\partial y}{\partial u} = \frac{1}{2}\times\left(-\frac{1}{2}\right) - \frac{1}{2}\times\left(\frac{1}{2}\right) = -\frac{1}{2}$$

Therefore the double integral becomes

$$\iint_R (x^2+2xy)\,dx\,dy = \int_{-1}^{1}\int_{-1}^{1}\left[\left(\frac{1}{2}(u+v)\right)^2 + 2\left(\frac{1}{2}(u-v)\right)\right]\frac{\partial(x,y)}{\partial(u,v)}\,du\,dv$$

$$= \int_{-1}^{1}\int_{-1}^{1}\left[\left(\frac{1}{2}(u+v)\right)^2 + 2\left(\frac{1}{2}(u+v)\right)\left(\frac{1}{2}(u-v)\right)\right]\frac{\partial(x,y)}{\partial(u,v)}\,du\,dv$$

$$= 2\int_0^{2x} d\phi \int_0^{\pi/2} \sin\theta\, d\theta \int_0^R \rho^2 d\rho = \frac{4\pi}{3}R^3$$

$$= \int_{-1}^{1}\int_{-1}^{1}\left(-\frac{3}{8}u^2 - \frac{1}{24}uv + \frac{1}{8}v^2\right)du\,dv = -\frac{1}{3}$$

There is *a* similar change of variables formula for triple integrals. Let T be *a* transformation from the S region in uvw-plane to the R region in xyz-plane:

$$T(u, v, w) = (x, y, z)$$

where x, y and z are related to u, v and w by the equations

$$x = g(u, v, w) \quad y = h(u, v, w) \quad z = k(u, v, w)$$

or, as we sometimes write,

$$x = x(u, v, w) \quad y = y(u, v, w) \quad z = z(u, v, w)$$

Again T is *a* C^1 transformation which means that g, h and k have continuous first-order partial derivatives. The Jacobian of T is determined by

$$\frac{\partial(x,y,z)}{\partial(u,v,w)} = \begin{vmatrix} \frac{\partial x}{\partial u} & \frac{\partial x}{\partial v} & \frac{\partial x}{\partial w} \\ \frac{\partial y}{\partial u} & \frac{\partial y}{\partial v} & \frac{\partial y}{\partial w} \\ \frac{\partial z}{\partial u} & \frac{\partial z}{\partial v} & \frac{\partial z}{\partial w} \end{vmatrix}$$

Similarly we have the following formula for triple integrals:

$$\iiint_R f(x,y,x)\,dA = \iiint_R f[g(u,v,w),h(u,v,w)]\left|\frac{\partial(x,y,z)}{\partial(u,v,w)}\right| du\,dv\,dw$$

Example:

1. Find the Jocobian of the spherical coordinates
2. Using the above results, show that the volume of *a* sphere with radius *R* is

$$V = \frac{4\pi}{3}R^3$$

Solution:

1. The Jocobian of the spherical coordinates is given by

$$\frac{\partial(x,y,z)}{\partial(\rho,\theta,\phi)} = \begin{vmatrix} \frac{\partial x}{\partial \rho} & \frac{\partial x}{\partial \theta} & \frac{\partial x}{\partial \phi} \\ \frac{\partial y}{\partial \rho} & \frac{\partial y}{\partial \theta} & \frac{\partial y}{\partial \phi} \\ \frac{\partial z}{\partial \rho} & \frac{\partial z}{\partial \theta} & \frac{\partial z}{\partial \phi} \end{vmatrix} = \begin{vmatrix} \sin\theta\cos\phi & \rho\sin\theta\cos\phi & -\rho\sin\theta\sin\phi \\ \sin\theta\sin\phi & \rho\cos\theta\sin\phi & \rho\sin\theta\cos\phi \\ \cos\theta & -\rho\sin\theta & 0 \end{vmatrix}$$

$= \rho^2\sin\theta$

2. The volume of the sphere is given by the triple integral

$\iiint_E d\,xd\,ydz$ where $E = \{(x, y, z)|x^2 + y^2 + z^2 \le R^2\}$

By using the spherical coordinate $\begin{cases} x = \rho\sin\theta\cos\phi \\ y = \rho\sin\theta\sin\phi \\ z = \rho\cos\theta \end{cases}$, we can transform the above integral into

$$\iiint_E d\,xd\,yd\,zd\,V = 2\int_0^{2x}\int_0^{\pi/2}\int_0^R \rho^2\sin\theta d\rho d\theta d\phi$$

$$= 2\int_0^{2x} d\phi \int_0^{\pi/2}\sin\theta d\theta \int_0^R \rho^2 d\rho = \frac{4\pi}{3}R^3$$

Chapter 4

Differential Operators

LANGUAGE AND CLASSIFICATION

In this chapter we will take a look at the language of partial differential equations. As in any technical subject, we shall need some standard terms in order to carefully describe the things we are working with.

Several terms are probably familiar already. Suppose that we have a partial differential equation in u. The *order* of the equation is the order of the highest partial derivative of u, the *number of variables* is simply the number of independent variables for u in the equation, and *the equation has constant coefficients* if u and the coefficients of all the partial derivatives of u are constant. If all the terms involving u are moved to the left side of the equation, then the equation is called *homogeneous* if the right side is zero, and *non homogeneous* if it is not zero. The system is linear if that left side involves u in only a "linear way" through some linear operator L[u]

Examples:

1. $4u_{xx} - 24\,u_{xy} + 11u_{yy} - 12u_x - 9u_y - 5u = 0$

   is a linear, homogeneous, second order partial differential equation in two variables with constant coefficients.

2. $u_t = x\,u_{xx} + y\,u_{yy} + u^2$

   is a nonlinear, second order partial differential equation in two variables and does not have constant coefficients.

3. $u_{tt} - u_{xx} = \sin(\pi\,t)$

   is a non-homogeneous equation.

In thinking of partial differential equations, we shall carry over the language that we used for matrix or ordinary differential equations as far as possible.

So, in partial differential equation, we consider linear equations

$$Lu = 0, \text{ or } u' = Lu,$$

Only now L is a linear operator on a space of functions. For example, it may be that $L(u) = u_{xx} + u_{yy}$, and corresponding notation for the heat equation

$$u' = Lu$$

is

$$u_t = u_{xx} + u_{yy}.$$

We shall be interested in a rather general second order, differential operator. In two variables, we consider the operator

$$L(u) = \sum_{p=1}^{2}\sum_{q=1}^{2} A_{pq} \frac{\partial^2 u}{\partial xp\, \partial xq} + \sum_{p=1}^{2} B_p \frac{\partial u}{axp} + cu.$$

There is an analogous formula for three variables.

We suppose that u is smooth enough so that

$$\frac{\partial^2 u}{\partial x_1\, \partial x_2} = \frac{\partial^2 u}{\partial x_2\, \partial x_1};$$

That is, we can interchange the order of differentiation. In this first consideration, the matrix A, the vector B, and the number C do not depend on u; we take the matrix A to be symmetric. Because of (19.2) we are free to arrange this, since the coefficient of the cross term is $A_{12} + A_{21}$, and we may assign A_{12} and A_{21} any way we wish, so long as the sum has the right value.

Example: Write an example in the matrix representation for constant coefficient equations:

$$L[u] = 4u_{xx} - 24\, u_{xy} + 11u_{yy} - 12u_x - 9_y - 5u.$$

Answer:

$$L[u] = (\partial/\partial x\ \partial/\partial y)\begin{pmatrix} 4 & -12 \\ -12 & 11 \end{pmatrix}\begin{pmatrix} \partial/\partial x \\ \partial/\partial y \end{pmatrix}u$$

$$+(-12\quad -9)\begin{pmatrix} \partial/\partial x \\ \partial/\partial y \end{pmatrix}u - 5u.$$

Many equations of interest have the form

$$L(u) = f$$

or

$$u' = L(u) + f.$$

In this section, u is a function on R^2 or R^3 and the equations are to hold in an open, connected region D of the plane. We will also assume that the boundary of the region is piece-wise smooth, and denote this boundary by ∂D. Just as in ordinary differential equations, in partial differential equations some boundary conditions will be needed to solve the equations. We will take the boundary conditions to be linear and have the general form

$$B(u) = a\,u + b\,u_n,$$

Where u_n is the derivative taken in the direction of a normal to the boundary of the region.

The techniques of studying partial differential operators and the properties of these operators change depending on the "type" of operator. These operators have been classified into three principal types. The classifications are made according to the nature of the coefficients in the equation which defines the operator. The operator is called an *elliptic* operator if the eigenvalues of A are non-zero and have the same algebraic sign. The operator is *hyperbolic* if the eigenvalues have opposite signs and is *parabolic* if at least one of the eigenvalues is zero.

The classification makes a difference both in what we expect from the solutions of the equation and in how we go about solving them. Each type is modeled on an important equation of physics.

The basic example of a parabolic equation is the one dimensional heat equation. Here, u(t,x) represents the heat on a line at time t and position x. One should be given an initial distribution of temperature which is denoted u(0,x), and some boundary conditions which arise in the context of the problem. For example, it might be assumed that the ends are held at some fixed temperature for all time. In this case, boundary conditions for a line of length L would be $u(t,0) = \alpha$ and $u(t,L) = \beta$. Or, one might assume that the ends are insulated. A mathematical statement of this is that the rate of flow of heat through the ends is zero:

$$\frac{\partial u}{\partial x}(t,0) = \frac{\partial u}{\partial x}(t,L) = 0$$

The manner in which u changes in time is derived from the physical principle which states that the heat flux at any point is proportional to the temperature gradient at that point and leads to the equation

$$\frac{\partial u}{\partial t}(t,x) = \frac{\partial^2 u}{\partial x^2}(t,x)$$

Geometrically, one may think of the problem as one of defining the graph of u whose domain is the infinite strip bounded in the first quadrant by the parallel lines x = 0 and x =

L The function u is known along the x axis between x = 0 and x = L, To define u on the infinite strip, move in the t direction according to the equation

$$\frac{\partial u}{\partial t}(t,x) = \frac{\partial^2 u}{\partial x^2}(t,x)$$

While Maintaining the Boundary Conditions.

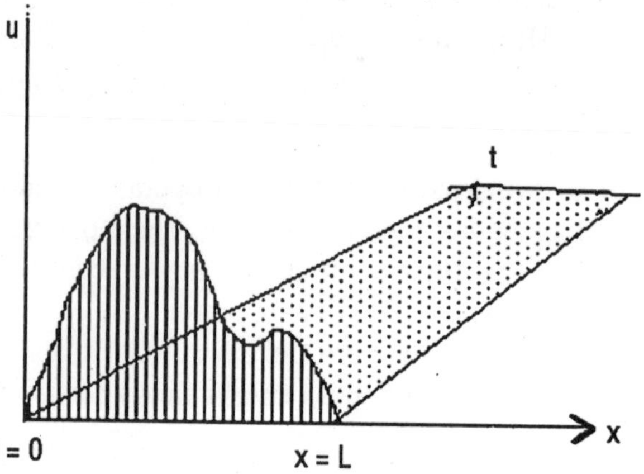

We could also have a source term. Physically, this could be thought of as a heater (or refrigerator) adding or removing heat at some rate along the strip. Such an equation could be written as

$$\frac{\partial u}{\partial t}(t,x) = \frac{\partial^2 u}{\partial x^2}(t,x) + Q(t,x).$$

Boundary and initial conditions would be as before. In order to rewrite this equation in the context of this course, we should conceive of the equation as L[u] = f, with appropriate boundary conditions. The operator L is

$$L[u] = \frac{\partial u}{\partial t}(t,x) - \frac{\partial^2 u}{\partial x^2}(t,x)$$

This is a parabolic operator according to the definition given above; in fact, the matrix A is given by

$$A = \begin{pmatrix} 0 & 0 \\ 0 & -1 \end{pmatrix}.$$

(A and Laplace operator, or Laplacian:

$$L[u] = \frac{\partial^2 u}{\partial x^2} + \frac{\partial^2 u}{\partial y^2}.$$

Laplace's equation states that

$$L[u] = 0$$

and Poisson's equation states that

$$L[u] = f(x,y)$$

for some given function f. A physical situation in which it arises is in the problem of finding the shape of a drum under force. Suppose that the bottom of the drum sits on the unit disc in the xy-plane and that the sides of the drum lie above the unit circle. We do not suppose that the sides are at a uniform height, but that the height is specified on the circle.

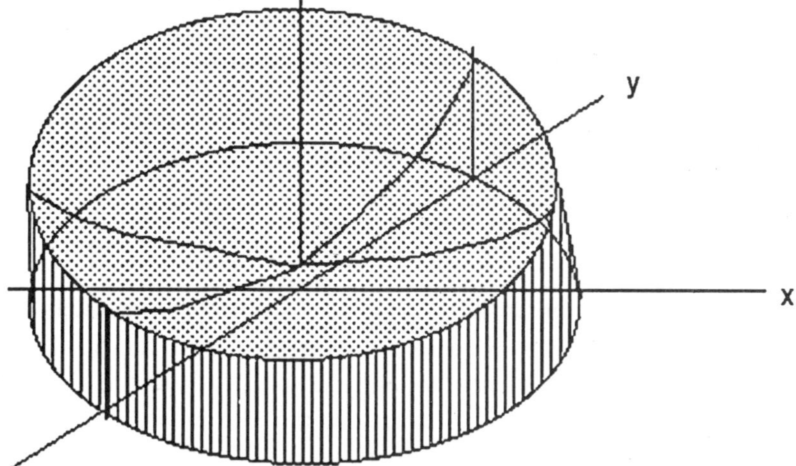

That is, we know u(x,y) for {x,y} on the boundary ot the drum. We also suppose that there is a force pulling down, or pushing up, on the drum at each point and that this force is not changing in time. An example of such a force might be the pull of gravity. The question is, what is the shape of the drum? As we shall see, the appropriate equations take the form: Find u if

$$\frac{\partial^2 u}{\partial x^2} + \frac{\partial^2 u}{\partial y^2} = f(x, y) \text{ for } x^2 + y^2 < 1$$

With

u(x,y) specified for $x^2 + y^2 = 1$.

Laplace's and Poisson's equations also arise in electromagnetism and fluid mechanics as the equations for potential functions. Still another place where you may encounter Laplace's equation is as the equation for a temperature distribution in 2 or 3 dimensions, when thermal equilibrium has been reached. The Laplace operator is elliptic by our definition, for the matrix A in is given by

$$A = \begin{pmatrix} 1 & 0 \\ 0 & 1 \end{pmatrix}.$$

Finally, the model for a hyperbolic equation is the one dimensional wave equation. One can think of this equation as describing the motion of a taunt string after an initial perturbation and subject to some outside force. Appropriate boundary conditions are given. To think of this as being a plucked string with the observer watching the up and down motion in time is not a bad perspective, and certainly gives intutive understanding. Here is another perspective, however, which will be more useful in the context of finding the Green function to solve this one dimensional wave equation:

$$\frac{\partial^2 u}{\partial t^2} = \frac{\partial^2 u}{\partial x^2} \text{ f (t, x) for } 0 < x < L,$$

$$u(t, x) = u\ (t, L) = 0 \text{ for } t > 0,$$

$$u(0, x) = g\ (x)$$

and $$\frac{\partial u}{\partial t}\ (0, x) = h(x) \text{ for } 0 < x < L.$$

As in example (a), the problem is to describe u in the infinite strip within the first quadrant of the xt-plane bounded by the x axis and the lines x = 0 and x = L. Both u and its first derivative in the t direction are known along the x axis. Along the other boundaries, u is zero. What must be the shape of the graph above the infinite strip?

To classify this as a hyperbolic problem, think of the operator L as

$$L[u] = \frac{\partial^2 u}{\partial t^2} - \frac{\partial^2 u}{\partial x^2}$$

and re-write it in the appropriate form for classification. The matrix A in (19.1) is given by

$$A = \begin{pmatrix} 1 & 0 \\ 0 & -1 \end{pmatrix}.$$

(A physical derivation of the wave equation and further discussion of it)

There are standard procedures for changing more general partial differential equations to the familiar standard forms, which we shall investigate in the next section.

These very names and ideas suggest a connection with quadratic forms in analytic geometry. We will make this connection a little clearer. Rather than finding geometric understanding of the partial differential equation from this connection we will more likely develop algebraic understanding. Especially, we will see that there are some standard forms. Because of the nearly error-free arithmetic that Maple is able to do, we will offer syntax in

Maple that enables the reader to use this computer algebra system to change second order, linear systems into the standard forms.

If presented with a quadratic equation in x and y, one could likely decide if the equation represented a parabola, hyperbola, or ellipse in the plane. However, if asked to draw a graph of this conic section in the plane, one would start recalling that there are several forms that are easy to draw:

$$a\,x^2 + b\,y^2 = c^2, \text{ and the special case } x^2 + y^2 = c^2,$$

$$a\,x^2 - b\,y^2 = c^2, \text{ and the special case } x^2 - y^2 = c^2,$$

or

$$y - a\,x^2 = 0 \text{ and } x - b\,y^2 = 0.$$

These quadratic equations represent the familiar conic sections: ellipses, hyperbolas and parabolas, respectively. If a quadratic equation is given that is not in these special forms, then one may recall procedures to transform the equations algebraically into these standard forms. This will be the topic of the next section.

The purpose for doing the classification is that the techniques for solving equations are different in the three classes, if it is possible to solve the equation at all. Even more, there are important resemblance among the solutions of one class; and there are striking differences between the solutions of one class and those of another class. The remainder of these notes will be primarily concerned with finding solutions to hyperbolic, second order, partial differential equations. As we progress, we will see the importance of the equation being a hyperbolic partial differential equation to use the techniques of these notes.

Before comparing the similarity in procedures for changing the partial differential equation to standard form with the preceeding arithmetic, we pause to emphasize the differences in geometry for the types: elliptic, hyperbolic, and parabolic.

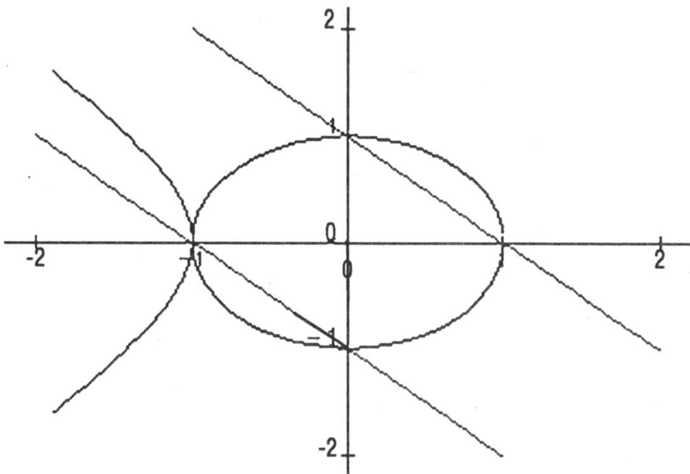

Here are three equations from analytic geometry:

$x^2 + y^2 = 4$ is an ellipse,

$x^2 - y^2 = 4$ is a hyperbola,

and

$x^2 + 2\,x\,y + y^2 = 4$ is a parabola.

Contains the graphs of all three of these. Their shapes and their geometry, are strikingly different. Even more, naively, one might say that the graph of the third of those above is not the graph of a parabola. Indeed. It does, however, meet the criteria: $b^2 - 4\,a\,c = 0$. One might think of the graph as that of a parabola with vertex at the "point at infinity."

The criterion for classifying second order, partial differential equations is the same: ask what is the character of $b^2 - 4\,a\,c$ in the equation

$$a\frac{\partial^2 u}{\partial x^2} + b\frac{\partial^2 u}{\partial x \partial y} + c\frac{\partial^2 u}{\partial y^2} + d\frac{\partial u}{\partial x} + e\frac{\partial u}{\partial y} + fu = 0.$$

We now present solutions for three equations that have the same start - the same initial conditions. However the equations are of the three types. There is no reason you should know how to solve the three equations yet. There is no reason you should even understand that solving the hyperbolic equation by the method of characteristics is appropriate. But, you should be able to check the solutions - to see that they solve the specified equations. Each equation has initial conditions

$$u(0,y) = \sin(y) \text{ and } ux(0,y) = 0.$$

The equations are

$$\frac{\partial^2 u}{\partial x^2} + \frac{\partial^2 u}{\partial y^2} = 0$$

Laplace's equation - elliptic;

$$\frac{\partial^2 u}{\partial x^2} - \frac{\partial^2 u}{\partial y^2} = 0$$

a form of the wave equation - hyperbolic; and

$$\frac{\partial^2 u}{\partial x^2} + 2\frac{\partial^2 u}{\partial x \partial y} + \frac{\partial^2 u}{\partial y^2} = 0$$

is a parabolic partial differential equation,

These have solutions cosh(x) sin(y), cos(x) sin(y), and cos(x–y) x/2 + sin(y–x) respectively. Figure 19.2 has the graphs of these three solutions in order.

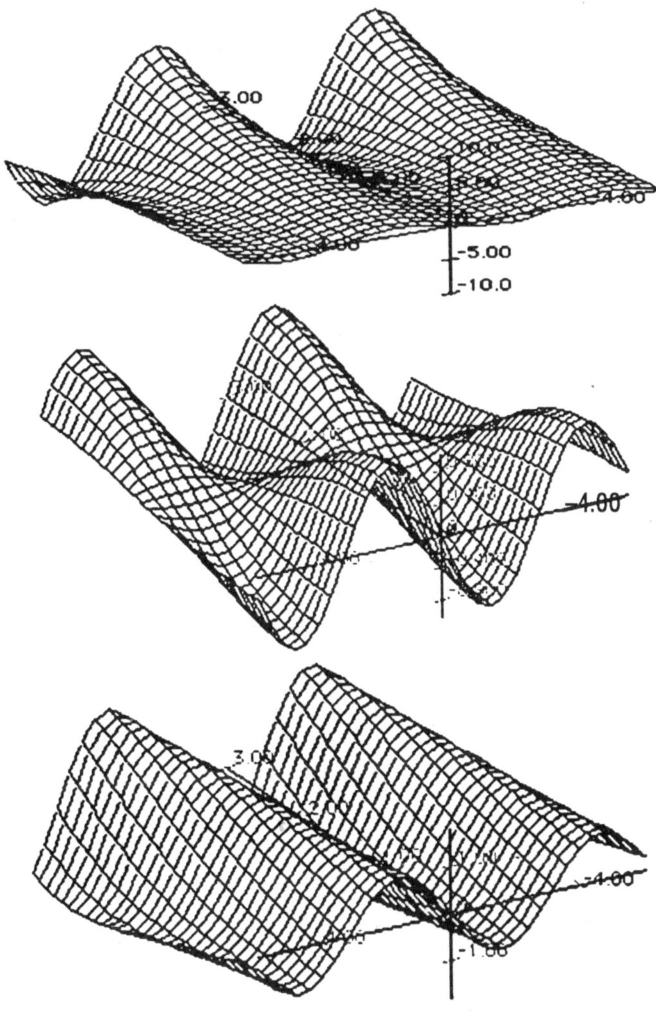

CANONICAL FORMS

In this section, we review the procedures to get second-order equations into standard forms, and compare these with the techniques to get second-order partial differential equations with constant coefficients into a few standard forms, called the *canonical forms*.

Performing these algebraic procedures corresponds to the geometric notions of translations and rotations.

For example, the equation

$$x^2 - 3y^2 - 8x + 30y = 50$$

Represents a hyperbola. To draw the graph of the hyperbola, one algebraically factors the equation or, geometrically, translates the axes:

$$(x-4)^2 - 3(y-5)^2 = 1.$$

Now, the response that this is a hyperbola with center {4,5} is expected. More detailed information about the direction of the major and minor axes could be made, but these are not notions that we will wish to carry over to the process of getting second order partial differential equations into canonical forms.

There is another idea more appropriate. Rather than keeping the hyperbola in the Euclidean plane where it now has the equation

$$x^2 - 3y^2 = 1$$

In the translated form, think of this hyperbola in the Cartesian plane, and do not insist that the x axis and the y axis have the same scale. In this particular case, keep the x axis the same size and expand the y axis so that every unit is the old unit multiplying by R(3). Algebraically, one commonly writes that there are new coordinates {x′,y′} related to the old coordinates by

$$x = x', \ R(3)\ y = y'.$$

The algebraic effect is that the equation is transformed into an equation in {x′,y′} coordinates:

$$x'^2 - y'^2 = 1.$$

Pay attention to the fact that it is a mistake to carry over too much of the geometric language for the form. For example if the original quadratic equation had been

$$x^2 + 3y^2 - 8x + 30y = 50$$

and we had translated axes to produce

$$x^2 + 3y^2 = 50,$$

and then rescaled the axes to get

$$x^2 + y^2 = 50$$

we have not changed an ellipse into a circle for a circle is a geometric object whose very definition involves the notion of distance. The process of changing the scale on the x axis and the y axis certainly destroys the entire notion of distance being the same in all directions.

Rather, the rescaling is an idea that is *algebraically* simplifying. Before we pursue the idea of rescaling and translating in second order partial differential equations in order to come up with canonical forms, we need to recall that there is also the troublesome need to rotate the axis in order to get some quadratic forms into the standard one. For example, if the equation is

$$xy = 2,$$

We quickly recognize this as a quadratic equation. Even more, we could draw the graph. If pushed, we would identify the resulting geometric figure as a hyperbola. We ask for more here since these geometric ideas are more readily transformed into ideas about partial differential equations if they are converted into algebraic ideas. The question, then, is how do we achieve the algebraic representation of the hyperbola in standard form?

One recalls from analytic geometry, or recognizes from simply looking at the picture of the graph of the equation, that this hyperbola has been rotated out of standard form. To see it in standard form, we must rotate the axes. One forgets the details of how this rotation is performed, but should know a reference to find the scheme. From here on, the following may serve as your reference!

Here is the rotation needed to remove the xy term in the equation

$$a\,x^2 + b\,xy + c\,y^2 + d\,x + e\,y + f = 0.$$

The new coordinates $\{x', y'\}$ are given by

$$\begin{pmatrix} x' \\ y' \end{pmatrix} = \begin{pmatrix} \cos(a) & \sin(a) \\ -\sin(a) & \cos(a) \end{pmatrix} \begin{pmatrix} x \\ y \end{pmatrix},$$

where

$$a = \begin{pmatrix} p/4 \text{ if } a = c \\ \frac{1}{2}\arctan\left(\frac{b}{a-c}\right) \text{if } a\ \pi\ c \end{pmatrix}.$$

And "quivalent way to write is to multiply both sides by the inverse of the matrix:

$$\begin{pmatrix} x \\ y \end{pmatrix} = \begin{pmatrix} \cos(a) & -\sin(a) \\ \sin(a) & \cos(a) \end{pmatrix} \begin{pmatrix} x' \\ y' \end{pmatrix}$$

Thus, substitute $x = x'\cos(a) - y'\sin(a)$ and $y = x'\sin(a) + y'\cos(a)$ into the equation, where a is as indicated above and the cross term, bxy, will disappear.

Given a general quadratic in two variables, there are three things that need to be done to get it into standard form: get rid of the xy terms, factor all the x terms and the y terms separately, and rescale the axes so that the coefficients of the x^2 term and the y^2 terms are the same. Geometrically this corresponds, as we have recalled, to a rotation, translation, and expansion, respectively. From the geometric point of view, it does not matter which is done first: the rotation and then the translation, or vice versa. Algebraically, it is better to remove the xy terms first, for then the factoring is easier.

Finally, it should also be remembered that the sign of $b^2 - 4ac$ can be used to predict whether the curve is a hyperbola, parabola, or ellipse.

What follows is a Maple program for removing the cross term. Using the program assures fast accurate computation for otherwise tedious calculations to determine the rotation for equations of the form

$$a x^2 + b xy + c y^2 + d x + ey + f = 0.$$

The coefficients are read in first.

```
* a:= ?: b:= ?:c:= ?:d:= ?:e:= ?:f:= ?:
```

The angle is determined and the rotation is performed.

```
*if a = c then alpha: = Pi/4
else alpha: = arctan(b/(a-c))/2 fi;
* si: = sin(alpha);
* co: = cos(alpha);
* x: = co*u – si*v; y: = si*u + co*v;
* q: = a*x ^ 2 + b*x*y + c*y^2 + d*x + e*y + f;
* simplify(");
* evalf(");
```

The purpose of the previous paragraphs recalling how to change algebraic equations representing two dimensional conic sections into standard form was to suggest that the same ideas carry over almost unchanged for the second degree partial differential equations. The techniques will change these equations into the canonical forms for elliptic, hyperbolic, or parabolic partial differential equations.

Here are the forms we want:

Elliptic Equations:

$$L[u] = \frac{\partial^2 u}{\partial x^2} + \frac{\partial^2 u}{\partial y^2} + a\frac{\partial u}{\partial x} + b\frac{\partial u}{\partial y} \pm k\,u,$$

Hyperbolic Equations:

$$L[u] = \frac{\partial^2 u}{\partial x^2} + \frac{\partial^2 u}{\partial y^2} \pm k\,u,$$

Parabolic Equation:

$$L[u] = \frac{\partial u}{\partial x} - \frac{\partial^2 u}{\partial y^2} \pm k\,u,$$

The choice for which form to use is determined by the techniques used to solve the equations.

We turn now to the techniques of arriving at standard forms for partial differential equations.

If one has a second order partial differential equation with constant coefficients that is not in the standard form, there is a method to change it into this form. The techniques are similar to those used in the analytic geometry. Having the standard form, one might then solve the equation. Finally, the solution should be transformed back into the original coordinate system.

We will illustrate the procedure for transformation of a second order equation into standard form. Consider the equation

$$11\frac{\partial^2 u}{\partial x^2}+4\sqrt{3}\frac{\partial^2 u}{\partial x\partial y}+7\frac{\partial^2 u}{\partial y^2}-5u=0.$$

In the original equation, if we think of the equation as

$$a\frac{\partial^2 u}{\partial x^2}+b\frac{\partial^2 u}{\partial x\partial y}+c\frac{\partial^2 u}{\partial y^2}+d\frac{\partial u}{\partial x}+e\frac{\partial u}{\partial y}+f\,u=0.$$

Then, a = 11, b = 4 R(3), c = 7, so that $b^2 - 4\,a\,c = -260$ and we identify this as an elliptic equation.

We would like to transform the equation into the form

$$\frac{\partial^2 u}{\partial x^2}+\frac{\partial^2 u}{\partial y^2}+cu=0$$

Introduce new coordinates (x,h) by rotation of axes so that in the transformed equation the mixed second partial derivative does not appear. Let

$$\begin{pmatrix} x \\ h \end{pmatrix}=\begin{pmatrix} \cos(a) & \sin(a) \\ -\sin(a) & \cos(a) \end{pmatrix}\begin{pmatrix} x \\ y \end{pmatrix}$$

or,

$$\begin{pmatrix} x \\ y \end{pmatrix}=\begin{pmatrix} \cos(a) & -\sin(a) \\ \sin(a) & \cos(a) \end{pmatrix}\begin{pmatrix} x \\ h \end{pmatrix}$$

Using the chain rule,

$$\frac{\partial}{\partial x}=\cos(a)\frac{\partial}{\partial x}-\sin(a)\frac{\partial}{\partial h}$$

and

$$\frac{\partial}{\partial y} = \sin(a)\frac{\partial}{\partial x} + \cos(a)\frac{\partial}{\partial h}.$$

It follows that

$$\frac{\partial^2}{\partial x^2} = \frac{\partial}{\partial x}\frac{\partial}{\partial x} = \left(\cos(a)\frac{\partial}{\partial x} - \sin(a)\frac{\partial}{\partial h}\right)\left(\cos(a)\frac{\partial}{\partial x} - \sin(a)\frac{\partial}{\partial h}\right)$$

so that

$$\frac{\partial^2}{\partial x^2} = \cos^2(a)\frac{\partial^2}{\partial x^2}2 - \sin(a)\cos(a)\frac{\partial^2}{\partial x \partial h} + \sin^2(a)\frac{\partial^2}{\partial h^2}$$

In a similar manner,

$$\frac{\partial^2}{\partial x \partial y} = \frac{\partial^2}{\partial x^2} + \left(\cos^2(a) - \sin(a)\right)\frac{\partial^2}{\partial x \partial h} - \sin^2(a)\cos(a)\frac{\partial^2}{\partial h^2},$$

and

$$\frac{\partial^2}{\partial y} = \sin^2(a)\frac{\partial^2}{\partial x^2}2\sin(a)\cos(a)\frac{\partial^2}{\partial x \partial h} + \cos^2(a)\frac{\partial^2}{\partial h^2}.$$

The original equation described u as a function of x and y. We now define v as a function of x and h by v(x,h) = u(x,y). The variables x and h are related to x and y as described by the rotation above:

$$\begin{aligned} v(x,h) &= u(x(x,h),\ y(x,h)) \\ &= u(\cos(a)\,x - \sin(a)\,h,\ \sin(a)\,x + \cos(a)\,h) \end{aligned}$$

Of course, we have not specified a yet. This comes next.

The equation satisfied by v is

$$[11c^2 + 4\sqrt{3}sc + 7s^2]\frac{\partial^2 v}{\partial x^2} + [-8sc + 4\sqrt{3}(c^2 - s^2)]\frac{\partial^2 v}{\partial x \partial h}$$

$$+[11s^2 + 4\sqrt{3}sc + 7c^2]\frac{\partial^2 v}{\partial h^2} - 5v,$$

Where we have used the abbreviations s = sin(a) and c = cos(a). The coefficient of the mixed partials will vanish if a is chosen so that

$$-8\sin(a)\cos(a) + 4\sqrt{3}(\cos^2(a) - \sin^2(a)) = 0$$

That is,

$$\tan(2a) = 3^{1/2}/2$$

This means

$$a = p/6, \sin(a) = \sqrt{\frac{1-\cos(2a)}{2}} \text{ and } \cos(a) = \sqrt{\frac{1+\cos(2a)}{2}}.$$

After substitution of these values, the equation satisfied by v becomes

$$13\frac{\partial^2 v}{\partial x^2} + 5\frac{\partial^2 v}{\partial h^2} - 5v = 0$$

This special example, together with the foregoing discussions of analytic geometry makes the following statement believable: Every second order partial differential equation with constant coefficients can be transformed into one in which mixed partials are absent.

It is left as an exercise to see that in the general case, b^2 - 4 a c is left unchanged by this rotation. Surely, Maple is to be used to do this calculation. We are now ready for the second step: to remove the first order term. For economy of notation, let us assume that the given equation is already in the form

$$\frac{\partial^2 u}{\partial x^2} - 4\frac{\partial^2 u}{\partial y^2} + 3\frac{\partial u}{\partial x} + u = 0$$

Define v by

$$v(x, y) = e^{-bx}u(x, y) = e^{bx}v(x, y),$$

Where b will be chosen so that the transformed equation will have the first order derivative removed. Differentiating u and substituting into the equation we get that

$$\frac{\partial^2 v}{\partial x^2} - 4\frac{\partial^2 v}{\partial y^2} + (2b+3)\frac{\partial v}{\partial x} + (b^2 + 3b + 1)v = 0.$$

If we choose b = – 3/2, we have

$$\frac{\partial^2 v}{\partial x^2} - 4\frac{\partial^2 v}{\partial y^2} - \frac{5}{4}v = 0.$$

Notice that this transformation to achieve an equation lacking the first derivative with respect to x is generally possible when the coefficient on the second derivative with respect to x is not zero, and is otherwise impossible. The same statements hold for derivatives with respect to y.

The final step is rescaling. We choose variables ξ and η by

ξ =\mu x and η =\nu y, where m and n are chosen so that in the transformed equation the coefficients of v ξ ξ, v η η and v are equal in absolute value. We have

$$\frac{\partial^2 u}{\partial x^2} = \mu^2 \frac{\partial^2 v}{\partial \xi^2} \text{ and} \frac{\partial^2 u}{\partial y^2} = v^2 \frac{\partial^2 v}{\partial \eta^2}.$$

Our equation becomes

$$\mu^2 \frac{\partial^2 v}{\partial \xi^2} - 4v^2 \frac{\partial^2 v}{\partial \eta^2} - \frac{5}{4} v = 0$$

The condition that

$$\mu^2 = 4\ ^2 = 5/2$$

will be satisfied if $= (5/2)^{1/2}$ and $= (5/8)^{1/2}$. Then, we obtain the standard form.

GENERALIZED ROTATIONS TO REMOVE CROSS TERMS

The removal of cross terms as typically presented is complicated by the necessity of finding the angle a of rotation and of working with the ensuing trigonometric functions. These ideas seem inappropriate for higher dimensions. An alternative approach is to address the problem from the perspective of linear algebra, where the ideas even generalize for large dimensions. We will need the linear algebra package to make this work.

```
> with(linalg);
```

We choose the constants as the coefficients for the partial differential equations

$$au_{xx} + b_{xy} + c_{yy} + d\ u_x + e\ u_y + f = 0.$$

```
> a: = 2: b: = 3: c: = –2: d: = 1: e: = 1: f: = 3/25:
```

Make a symmetric matrix with a, b/2, and c as the entries.

```
> A: = matrix([[a,b/2],[b/2,c]]):
```

It is a theorem in linear algebra that symmetric matrices have a diagonal Jordan form.

```
> jordan(A,P);
```

We want K to be a matrix of eigenvectors.

```
> K:= inverse(P);
```

In order to make this process work, we ask that the eigenvectors - which form the columns of K - should have norm 1. This will cause the eigenvectors to be orthonormal.

```
> N1:=norm([K[1,1],K[2,1]],2): N2:=norm([K[1,2],K[2,2]],2):
> L:=matrix([[K[1,1]/N1,K[1,2]/N2],[K[2,1]/N1,K[2,2]/N2]]);
```

It will now be the case that (transpose(L) * A * L) is the Jordan form.

```
> evalm(transpose(L) &* A &* L);
```

To remove the cross-product term, we now perform a change of variables defining s and t.

```
> s:=(x,y)->evalm(transpose(L) &* vector([x,y]))[1]:
t:=(x,y)->evalm(transpose(L) &* vector([x,y]))[2]:
```

If we define v as indicated below, then it will satisfy a PDE with the cross-term missing and with u satisfying the original partial differential equation.

```
> u: = (x,y) – >v(s(x,y),t(x,y)):
> a*diff(u(x,y),x,x) + b*diff(u(x,y),x,y) + c*diff(u(x,y),y,y) +
    d*diff(u(x,y),x) + e*diff(u(x,y),y) + u(x,y):
> simplify(");
> collect(",D[1,2](v)(s(x,y),t(x,y)));
> pde: = collect(
    collect(
      collect(
        collect(",D[1](v)(s(x,y),t(x,y)))
         ,D[2](v)(s(x,y),t(x,y)))
           ,D[2,2](v)(s(x,y),t(x,y)))
             ,D[1,1](v)(s(x,y),t(x,y)));
```

Thus, we solve this simpler system and create u that satisfies the original equation.

In the chapters that came before, an understanding of the adjoints of linear functions was critical in determining when certain linear equations would have a *solution*...and even in computing the solutions for some cases. It is, then, no surprise that we shall be interested in the computation of adjoints in this setting, too.

In a general inner product space, the adjoint of the linear operator L is defined as the operator L* such that for all u,v,

$$< L(u), v > = < u, L^*(v) >.$$

For ordinary differential equations boundary value problems, the dot product came with the problem in a sense: it was an integral over an appropriate interval on which the functions were defined.

For partial differential equations with boundary conditions, the inner product will be of the standard kind:

$$\{f, g\} := \int_a^b sf(x)\overline{g(x)}\, dx.$$

But now the integral is multidimensional and the variables run over the *region* of interest. For ordinary differential equations, integration-by-parts played a key role in deciding the appropriate boundary conditions to impose so that the formal adjoint would be the real adjoint. Now, Green's identities provide the appropriate calculus.

In fact, Green's second identity can be used to compute adjoints of the Laplacian. We will see that the divergence theorem is useful for the more general second-order, differential. operators.

Example: Consider

$$\nabla^2 u = f,$$

$$u(x,0) = u(x,b) = 0$$

$$\partial u(0,y)/\partial x = \partial u(a,y)/\partial x = 0$$

On an interval [0,a] x [0,b] in the plane. This problem invites consideration of the operator L defined on a manifold as given below:

$$L(u) = \nabla^2 u$$

and

$$M = \{u: u(x,0) = \partial u(x,b) = 0, \partial u(0,y)/\partial y = \partial u(a,y)/\partial y = 0\}.$$

The second identity presents a natural setting in which the operator L is self-adjoint in the sense that L = L*. Let u and v be in M:

Integral$_D$[[v L(u) – L(v) u] dA] = Integral ∂_D[v $\partial u/\partial\eta - \partial v/\partial\eta u$] ds = 0.

This last equality follows because, on the boundary, all of u,

$\partial u/\partial\eta$, v, and $\partial v/\partial\eta = 0$.

In order to discuss adjoints of more general second order partial differential equations, let A, B, C, and c be scalar valued functions. Let b be a vector valued function. Let L(u) be given by

$$L(u) = A\partial^2 u/\partial x^2 + 2B\partial^2 u/\partial x\partial y + C\partial^2 u/\partial y^2 + \langle b, \nabla u\rangle + cu.$$

Definition: The FORMAL ADJOINT is given by

$$L^*(v) = \partial^2(Av)/\partial x^2 + 2\partial^2(Bv)/\partial x\partial y + \partial^2(Cv)/\partial y^2 - \nabla^*(bv) + cv.$$

Take A, B, and C to be constant. What would it mean to say that L is formally self-adjoint? That L = L* (formally)? Then $\langle b, \nabla u\rangle$ must be $-\nabla^* bu = \langle -b, \nabla u\rangle - \nabla u * b$.

Thus, 2 < b, ∇u > = – u (∇*b) for all u. Since this must hold for all u, it must hold in the special case that u = 1 identically, which implies that ∇*b = 0. Taking u(x,y) to be x, or to be y gets that each of b1 and b2 = 0. Hence, if L is formally self adjoint, then b = 0.

Examples:

1. Let L[u] = 3∂²u/ ∂x²) + 5 ∂²u/ ∂y²). The formal adjoint of L is L. Note that
   L[u] v – u L[v] = ∂/ ∂x (3[∂u/ ∂x v – u ∂v/ ∂x]) + ∂/ ∂y (5[∂u/ ∂y v – u ∂v/ ∂x]
   = ∇.(3[∂u/ ∂x v – u ∂v/∂ x] 5[∂u/ ∂y v – u ∂v/ ∂x])
2. Let L[u] = 3 F(∂²u,∂x²) + 5 F(∂²u,∂y²) + 7 F(∂u,∂x) + 11 F(∂u,∂y) + 13 u. The formal adjoint of L is
   L*[v] = 3 F(∂²v,∂x²) + 5 F(∂²v,∂y²) - 7 F(∂v,∂x) – 11 F(∂v,∂y) + 13 v. Note that
   L[u] v – u L*[v] = F(∂,∂ x) (3[F(∂u,∂x) v – u F(∂v,∂x)] + 7 uv) + F(∂, ∂y) (5[F(∂u,∂y) v – u F(∂v,∂x)] + 11 uv)
   = ∇.(3[F(∂u,∂x) v – u F(∂v,∂x)] +7 uv, 5[F(∂u,∂y) v – u F(∂v,∂x)] + 11 uv).
3. Let L[u] = e^x F(∂²u,∂x²) + 5 F(∂u,∂y) + 3u. The formal adjoint of L is L* given by
   L*[v} = e^x F(∂²v,∂x²) + 2e^x F(∂v,∂x) – 5F(∂v,∂y) + (e^x + 3) u. Note that
   L[u] v – u L*[v] = F(∂,∂x) (e^x v F(∂u,∂x) – u F(∂e^xv, ∂x)) + F(∂, ∂y) (5uv)
   = ∇.(e^x v F(∂u, ∂x) – u F(∂e^xv, ∂x), 5uv).

THE CONSTRUCTION OF M

We now come to the important part of the construction of the real adjoint: how to construct the appropriate adjoint boundary conditions. We are given L and M; we have discussed how to construct L*. We now construct M*. To see what is M* in the general case, the divergence theorem is recalled:

òòD ∇*F dx dy = ò∂D < F, η> ds.

The hope, then, is to write v L(u) – L*(v) u as ∇*F for some suitable chosen F.

Theorem: If L is a second order differential operator and L* is the formal adjoint, then there is F such that vL(u) – L*(v)u = ∇.F.

Here's how to see that. Note that

(v A∂²u/∂x² + v C∂²u/∂y²) – (u ∂²(Av)/∂x² + u∂ ²(Cv)/∂y²)

= ∂/∂x(vA∂u/∂x – u∂(Av)/∂x) + ∂/∂y (vC∂u/∂y – u∂(Cv)/∂y)

= ∇ *{v(A∂ u/∂x, C∂ u/∂y) – u(∂Av/∂x, ∂Cv/∂y)}.

Also, v < b, ∇u > + u ∇*(bv)

= v b1 ∂u/∂x + b2 ∂u/∂y v + u ∂(b1v)/∂x + u∂ (b2v)/∂y

= ∇ *(vbu).

COROLLARY: òòD [vL(u) – uL*(v)]

= òòD ∇*{v(A∂u/∂x,C∂u/∂y) – u(∂(Av)/∂x,∂(Cv)/∂y) + vbu)

=ò∂ D [v{A∂u/∂x,C∂u/∂y}– u{[[patialdiff]](Av)∂x,∂(Cv)/∂y} +vbu}]*η ds.

Examples:

1. Let L[u] be as in example 1 above for {x,y} in the rectangle D = [0,1]x[0,1] and M = {u: u = 0 on ∂D}. Then, according to Example 1, L = L* and

   òòD [vL(u) – uL*(v)] dA=

   òD < {3 [F(∂u,∂x) v – u F(∂v,∂x)], 5 [F(∂u,∂y) v – u F(∂v,∂y)]}, η> ds.

   Recalling that the unit normal to the faces of the rectangle D will be {0.–1}, {1,0}, {0,1}, or {–1,0} and that u = 0 on ∂D, we have that

   òòD [vL(u) – uL*(v)] dA=

   = – I(0,1,) 5F(∂u,∂y) v dx + 3 I(0,1,)F(∂u,∂x) v dy

   + 5 I(1,0,) F(∂u,∂y) v |dx| – 3 I(1,0,) F(∂u,∂y) v |dy|.

   In order for this integral to be zero for all u in M, it must be that v = 0 on ∂D. And M = M*. Hence, {L, M} is (really) self adjoint.

2. Let L[u] be as in Example 2 above and M = {u: u(x,0) = u(0,y) = 0, and F(∂u, ∂x)(1,y) = F(∂u, ∂y)(x,1) = 0, 0 < x < 1, 0 < y < 1}. Using the results from above,

   òòD [L(u)v – uL*(v)] dA =

   – I(0,1,) 5 F(∂u, ∂y) v dx + I(0,1,) [–3 u F(∂v, ∂x) + 7uv] dy

   + I(1,0,) [5u F(∂v, ∂x) + 11 uv] |dx| – I(1,0,)[3 F(∂u, ∂x) v] |dy|.

   It follows that M* = {v: v(x,0) = 0, v(1,y) = F(3,7) F(∂v, ∂x) (1,y), v(x,1) = F(5,11) F(∂v, ∂x) (x,1), and v(0,y) = 0}.

3. Let L[u] be as in Example 3 above for [x,y} in the first qaudrant. Let M = {u: u = 0 on ∂D}. Then

   òòD [vL(u) – uL*(v)] dA =ò ∂D <{e^x v F(∂u,∂x) – u F(∂e^xv, ∂x), 5uv}, η> ds = – I(0,*,)v F(∂u,x) dy.

   Thus, M* = {v: v(0,y) = 0 for y > 0}.

THE FREE GREEN FUNCTION AND THE METHOD OF IMAGES

In this section we shall construct Green functions for the problems of the type: Find u such that

$$\nabla^2 u = f \text{ on } D,$$

$$u = g \text{ on the boundary of } D.$$

This partial differential equation is known as Poisson's equation, and its solutions may be thought of as the electrostatic potential in the presence of a charge distribution ρ, related to f by

$$f(x) = -4\pi\rho.$$

Other quantities termed "potential" which arise in other contexts also solve. The other conditions in the boundary-value problem specify the value of the potential on the surface of the region. In most of our examples it will be 0, which means that the boundary is grounded.

For physical reasons we expect that knowing the charge in the interior and the potential on the boundary of a region is sufficient to determine the potential in the interior uniquely. Indeed, a uniqueness theorem guarantees precisely the problem is well posed in this sense.

In order to simplify notation, we will let P or Q represent points in the plane. For example, P might represent {x,y} and Q represent {a,b}. We indicate that in two dimensions,

$$\nabla^2 G(P, Q) = G_{xx} + G_{yy},$$

instead of partials with respect to a and b, by writing $\nabla_p{}^2G(P,Q)$.

The function G is to be constructed to have these properties:

$$\nabla_p\, p^2\, G(P,Q) = (P,Q)$$

and

$$G(.,Q) = 0 \text{ on the boundary of D}$$

Dirichlet type boundary conditions: Having such a G, the following applications of Green's identities show that we can determine the solution to the partial differential equation:

$$\iint G(P,Q)f(P)d_pA = \iint G(P,Q)\nabla_p{}^2(P)\, d_pA$$

$$= \iint \nabla_p.[G(P,Q)\nabla_p u - \nabla_p G(P,Q)\, d_pA$$

$$+ \iint \nabla_p{}^2 G(P,Q)u(P)d_pA$$

$$= -\int \partial G/\partial\eta(P,Q)u(P)d_ps + \iint \delta(P,Q)u(P)d_pA$$

$$= -\int \partial G/\partial\eta(P,Q)g(P)d_ps + \iint \delta(P,Q)u(P)d_pA\,.$$

Thus, having such a G and knowing f and g, we have a formula for u which provides a solution to the problem:

$$u(Q) = \iint G(P,Q)f(P)d_pA + \int \partial G/\partial\eta(P,Q)g(P)d_ps$$

How is such a G constructed? We will do it in two pieces. We construct G = F + R where F is the *free Green function* (also known as the *fundamental* or *singular* part) and satisfies:

$$\nabla^2 G = \delta$$

on all of R^2, with F (.,Q) independent of (in polar coordinates).

R is the *regular* part and satisfies

$$\nabla^2 R = 0 \text{ on } D$$

$$R = -f \text{ on } \partial D$$

Let us recall that whenever we have a linear, inhomogeneous problem, the general solution is given by any particular solution, plus the general solution of the related homogeneous problem. Equation (19.4) is the homogeneous equation associated with (19.3). It has no delta function, so its solutions will be completely regular. (Even if the boundary conditions were rough this is guaranteed by a general theorem for Laplace's equation.)

In other words, the singular parts of any two Green functions for the same equation, but different boundary conditions, will be the same.

FINDING THE FREE GREEN FUNCTION

We begin by constructing F. Recall the formula for the Laplace operator ∇^2 in polar coordinates:

$$\nabla^2 u = \frac{1}{r}\frac{\partial(r\partial u/\partial r)}{\partial r} + \frac{1}{r^2}\frac{\partial^2 u}{\partial \Theta^2}.$$

In seeking F such that

$$\nabla^2 f = \delta,$$

We recall that, in the sense of distributions, δ ({x,y},{a,b}) = 0 unless {x,y} = {a,b}. Also, δ is radially symmetric. Thus, supposing initially that Q = {a,b} is at the origin, we have

$$0 = \nabla^2 F = 1/r\ \partial(r\partial F/\partial r + 1/r^2\ \partial^2 F/\partial \Theta^2$$

$$= 1/r\ \partial(r\partial F/\partial r)/\partial r.$$

This last equality is because F is independent of θ. Thus r F_r is constant and F(r) = A ln(r) + B for some A and B. It remains to find A and B.

To this point we have not used information about F at the origin, only at {x,y} different from {0,0}. Information about F at {a,b} comes through the integral. If we integrate the delta function over any disk with radius c > 0 we get 1, so:

$$1 = \iint \nabla^2 F \, dA = \iint \nabla . \nabla F \, dA = \int \nabla F . \eta \, ds$$

by the divergence theorem. Since F is radially symmetric, we can calculate the last integral as

$$= \int A / r \, ds = 2\pi A.$$

Thus, $A = 1/(2\pi)$ and B is undetermined. We may as well choose it to be 0 at this stage. Of course there is no singularity in the constant function, so its derivatives do not contribute to the delta function. When Q is not at the origin, we simply shift the calculation by Q:

$$F(P,Q) = \ln[|P - Q|]/(2\pi).$$

In coordinates

$$F(\{x,y\},\{a,b\}) = \ln[(x - a)^2 + (y - b)^2]/(4)$$

(The factor of 1/2 comes because $\ln(r^a) = a \ln(r)$).

The function F is the Green function for the following problem:

$$\nabla^2 u = f \text{ on all } R^2, \text{ for } u \in L^2.$$

This problem is in the first alternative, as a consequence of a theorem of Liouville for solutions of Laplace's equation (known as *harmonic functions*), which implies that the only square-integrable solution is 0. See exercise 3 for why the assumption that u is square integrable is necessary. Some mild assumptions on f are needed, for instance $|f|$ could be assumed bounded by any large constant times $(1 + r)^{-3}$.

THE METHOD OF IMAGES

We next discuss a method for finding the Green function on regions other than R^n. This method may apply if the region is highly symmetric:

Example: Now let us solve the problem (19.1) on a new domain, the *upper half plane*, where x may have any value but $y > 0$. We need to find a Green function which = 0 when $y = 0$.

Solution. We may as well imagine that the problem we wish to solve is (19.1) for $y > 0$, with the boundary condition

$$u(x,0) = 0,$$

For it has the same Green function as other with more complicated Dirichlet boundary data on the x-axis, as shown above.

Here are two ways to derive the Green function.

I. The problem (19.1) is well-posed problem, owing to the uniqueness theorem. Thus if we can find a solution by imagining a different situation which is easier to solve, but which satisfies the partial differential equation and the boundary condition, it must be the right solution. The different situation will be one on all of R^2, like the problem we just solved, but where the boundary condition happens to be satisfied. Here is how:

Since f(x,y) is physically meaningful only for y > 0, we are free to define it however we will when y < 0. Let us choose to define it by the *odd extension*, so f(x,– y) = – f(x,y). We can solve this modified problem with the free Green function, as

$$\nabla^2 u = \frac{\partial^2 u}{\partial r^2} + \frac{2}{r}\frac{\partial u}{\partial r} + \frac{1}{r^2 \sin^2(\theta)}\frac{\partial^2 u}{\partial \varphi^2} + \frac{1}{r^2}\frac{\partial^2 u}{\partial \theta^2}.$$

The last step here came from rewriting the integral over y < 0 using the odd symmetry. Notice that if y = 0, this integral vanishes, while for y > 0, the extended function f has the same values as posited in the problem. Because of uniqueness, it is the right solution to the problem on the upper half plane. The Green function for the half-plane problem is

$$G(P,Q) = F(P,Q) - F(P^R, Q),$$

where $P^R = \{x,-y\}$ is the point found by reflecting P across the x-axis. It is as if there were an equal and opposite image source for the problem, located in the lower half plane.

II. Here is another way to derive the solution and the Green function. As remarked above, the Green function for this problem is of the form

$$G(P,Q) = F(P,Q) + R(P,Q),$$

where R is a regular function, solving Laplace's equation on the domain D where the problem resides, in this case the upper half plane. Can we find such a solution, such that G will satisfy a zero Dirichlet boundary condition? Yes, because the function F(S,Q) solves

$$\nabla^2 F(S,Q) = 0$$

(in the Q variable) for any S which is not in the domain D, since the delta function is zero everywhere except at the singularity. Taking $S = P^R$ and $R(P, Q) = -F(P^R, Q)$ gives a Green function which is 0 when y = 0, as required.

Let's use the method of images to solve some more problems. First, we would like to note that the idea of exploiting symmetries together with the uniqueness theorem is not logically connected to the method of Green functions, but can be used in conjunction with any method of solution. Indeed, we used the same ideas in our discussion of d'Alembert's solution of the wave equation.

Example: We continue here to solve Poisson's equation, but now on the strip D = {x unrestricted, 0 < y < 1}. Dirichlet boundary conditions are imposed on the strip.

Solution. Have you ever been in a barber shop or hair dresser where you sit between two mirrors? You see an infinite number of reflections of yourself, alternating as to whether they face toward you or away from you. If we want to place image Green functions in the plane to match both boundary conditions, we need an infinite number of them. The position reflected from height $y < 1$ through $y = 1$ is $2 - y$ (when $y = 0$ the reflection is at 2 and when $y = 1$ it meets its reflection). The next reflection will be at $2 + y$, the next one at $4 - y$, and so forth. We need to alternate signs to get all the cancellations. The result ought to be: $G(\{x,y\},Q) = \sum_{n>-\infty}^{\infty} F(\{x,y+2n\},Q) - \sum_{n>-\infty}^{\infty} F(\{x,+2n-y\},Q.$ There is a subtlety here, however, which is that the infinite sums do not converge separately; this expression is off the form infinity – infinity.

Remember, however, that we can subtract a regular solution of Laplace's equation

$$\nabla^2 R = 0$$

to the free Green function, so long as we end up satisfying the boundary condition. There is more than one way to do this, but the simplest one to use here (see exercise) is

$$F_n(P,Q) = F(P,Q) - \frac{y - b}{2\pi n},$$

and to sum the images in the following way:

$$G(\{x,y\},Q) = \sum_{n>-\infty}^{\infty} F_n(\{x,y+2n\},Q) - F_n(\{x,2n-y\},Q).$$

(It should be easy to see that $c\,(y - b)$ solves Laplace's equation for any b,c.)

Example: We continue to solve Poisson's equation, again on the upper half-plane $y > 0$, but this time with Neumann-type boundary conditions,

$$u_y(x,0) = 0.$$

Solution. The Green function we need this time uses the even reflection through $y = 0$. The Green function with the correct boundary condition is:

$$G(P, Q) = F(P,Q) + F(P^R,Q).$$

In some ways things are simpler in three dimensions for this problem. Let us now seek the three-dimensional free Green function with the same technique as we used above. We begin by looking for a solution of Laplace's equation which depends only on the radial coordinate in the spherical coordinate system. In this system, the Laplace operator has the form

$$\nabla^2 u = \frac{\partial^2 u}{\partial r^2} + \frac{2}{r}\frac{\partial u}{\partial r} + \frac{1}{r^2 \sin^2(\theta)}\frac{\partial^2 u}{\partial \varphi^2} + \frac{1}{r^2}\frac{\partial^2 u}{\partial \theta^2}.$$

So if the point Q is put at the origin, the free Green function will satisfy the ordinary differential equation

$$F'' + 2\,F'/r = 0$$

in r for all $r > 0$. The solutions of this are of the form $F = A/r + B$.

Again, we may choose $B = 0$ in most circumstances, and use the divergence theorem to evaluate A:

$$1 = \int_0 \nabla^2 F\, d^3x = \int_0 \nabla.\nabla F\, d^3x = \int_0 \nabla F\,.\,n\, d^2x$$

(see above). Since F is radially symmetric, we can integrate over a ball of radius r and calculate the last integral as

$$-\int \frac{A}{r^2} d^2x = 4\pi A.$$

We conclude that in three dimensions,

$$F(P,Q) = -\,1/(4\,\pi|P-Q|).$$

For those of you who have an interest in electromagnetism, this may be familiar as the electric potential of a unit charge (with a certain choice of physical units and a sign convention which may be the opposite of what you will find in a physics class). If there is a continuous charge distribution $\rho(P)$, then the electrostatic potential in free space is the solution of

$$\nabla^2 u = -\,4\,\pi\,\rho$$

i.e.,

$$u(Q) = \int \frac{\rho(P)}{|P-Q|} d^2x = -\int \frac{f(P)}{4\pi|P-Q|} d^3x.$$

You can think of this as a continuous superposition of the potentials due to point charges, distributed according to the function f.

The method of images works in three dimensions just as in two dimensions:

Example: Now let us solve Poisson's equation on the *upper half space*, where x and y may have any value but $z > 0$. We need to find a Green function which = 0 when $z = 0$.

Solution. If $P = \{a,b,c\}$ and $Q = \{x,y,z\}$, then the Green function is:

$$G(\{x,y,z\},\ \{a,b,c\}) = \frac{1}{4\pi\sqrt{(x-a)^2 + (y-b)^2 (z+c)^2}} - \frac{1}{4\pi\sqrt{(x-a)^2 + (y-b)^2 (z-c)^2}}.$$

There is another remarkable situation in which the method of images is useful. For motivation, suppose we work the problem in reverse, but trying to locate a new region on which we can solve with zero Dirichlet boundary conditions. For example, if we look at

$$G(P,Q) := F(P,Q) - q\,F(P,Q'),$$

where the "charge" $q > 1$, then G will be negative for P sufficiently near Q and positive for P sufficiently near Q', since the Free Green function diverges to - infinity at its singularity. The function G must also be positive very far away from Q and Q', for the assumption that $q > 1$ means that the second contribution is more important at large distances: Physically speaking, a probe at a large distance will detect the net charge, $q - 1$. Clearly, $G = 0$ on some region surrounding Q but not Q'.

In order to determine this region, let us choose the x-axis so that

$$Q = \{a_1, 0, 0\} \text{ and } Q' = \{a_2, 0, 0\}.$$

The region must be rotationally symmetric about the x-axis, so to determine it we can set $z = 0$. The coordinates of the points in the y - z plane where $G = 0$ thus satisfy

$$\frac{1}{4\pi\sqrt{(x-a_1)^2+y^2}} = \frac{q}{4\pi\sqrt{(x-a_2)^2+y^2}}$$

This is equivalent to

$$(x-a_1)^2 + y^2 = \frac{(x-a_2)^2+y^2}{q^2},$$

when we take reciprocals of both sides and square. Collecting terms, we get

$$(x^2+y^2)\left(1-\frac{1}{q^2}\right)-2x\left(a_1-\frac{a_2}{q^2}\right)+\left(a_2^1-\frac{a_2}{q^2}\right).$$

This is the equation of a circle with a center somewhere on the x-axis! If we do not specialize to the x-y plane and points Q and Q' on the x-axis, we must conclude:

$$F(P, Q) - q\,F(P, Q') = 0$$

precisely on some sphere when $q > 1$.

Armed with this knowledge, it is a matter of algebra to go the other way and find the values of Q' and q which produce the correct Green function for any particular sphere. Supposing that the sphere is centered at the origin and has radius R. If $Q = (a_1,0,0)$ we find:

$$a_2 = R^2/a_1$$

$$q = -R/a_1.$$

If Q is not located on the x-axis, we can find q and Q′ simply by rotating this relationship: $Q' = Q\,R^2/|Q|^2$, and $q = R/|Q|$.

Theorem: The Green function for Poisson's equation in the sphere of radius R (or, in two-dimensions, the circle of radius R), with Dirichlet boundary conditions, is

$$G(P,Q) = F(P,Q) - \frac{R}{Q} F\left(P \frac{QR^2}{|Q|^2}\right),$$

where F is the free Green function

$$F(P,Q) = \frac{-1}{4\pi|P-Q|}(3D).$$

or

$$F(P,Q) = \frac{-1}{2\pi} \text{In } (|P-Q|)(2D).$$

Chapter 5

Linearity

This chapter is concerned with some of the most important methods of applied mathematics, without which many technological developments of the nineteenth and twentieth centuries would have been impossible. When you have finished this course you will be able to solve most of the partial differential equations and integral equations that you are likely to encounter in engineering, and, just as importantly, you will understand what to expect from the solutions, which might emerge in a mysterious fashion from a computer at your workplace. The leitmotifs are the notions of linearity and orthogonality. By this stage in your education, you are familiar with vectors. You have added them by the head-to-tail construction, or by adding their components. You also know how to take their scalar (dot) product, producing a number equal to the product of the two lengths and the cosine of the angle between them:

$$\nabla \cdot \nabla = |\nabla||\nabla| \cos(\theta)$$

One of the most powerful and astonishing ideas in modern mathematics is that in many regards functions can be treated just like vectors. There is a very close analogy between sets of functions and spaces of vectors, which will guide us to striking solutions for several of the most important differential equations you will use in science and engineering, especially the wave equation, the heat equation, and the potential equation.

Mathematicians have a way of making analogies without the mushiness that so often afflicts analogies in so many other contexts. It is called *abstraction*. While abstraction may sometimes seem like theory divorced from reality, when used properly it is a wonderful, and very practical, tool for solving problems.

This is why we will begin by making vectors abstract. Let us recall some facts about everyday vectors. If we have a bag full of vectors, we can scale them and add them together. We call the result a *linear combination*:

$$\alpha_1 v_1 + \alpha_2 v_2 + \cdots + \alpha_n v_n$$

we'll normally use Greek letters for scalars (= ordinary real or complex numbers). It doesn't matter how many are in the combination, but unless we explicitly state otherwise, we will assume that it is only a finite sum, a *finite linear combination*. Of course, we can make linear combinations of functions, too:

$$\alpha_1 f_1(x) + \alpha_2 f_2(x) + \cdots + \alpha_n f_n(x),$$

and the result is another function. In this way, the set of all functions is a vector space.

Definition: More formally, a *vector space* over the complex numbers is a set of entities, abstractly called vectors, for which

1. Any finite linear combination of vectors is a member of the same set
2. The usual commutative rule holds for addition: v + w = w + v,
3. Just for consistency, the usual commutative, associative, and distributive laws hold for vector addition and multiplication by scalars. In other words

$$\alpha v + \alpha w = \alpha(v + w),$$

$$\alpha v + \beta v = (\alpha + \beta)v,$$

$$(\alpha\beta)v = \alpha(\beta v).$$

$$\text{Also,} 1v = v.$$

In practice the rules in 3 are obvious and not very interesting. From these rules, you can show some other properties, such as:

There is a special element, which will be called the zero vector, equal to the scalar 0 times any vector whatsoever. It has the property that for any vector v, 0 + v = v.

For any vector v, there is a negative vector, called -v, such that v + (– v) = 0 (the zero vector).

Certain pretty reasonable conventions will be made, such as writing v - w instead of v + (– w). Great - all the usual stuff works, right? Well, not quite. We *don't* assume abstractly that we can multiply vectors by one another. Things like the dot product and the cross product work for some vector spaces but not for others. The other deep issue lurking in the shadows is infinite linear combinations, i.e., infinite series. Fourier liked them, and we'll see lots of them later. But if the vectors are functions, perhaps the infinite series converges for some x and not for others. For instance,

$$\sum_{n=1}^{\infty} \sin(nx) = 0$$

when x is a multiple of π, but it is certainly an improper sum when x = π/2, and it is not immediately clear what happens for other values of x. What does the series mean then?

Examples:

1. The usual two-dimensional or three-dimensional vectors. It makes little difference whether they are thought of as column vectors such as $\frac{1}{2}$ or as row vectors (2,3). The set of all such vectors will be denoted C^2 or C^3 (assuming complex entries are allowed - otherwise R^2 or respectively R^3). Mathematica and Maple can perform all of the usual vector operations in a straightforward way.
2. The set of complex numbers. Here there is no difference between a vector and a scalar, and you can check all the properties pretty easily. We call this a one-dimensional vector space.
3. The set C^n of n numbers in a list. These are manipulated just like 2– or 3– vectors, except that the number of components is some other fixed number, n. For instance, with C^4, we might have elements such as (1,2,3,4) and (1,0,–1,–2), which can be added and multiplied as follows:

   $(1,2,3,4) + (1,0,-1,-2) = (2, 2, 2, 2)$

   $(1,2,3,4).\ (1,0,-1,-2) = -10$, etc.
4. The set of continuous functions of a variable x, $0 < x < 1$. The rather stupid function $f_0(x) = 0$ for all x plays the role of the zero element.
5. A smaller vector space of functions. Instead of simply listing n numbers, let us multiply them by n different functions, to define another vector space of functions. For example, for some fixed n, consider

   $a_1 \sin(x) + a_2 \sin(2x) + \ldots + a_n \sin(nx)$,

   where the numbers a_k can take on any value. Notice that this vector space is a part of the one of example 4. In other words, it is a *subspace*.
6. The set of, say, 2 by 3 matrices. Addition means addition component-wise:

$$\begin{bmatrix} 1 & 0 & i \\ -1 & \pi & 2+i \end{bmatrix} + \begin{bmatrix} 2 & i & 0 \\ -1 & -\pi & 2-1 \end{bmatrix} = \begin{bmatrix} 3 & -i & i \\ -2 & 0 & 4 \end{bmatrix}$$

   and scalar multipication affects all componets:

$$5\begin{bmatrix} 1 & 0 & i \\ -1 & \pi & 2+i \end{bmatrix} = \begin{bmatrix} 5 & 0 & 5i \\ -5 & 5\pi & 10+5i \end{bmatrix}$$

7. The set of 3-component vectors (x,y,z) such that $x - 2y + z = 0$. This is a plane through the origin.

Definitions: A set of vectors $\{v_1, \ldots, v_n\}$ is *linearly independent* if it is impossible to write any of the vectors in the set as a linear combination of the rest. *Linearly dependent* is the opposite notion. The *dimension* of a vector space V is the largest number n of linearly independent vectors $\{v_1,..., v_n\}$ which are contained in V. The *span* of a set $\{v_1,..., v_n\}$ is the vector space V obtained by considering all possible linear combinations from the set. We also say that $\{v_1,..., v_n\}$ *spans* V. A set $\{v_1,..., v_n\}$ is a *basis* for a finite-dimensional vector space V if $\{v_1,..., v_n\}$ is linearly independent and spans V.

Notice that the only way that two vectors can be linearly dependent is for them to be proportional (parallel) or for one of them to be 0. If the set has the same number of vectors as each vector has components, which frequently is the case, then there is a calculation to test for linear dependence. Array the vectors in a square matrix and calculate its determinant. If the determinant is 0, they are dependent, and otherwise they are independent. For example, consider the vectors {(1,2,3), (4,5,6),(7,8,9)}, which are not obviously linearly dependent. A calculation shows that

$$\det\begin{vmatrix} 1 & 2 & 3 \\ 4 & 5 & 6 \\ 7 & 8 & 9 \end{vmatrix} = 0$$

Indeed, we can solve for one of these vectors as a linear combination of the others:

$$(7,8,9) = (-1)\,(1,2,3) + 2\,(4,5,6)$$

Many of our vector spaces will be infinite-dimensional. For example, {sin(nx)} for n = 1, 2,..., is an infinite, linearly independent set of continuous functions - there is no way to write sin(400 x), for example, as a linear combination of sine functions with lower frequencies, so you can see that each time we introduce a sine function of a higher frequency into the list, it is independent of the sines that were already there. An infinite linearly independent set is a basis for V is every element of V is a limit of finite linear combinations of the set - but we shall have to say more later about such limits. A vector space has lots of different bases, but all bases contain the same number of items.

For practical purposes, the dimension of a set is the number of degrees of freedom, i.e., the number of parameters it takes to describe the set. For example, C^n has dimension n, and the set of 2 by 3 matrices has six elements, so its dimension is 6.

Model problem: Show that the plane x + y + z = 0 is a two-dimensional vector space.

The verification of the vector space properties will be left to the reader. Here is how to show that the dimension is 2:

Solution. The solution can't be 3, since there are vectors in R^3 which are not in the plane, such as (1,1,1). On the other hand, here are two independent vectors in the plane: (1,–1,0) and (1,1,–2).

Further observations. A general vector in the plane can be written with 2 parameters multiplying these two vectors, and it is not hard to find the formula to express the general vector this way. First write the general vector in the plane as (x,y,–x–y) (by substituting for z – notice 2 parameters for 2 dimensions). It is a straightforward exercise in linear algebra – with or without software - to solve for and such that

$$(x,y,-x-y) = (1,-1,0) + (1,1,-2).$$

We can solve this equation with the choices = (x–y)/2 and = (x + y)/2. Hence the two vectors we found in the solution are a basis for the plane. Any two independent vectors in the plane form a basis.

Definition: A *linear transformation* is a function on vectors, with the property that it doesn't matter whether linear combinations are made before or after the transformation. Formally,

$$F(\alpha_1 v_1 + \alpha_2 v_2 + \alpha_3 v_3 + \alpha_4 v_4)$$
$$= \alpha_1 F(v_1) + \alpha_2 F(v_2') + \alpha_3 F(v_3) + \alpha_4 F(v_4)$$

Linear transformations are also called *linear operators*, or just operators for short. You know plenty of examples:

Examples:

1. Matrices. If M is a matrix and v, w etc. are column vectors, then

$$M(\alpha_1 v_1 + \alpha_2 v_2 + \alpha_3 v_3 + \alpha_4 v_4) = \alpha_1 M v_1 + \alpha_2 M v_2 + \alpha_3 M v_3 + \alpha_4$$

Think of rotation and reflection matrices here. If you put a bunch of vectors head-to-tail and rotate the assemblage, or look at it in a mirror, you get the same effect as if you first rotate or reflect the vectors, and then put them head-to-tail.

It may be less geometrically obvious when the matrix distorts vectors in a trickier way, or there are more than three dimensions, but it is still true. In this example, algebraic intuition might be more convincing than geometric intuition

2. Derivatives and integrals. As we know,

$$\frac{d}{dx}(\alpha_1 f_1(x) + \alpha_2 f_2(x)) = \alpha_1 f_1'(x) + \alpha_2 f_2'(x)$$

$$\int_0 ((\alpha f(y) + \beta g(y))dy = \left(\alpha \int_0 f(y)dy + \beta \int_0 g(y)dy\right)$$

3. This may seem silly at first, but the *identity operator* Id, which just leaves functions alone is a linear transformation:

$$\text{Id}(\alpha_1 f_1 + \alpha_2 f_2) = \alpha_1 \text{ Id } f_1 + \alpha_2 \text{ Id } f_2,$$

Since both sides are just round-about ways of writing $\alpha_1 f_1 + \alpha_2 f_2$. The identity operator is a useful bit of notation for much the same reason as the identity matrix,

$$\text{Id} = \begin{bmatrix} 1 & 0 & 0 \\ 0 & 1 & 0 \\ 0 & 0 & 1 \end{bmatrix},$$

the effect of which on any vector is to leave it unchanged.

A linear transformation is a function defined on vectors, and the output is always a vector, but *the output need not be the same kind of vector as the input*. You should be familiar with this from matrices, since a 2 by 3 matrix acts on 3-vectors and produces 2-vectors. Similarly, the operator D acts on a vector space of functions assumed to be differentiable and produces functions which are not necessarily differentiable.

Example: More exotic possibilities are possible, such as the operator which acts on 2 by 2 matrices by the rule:

$$F\left(\begin{bmatrix} m_{11} & n_{12} \\ m_{21} & m_{22} \end{bmatrix}\right) = m_{11}\sin(x) - 3m_{22}\sin(2x).$$

The whole theory we are going to develop begins with the analogy between Examples. We can think of the *linear operator* of differentiation by x as a kind of abstract matrix, denoted D. If we also think of the functions f and g as entities unto themselves, without focusing on the variable x, then the expression in the first part of Example:

$$D(\alpha_1 f_1 + \alpha_2 f_2) = \alpha_1 D\, f_1 + \alpha_2 D\, f_2.$$

The custom with linear transformations, as with matrices, is to do without the parentheses when the input variable is clear. It is tempting to manipulate linear operators in many additional ways as if they were matrices, for instance to multiply them together. This often works. For instance $D^2 = D\,D$ can be thought of as the second derivative operator, and expressions such as $D^2 + D + 2$ Id make sense. In passing, notice that if S is the integration operator of the second part of Example 2, then D S f = f, so D is an inverse to S in the sense that D S = Id.

Linear ODE's. In your course on ordinary differential equations, you studied linear differential equations, a good example of which would be

$$D^2y + Dy + 2y = 0$$

More specifically, this is an example of a "linear, homogeneous differential equation of second order, with constant coefficients". We can picture this equation as one for the *null space* of a *differential operator* A:

$$Ay := (D^2 + D + 2\,\text{Id})\, y = 0.$$

By definition, the *null space* N(A) is the set of all vectors solving this equation; some texts refer to it as the *kernel* of A. There is no difference. You may remember that the null space of a matrix is always a linear subspace, and the same is true for the null space of any linear operator. E. g., the plane of Example is the null space of the matri

$$M = \begin{bmatrix} 1 & -2 & 1 \\ 1 & -2 & 1 \\ 1 & -2 & 1 \end{bmatrix}.$$

What does this abstract statement about subspaces mean in a concrete way for the problem of solving a linear homogeneous problem? It is the famous *superposition principle*: If I can find two (or more) solutions of a linear homogeneous equation, then any linear combination of them is also a solution.

For matrices the general solution of the homogeneous equation is the set of linear combinations of a finite number of particular solutions. This will also be true for linear ordinary differential equations (the situation is a little more complicated for linear partial differential equations). Indeed, the number of independent functions in the null space of an ordinary differential operator is equal to its order (the highest power of D). This is true even when the coefficients are allowed to depend on x, so long as there are no singularities (such as values of x where the coefficients become 0 or infinity).

Let us illustrate the situation with the two operators mentioned above.

Example: Find the null space of M as given in, which is the same as finding the general solution of M v = 0.

In a linear algebra class you probably learned several techniques for finding the solution vectors v, and the way you describe the null space may depend on the technique chosen.

Solution. The nullspace can be found with software as a formula relating x, y, and z:

$$x = 2y - z$$

Equivalently this is the plane satisfying $x - 2y + z = 0$. This is a plane passing through the origin, so it must be true that any linear combination of vectors in this plane again lies in the plane. (Why is it important that the plane passes through the origin?) A second way to describe the plane is to find a basis, i.e., two independent vectors such as

$$v_1 = (1,1,1) \text{ and } v_2 = (1,0,-1),$$

which are both in the plane and which form a basis for it.

Example: Find the general solution of

$$Au(x) = 0.$$

Solution. The method here is to guess a solution of the form e^{mx}, and substitute to find what m must be. Usually, we find two possible values of m, and in a class on ordinary differential equations we learned that the general solution can be obtained from any two linearly independent solutions (for second order ODE's). If u(x) is of the form e^{mx}, then A u = $(m^2 + m + 2)$ u, so if u is not the zero function, we must have

$$m = \frac{1}{2} \pm \frac{\sqrt{-7}}{2} = -\frac{1}{2} \pm \frac{\sqrt{7}i}{2}.$$

Thus, two linearly independent solutions to (1.3) are

$$u + (x) = \exp\left(\left(-\frac{1}{2} + \frac{\sqrt{7}j}{2}\right)x\right) \text{ and } u(x) = \exp\left(\left(-\frac{1}{2} - \frac{\sqrt{7}j}{2}\right)x\right).$$

As these are complex valued - remembering Euler's formula that

exp(i gamma) = cos(gamma) + i sin(gamma)

-they are not necessarily in the most convenient form. However, they can be recombined easily into a pair of independent real functions,

The general solution is the span of these two functions, i.e.,

$\alpha_1 u_1(x) + \alpha_2 u_2(x)$.

Finally, let us recall the solution of nonhomogeneous linear problems, and observe how similarly it looks for matrix equations and differential equations.

Example: Find the general solution of

[1]

M v = [1]

[1]

Solution. The first step is to find a *particular* solution A bit of trial and error leads to, perhaps,

$$v_p = \begin{bmatrix} 1 \\ 0 \\ 0 \end{bmatrix}.$$

There are many other, equally good choices for v_p. The general solution is the set of vectors of the form

$$v = v_p + \sum_j \alpha_j v_j$$

So we see that the homogeneous solution is just added onto v_p. The matrix M annihilates the terms $\alpha_j v_j$, so

$$Mv = Mv_p = \begin{bmatrix} 1 \\ 1 \\ 1 \end{bmatrix}$$

The explicit answer is

$$Mv = Mv_p = \begin{bmatrix} 1 \\ 1 \\ 1 \end{bmatrix}.$$

Compare with the following:

Example: Find the general solution of

$$A\ u = \exp(2x).$$

Solution. Again, the first step is to find a particular solution. Trial and error with the guess $u_p(x) = C \exp(2\ x)$ leads to

$$A\ u_p = (4 + 2 + 2)\ C \exp(2x),$$

showing us that we must choose $C = 1/8$, making

$$u_p = \exp(2x)/\ 8.$$

The general solution is

$$u(x) = \exp(2x)/\ 8 + a_1\ u_1(x) + a_2\ u_2(x).$$

where $u_{1,2}$ are as given above.

THE GEOMETRY OF FUNCTIONS

The other familiar vector operation we shall use, besides sums and scalar multiples, is the dot product, which we abstractly call an *inner product.* (We won't be concerned with analogues of the cross product, although you may see these in other courses.) You probably learned about the dot product as connected with trigonometry - the dot product of two vectors is given as the product of their lengths times the cosine of the angle between them:

$$\nabla\nabla = |\nabla|\ |\nabla|\cos\ (\angle\ \nabla,\nabla)$$

Later you learned that this could be conveniently calculated from the coordinates of the two vectors, by multiplying given components together and summing: If the components of the vector v are $\upsilon_1,\ldots,\ \upsilon_n$ and those of w are $\omega_1,\ldots,\ \omega_n$., then

$$vw = \sum_k \upsilon_k \overline{\omega_k}$$

From the abstract point of view it is best not to begin with angle, since we don't have a good intuitive picture of the angle between two functions or matrices. Instead, we will make sense of the "angle" between functions from the connection between these formulae.

In the abstract setting, we shall denote the inner product rather than v.w. Given an inner product, we may speak about the length, defined by $||v|| = (v.v)^{1/2}$

We shall usually call this the *norm* rather than the length.

An abstract inner product will share most of the properties you learned about for the ordinary dot product:

1. The inner product is a mapping taking pairs of vectors and producing scalars

2. *Linearity*:

$$(\alpha_1 W_1 + \alpha_2 w_2, v) = \alpha_1 (W_1 v) + \alpha_2 (w_2, v),$$

Note: some texts define an inner product to be linear on the right side rather than the left side. This makes no practical difference, but if there are complex quantities, you should be careful to be consistent with whichever convention you follow.

3. *Symmetry*:

$$(v, w) = \overline{(w, v)}$$

(The bar denotes complex conjugate, in case these are complex numbers.)

4. *Positivity*:

$(v,v) \geq 0$, and $(v, v) = 0$ only if v is the zero vector.

5. The *Cauchy, Schwarz*, or *Buniakovskii inequality* (depending on your nationality):

$$|(v, w)| \leq ||v|| \, ||w||.$$

6. The *triangle inequality*:

$$||v + w|| \leq ||v|| + ||w||.$$

We need Property 5 if an inner product is to correlate with Equation, and we need Property 6 if the length

$$||v|| = \sqrt{(v, v)}$$

is to make sense geometrically. We wil! shortly define an inner product for functions. The reason we can speak of the geometry of functions is that properties 5 and 6 follow automatically from Properties 1-4.

Since this is not at all obvious, I shall prove it below.

Definition: Given an abstract vector space, an *inner product* is any mapping satisfying properties 1-4 above. A vector space with an inner product is called an *inner product space*.

Examples:

1. The usual dot product.
2. A modified dot product on 2-vectors. A modification of the usual dot product can also produce an inner product defined by

$$(v, w)_A := \sum_{jk=1}^{2} \upsilon_j A_k \overline{w_k}$$

where ·

$$A_k = \begin{bmatrix} 2 & 1 \\ 1 & 2 \end{bmatrix}.$$

Actually, we could consider the vector space of n-vectors and let A be any positive n by n matrix.

(By definition, a *positive matrix* is one for which the inner product so defined satisfies Property 4 of the inner product.)

3. The standard inner product for functions. Consider the set of functions which are continuous on an interval $a \le x \le b$. Then the *standard inner product* on them is the integral:

$$(f, g) := \int_b^a f(x)\overline{g(x)}\, dx.$$

Another name for this is the L^2 inner product. We can use it for functions with discontinuities and even singularities, so long as the singularities are not too strong.

The most general set of functions, defined on a set Ω, for which this inner product is defined is known as the set of *square-integrable* functions, denoted $L^2(\Omega)$.

Some fancier examples of inner products.

Theorem: If V is an inner product space, then the CSB inequality and the triangle inequality hold.

Proof: Having the CSB inequality in hand, we may now define a strange but useful idea - the angle between two functions. Consider, for example, the interval $0 \le x \le L$ and the functions $f(x) = 1$ and $g(x) = x$. With the standard inner product, we first calculate their "L^2" norms:

$$\|1\| := \sqrt{\int_0^x 1^2 dx} = \sqrt{L}, \text{and } \| x \| := \sqrt{\int_0^x x^2 dx} = \sqrt{\frac{L^3}{3}}$$

Since their inner product is

$$(1, x) = \int_0^x x\,dx = \frac{L^3}{3}.$$

the cosine of the angle between the two functions must be

$$\cos(\theta) = \frac{\frac{L^2}{2}}{\sqrt{L}\sqrt{\frac{L^3}{3}}} = \frac{\sqrt{3}}{2}$$

Thus the "angle" between the functions 1 and x is π/6 radians. The most useful angle to deal with, however, is a right angle:

Definition: Two functions f and g are said to be *orthogonal* if $\langle f,g \rangle = 0$. A set of functions $\{f_j\}$ is *orthogonal* if $\langle f_j,f_k \rangle = 0$ whenever j differs from k. The set is said to be *orthonormal* if it is orthogonal and $\|f_j\| = 1$ for all j.

With the Kronecker delta symbol, $\delta_{jk} = 0$ when j differs from k, and $\delta_{kk} = 1$, orthonormality can be expressed as $\langle f_j,f_k \rangle = \delta_{jk}$.

Examples:

1. Integral tables, mathematical software, integration by parts (twice), substitution with the cosine angle-sum rule, and rewriting trigonometric functions as complex exponentials can all be used to evaluate integrals such as

$$\int_0^x \sin\left(\frac{m\pi x}{L}\right)\sin\left(\frac{n\pi x}{L}\right)dx.$$

Any or all of these methods will lead to the same conclusion, *viz.*:

$$\int_0^x \sin\left(\frac{m\pi x}{L}\right)\sin\left(\frac{n\pi x}{L}\right)dx = \frac{L}{2}\delta_{m\pi}$$

The set of functions

$$\left\{\sin\left(\frac{m\pi x}{L}\right)\right\}_{m=1}^{\infty}$$

Is orthogonal on the interval [0,L], and to turn it into an orthonormal set, we normalize the functions by multiplying by the appropriate constant:

$$\left\{\sqrt{\frac{2}{L}}\sin\left(\frac{m\pi x}{L}\right)\right\}_{m=1}^{\infty}$$

2. Similarly,

$$\left\{\sqrt{\frac{2}{L}}\cos\left(\frac{m\pi x}{L}\right)\right\}_{m=1}^{\infty}$$

is orthonormal on the interval [0,L], and we can even include another function, the constant:

$$\left\{\sqrt{\frac{1}{L}}\right\}\cup\left\{\sqrt{\frac{2}{L}}\cos\left(\frac{m\pi x}{L}\right)\right\}_{m=1}^{\infty}$$

3. We can mix the previous two sets to have both sines and cosines as long as we leave out all of the odd coefficients:

$$\left\{\sqrt{\frac{1}{L}}\right\}\cup\left\{\sqrt{\frac{2}{L}}\cos\left(\frac{2m\pi x}{L}\right)\right\}_{m=1}^{\infty} \text{È} \left\{\sqrt{\frac{2}{L}}\sin\left(\frac{2n\pi x}{L}\right)\right\}_{n=1}^{\infty}$$

is also an orthonormal set. This one is the basis of the usual Fourier series, and is perhaps the most important of all our orthonormal sets. By the way, we do not claim that the functions in equqtionsare orthogonal to functions in the other sets, but only separately among themselves. For instance,

$\sqrt{\frac{2}{L}}\sin\left(\frac{nx}{L}\right)$ is not orthogonal to $\sqrt{\frac{1}{L}}$ on the interval [o, L].

4. Recall that by Euler's formula, exp(i α):= $e^i\,\alpha$:= cos(α) + i sin (α). The complex trigonometric functions exp $\left(\frac{2mnx}{L}\right)$

have many useful properties, including

$$\left|\exp\left(\frac{2\pi inx}{L}\right)\right| = 1 \text{ for all x;}$$

and

$$\left\{\frac{1}{\sqrt{L}}\exp\left(\frac{2\pi inx}{L}\right)\right\}_{n>-\infty}^{\infty}$$

is an orthonormal set on [0,L].

For later purposes, we observe here that the sets of functions are each orthogonal on *any* interval [a,b] with L = b – a.

Before finishing this section we need two more notions about vectors and functions, thought of as abstract vectors.

The first is distance. With the standard inner product, we would like to define the distance between two functions f and g as the L^2 norm of f – g:

$$\|f - g\| = \sqrt{\int_b^a |f(x) - g(x)|^2 \, dx}.$$

This turns out to be a familiar quantity in data analysis, called the *root-mean-square, or r.m.s., deviation.*

It is a convenient way of specifying how large the error is when the true function f is replaced by a mathematical approximation or experimentally measured function g. It is always positive unless f = g *almost everywhere*.

Definition: *Almost everywhere* is a technical phrase meaning that f and g differ for sufficiently few values of x that all integrals involving f have the same values as those involving g. Such a negligible set is called a *null set*. For most practical purposes we may regard f and g as the same functions, and write

$$f = g \text{ a.e.}$$

A typical example is where a function is discontinuous at a point and some arbitrary decision is made about the value of the function at the discontinuity. Different choices are still equal a.e. Much more exotic possibilities than this typical example can occur, but will not arise in this course.

The second notion we generalize from ordinary vectors is that of projection. Suppose that we have a position vector in three dimensions such as v = (3, 4, 5)

Example: We wish to find the vector in the plane spanned by (1,1,1) and (1,–1,0), which is the closest to v = (3,4, 5).

Solution. We solve this problem with a projection. The projection of a vector v onto v_1 is given by the formula:

$$Pv_1 \, v = \frac{(v, v_1) v_1}{\| v_1 \|^2}$$

Notice that this points in the direction of v_1 but has a length equal to ||v|| cos(theta), where theta is the angle between v and v_1.

The length of v_1 has nothing to do with the result of this projection - if we were being very careful, we would say that we were projecting v onto the *direction determined by* v_1, or onto the *line through* v_1. For similar reasons we notice that the vector v_1 could be normalized to have length 1, so that the denominator can be ignored - it is 1. In our example,

$$P_{\begin{bmatrix}1\\1\\1\end{bmatrix}}\begin{bmatrix}3\\4\\5\end{bmatrix} = \frac{(12)\begin{bmatrix}1\\1\\1\end{bmatrix}}{3} = \begin{bmatrix}4\\4\\4\end{bmatrix}.$$

(Here we write the vectors as column vectors because the projection operator is equivalent to a 3 by 3 matrix multiplying them.)

If the basis for the plane consists of orthogonal vectors v_1 and v_2, as in our example, then the projection into the plane is just the sum of the projections onto the two vectors:

$$P\,\{v_1, v_n\}\ v = \sum_{n=1}^{2} \frac{(v\ v_n)v_n}{\|v_n\|^2}.$$

In our example,

$$P\,\{v_1\ v_n\}\ v = \begin{bmatrix}4\\4\\4\end{bmatrix} + \begin{bmatrix}-1/2\\1/2\\0\end{bmatrix} = \begin{bmatrix}7/2\\9/2\\4\end{bmatrix}.$$

Example: We wish to find a) the vector in the plane spanned by (1,1,1) and (1,2,3), which is the closest to v = (3, 4, 5) and b) the vector in the plane closest to (1,–1,0).

Solution. This is similar to the previous problem, except that the vectors defining the plane are not orthogonal.

We need to replace them with a different pair of vectors, which are linear combinations of the first, but which are orthogonal. The formula is definitely wrong if the vectors v_n are not orthogonal.

After finding the new pair of vectors, however, the solution will be as before - just sum the projections onto the orthogonal basis vectors.

There is more than one good choice for the pair of orthogonal basis vectors. If we solve the vector equation

$$\begin{bmatrix}1\\1\\1\end{bmatrix} \cdot \left(\begin{bmatrix}1\\2\\3\end{bmatrix} - \alpha\begin{bmatrix}1\\1\\1\end{bmatrix}\right) = 0$$

for α we get $\alpha = 2$, so a suitable second vector which is orthogonal to (1,1,1) is (–1,0,1).

The projection of (3,4,5) into the plane is the sum of its projections onto these vectors, i.e.,

$$\frac{3+4+5}{3}\begin{bmatrix}1\\1\\1\end{bmatrix}+\frac{-3+0+5}{2}\begin{bmatrix}-1\\0\\1\end{bmatrix}=\begin{bmatrix}3\\4\\5\end{bmatrix}.$$

Perhaps it looks strange to see that the projection of the vector (3,4,5) is itself, but the interpretation is simply that the vector was already in the plane before it was projected. In general a vector will be moved (and shortened) when it is projected into a plane, and we can see this when we project (1,–1,0):

$$\frac{1-1+0}{3}\begin{bmatrix}1\\1\\1\end{bmatrix}+\frac{-1+0+0}{2}\begin{bmatrix}-1\\0\\1\end{bmatrix}=\begin{bmatrix}1/2\\0\\-1/2\end{bmatrix}.$$

Now that we have a vector in the plane, if we project again, it won't move. Algebraically, projections satisfy the equation $P^2 = P$.

We shall make these same calculations in function space to find the best mean-square fit of a function f(x) by a nicer expression or polynomials. The formula simply replaces the dot product with the standard inner product:

$$P_{\{g(x)\}}\,(f(x)) := \frac{\langle f,g\rangle}{\|g\|^2}\,g(x)$$

For example, if we wish to find the multiple of sin(3x) which is closest to the function x on the interval 0 < x <, we find:

$$P_{\{\sin(3x)\}}(x)) := \frac{\int_0^x \tilde{x}\sin(3\tilde{x})\,d\tilde{x}}{\int_0^x (\sin(3\tilde{x}))^2\,d\tilde{x}}\sin(3x) = \frac{2}{3}\sin(3x).$$

The projection of a vector/function onto its own direction will be the same as the vector/function itself.

Now consider what we mean when we project a function onto the constant function 1. This should be the best approximation to f(x) consisting of a single number. What could this be but the average of f? Indeed, the projection formula gives us

$$P_{\{1\}}\,(f(x)) := \frac{\langle f,1\rangle}{\|1\|^2}1 = \frac{\int_a^b f(x)\,dx}{b-a},$$

which is familiar as the average of a function.

Example: Consider the set of functions on the interval $-1 \le x \le 1$ We wish to find the function in the span of 1, x, and x^2, which is the closest to $f(x) = \cos(\pi\, x/2)$. In other words, find the best quadratic fit to the function f in the mean-square sense.

Solution. The calculations can be done with Mathematica or Maple if you prefer. First let us ask whether the three functions are orthogonal. Actually, no.

The function x is orthogonal to each of the other two, but 1 and x^2 are not orthogonal. We can see this immediately because x is odd, while 1 and $x^2$are both even, and both positive.

A more suitable function than x^2 would be x^2 minus its projection onto the direction of 1, that is, x^2 minus its average, which is easily found to be 1/3. The set of functions {1, x, and $x^2 - 1/3$} is an orthogonal set, as you can check.

Then we can project cos(¹x/2) onto the span of the three functions 1, x, and $x^2 - 1/3$:

$$P_{\{1\}}\left(\cos(\pi x/2)\right) = \frac{1}{2}\int_{-1}^{1}\cos(\pi x/2)\,dx = \frac{2}{\pi}.$$

$P_{\{x\}}\left(\cos(\pi x/2)\right) = 0$ (the cosine is even and x is odd).

$$P_{\{x^2-1/3\}}\left(\cos(\pi x/2)\right) = \frac{\int_{-1}^{1}\left(t^2-1/3\right)\cos(\pi t/2)\,dt}{\int_{-1}^{1}\left(t^2-1/3\right)^2 dt}$$

$$= \frac{15\left(24-2\pi^2\right)}{2\pi^3}\left(x^2-1/3\right)$$

The best quadratic approximation to $\cos(\pi\, x/2)$ on the interval $-1 < x < 1$ is the sum of these three functions. Here is a graph showing the original function and its approximation:

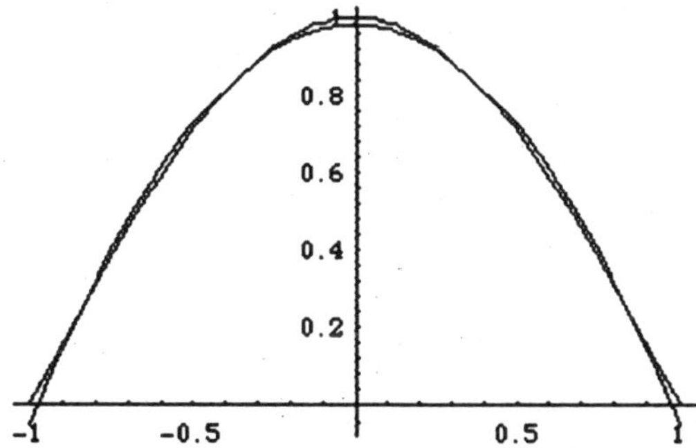

For comparison, here is a plot which shows the Taylor approxim ation as well as the original and the best quadratic:

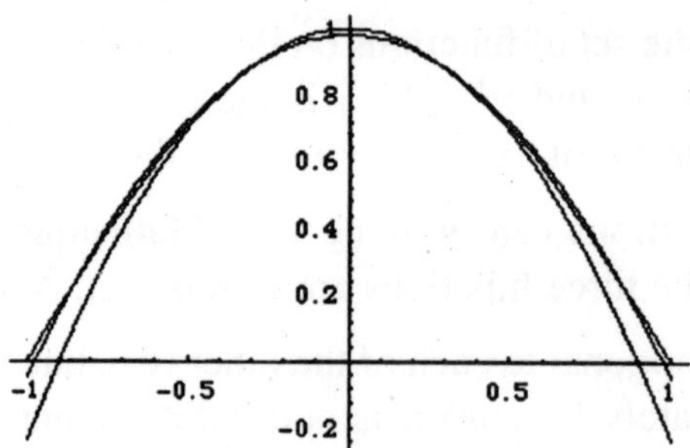

The general algorithm for finding the best approximation to a function is as follows. Suppose that we want to find the best approximation to f(x), $a \le x \le b$, of the form $a_1g_1(x) + a_2g_2(x) + ... + a_{1n}g_n(x)$, where $g_1...g_n$ are some functions with nice properties - they may oscillate with definite frequencies, have simple shapes, etc. They could be chosen to capture the important features of f(x), while possibly simplifying its form or filtering out some noise.

Step 1. Replace $g_1...g_n$ by an orthogonal set with the same span. Let's call the orthogonal set $h_1...h_n$.

Step 2. Project f onto each of the h_k.

Step 3. Sum the projections. If P denotes the span of $g_1...g_n$, then

$$\text{Proj}_p(f) = \sum_{k=1}^{n} \frac{(f, h_k)}{\| h_k \|^2} h_k(x)$$

Perhaps the most important functional approximation uses the Fourier functions as we shall learn to call them. The coefficients of these functions in the projection are called *Fourier coefficients*, and the approximation is as follows:

$$f(x) \cong a_0 + \sum_{m=1}^{M} a_m \cos\left(\frac{2\pi mx}{l}\right) + \sum_{n=1}^{N} b_n \sin\left(\frac{2\pi nx}{l}\right).$$

The right side should be the projection of f(x) on the span of the sines and cosines (including the constant) on the right. To get the coefficients we use the analogue of the formula. For example,

$$P_{\text{coc}\left(\frac{2\pi mx}{L}\right)} f = \frac{\left\langle f, \cos\left(\frac{2\pi mx}{L}\right)\right\rangle \cos\left(\frac{2\pi mx}{L}\right)}{\left\|\cos\left(\frac{2\pi mx}{L}\right)\right\|^2}$$

$$= \frac{2}{L}\left(\int_0^x \cos\left(\frac{2\pi mt}{L}\right) f(t)\, dt\right) \cos\left(\frac{2\pi mx}{L}\right)$$

In other words, the formula for the coefficients a_m must be:

$$a_m = \frac{2}{L}\int_0^x \cos\left(\frac{2\pi mt}{L}\right) f(x)\,dx,\ m = 1,2,...$$

$$a_0 = \frac{1}{L}\int_0^x f(x)\,dx,\ m = 1,2,...$$

As mentioned above, the projection of f onto a constant function is its average.

Finally, the analogous calculation for the sines gives:

$$b_n = \frac{2}{L}\int_0^x \sin\left(\frac{2\pi nt}{L}\right) f(x)\,dx,\ n = 1,2,...$$

Formulae will be the basis of the Fourier series in the next section.

The notions of best approximation and projection in function space are at the heart of filtering theory. A filtered signal is nothing other than a function that has been projected in some way to remove certain frequencies or some undesired characteristics such as noise.

Constructing orthonormal sets. It is often convenient to have orthonormal, or at least orthogonal sets. These are analogous to the usual basis vectors for the plane or for 3-space (denoted R^3), but you may recall that there are many choices of orthogonal bases. For instance, you may have unit vectors oriented along the x,y, and z axes, but someone else may have unit vectors oriented along a rotated set of axes. Although we shall first concentrate on the set (2.8) as a basis for a vector space of functions, the other sets of orthonormal functions (2.5) – (2.7) and (2.9) will be useful later for the same purpose. The choice among these sets is analogous to the choice of different bases for R^3.

But how do we come up with a basis in the first place? Suppose you are given several vectors, such as $v_{1,2,3}$ = (1,0,0), (1,1,0), and (1,1,1), and you want to recombine them to get an orthonormal, or at least orthogonal set. You can do this by projecting away the parts of the vectors orthogonal to each other. The systematic way of doing this is called the Gram-Schmidt procedure, and it depends a great deal on the order in which it is done.

Example: Find an orthonormal set with the same span as $v_{1,2,3}$ = (1,0,0), (1,1,0), and (1,1,1), beginning with $w_1 = v_1$ = (1,0,0). (We rename it because it is the first vector in a new set of recombined vectors.)

Solution. The next vector v_2 is not orthogonal to v_1, so we subtract off the projection of v_2 onto the direction of v_1:

$$w_2 = v_2 - Pw_1\, v_2 = \begin{bmatrix}1\\1\\0\end{bmatrix} - \begin{bmatrix}1\\0\\0\end{bmatrix} = \begin{bmatrix}0\\1\\0\end{bmatrix}.$$

For w_3, we begin with v_3 and project away the parts in the plane spanned by v_1 and v_2, which is the same as the plane spanned by w_1 and w_2.

We find the standard basis vector

$$w_3 = \begin{bmatrix} 0 \\ 0 \\ 1 \end{bmatrix}$$

Example: Find an orthonormal set with the same span as $v_{1,2,3} = (1,0,0), (1,1,0)$, and $(1,1,1)$, beginning with v_3.

Solution. Fist let's turn v_3 into a unit vector:

$$\tilde{w}_1 = \frac{1}{\sqrt{3}} \begin{bmatrix} 1 \\ 1 \\ 1 \end{bmatrix}.$$

For the second unit vector we could take v_2 and project away the part pointing along v_3, getting

$$\begin{bmatrix} 1 \\ 1 \\ 0 \end{bmatrix} - \frac{2}{3}\begin{bmatrix} 1 \\ 1 \\ 1 \end{bmatrix} = \begin{bmatrix} \frac{1}{3} \\ \frac{1}{3} \\ -\frac{2}{3} \end{bmatrix},$$ which we can nomalizas

which we can normalize as

$$\tilde{w}_2 = \frac{1}{\sqrt{6}} \begin{bmatrix} 1 \\ 1 \\ -2 \end{bmatrix},$$

Finally, taking v_1, projecting out the components in these directions, and normalizing, gives us the vector

$$\tilde{w}_3 = \frac{1}{\sqrt{2}} \begin{bmatrix} 1 \\ -1 \\ 0 \end{bmatrix}.$$

Example: Construction of the Legendre polynomials. Let us consider the interval -1 x 1, and find a set of orthonormal functions which are more mundane than the trigonometric functions, namely the polynomials.

We begin with the power functions 1, x, x^2, x^3,.... Some of these are orthogonal because some are even functions and others are odd, but they are not all orthogonal to one another. For instance,

$$\{1, x_2\} = \int_{-1}^{1} x^2 \, dx = \frac{2}{3}$$

Let us denote the set of orthogonal polynomials we get using the Gram-Schmidt procedure on the power functions (in this order) p_0, p_1, p_2,.... These are important in approximation theory and differential equations, and are known as the *normalized Legendre polynomials*. Beginning with the function $x^0 = 1$, after normalization we find

$$P_0(x) = \sqrt{\frac{1}{2}}$$

The next power is $x^1 = x$. Since x is already orthogonal to any constant function, all we do is normalize it:

$$P_1(x) = \sqrt{\frac{3}{2}}\, x$$

To make x^2 orthogonal to p_0, we need to subtract a constant: $x^2 - 1/3$. Because of symmetry, it is already orthogonal to p_1, so we don't worry about p_1 yet, and just normalize:

$$P_2(x) = \sqrt{\frac{5}{2}}\left(\frac{3}{2}x^2 - \frac{1}{2}\right)$$

Similarly, when orthogonalizing x^3 we need to project out x but not 1 or x^2. We find $x^3 - 3x/5$, or, when normalized:

$$p3(x) = \sqrt{\frac{7}{2}}\left(\frac{5}{2}x^3 - \frac{3}{2}x\right) \text{ etc.}$$

By the way, Legendre polynomials are traditionally not normalized as we have done, but rather are denoted $P_k(x)$ and scaled so that $P_k(1) = 1$. The normalization for Legendre polynomials of arbitrary index is such that

$$p_n(x) = (n + 1/2)^{1/2} \, P_n(x).$$

Most special functions are known to Mathematica and Maple, so calculations with them are no more difficult than calculations with sines and cosines.

Chapter 6

Linear Ordinary Differential Equations

In this chapter we shall employ what has been developed to give a brief overview of the theory of linear ordinaryi differential equaqtions of nonlinear differential equations as well as bifurcation theory for periodic orbits and many other facets where linearization techniques are of importance. The results are also of interest in their own right.

Preliminaries

Let $I \cap \mathbb{R}$ be a real interval and let

$$A : I \to £(\mathbb{R}^N, \mathbb{R}^N)$$

$$f : I \to \mathbb{R}^N$$

be continuous functions. We consider here the system of ordinary differential equations

$$u' = A(t)u + f(t), t \in I,$$

and

$$u' = A(t)u, t \in I.$$

Using earlier results we may establish the follwing basic proposition.

Proposition: For any given *f*, initial value problem for (1) are uniquely solvable and solutions are defined on all of I.

Remark: More generally we may assume that *A* and *f* are measurable on *I* and locally integrable there, in which case the conclusion of Proposition I still holds. We shall not go into details for this more general situation, but leave it to the reader to present a parallel development.

Proposition: The set of solutions of is a vector space of dimension *N*.

Proof: That the solution set forms a vector space is left as an exercise. To show that the dimension of this space is *N*, we employ the uniqueness principle above. Thus let $t_0 \in I$, and let $u_k(t), k = 1, ..., N$ be the solution such that

$$u_k(t_0), = e_k, e_k^i = \delta_{ki} \text{ (Kronecker delta).}$$

It follows that for any set of constants $a_1,..,a_N$,

$$u(t) = \sum_{1}^{N} a_i u_i(t)$$

is a solution. Further, for given $\xi \in \mathbb{R}^N$, the solution u such taht $u(t_0) = \xi$ is given with $a_i = \xi^i, i = 1,\ldots,N$

Let the $N \times N$ matrix function Φ be defined by

$$\Phi(t) = (u_j^i(t)), 1 \le i, j \le N,$$

i.e. the columns of Φ are solutions of (2). Then (4) takes the form

$$u(t) = \Phi(t)a, = a = (a^1,\ldots,a^N)^T.$$

Hence for given $\xi \in \mathbb{R}^N$, the solution u such that $u(t_1) = \xi, t_1 \in I$, is provided that $\Phi(t_1)$ is a nonsingular matrix, in which case a may be uniquely determined. That this matrix in never singular, provided it is nonsingular at some point, is known as the Abel-Liouville lemma, whose proof is left as an exercise below.

***Lemma**:* If g(t) = detF(t), then gsatisfies

$$g(t) = g(t_0)e^{\int_{t_0}^{t} traceA(s)} ds.$$

Hence, if Φ is defined, where $u_1,\ldots,uN$ are solutions, then $\Phi(t)$ is nonsingular for all $t \in I$ if and only if $\Phi(t_0)$ is nonsingular for some $t_0 \in I$.

Fundamental Solutions

A nonsingular $N \times N$ matrix function Ψ whose columns are solutions of the equation is called a fundamental matrix solution or a fundamental system. Such a matrix is a nonsingular solution of the matrix differential equation

$$\Psi' = A(t)\Psi.$$

The following proposition characterizes the set of fundamental solutions; its proof is again left as an exercise.

Proposition: Let F be a given funamental matrix solution. Then every other fundamental matrix solution has the form $\Psi' = \Psi C$, where C is a constant nonsingular $N \times N$ matrix. Furthermore the set of all solution is given by

$$\{\Phi c : c \in \mathbb{R}^N\},$$

where Φ is a fundamental system.

Variation of Constants

It follows from Proposition that all solutions are given by

$$\{\Phi(t)c + u_p(t) : c \in \mathbb{R}^N\},$$

where Φ is a fundamental system and u_p is some particular solution. Hence the problem of finding all solutions is solved once a fundamental system is know and some particular solution has been found. The following formula, known as the variation of constants formula, shows that a particular solution may be obtained from a fundamental system.

Proposition: Let F be a fundamental matrix solution of (2) $t_0 \in I$. Then

$$u_p(t) = \Phi(t)\int_{t_0}^{t} \Phi^{-1}(s) f(s) ds$$

is a solution of (1). Hence the set of all solutions of (1) is given by

$$\{\Phi(t)\left(c + \int_{t_0}^{t} \Phi^{-1}(s) f(s) ds\right) : c \in \mathbb{R}^N\},$$

where Φ is a fundamental system.

Constant Coefficient Systems

In this section we shall assume that the matric A is a constant matrix and thus have that solutions are dfined for all $t \in \mathbb{R}$. In this case a fundamental matrix solution Φ is given by

$$\Phi(t) = e^{tA}C,$$

where C is a nonsingular constant $N \times N$ matrix and

$$e^{tA} = \sum_{0}^{\infty} \frac{t^n A^n}{n!}.$$

Thus the solution u with $u(t_0) = \xi$ is given by

$$u(t) = e^{(t-t_0)A}\xi.$$

To compute e^{tA} we use the (complex) Jordan canonical form J of A. Since A and J are similar, there exists a nonsingular matrix P such that $A = PJP^{-1}$ and hence $e^{tA} = Pe^{tJ}P^{-1}$. We therefore compute e^{tJ}. On the other hand J has the form

$$J = \begin{pmatrix} J_0 & & & \\ & J_1 & & \\ & & \ddots & \\ & & & J_s \end{pmatrix}$$

where

$$J_0 = \begin{pmatrix} \lambda_1 & & & \\ & \lambda_2 & & \\ & & \ddots & \\ & & & \lambda_q \end{pmatrix}$$

is a q × q diagonal matrix whose entries are the simple (algebraically) and semisimple eigenvalues of A, repeated according to their multiplicities, and for $1 \le i \le s$,

$$J_i = \begin{pmatrix} \lambda_{q+i} & 1 & & & \\ & \lambda_{q+i} & 1 & & \\ & & \ddots & \ddots & \\ & & & \lambda_{q+i} & 1 \end{pmatrix}$$

is a $q^i \times q^i$ matrix, with $\quad q + \sum_1^s q_i = N$

by the laws of matrix matrix multiplication it follows that

$$e^{tJ} = \begin{pmatrix} e^{tJa} & & & \\ & e^{tJ_1} & & \\ & & \ddots & \\ & & & e^{tJ_s} \end{pmatrix},$$

and

$$e^{tJ_0} = \begin{pmatrix} e^{\lambda_1 t} & & & \\ & e^{\lambda_2 t} & & \\ & & \ddots & \\ & & & e^{\lambda_q t} \end{pmatrix}.$$

Futher, since $J_i = \lambda_{q+i} I_{r_i} + z_i$, where I_{r_i} is the $r_i \times r_i$ identity matrix and Z_i is given by

$$Z_i = \begin{pmatrix} 0 & 1 & & & \\ & 0 & 1 & & \\ & & \ddots & \ddots & \\ & & & & 1 \\ & & & & 0 \end{pmatrix}$$

we obtain that

$$e^{tJ_i} = e^{t\lambda_{q+i}I_{r_i}} e^{tZ_i} = e^{t\lambda_{q+i}} e^{tZ_i}.$$

An easy computation now shows that

$$e^{tZ_i} = \begin{pmatrix} 1 & t & \frac{t^2}{2!} & \cdots & \frac{t^{r_i-1}}{r_{i-1}!} \\ 0 & 1 & t & \cdots & \frac{t^{r_i-2}}{r_{i-2}!} \\ \vdots & \vdots & \vdots & \cdots & \vdots \\ 0 & 0 & 0 & \cdots & 1 \end{pmatrix}.$$

Since P is a nonsingular matrix $e^{tA}P = Pe^{tJ}$ is a fundamental matrix solution as well. Also, since J and P may be complex we obtain the set of all real solutions as

$$\{\operatorname{Re} Pe^{tJ}c, \operatorname{Im} Pe^{tJ}c : c \in \mathbb{C}^N\}.$$

The above considerations have the following proposition as a consequence.

Proposition: Let A be an $N \times N$ constant matrix and consider the differential equation

$$u' = Au.$$

Then:

1. All solutions u satisfy $u(t) \to 0,\ as\ t \to \infty$, if and only $\operatorname{Re}\lambda < 0$, for all eigenvalues l of A.
2. All solutions u are bounded on $[0, \infty)$, if and only if $\operatorname{Re}\lambda \leq 0$, for all eigenvalues l fo A and those with zeor real part are semisimple.

Floquet Theory

Let $A(t), t \in \mathbb{R}$ be an $N \times N$ continuous matrix which is periodic with respect to t of period T, i.e., $A(t+T) = A(t), -\infty < t < \infty$, and consider the differential equation

$$u' = A(t)u.$$

We shall associate to (12) a constant coefficient system which determines the asymptotic behavior of solutions.

Proposition: Let $\Phi(t)$ be a fundamenetal matrix solution, then so is $\Psi(t) = \Phi(t+T)$.

Proof: Since Φ is a fundamental matrix it is nonsingular for all t, hence Ψ is nonsingular, Further

$$\begin{aligned}\Psi'(t) = \Phi'(t+T) &= A(t+T)\Phi(t+T)\\ &= A(t)\Phi(t+T)\\ &= A(t)\Phi(t).\end{aligned}$$

It follows by our earlier consideration that there exists a nonsingular constant matrix Q such that

$$\Phi(t+T) = \Phi(t)Q$$

Since Q is nonsingular, there exists a matrix R (see exercises at the end of this chapter) such that

$$Q = e^{TR}.$$

Letting $C(t) = \Phi(t)e^{-tR}$ we compute

$$\begin{aligned}C(t+T) &= \Phi(t+T)e^{-(t+T)R}\\ &= \Phi(t)Qe^{-TR}\,e^{-tR}\\ &= \Phi(t)e^{-tR} = C(t).\end{aligned}$$

We have proved the following proposition.

Proposition: Let $\Phi(t)$ be a fundamental matrix solution, then there exists a nonsingular periodic (of peiod T) matrix C and a constant matrix R such that

$$\Phi(t) = C(t)e^{-tR}.$$

From this representation we may immediately deduce conditions which guarantee the existence of nontrivial T-periodic and mT-periodic (subharmonics) solutions.

Corollary: For any positive integer m has a nontrivial mT– periodic solution if and only if $\Phi^{-1}(0)\Phi(T)$ has an *m–th* root of unity as an eigenvalue, where F is a fundamental matix solution.

Proof: The properties of fundamental matrix solutions guarantee that the matrix $\Phi^{-1}(0)\Phi(T)$ is uniquely dtermined by the equation and Proposition implies that

$$\Phi^{-1}(0)\Phi(T) = e^{TR}.$$

On the other hand a solution u of (12) is given by

$$u(t) = C(t)e^{tR}d,$$

where $u(0) = c(0)d$. Hence u is peiodic of peiod mT if and only if

$$u(mT) = C(mT)e^{mTRd} = C(0)e^{mTRd} = c(0)d,$$

Which is the case if and only if $e^{mTRd} = (e^{TR})^m$ has 1 as an eigenvalue.

Let us apply these results to the second order scalar equation

$$y'' + p(t)y = 0,$$

(Hill's equation) where $p: \mathbb{R} \to \mathbb{R}$ is a T-periodic. Equation may be rewritten as the system

$$u' = \begin{pmatrix} 0 & 1 \\ -p(t) & 0 \end{pmatrix} u.$$

Let y_1 be the solution of (14) such that

$$y_1(0) = 0, y_1'(0) = 0,$$

and y_2 be the solution of (14) such that

$$y_2(0) = 0, y_2'(0) = 1,$$

Then

$$\Phi(t) = \begin{pmatrix} y_1(t) & y_2(t) \\ y_1'(t) & y_2'(T) \end{pmatrix}$$

will be a fundamental solution of (15) and

$$\det \Phi(t) = e^{\int_0^t \text{traceA}(s)ds} = 1$$

by the Abel -Liouville formula. Hence, or equivalently, will have a mT–periodic solution if and only if $\Phi(T)$ has an eigenvalues of $\Phi(T)$ are solutions of the equation

$$\det \begin{pmatrix} y_1(T) - \lambda & y_2(T) \\ y_1'(T) & y_2'(T) - \lambda \end{pmatrix} = 0,$$

or

$$\lambda^2 - a\lambda + 1 = 0,$$

where

$$a = y_i(T) + y_2'(T).$$

Therefore

$$\lambda = \frac{a \pm \sqrt{a^2 - 4}}{2}.$$

$$u' = Au \mid g(t,u)$$

and show how to construct Lyapunov functionals to test the stability of the trivial solution of this system.

The type of Lyapunov functional we shall be looking for are of the form

$$v(x) = x^T Bx,$$

where B is a constant $N \times N$ matrix, i.e. we are looking for v as a quadratic form. If u is a solution of then

$$\frac{dv^*}{dt} = \frac{dv(t,u(t))}{dt}$$

$$= u^T (A^T B + BA)u + g^T (t,u)Bu + u^T Bg(t,u).$$

Hence, given, A if B can be found so that $C = A^T B + BA$ has certain definniteness properties, then the results of the trivial solution. To proceed along these lines we need some linear algebra results.

Proposition: Let A be a constant $N \times N$ matrix having the property that for any eigenvalue λ of A, $= \lambda$ is not an eigenvalue.

Then for any $N \times N$ matrix C. there exists a unique $N \times N$ matrix B such that $C = A^T B + BA\cdot$

Proof: On the space of $N \times N$ matrices define the bounded linear operator L by

$$L(B) = A^T B + BA\cdot$$

Then L may be viewed as a bounded linear operator of $\mathbb{R}^{N\times N}$ to itself, hence it will be a bijection provided it does not have 0 as an eigenvalue, Once we show the latter, the result follows. Thus let μ be an eigenvalue of L, i.e., there exists a nonzero matrix B such that

$$L(B) = A^T B +, \mu B$$

Hence

$$A^T B + B(A - \mu I) = 0$$

From this follows

$$B(A-\mu I)^n = (-A^T)^n B,$$

for any integer $n \geq 1$, hence for any polynomial p

$$Bp(A-\mu I)^n = p(-A^T)B.$$

Since, on the other hand if F and G are two matrices with no common eigenvalues, there exists a polynomial p such that $p(F) = I, p(G) = 0$, implies that $A-\mu I$ and $_-A^T$ must have a common eigenvalue.

From which follows taht μ is the sum of two eigenvalues of A, which, by hypothesis connot equal 0.

This proposition has the following corollary.

Corollary: Let A be a constant $N \times N$ matrix. Then for any $N \times N$ matrix C, there exists $\mu > 0$ and a unique $N \times N$ matrix B such that $2\mu B + C = A^T B + BA$.

Proof: Let

$$S = \{\lambda \in \mathbb{C} : \lambda = \lambda_1 + \lambda_2\},$$

where λ_1 and λ_2 are eigenvalues of A. Since S is a finite set, there exists $r_0 > 0$. such that $\lambda(\neq 0) \in S$ implies that $|\lambda| > r_0$. Choose $0 < \mu \leq r_0$ and consider the matrix $A_1 = A - \mu I$.

We may now apply Proposition 21 to the matrix A_1 and find for a given matrix C, a unique matrix B such that $C = A_1^T B + BA_1$, i.e. $2\mu BC = A^T B + BA$.

Corollary: Let A be a constant $N \times N$ matrix haivng the property that all eigenvalues λ of A have negative real parts. Then for any negative definite $N \times N$ matrix B such that $C = A^T B + BA$.

Proof: Let C be a negative definite matrix and let B be given by Proposition, which may be applied since all eigenvalues of A have negative real part. Let $v(x) = x^T Bx$, and let u be a solution of $u' = Au, u(0) = x_0 \neq 0$ Then

$$\frac{dv^*}{dt} = \frac{dv(u(t))}{dt} = u^T (A^T B + BA)u$$

$$= u^T Cu \leq -\mu |u|^2,$$

since C is negative definite. Since $\lim_{t\to\infty} u(t) = 0$ (all eigenvalues of A have negative real part!), it follows that $\lim_{t\to\infty} v(u(t)) = 0$.

We also have

$$v(u(t)) \leq v(x_0) - \int_0^t \mu\,|\,u(s)\,|^2\, ds,$$

from which follows that $v(x_0) > 0$. Hence B is positive definite.

The next corollary follows from stability theory equations and what has just been discussed.

Corollary: A necessary and sufficient condition that an $N \times N$ matrix A have all of its eigenvalues with negative real part is that there exists a unique positive definite matrix B such that

$$A^T B + BA = -I.$$

We next consider the nonlinear problem with

$$g(t,x) = o(|\,x\,|),$$

uniformly with respect to $t \in [t_0, \infty)$, and show that for certain types of matrices A the trivial solution of the perturbed system has the same stability property as that of the unperturbed problem. The class of matrices we shall consider is the following.

Definition: We call an $N \times N$ matrix A critical if all its eigenvalues have non-positive real part and there exists at least one eigenvalue with zero real part. We call it noncritical otherwise.

Theorem: Assume A is a noncritical $N \times N$ matrix and let g satisfy. Then the stability behavior of the trivial solution of is uniformly asymptotically stable if all eigenvalues of A have negative real part and it is unstable if A has an eigenvalue with positive real part.

Proof: Assume all eigenvalues of A have negative real part. By the above exists a unique positive definite matrix B such that

$$A^T B + BA = -I.$$

Let $v(x) = x^T Bx.$ Then v is positive definite and if u is a solution it satisfies

$$\frac{dv^*}{dt} = \frac{dv(u(t))}{dt}$$

$$= -[u(t)\,|^2|\, g^T(t,u)Bu + u^T Bg(t,u).$$

Now

$$|\, gT(t,u)bu + u^T Bg(t,u)\,| \leq 2\,|\, g(t,u)\,||\, B\,||\, u\,|.$$

Choose $r > 0$ such that $|\,x\,| \leq r$ implies

$$|g(t,u)| \leq \frac{1}{4}|B|^{-1}|u|,$$

then

$$\frac{dv^*}{dt} \leq -\frac{1}{2}|u(t)|^2,$$

as long as $|u(t)| \leq r$.

Let A have an eigenvalue with positive real part.

Then there exists $\mu > 0$ such that $2\mu < |\lambda_i + \lambda_j|$, for all eigenvalues λ_i, λ_j of A, and a unique matrix B such that

$$A^T B + BA = 2\mu B = I,$$

as follows from Corollary 22. We note that B connot be positive definite nor positive semidefinite for otherwise we must have, letting $v(x) = x^T Bx$,

$$\frac{dv^*}{dt} = 2\mu v^*(t) - |u(t)|^2,$$

or

$$e^{-2\mu t} v^*(t) - v^*(0) = -\int_0^t e^{-2\mu s} |u(s)^2 ds$$

i.e.

$$0 \leq v^*(0) - \int_0^t e^{-2\mu s} |u(s)|^2 \, ds$$

for any solution, u, contradicting the fact that solution u exist for which

$$\int_0^t e^{-2\mu s} |u(s)|^2 \, ds$$

becomes unbounded as $t \to \infty$. Hence there exists $x_0 \neq 0$, of arbitrarily small norm, so that $v(x_0) < 0$. Let u be a solution.

If the trivial solution were stable, then $|u(t)| \leq r$ for some $r > 0$. Again letting $v(x) = x^T Bx$ we obtain

$$\frac{dv^*}{dt} = 2\mu v^*(t) - |u(t)|^2 + g^T(t,u) \quad Bu + u^T Bg(t,u)$$

We can choose r so small that

$$2|\overline{g(t,u)}||B||u| \leq \frac{1}{2}|u|^2, |u| \leq r,$$

hence

$$e^{-2\mu t} v^*(t) - v^*(0) \leq -\frac{1}{2}\int_0^t e^{-2\mu s} |u(s)|^2 \, ds,$$

i.e. $$v^*(t) \leq e^{-2\mu t} v^*(0) \to -\infty,$$

contradicting that $v^*(t)$ is bounded for bounded u. Hence u connot stay bounded, and we have instability.

It it is the case that the matrix A is a critical matrix, the trivial solution of the linear system may still be stable or it may be unstable.

In either case, one may construct examples, where the trivial solution of the perturbed problem has either the same or opposite stability behavior as the unperturbed system.

Chapter 7

System of Differential Equations

Suppose we want to solve the system of first order ODEs,

$$\dot{Y} = F(t, Y), Y(a) = A.$$

Here **Y** and **A** are n-vectors and **F** is a nonlinear vector-valued function on $R \times R^n$:

$$Y = \begin{bmatrix} Y_1 \\ Y_2 \\ \cdots \\ Y_n \end{bmatrix}, \ A = \begin{bmatrix} A_1 \\ A_2 \\ \cdots \\ A_n \end{bmatrix}, \ F = \begin{bmatrix} F_1(t, Y_1, Y_2, \cdots, Y_n) \\ F_2(t, Y_1, Y_2, \cdots, Y_n) \\ \cdots \\ F_n(t, Y_1, Y_2, \cdots, Y_n) \end{bmatrix},$$

To insure existence and uniqueness of a solution and hence establish effective numerical procedures, we require that $F(t, Y)$ and $\partial F/\partial Y$ be continuous in the box B: $|t-a| \le a \| Y{-}A \| \le \beta$, where α and β are positive numbers and $\| \ \|$ is the vector Euclidean norm. If $\|F(t, Y)\| \le M$ for all (t, Y) in B and if $h$is the smaller of α and β/M, then the IVP has a unique solution for $|t - a| \le h$. Weaker conditions do exist, but these will suffice for our purposes.

However, problems often arise in different forms. For example, one is often interested in solving the second order equation,

$$\ddot{y} = g(t, y, \dot{y}),$$

with initial conditions $y(a) = A_1, \ \dot{y}(a) = A_2$,. Here there is one unknown function $y(t)$ from which we can, *in principle*, obtain $\dot{y}(t)$ A new problem can be formed that involves two unknown quantities, $Y_1(t)$ and $Y_2(t)$. The idea is to identify *independently* $y(t)$ by Y_1 (t) and $\dot{y}(t)$ by $Y_2(t)$. Some manipulation then shows that $Y_1(t)$ and $Y_2(t)$ satisfy the system,

$$\dot{Y}_1 = Y_2,$$

$$\dot{Y}_2 = g(t, Y_1, Y_2),$$

with

$$Y_1(a) = A_1,\ Y_2(a) = A_2.$$

Then $y(t) = Y_1(t)$ is the solution of the original problem. This can be put into the form of equation by defining

$$Y = \begin{bmatrix} y \\ \dot{y} \end{bmatrix},\ F = \begin{bmatrix} Y_2 \\ g(t, Y_1, Y_2) \end{bmatrix},\ A = \begin{bmatrix} A_1 \\ A_2 \end{bmatrix}.$$

More generally, an th order equation in one unknown,

$$y^{(n)} = g(t, y, \dot{y}, \ldots, y^{(n-1)}),$$

$$y(t_0) = A_1, \dot{y}(t_0) = A_2, \ldots, y^{(n-1)} = A_n,$$

can be put into the form of equation via

$$Y = \begin{bmatrix} y \\ \dot{y} \\ \cdots \\ y^{(n-2)} \\ y^{(n-1)} \end{bmatrix}, \quad F = \begin{bmatrix} Y_2 \\ Y_3 \\ \cdots \\ Y_n \\ g(t, Y_1, Y_2, \ldots, Y_n \end{bmatrix}, \quad A = \begin{bmatrix} A_1 \\ A_2 \\ \cdots \\ A_n \end{bmatrix},$$

The general procedure here is to introduce unknowns to handle derivatives up to one less than the highest appearing in the equation.

Example: Conversion of a higher order system to an equivalent system of first ordetions

Consider the simplified equations of planetary motion,

$$\ddot{r} - r(\dot{\theta}) = -\frac{K}{r^2},$$

$$r\ddot{\theta} + 2\dot{\theta}\dot{r} = 0$$

with the initial conditions $r(0) = r_0$, $\dot{r}(0) = \dot{r}_0 \theta(0) = \theta_0$, and $\dot{\theta}(0) = \dot{\theta}_0$. Letting $Y_1 = r$, $Y_2 = \dot{r}$ and $Y_4 = \dot{\theta}$, we derive the equivalent first order system,

$$\dot{Y}_1 = Y_2,$$

$$\dot{Y}_2 = Y_1 Y_4 - \frac{K}{Y_1^2},$$

$$\dot{Y}_3 = Y_{4,}$$

$$\dot{Y}_4 = -\frac{2Y_4Y_2}{Y_1},$$

with initial conditions $Y_1(0) = r_0\ Y_2(0) = \dot{r}_0, Y_3(0) = \theta_0$ and $Y_4(0) = \dot{\theta}_0$.

This is the usual way to go from a higher order system to a system of first order equations, but there are many other ways to do it; see Exercise 1.3 and Example 7.

Example: A nonstandord way to convert a second order ODE to an equivalent sya first order ODE*s*.

The following problem arises in the study of nonlinear mechanics:

$$\ddot{y} + g(y)\dot{y} + y = 0$$

The change of variables, $Y_1 = y$, $Y_2 = \dot{y}$, yields the system

$$\dot{y}_1 = Y_2,$$

$$\dot{y}_2 = -g(Y_1)Y_2 - Y_1.$$

However, in many problems the function $g(x)$ is simple in form, but only smooth in pieces. If we have an analytical expression for an indefinite integral $G(x)$ of the function g (x), that is $d/dx\ (G(x)) = g(x)$, then we note that

$$\ddot{y} + g(y(t))\dot{y}(t) + y(t) = \frac{d}{dt}[\dot{y}(y)G(y(t))] + y(t).$$

So we use the change of variables suggested by Lienard,

$$Y_1(t) = y(t),\ Y_2(t) = \dot{y}(t) + G(y(t)),$$

to get the system

$$\dot{Y}_1 = Y_2 - G(Y_1)$$

$$\dot{Y}_2 = -Y_1.$$

This system has smoother functions than the standard system making it easier to solve numerically.

NUMERICAL SOLUTION METHODS

We begin with numerical methods for solving a scalar version of equation, i.e., the case for $n = 1$:

$$\dot{y} = f(t, y),\ y(a) = A.$$

The methods we develop for solving of equation can easily be extended to systems of first order differential equations and to higher order differential equations. The methods are referred to as discrete variable methods and generate a sequence of approximate values for $y(t)$, $y_1, y_2, y_3, \ldots$, at points $t_1, t_2, t_3, \ldots$. No attempt is made to approximate the exact solution, $y(t)$, over a continuous range of the independent variable t.

In our development, we will assume a constant spacing $h$between $t$points. In realistic implementations of these methods, however, is chosen to satisfy a user-specified accuracy request. The expression $y(t_i)$ will always be used to denote the solution to the equation at $i = t_i$, and $y_i$will always be used for an approximation to $y(t_i)$.

Errors enter into the numerical solution of IVPs from two sources. The first is *discretization error* and depends on the method being used. The second is *computational error* which includes such things as *roundoff error*, the error in evaluating implicit formulas, etc. In general, roundoff error can be controlled by carrying enough significant figures in the computation. The control of other computational errors again depends on the method being used.

There are two measures of discretization error commonly used in discussing the accuracy of numerical methods for solving IVPs. The first is *true* or *global* error. For any $t = t_{i+1}$, global error is simply the difference between the true solution and our numerical approximation to it:

$$e_{i+1} = y\ (t_{i+1}) - y_{i+1}).$$

Even though this is the error in which we are usually interested, it is a relatively difficult and expensive to estimate. The other measure of error is *local error*. It is the error incurred in taking a single step using a numerical method. If we let $u(t)$ be the solution to the IVP,

$$\dot{u} = f(t,u),\ u(t_i) = y_i,$$

then the local error at $t = t_{i+1}$ is given by

$$d_{i+1} = u\ (t_{i+1}) - y_{i+1}.$$

Most codes for solving IVPs estimate the local error at each step and attempt to adjust accordingly. Control of local error controls global error indirectly; this, of course, depends on the stability of the problem itself. Most problems are at least moderately stable, and the global error is comparable to the error tolerance. Also, the cost of estimating global error is twice or more the cost of the integration itself.

ONE-STEP METHODS

A differential equation has no "memory". That is, the values of $y(t)$ for t before $t_i$do not directly affect the values of $y(t)$ for t after t_i. Some numerical methods have memory, and some do not. We shall first describe a class of methods known as *one-step* methods. They have no memory; given y_i there is a recipe for y_{i+1} that depends only on information at t_i.

Suppose we want to approximate the solution to the equation on the interval $[a, b]$. Let the t points be equally spaced; so for some positive integer n and $h = (b-a)/n$, $t_i = a + ih$, $i = 0, 1, \ldots, n$. If $a < b$, h is positive and we are integrating forward; if $a > b$, is negative and we are integrating backwards. The latter case could occur if we were solving for the initial point of a solution curve given the terminal point. A general one-step method can then be written in the form

$$y_{i+1} = y_i + h\,\Delta(t_i, y_i), \qquad y_0 = y\,(t_0),$$

where Δ is a function that characterizes our method. We seek accurate algorithms of equation. By this we mean algorithms for which the true solution, $y(t)$.

$$y(t_{i+1}) = y(t_i) + h\Delta(t_i, y(t_i)) + h\tau_i$$

with τ_i"small." The quantity $h\tau_i$ is called the *local (truncation) error* of the method. The method of equation is said to be of *order p* if for all $a \le t_i \le b$, and for all sufficiently small , there are constants C and p such that

$$|\tau i| = Ch^p$$

This can be interpreted as meaning that $|\tau_i|$ goes to zero no slower than Ch^p. Hereafter, we shall write terms like this as $\tau_i = O\,(h^p)$. The constant C depends, in general, on the solution $y(t)$, its derivatives, and the length of the interval over which the solution is to be found, but is independent of h.

Note that, the order of our method is p even though the order of the local (truncation) error is $p + 1$, because these errors tend to accumulate as the integration proceeds. The order of a method may be viewed as a measure of how fast the error in the computed solution goes to zero at a fixed point as more and more steps are taken, i.e., as h approaches zero. Our goal is to find functions Δ that are inexpensive to evaluate, yet of as high order p as possible. In what follows, different Δ functions are displayed, giving rise to the Taylor series methods and the Runge-Kutta methods.

TAYLOR SERIES METHODS

Perhaps the simplest one-step methods of order p are based on Taylor series expansion of the solution $y(t)$. If is $y(p+1)\ (t)$ continuous on $[a, b]$, then Taylor's formula gives

$$y(t_{i+1}) = y(t_i) + h\left[\dot{y}(t_i) + \cdots + y^{(p)}(t_i)\frac{h^{p-1}}{p!}\right] + y^{(p+1)}(\zeta_i)\frac{h^{p+1}}{(p+1)!},$$

where $t_i \le \zeta_i \le t_{i+1}$. The continuity of $y^{(p+1)}(t)$ implies that it is bounded on $[a, b]$ and so

$$y^{(p+1)}(\zeta_i)\frac{h^{p+1}}{(p+1)!} = O(h^{p+1}) = hO(h^p)$$

Using the fact that $\dot{y} = f(t,y)$, the equation can be written in the form

$$y(t_{i+1}) = y(t_i) + h\left[f(t_i, y(t_i)) + \cdots + f^{(p-1)}(t_i, y)(t_i))\frac{h^{p-1}}{p!}\right] + hO(h^p)$$

where the total derivatives of f are defined recursively by

$$f^{(1)}(t, y = f_t(t, y) + f_y(t, y)f(t, y),$$

$$f(k)(t, y) = f_t^{(k-1}(t, y) + f_y^{(k-1)}(t, y)f(t, y),\ k = 2, 3, \ldots..,$$

Comparison of equations shows that to obtain a method of order p, we can let

$$\Delta\ (t, y(t_i)) = f(t_i, y(t_i)) + \cdots + f^{(p-1)}(t_i, y(t_i))\frac{h^{p-1}}{p!}.$$

This choice leads to a family of methods known as the Taylor series methods, given in the following algorithm.

ALGORITHM

Taylor Series Algorithm

To obtain an approximate solution of order p to the IVP on $[a, b]$, let $h = (b - a)/n$ and generate the sequences

$$y_{i+1} = y_i + h\left[f(t_i, y_i) + \cdots + f^{(p-1)}(t_i, y_i)\frac{h^{p-1}}{p!}\right],$$

$$t_{i+1} = t_i + h,\ i = 0, 1, \ldots, n-1,$$

where $t_0 = a$ and $y_0 = A$.

THE EULER METHOD

Example: The Taylor method of order $p = 1$ is known as Euler's method:

$$y_{i+1} = y_i + hf(t_i, y_i),$$
$$t_{i+1} = t_i + h.$$

To illustrate it, we approximate the solution to the IVP,

$$\dot{y} = y,\ y(0) = 1,$$

at $t = 1.0$ with step $h = 0.25$. Here $f(t, y) = y$, The equation simplifies to

$$y_{i+1} = y_i + 0.25y_i,$$
$$t_{i+1} = t_i + h.$$

Starting with $t_0 = 0.0$ and $y_0 = 1,0$, we compute, truncating results to four decimal places:

$$y_1 = y_0 + 0.25y_0 = (1.0) + (0.25)(1.0) = 1.25,$$

$$t_1 = t_0 + h = 0.0 + (0.25) = 0.25;$$

$$y_2 = y_1 + 0.25y_1 = (1.25) + (0.25)(1.25) = 1.5625,$$

$$t_2 = t_1 + h = 0.25 + (0.25) = 0.50;$$

$$y_3 = (1.5625) + (0.25)(1.5625) = 1.9531,$$

$$t_3 = t_2 + h = 0.50 + (0.25) = 0.75;$$

$$y_4 = (1.9531) + (0.25)(1.9531) = 2.4414,$$

$$t_4 = t3 + h = 075 + (0.25) = 1.00.$$

The exact solution is $y(t) = e^t$, so $y(1) = 2.7183$, correct to four decimal places, and the magnitude of the true or global error in our approximation is $|y_4 - y(1)| = 0.2768$. The approximate and exact solutions are represented graphically below in Figure 2, where the approximating values y_1, y_2, y_3, and y_4 have been joined by straight line segments.

We note from the last column of Table 1 that $|y_n - e| = O(h)$; in fact, for ,. This is consistent with the fact that Euler's method is of order $p = 1$.

If we repeat the above calculations using a Taylor method of order 2, we obtain

$$y_{i+1} = y_i + h\left(y_i + \frac{h}{2}y_i\right),$$

$$t_{i+1} = t_i + h,$$

and the results in Table 2. From Table 2, we see that for $n \geq 16$ $|y_n - e| = 0.45h^2$,.

What would happen if we were to increase the order even more? For $p = 3$ and $p = 4$ the formulas become

$$\text{For } p = 3,\ y_{i+1} = y_i\left(1 + h + \frac{h^2}{2} + \frac{h^3}{6}\right),$$

$$\text{For } p = 4,\ y_{i+1} = y_i\left(1 + h + \frac{h^2}{2} + \frac{h^3}{6} + \frac{h^4}{24}\right),$$

and we obtain using $h = 0.25$ the results that appear in Table 3. Note that the same order of accuracy is obtained for $p = 4$ with $h = 0.25$ as for $p = 2$ and $h = 7.8125 \times 10^{-3}$. Of course, higher total derivatives $f^{(n)}$ are easy to compute for $f(t, y) = y$.

RUNGE-KUTTA METHODS

Runge-Kutta methods are designed to approximate Taylor series methods, but have the advantage of not requiring explicit evaluations of the derivatives of $f(t, y)$. The basic idea is to use a linear combination of values of $f(t, y)$ to approximate $y(t)$.

This linear combination is matched up as closely as possibly with a Taylor series for $y(t)$ to obtain methods of the highest possible order p. Euler's method is an example using one function evaluation.

We illustrate the development of Runge-Kutta formulas by deriving a method using two evaluations of $f(t, y)$ per step; the technique employed in the derivation extends easily to the development of all Runge-Kutta type formulas. Given values t_i, y_i, choose values t_i, y_i and constants α_1, α_2, so as to match with the Taylor expansion,

$$y(t_{i+1}) = \left[f(t_i, y_i) + f^{(1)}(t_i, y_i)\frac{h}{2} + f^{(2)}(t_i, y_i)\frac{h^2}{6} \cdots \right],$$

as closely as possible. In what follows all arguments of f and its derivatives will be suppressed when they are evaluated at (t_i, y_i). It will also be convenient to express t_i, y_i, as

$$\hat{t}_i = t_i + h\beta_i,$$

$$\hat{y}_i = y_i + \beta_2 hf(t_i, y_i).$$

So the object is to match

$$R = \alpha_1 f + \alpha_2 f(t_i + \beta_1 h, y_i + \beta_2 hf)$$

$$= (\alpha_1 + \alpha_2) f + \alpha_2 h(\beta_2 ff_y + \beta_1 f_t) + \frac{\alpha_2 h^2}{2}\left(\beta_2^2 f^2 f_{yy} + 2\beta_1\beta_2 ff_{ty} + \beta_1^2 f_{tt}\right) + O(h^3)$$

with the Taylor expansion

$$T = f + \frac{h}{2} f^{(1)} + \frac{h^2}{6} f^{(2)} + O(h^3)$$

$$= f + \frac{h}{2}(ff_y + f_t) + \frac{h^2}{6}(f^2 f_{yy} + 2ff_{ty} + f_{tt} + f_t f_y + ff_y^2) + O(h^3).$$

Equating coefficients of like powers of in the above expressions for R and T, we are able to obtain agreement in terms involving h^0 and h^1:

$$h^0: \quad \alpha + \alpha 2 = 1,$$

$$h^1: \quad \alpha_2\beta_2 = \alpha_2\beta_1 = \frac{1}{2}.$$

If we choose $\alpha_2 = \gamma$, an arbitrary parameter, these equations can be solved exactly to give

$$\alpha_2 = \gamma,$$

$$\alpha_1 = 1 - \gamma,$$

$$\beta_1 = \beta_2 = \frac{1}{2\gamma}, \qquad \gamma \neq 0.$$

Combining all of this gives a one-step method of order $p = 2$ if $\gamma \neq 0$ and f is sufficiently smooth. We state this in the following algorithm.

ALGORITHM

Runge- Kutta algorithm of order

To obtain an approximate solution of order $p = 2$ to the IVP, let $h = (b - a)/n$ and generate the sequences

$$y_{i+1} = y_i + h\left[(1-\gamma)f(t_i, y_i) + \gamma f\left(t_i + \frac{h}{2\gamma} f(y_i + \frac{h}{2\gamma} f(t_i,\right)\right]$$

$$ti+1 = ti + h, \; i = 0, 1, ..., n - 1,$$

where $\gamma \neq 0$, $t_0 = a$, $y_0 = A$.

Euler's method is the special case, $\gamma = 0$, and has order 1; the improved Euler method has $\gamma = 1/2$ and the Euler-Cauchy method has $\gamma = 1$.

Example: An illustration of the Euler Cauchy method.

Approximate the solution to $\dot{y} = y^2 + 1, y(0) = 0$ at $t = 0.1, 0.2, ..., 1.0$ using the Euler-Cauchy method with $h = 0.1$.

The recurrence relation for y_{i+1} is

$$y_{i+1} = y_i + hf\left(t_i + \frac{h}{2}, y_i + \frac{h}{2} f(t_i, y_i)\right)$$

$$= y_i + \left[\left(h1 + (y_i + \frac{h}{2}(1 + y_i^2)\right)^2\right]$$

and the resulting approximations are given in Table 4. The IVP has the solution $y(t) = \tan t$. The approximate and exact solutions are represented graphically in Figure 3 where the approximating values $i = 0, 1, ..., 10$ have been joined by straight line segments.

If we increase n and tabulate the solution at $t = 1.0$, we obtain the results in Table 5, and we see that the error, $| y_n - \tan(1.0) | \approx 1.7h^2$ as h approaches zero.

A basic assumption in the derivation of the family of Runge-Kutta formulas was that the solution $y(t)$ had three continuous derivatives. What if a formula of order 2 is used to solve an initial value problem whose solution has only two continuous derivatives, but not three. Examination of the local truncation error shows that the formula is then of order 1 and convergence is $O(h)$ and not $O(h^2)$. The point here is that higher order procedures can be used on problems whose solutions are not sufficiently smooth, but their rate of convergence may be reduced.

Example: An IVP whose solution is not smooth.

A common example of problems whose solutions will not be smooth are those where the coefficients have a jump discontinuity at some point in the range of integration. In solving such problems numerically, integration should not be performed across the discontinuity. For example, suppose we are solving the problem

$$\ddot{y} + y = f(t), 0 \le t \le 2,$$

$$y(0) = 0, \ \dot{y}(0) = 1$$

$$f(t) = \begin{cases} 1, \text{ for } 0 \le t \le 1, \\ 0, \text{ for } 1 \le t \le 2, \end{cases}$$

A good procedure would be to integrate from $t = 0$ to $t = 1$ and then from $t = 1$ to $t = 2$. On each subinterval, the differential equation has smooth solutions and convergence rates will be as advertised.

A major limitation of Runge-Kutta formulas is the amount of work required; work is measured in terms of the number of times the function f is evaluated. For higher order formulas, the work goes up dramatically; p evaluations per step lead to procedures of order p for $p = 1, 2, 3$, and 4, but not for 5; 6 evaluations are required for a formula of order 5, 7 for order 6, 9 for order 7, 11 for order 8, etc. For this reason, fourth order procedures are quite common. As in the second order case where the parameter γ was arbitrary, there is a family of fourth order formulas that depend on several parameters. One choice leads to the so-called *classical* formulas.

Algorithm: Classical Runge Kutt Formulas

To obtain an approximate solution of order $p = 4$ to the IVP on $[a, b]$, let $h = (b - a)/n$ and generate the sequences

$$y_{i+1} = y_i + \frac{h}{6}(k_1 + 2k_2 + 2k_3 + k_4),$$

$$t_{i+1} = t_i + h, \qquad i = 0, 1, \ldots, n - 1$$

where

$$k1 = f(t_i, y_i),$$

$$k_2 = f(t_i + h/2, y_i + (h/2)k_1),$$

$$k_3 = f(t_i + h/2, y_i + (h/2)k_2),$$

$$k_4 = f(t_i + h, y_i + hk_2),$$

and , $t_0 = a\, y_0 = A.$

As we pointed out earlier, the methods developed extend readily to systems of first order IVPs. As an illustration, the formulas in the case of an -dimensional system become

$$\mathrm{Y}_{i+1} = \mathrm{Y}_i + \frac{h}{6}(\mathrm{K}_1 + 2\mathrm{K}_2 + 2\mathrm{K}_3 + \mathrm{K}_4),$$

$$t_{i+1} = t_i + h \text{ , } i = 0, 1, \ldots, n-1,$$

where

$$\mathrm{K}_1 = \mathrm{F}(t_i, \mathrm{Y}_i)$$

$$\mathrm{K}_2 = \mathrm{F}\left(t_i + \frac{h}{2}, \mathrm{Y}_i + \frac{h}{2}\mathrm{K}_1\right)$$

$$\mathrm{K}_3 = \mathrm{F}\left(t_i + \frac{h}{2}, \mathrm{Y}_i + \frac{h}{2}\mathrm{K}_2\right)$$

$$\mathrm{K}_4 = \mathrm{F}(t_i + h, \mathrm{Y}_i + h\,\mathrm{K}_3).$$

Example: Classical Runge Kutta method applied to a system of IVP*s*.

Consider the system of first order IVPs,

$$\dot{Y}_1 = Y_2,$$

$$\dot{Y}_1 = Y_1 + t,$$

$$Y_1(0) = 1,\ Y_2(0) = 1,$$

where, in terms of our previous notation,

$$\mathrm{Y} = \begin{bmatrix} Y_1 \\ Y_2 \end{bmatrix}, \mathrm{F} = \begin{bmatrix} F_1 \\ F_2 \end{bmatrix} = \begin{bmatrix} Y_2 \\ Y_1 + t \end{bmatrix}, \mathrm{A} = \begin{bmatrix} 1 \\ 1 \end{bmatrix}.$$

So,

$$\mathrm{K}_1 = \mathrm{F}(t, \mathrm{Y}) = \begin{bmatrix} Y_2 \\ Y_1 + t \end{bmatrix},$$

which implies that

$$Y + (h/2)\,K_1 = \begin{bmatrix} Y_1 \\ Y_2 \end{bmatrix} + (h/2)\begin{bmatrix} Y_2 \\ Y_1 + t \end{bmatrix} = \begin{bmatrix} Y_1 + (h/2)Y_2 \\ Y_2 + (h/2)(Y_1 + t) \end{bmatrix},$$

and therefore

$$K_2 = F(t + (h/2),\, Y + (h/2)\,K_1) = \begin{bmatrix} Y_2 + (h/2)(Y_1 + t) \\ Y_1 + (h/2)Y_2 + t + (h/2) \end{bmatrix},$$

etc. At the $(i + 1)$ st step, the values of t, Y_1, and Y_2 are assumed to be evaluated at the (t_i, Y_i).

SOME IMPLEMENTATION ISSUES

A too small value of h means that we are doing unnecessary computation that could lead to roundoff error (error induced because the arithmetic is not exact); a too large value of means that we are probably not h meeting the desired accuracy requirements (errors induced by discretization).

Although we will not develop the details here, something can be said about the qualitative behavior of global error as a function of for a one-step method of order p. If the error in evaluating $\Delta\,(t_i, y_i)$ is ε_i with $|\varepsilon_i| \le \varepsilon$ for all and the error in forming $h\Delta\,(t_i, y_i)$ is ρ_i with $|\rho i| \le \rho$ for all i, then it can be shown that

$$|\, y(t_i) - y_i \,| \le |\, y(t_0) - y_0 \,|\, e^{L(b-1)} + \frac{e^{L(b-a)}}{L}\left(\frac{Ch^p}{2} + C + \frac{\rho}{r}\right),$$

where C and L are constants depending on y and its derivatives. A graph has the qualitative behavior shown in Figure 3.

As we said earlier, most zcodes estimate local error at each step and attempt to adjust h accordingly. A widely used procedure for estimating local error follows. Suppose we have computed approximations of order p and $p + 1$ at t_{i+1}, y_{i+1}^{p} and y_{i+1}^{p+1} respectively. Then using the equation, the local error in the p th order approximation can be written as

$$d_{i+1} = u(t_{i+1}) - y_{i+1}^{p} = \left[u(t_{i+1}) - y_{i+1}^{p+1}\right] + \left[y_{y_{i+1}}^{p+1} - y_{i+1}^{p}\right]$$

If h is sufficiently small, the term $\left[u(t_{i+1}) - y_{i+1}^{p+1}\right]$ can be neglected and we can use the computed value $\left[y_{y_{i+1}}^{p+1} - y_{i+1}^{p}\right]$ as an estimate of the error in the p th order formula.

This sort of approximation has been validated by extensive numerical experimentation over the years.

THE CODES RKSUITE

Rksuite: is an excellent collection of codes based on Runge-Kutta methods for the numerical solution of an IVP for a first order system of ODEs and is available in the public domain. It supersedes some very widely used codes written by the authors and their coauthors, namely the RKF45 code and its descendant DDERKF in the SLATEC library and DO2PAF and associated codes in the NAG Fortran library. RKSUITE is written in standard FORTRAN 77 and is distributed in source form.

The advanced algorithms provide more functionality than is found in earlier codes, including new more efficient formulas, interpolation, automatic selection of the initial step size, a stiffness diagnostic, and global error assessment. The advanced software design includes a novel interface permitting both interval- and step-oriented solutions and is highly portable. The documentation includes detailed instructions for the effective use of the codes, a collection of templates illustrating the use of the codes for common tasks, output from running the templates in a number of variations, and comprehensive instructions for installing the codes.

Rksuite: implements three Runge-Kutta pairs: (2,3), (4,5), and (7,8). The (4,5) pair, for example, uses both a 4th and a 5th order approximation to estimate the error in the 4th order formula; using extrapolation, it then produces a formula of order 5. Similarly, the (2,3) pair produces a formula of order 3 and the (7,8) pair a formula of order 8.

The documentation for RKSUITE provided in RKSUITE.DOC is especially recommended. It is fairly short, carefully written document, explaining some of the internal workings of the suit of codes. The reader should pay particular attention to the description of*error control* and the effect on delivered accuracy. The documentation also contains an excellent set of references that describe the formulas and algorithms used in RKSUITE; references are also included that describe the design, implementation and testing of codes based on explicit Runge-Kutta formulas.

A copy of RKSUITE, together with sample drivers, can be obtained from "netlib" using anonymous ftp as follows:

1. Ftp netlib.att.com (login as anonymous and use your e-mail address as password)
2. Bin (set binary transfer mode)
3. Cd netlib/ode/rksuite (change to directory containing VODE codes)
4. Mget * (gets the compressed vode file)
5. Quit (exits ftp)
6. Uncompress *.Z (uncompress file *)

The file "templates" contains sample drivers for the codes in rksuite.

For a complete listing of what is available from "netlib", mail netlib @ornl.gov and type send index in the body of your message. The Internet address "netlib@ornl.gov" refers to a gateway machine at Oak Ridge National Laboratory in Oak Ridge, Tennessee. This address should be understood on all major networks.

RKSUITE EXAMPLE

An important equation of nonlinear mechanics is van der Pol's equation,

$$\ddot{y} + \varepsilon(y^2 - 1)\dot{y} + y = 0$$

for $\varepsilon > 0$. For any initial conditions, the solution of this equation converges to a unique periodic solution, called a stable limit cycle. This code uses RKSUITE to solve this problem with $\varepsilon = 1$, $y(0) = 1$, $\dot{y}(0) = 0$, for $0 \leq t \leq 12$.

The code is followed by a tabulation of the solution at $t = 1, 2, \ldots, 12$ and a plot of the solution in the phase plane: $\dot{y}$ vs. y.

To put this problem in a form suitable for RKSUITE, we let $Y_1 = y$ $Y_2 = \dot{y}$, to obtain the equivalent system:

$$\dot{Y}_1 = Y_2,$$

$$\dot{Y}_2 = -\varepsilon\,(y_1^2 - 1)Y_2 - Y_1,$$

$$Y_1(0) = 1,\ Y_2(0) = 0.$$

The system of ODEs is implemented in the subroutine F. The code uses the (4,5) Runge-Kutta pair.

MULTI-STEP METHODS

The Taylor Series and explicit Runge-Kutta methods that we have discussed so far have no memory: the value of $y(t)$ for before t_i do not directly affect the values of $y(t)$ for t after t_i. Other methods take advantage of previously computed solution values and are referred to as multistep methods. The Adam's formulas for non-stiff problems and the Backward Differentiation Formulas for stiff problems furnish important and widely-used examples of multi-step methods.

THE ADAMS-BASHFORTH AND ADAMS-MOULTON FORMULAS

On reaching a mesh point t_i with approximate solution $y_i \cong y(t_i)$, there are (usually) available approximate solutions $yi+1-j \cong y(t_{i+1})$ for $j = 2,3,\ldots, p$. From the differential equation itself, approximations to the derivatives $\dot{y}(t_{i+1-j})$ can be obtained from

$$\dot{y}t_{i+1-j} = f(t_{i+1-j,}y(t_{i+1-j})) \cong f(t_{i+1-j}, y_{i+1-j}) = f_{i+1-j}.$$

This information can be exploited for solution values prior to the current point t_i by using the integrated form of the differential equation:

$$y(t_{i+1}) = y(t_i) + \int_{t_i}^{t_{i+1}} \dot{y}(t)dt = y(t_i) + \int_{t_i}^{t_{i+1}} f(t, y)(t))dt.$$

The Adams Bashforth formula of order p is obtained by integrating the polynomial $P(t)$that interpolates f_{i+1-j} at t_{i+1-j} for $j = 1, 2,..., p$, in place of f:

$$y_{i+1} = y_i + h\sum_{j=1}^{p} \alpha_{pj} f_{i+1-j},$$

Formula involves only one evaluation of f per step. An attractive feature of the approach is the underlying polynomial approximation $P(t)$ to $\dot{y}(t)$ because it can be used to approximate $y(t)$ between mesh points:

$$y(t) \cong y_i \int_{t_i}^{t} P(t)dt.$$

The lowest order Adams-Bashforth formula arises from interpolating the single value $f_i = f(t_i, y_i)$ by $P(t)$. The interpolating polynomial is constant so its integration from t_i to t_{i+1} results (t_i, y_i) in $hf(t_i, y_i)$ and the first order Adams-Bashforth formula (AB1):

$$y_{i+1} = y_i + hf(t_i, y_i).$$

This is just the familiar forward Euler formula. For constant step size h, the second order Adams-Bashforth formula (AB2) is also easily found to be

$$y_{i+1} = y_i + h\,[3/2)f(t_i, y_i) - (1/2)f(t_{i-1}, y_{i-1})].$$

The implicit Adams-Moulton formulas arises when the polynomial $P(t)$ interpolates f_{i+1-j} for $j = 0, 1,..., p-1$:

$$y_{i+1} = y_i + h\sum_{j=1}^{p-1} \hat{\alpha}_{p,j} f_{i+1-j}.$$

When $j = p-1$, the right hand side contains the term $fi+1 = f(t_{i+1}, y_{i+1})$, and we see that y_{i+1} is defined only implicitly by this formula. The solution is accomplished by first "predicting" the result using the explicit Adams-Bashforth formula, and then "correcting" it using the implicit formula; we then proceed by "simple" or "functional" iteration. If L is a bound on $|\partial f/\partial y|$ and the step size h is small enough so that for some constant ρ,

$$|\,h\hat{\alpha}_{k,0}\,|\,L \le \rho < 1,$$

then the equation has a unique solution y_{i+1} and the error is decreased by a factor of ρ at each iteration. For "small" step sizes h, the iteration converges very quickly.

Note that for a formula of order p, both the Adams-Bashforth and Adams-Moulton formulas interpolate the function f on pt-points. The t-points overlap and are illustrated graphically in Figure.

The lowest order Adams-Moulton formula involves interpolating the single value $f_{i+1} = f(x_{i+1}, y_{i+1})$ and an easy calculation leads to the formula

$$y_{i+1} = yi + hf(t_{i+1}, y_{i+1}),$$

which defines y_{i+1} implicitly. The resulting formula is called the backward Euler formula. From its definition it is clear that it has the same accuracy as the forward Euler method; its advantage is vastly superior stability. The second order Adams-Moulton method also does not use previously computed solution values; it is called the trapezoidal rule because it generalizes the trapezoidal rule for integrals to differential equations:

$$y_{i+1} = y_i + \frac{h}{2}[f(t_{i+1}, y_{i+1}) + f(t_i, y_i)].$$

The third order formula is more typical because it does involve a previously computed value. When the step size is a constant h, it is

$$y_{i+1} = y_i + h\,[(5/12)f(t_{i+1}, y_{i+1}) + (8/12)f(t_i, y_i) - (1/12)f(t_{i-1}, y_{i-1})].$$

The Adams-Moulton formula of order p is more accurate than the Adams-Bashforth formula of the same order, so that it can use a larger step size; the Adams-Moulton formula is also more stable. A modern code based on such methods is more complex than a Runge-Kutta code because it must cope with the difficulties of starting the integration and changing the step size. With enough "memorized" values, however, we can use whatever order formula we wish in the step from t_i. Modern Adams codes attempt to select the most efficient formula at each step as well as to choose an optimal step size h to achieve a user-specified accuracy.

Some general rules-of-thumb about how to choose between Runge-Kutta methods and Adams methods for solving nonstiff problems are given below:

1. Generally, Adams methods are superior if output at many points is needed.
2. If function evaluations are expensive, Adams methods are preferred.
3. If function evaluations are inexpensive and moderate accuracy is required, Runge-Kutta methods are generally best.
4. If storage is at a premium, Runge-Kutta methods win.
5. If accuracy over a wide range of tolerances is needed, the variable order Adams methods will outperform the fixed order Runge-Kutta methods.

Recent developments in Runge-Kutta methods have shifted these boundaries somewhat; RKSUITE, for example, has an interpolation capability that makes it more efficient than the previous generation of Runge-Kutta codes and the (7,8) solution pair is very efficient at stringent error tolerances.

Stiff Problems: Backward Differentiation Formulas

A problem is stiff if the numerical solution has its step size limited more severely by the stability of the numerical technique than by the accuracy of the technique. Frequently, these problems occur in systems of differential equations that involve several components that are decaying at widely differing rates.

A simple scalar example of stiffness is given by:

$$\dot{y} = \lambda[y - F(t)] + \dot{F}(t), \qquad \lambda \ll 0,$$

where $F(t)$ is a smooth, slowly varying function. The solution,

$$y = [y_0 - F(0)]e^{\lambda t} + F(t),$$

has a component $[y_0 - F(0)]e^{\lambda t}$ that will be insignificant compared to $F(t)$ for sufficiently large. The numerical method, however, will always have its step size h limited by the magnitude of λh for the entire integration.

Example: Gear's example.

In Gear's example above, choose $F(t) = t$. In Table 7 we tabulate the cost of integrating the IVP,

$$\dot{y} = \lambda(y - t) + 1, \qquad 0 \leq t \leq 10,\ y(0) = 1,$$

using Rksuite for $\lambda = -10, -20, -30, -100$. The solution is $y(t) = e^{\lambda t} + t$. Relative and absolute error tolerances were taken to be 10^{-6}.

Note that the smaller the solution component $e^{\lambda t}$ becomes, the harder RKSUITE works. Using a code designed specifically for stiff problems such as the code VODE presented in the next section, the number of function evaluations would have been approximately 120 for the range of λ -values considered.

A class of multi-step formulas which are highly effective in solving stiff problems are based on numerical differentiation. Again, we start by interpolating the previously computed solution values $y_i, y_{i-1}, \ldots, y_{i-p}$ as well as the new one y_{i+1} by a polynomial $P(t)$. The derivative of the solution at t_{i+1} is then approximated by $\dot{P}(t_{i+1})$. The approximation is related to the differential equation by insisting that it satisfy the differential equation at t_{i+1}:

$$\dot{P}(t_{i+1}) = f(t_{i+1}, P(t_{i+1})) = f(t_{i+1}, y_{i+1}).$$

Substituting for $\dot{P}(t_{i+1})$ in this equation, we obtain the family of backward differentiation formulas, the BDFs:

$$\bar{\alpha}_0 y_{i+1} + \bar{\alpha}_1 y_i + \cdots + \bar{\alpha}_p y_{i+1-p} = hf(t_{i+1}, y_{i+1}).$$

These formulas were popularized by Gear, and are sometimes known as Gear's formulas. They are implicit like the Adams-Moulton formulas, but not as accurate for formulas of the same order and not stable for orders 7 and up. However, at the orders for which the formulas are stable, they are much more stable than the

Adams-Moulton formulas. The formulas cannot be evaluated by simple iteration because this restricts the step size just as much as stability does for much less stable formulas. In practice, a modified Newton iteration (Linear Algebra Chapter) is used to solve the nonlinear algebraic equations for; this requires approximating partial derivatives and solving systems of linear equations.

As with Adams formulas, modern codes based on the BDFs vary the formula (the order used) as well as the step size. The solution of problems that are quite stiff are completely impractical with a method intended for non-stiff problems, such as an explicit Runge-Kutta method or an Adams-Moulton method evaluated by simple iteration

The simplest BDF is when $P(t)$ is the straight line interpolating y_{i+1} and y_i. The derivative at t_{i+1} is the constant slope of this line and setting it to $f(t_{i+1}, y_{i+1})$ results in

$$\frac{y_{i+1} - y_i}{h} = f(t_{i+1}, y_{i+1}).$$

Once again we have derived the backward Euler formula! Although this case results in a one-step formula, the higher order BDFs do involve previously computed solution values. For example, when the step size is a constant h, the backward differentiation formula of order two is

$$y_{i+1} = (4/3)y_{i.} - (1/3)y_{i-1} + h(2/3)f(t_{i+1}, y_{i+1}).$$

A code providing a highly efficient implementation of the Adam's formulas and the BDF formulas is given in the next section.

THE CODE VODE

VODE is a relatively new initial value ODE solver for stiff and non stiff problems and was written by P.N. Brown (LLNL), G.D. Byrne (currently at SMU), and A.C. Hindmarsh (LLNL). It uses variable coefficient Adams-Moulton and Backward Differentiation Formula (BDF) methods. The initial step size is selected internally and a method of order 1 is used to start the integration; the order is increased as sufficient data become available. The setting, MF = 10, is used to request a solution using the Adams method and setting of MF = 21, 22, 24, or 25 requests the BDF method.

A copy of the code may be obtained from "netlib" as follows:

1. Ftp netlib.att.com (login as anonymous and use your e-mail address as password)
2. Bin (set binary transfer mode)
3. Cd netlib/ode (change to directory containing VODE codes)
4. Get vode.f.Z (gets the compressed vode file)
5. Quit (exits ftp)
6. Uncompress *.Z (uncompress file *)

The reader should consult the SIAM article for a discussion of VODE. The article also contains a couple of interesting test cases and a good set of references for further reading. The code itself is well documented; leading comments contain a sample driver and details about the code input and output parameters.

VODE Example

The following IVP due to Robertson arises in the study of chemical kinetics:

$$\dot{Y}_1 = -0.04Y_1 + 10^4 Y_2 Y_3,$$

$$\dot{Y}_2 = 0.04Y_1 - 10^4 Y_2 Y_3 - 3\times 10^7 Y_2^2,$$

$$\dot{Y}_3 = 3\times 10^7 Y_2^2,$$

$$Y_1(0) = 1, Y_2(0) = 0, Y_3(0) = 0.$$

The problem is stiff and the code below uses VODE to solve it on the interval, $0 \le t \le 4 \times 10^{10}$. For an explanation of the various input parameters, see the prologue to VODE. Since the solution of a linear system is required in the BDF method, the Jacobian of **F** must be available.

VODE allows the Jacobian to be supplied by the user (MF = 21 or 24) or to be generated internally by the code (MF = 22 or 25). It is always preferable to supply the Jacobian if possible. Since

$$F = \begin{bmatrix} -0.04Y_1 + 10^4 Y_2 Y_3 \\ 0.04Y_1 - 10^4 Y_2 Y_3 - 3\times 10^7 Y_2^2 \\ 3\times 10^7 Y_2^2 \end{bmatrix},$$

the Jacobian is

$$J = (\partial F_i / \partial Y_j) = \begin{bmatrix} -0.04 & 10^4 Y_3 & 10^4 Y_2 \\ 0.04 & -10^4 Y_3 - 6\times10^7 Y_2 & -10^4 Y_2 \\ 0 & 6\times10^7 Y_2 & 0 \end{bmatrix}$$

F is implemented in the subroutine FEX and J in the subroutine JEX.

We note that although the Jacobian was not difficult to calculate here, it can be tedious in many applications; modern symbolic packages, such as MAPLE, MATHEMATICA, etc. make it easier to supply J.

This code is followed by two tabulations of the solution at $t = 4 \times 10^{-1}, 4 \times 10^0, 4 \times 10^1, \ldots, 4 \times 10^{10}$: the first for a supplied Jacobian (MF=21) and the second for an internally generated Jacobian (MF=22).

Both methods produce acceptably accurate solutions, but more work is required for the internally-generated Jacobian case. The quantity SUM = $Y_1 + Y_2 + Y_3$ is also printed out. Since $\dot{Y}_1(t) + \dot{Y}_2(t) + \dot{Y}_3(t) = 0$, SUM should have the value "1." Note that this condition is necessary for accuracy, but not for sufficient. The output is followed by a combined plot of $Y_1(t)$, $Y_2(t)$, and $Y_3(t)$ vs.t for $0 \le t \le 4 \times 10^{10}$.

Chapter 8

First Order Linear Equations

LINEAR DIFFERENTIAL EQUATIONS

A differential equation is linear in *a* set of one or more of its dependent variables if and only if each terms of the equation which contains *a* variable of the set or any of their derivatives is of the first degree in those variables and their derivatives

A differential equation which is not linear in some dependent variable is said to be nonlinear in that variable. *A* differential equation which is not linear in the set of all of its dependent variables is simply said to be nonlinear. For example

Differential Equation	*Linearity*
1. $y'' + 4xy' + 2y = \cos x$ 2. $y'' + 4yy' + 2x = \cos x$	Is linear, ordinary and order is 2 Is nonlinear ($\therefore$ yy')
3. $\dfrac{\partial^2 u}{\partial x^2} + \dfrac{\partial v}{\partial t} + u + v \ \sin u$	Is linear in v but nonlinear in u ($\therefore$ $\sin u$). The equation is nonlinear
4. $\dfrac{\partial^2 u}{\partial x^2} + \dfrac{\partial v}{\partial t} + u + v \ \sin u$	Is linear in each of the dependent variables x and y. But it is nonlinear in the set of $\{x, y\}$. The equation is nonlinear

FIRST-ORDER LINEAR DIFFERENTIAL EQUATION

By definition, *a* linear first-order differential equation in y cannot contain products, powers or other nonlinear combinations of y or y '. Have its most general form is

$$F(x)\frac{dy}{dx} + G(x)y = H(x)$$

or it appears in the more usual form

$$\frac{dy}{dx} + P(x)y = Q(x)$$

1. If $P(x) \equiv 0$, it is solvable by direct integration, or
2. If $Q(x) \equiv 0$, the equation is separable.

Theorem: The equation $\dfrac{dy}{dx} + P(x)y = Q(x)$ has $\exp\left(\int P(x)\,dx\right)$ as an integration factor.

Proof: When Equation is multiplied by $\exp\left(\int P(x)\,dx\right)$, it can be written in the form

$$\frac{d}{dx}\left(y\exp\left(\int P(x)\,dx\right)\right) = Q(x)\exp\left(\int P(x)\,dx\right)$$

which has solution

$$y = \exp\left(-\int P(x)\,dx\right)\int P(x)\exp\left(\int P(x)\,dx\right)dx + c\exp\left(-\int P(x)\,dx\right)$$

So, we have the following steps to solve *a* linear first-order differential equation

$$\frac{dy}{dx} + P(x)y = Q(x)$$

The following procedures are often helpful in finding solution to the equations:

1. Compute the integrating factor.
2. Multiply the right-hand side of the given equation by this factor and write the left-hand side as the derivative of y times the integrating factor.
3. Integrating and solving the equation for y.

SPECIAL FIRST-ORDER EQUATIONS

BERNOULLI EQUATION

A Bernoulli differential equation has the form

$$\frac{dy}{dx} + P(x)y = Q(x)y^n$$

Clearly, for $n = 0$ and 1, the equation is linear; for other values of n, it is nonlinear. However, changing the dependent variable $z = y^{1-n}$ converts it into the linear equation

$$\frac{dz}{dx} + (1-n)P(x)z \;\; (1-z)\,Q(x)$$

Proof: Multiply Eq.(14.3) by $(1-n)^{-n}$ to get

$$(1-n)y^{-n}y' + (1-n)P(x)y^{1-n} = (1-n)\,Q(x)$$

Note that $z = y^{1-n}$ and $z' = (1-n)y^{-n}\, y'$

After Equation is solved for z, a general solution of Equation can be found by substituting y^{-n} for z. Note however, if $n > 0$ Equation also has the suppressed solution $y = 0$.

Example: Solve the following differential equations by integration factor method:

1. $x\dfrac{dy}{dx} - ky = x^2$
2. $\dfrac{dy}{dx} + y\tan x = \sec x$

Solution:

1. The ODE $x\dfrac{dy}{dx} - ky = x^2$ becomes $\dfrac{dy}{dx} + \left(\dfrac{-k}{r}\right)y = x$. The integration factor is $\exp\left(\int -\dfrac{kdx}{x}\right) = x^{-k}$

   The ODE can be reduced to

$$\frac{dy}{dx} - kx^{-k-1}y = x^{1-k} \Rightarrow \left(\frac{d}{dx}\right)\left(x^{-k}y\right) = x^{1-k}$$

$$\therefore \quad x^{1-k}y = \int x^{1-k}dx = \frac{x^{2-k}}{2-k} + C$$

$$\therefore \quad y = \frac{x^2}{2-k} + Cx^k$$

2. The integration factor is $\exp\left(\int \tan x dx\right) = \sec x$ and the ODE can be reduced to

$$\sec x\frac{dy}{dx} + y\tan x \sec x = \sec^2 x \Rightarrow \frac{d}{dx} + (y\sec x) = \sec^2 x$$

$$\therefore \quad y\sec x = \int \sec^2 x dx = \tan x + C$$

$$\therefore \quad y = \sin x + C\cos x$$

Example: Solve the Bernoulli equation $2xy\dfrac{dy}{dx} - y^2 = x^2$

Solution: The ODE $2xy\,\dfrac{dy}{dx} - y^2 = x^2$ becomes $2y\dfrac{dy}{dx} - \dfrac{1}{x}y^2 = x$.

By changing the variable $z = y^2$, we have $\frac{dz}{dx} = \frac{dz}{dy}\frac{dy}{dx} = 2y\frac{dy}{dx}$.

Hence $\frac{dz}{dx} + \left(-\frac{1}{x}\right)z = x$ and its integration factor is $\exp\int\left(-\frac{1}{x}dx\right) = \frac{1}{x}$

Therefore, we obtain

$$\frac{1}{x}\frac{dz}{dx} + \left(-\frac{1}{x^2}\right) = 1 \Rightarrow \frac{d}{dx}\left(\frac{z}{x}\right) = 1 \Rightarrow \frac{z}{x} = x + C$$

$\therefore\ x^2 + Cx$ and $y = \pm\sqrt{x^2 + Cx}$

RICCATI EQUATION

A Riccati differential equation has the form

$$\frac{dy}{dx} = P(x)y^2 + Q(x)y + R(x)$$

Clearly, if $R(x) = 0$, it becomes Bernoulli equation. If $R(x) \neq 0$, however, *a* general solution can be found whenever one specific solution $y = u(x)$ is known. Then, substitution $y = u + \frac{1}{z}$ will transform the above equation into *a* linear first-order equation in z.

Proof:

By hypothesis, if $u(x)$ is *a* particular solution of the Riccati differential equation

$$\frac{dy}{dx} = P(x)y^2 + Q(x)y + R(x)$$

then

1. $\frac{du}{dx} = P(x)u^2 + Q(x)u + R(x)$

   By using the substitution $y = u + \frac{1}{z}$, we have

2. $\frac{dy}{dx} = \frac{d}{dx}\left(u + \frac{1}{z}\right) = \frac{du}{dx} - \frac{1}{z^2}\frac{dz}{dx}$

Substituting (2) into the Riccati equation, we obtain

$$\frac{du}{dx}-\frac{1}{z^2}\frac{dz}{dx} = P(x)\left(u+\frac{1}{z}\right)^2+Q(x)\left(u+\frac{1}{z}\right)+R(x)$$

$$= P(x)\left(u^2+\frac{2u}{z}+\frac{1}{z^2}\right)+Q(x)\left(u+\frac{1}{z}\right)+R(x)$$

$$= P(x)u^2\frac{2u}{z}P(x)+\frac{1}{z^2}P(x)+Q(x)u+\frac{1}{z}Q(x)+R(x)$$

$$= \left(P(x)u^2+Q(x)u+R(x)\right)$$

$$+\left(\frac{2u}{z}P(x)+\frac{1}{z^2}P(x)+\frac{1}{z}Q(x)\right)$$

Hence by (1), the equation is simplified into

$$-\frac{1}{z^2}\frac{dz}{dx} = \frac{2u}{z}P(x)+\frac{1}{z^2}P(x)+\frac{1}{z}Q(x) \Rightarrow \frac{dz}{dx} = -\ 2uzP(x)-zQ(x)$$

$$\therefore \quad \frac{dz}{dx}-(2uP(x)+Q(x)) = -P(x)$$

which is a linear first-order differential equation in z.

Example: Solve the Riccati's equation:

$$x\frac{dy}{dx}-3y+y^2 = 4x^2-4x$$

Solution: It is obvious that $y = 2x$ is *a* particular solution of the above ODE.

By changing the variable $y = 2x+\frac{1}{z}$, we have $\frac{dy}{dx} = 2-\frac{1}{z^2}\frac{dz}{dx}$. Hence the ODE is reduced to

$$x\left(2-\frac{1}{z^2}\frac{dz}{dx}\right)-3\left(2x+\frac{1}{z}\right)+\left(2x+\frac{1}{z}\right)^2 = 4x^2-4x$$

$$\Rightarrow 2x-\frac{x}{z^2}\frac{dz}{dx}-6x-\frac{3}{z}+4x^2+\frac{4x}{z}+\frac{1}{z^2} = 4x^2-4x$$

$$\Rightarrow -\frac{x}{z^2}\frac{dz}{dx} - \frac{3}{z} + \frac{4x}{z} + \frac{1}{z^2} = 0$$

$$\Rightarrow \frac{dz}{dx} + \left(4 - \frac{3}{x}\right) z = -\frac{1}{x}$$

Therefore, we obtain

$$\frac{d}{dx}(-4x\, e^{4x} z) = -4e^{4x} \Rightarrow -4xe^{4x} z = -e^{4x} + C$$

$$\therefore \quad \frac{1}{4x} - \frac{C}{4}e^{-4x} \text{ and } y = 2x + \frac{1}{\frac{1}{4x} - \frac{C}{4}e^{-4x}}$$

SOLVING SECOND ORDER DIFFERENTIAL EQUATION BY REDUCTION OF ORDER

Let us consider a second-order differential equation in the form

$$\frac{d^2 y}{dx^2} = F(y, y')$$

where $F(y, y')$ is a continuous function. To solve the differential equation of this type, we can at least reduce its order by one

Let $p = \frac{dy}{dx}$, then

$$\frac{d^2 y}{dx^2} = \frac{dp}{dx} = \frac{dp}{dy}\frac{dy}{dx} = p\frac{dp}{dy}$$

Therefore, the original differential equation Equation is reduced to *a* first-order differential equation

$$p\frac{dp}{dy} = F(y, p)$$

The above method is known as the reduction of order.

Example: Solve the differential equations $\frac{d^2 y}{dx^2} = y\frac{dy}{dx}$ by reduction of order:

Solution: Let $p = \frac{dy}{dx}$, we have $\frac{d^2y}{dx^2} = p\frac{dy}{dp}$ and the ODE becomes

$$p\frac{dp}{dy} = yp \Rightarrow dp = ydy \Rightarrow \int dp = \int ydy$$

$$\therefore\ p = \frac{1}{2}y^2 + \frac{C^2}{2} \Rightarrow \frac{dy}{dx} = \frac{1}{2}(y^2 + C^2)$$

$$\therefore\ \int\frac{dy}{C^2 + y^2} = \int\frac{1}{2}dx \Rightarrow \frac{1}{C}\tan^{-1}\left(\frac{y}{C}\right) = \frac{1}{2}x + C$$

$$\therefore\ y = c_1\tan\left(\frac{c_1}{2}x + c_2\right)$$

THEORY OF SOLUTIONS OF LINEAR DIFFERENTIAL EQUATIONS

LINEAR DIFFERENTIAL EQUATION

An nonhomogeneous nth-order linear differential equation has the form

$$a_n(x)y^{(n)} + a_{n-1}(x)y^{(n-1)} + \cdots + a_1(x)y' + a_0(x)y = f(x)$$

where $f(x)$ and the coefficients $a_0(x)$, $a_1(x)$,..., $a_n(x)$ depend solely on the variable x. In other words, they do not depend on y or on any derivative of y. Otherwise, if $f(x) = 0$ the equation becomes

$$a_n(x)y^{(n)} + a_{n-1}(x)y^{(n-1)} + \cdots + a_1(x)y' + a_0(x)y = 0$$

which is homogeneous nth-order differential equation.

Definition: A set of functions $y_1(x)$, $y_3(x)$,..., $y_n(x)$ is said to be linear dependent if there exist c_1, c_2,..., c_n constants, not all z ero, such that

$$c_1y_1(x) + c_2y_2(x) + \cdots + c_ny_n(x) \equiv 0$$

If there is not the case, the functions are said to be linear dependent.

For the homogeneous nth-order linear differential equation, we have the following theorem:

Theorem: The nth-order homogenous differential equation always has n linearly independent solutions.

If $y_1(x)$, $y_3(x)$,..., $y_n(x)$ are n independent solutions of the equation, then $y(x) = c_1y_1(x) + c_2y_2(x) + \cdots + c_ny_n(x)$ is the general solution of the equation Equation.

WRONSKIAN

The Wronskian of *a* set of functions $\{y_1(x), y_2(x), \ldots, y_n(x)\}$, having the property that each functions possesses derivatives, is the determinant

$$W(y_1, y_2, \ldots, y_n) = \begin{vmatrix} y_1 & y_2 & \cdots & y_n \\ y_1' & y_2' & \cdots & y_n' \\ y_1'' & y_2'' & \cdots & y_n'' \\ \vdots & \vdots & \vdots & \vdots \\ y_1^{(n-1)} & y_2^{(n-1)} & \cdots & y_n^{(n-1)} \end{vmatrix}$$

Theorem: If the Wronskian of *a* set of *n* functions $\{y_1(x), y_2(x), \ldots, y_n(x)\}$ is nonzero at *a* certain point x_0, then the set of functions is linear independent at the point x_0.

Proof: Let us assume that there exists *n* real constants $c_1, c_2, \ldots, c_n$ such that

1. $c_1 y_1(x_0) + c_2 y_2(x_0) + \cdots + c_n y_n(x_0) = 0$

   Differentiating (1) successively w.r.t. x for $(n-1)$ times. We have, together with (1), a system of *n* homogeneous equations in $c_1, c_2, \ldots, c_n$:

2. $$\begin{cases} c_1 y_1(x_0) + c_2 y_2(x_0) + \cdots + c_n y_n(x_0) = 0 \\ c_1 y_1'(x_0) + c_2 y_2'(x_0) + \cdots + c_n y_n'(x_0) = 0 \\ c_1 y_1''(x_0) + c_2 y_2''(x_0) + \cdots + c_n y_n''(x_0) = 0 \\ \vdots \\ c_1 y_1^{(n-1)}(x_0) + c_2 y_2^{(n-1)}(x_0) + \cdots + c_n y_n^{(n-1)}(x_0) = 0 \end{cases}$$

By hypothesis, the Wronskian of *a* set of *n* functions $\{y_1(x), y_2(x), \ldots, y_n(x)\}$ is nonzero at a certain point x_0, i.e.

$$W(x_0) = \begin{vmatrix} y_1(x_0) & y_2(x_0) & \cdots & y_n(x_0) \\ y_1'(x_0) & y_2'(x_0) & \cdots & y_n'(x_0) \\ y_1''(x_0) & y_2''(x_0) & \cdots & y_n''(x_0) \\ \vdots & \vdots & \vdots & \vdots \\ y_1^{(n-1)}(x_0) & y_2^{(n-1)}(x_0) & \cdots & y_n^{(n-1)}(x_0) \end{vmatrix} \neq 0$$

Therefore the system of homogenous equations (2) will have *a* trivial solution, i.e $c_1 = c_2 = \ldots = c_n = 0$.

Hence the set of *n* functions $\{y_1(x), y_2(x), \ldots, y_n(x)\}$ is linearly independent at $x = x_0$.

Example: Find the Wronskian of the following functions

1. {sin 3x, cos 3x)
2. x, x^2, x^3)

Solution: 1. The Wronskian is $W(x) = \begin{vmatrix} \sin 3x & \cos 3x \\ 3\cos 3x & -3\sin 3x \end{vmatrix} = -\sin^2 3x - 3\cos^2 3x = -3$.

2. The Wronskian is

$$W(x) = \begin{vmatrix} x & x^2 & x^3 \\ 1 & 2x & 3x^2 \\ 0 & 2 & 6x \end{vmatrix} = 2x^{3.}$$

Example: Show that the functions $\{1 - x, 1 + x, 1 - 3x\}$ are linear dependent for all values of x.

Solution: The Wronskian of the functions $\{1 - x, 1 + x, 1 - 3x\}$ is

$$W(x) = \begin{vmatrix} 1-x & 1+x & 1-3x \\ -1 & 1 & -3 \\ 0 & 0 & 0 \end{vmatrix} = 0 \text{ for all values of } x$$

Therefore the functions $\{1 - x, 1 + x, 1 - 3x\}$ are linear dependent for all values of x.

LINEAR HOMOGENEOUS SECOND-ORDER DIFFERENTIAL EQUATIONS

In particular, for *a* homogeneous second-order linear equation of the form

$$\frac{d^2y}{dx^2} + P(x)\frac{dy}{dx} + Q(x)y = 0$$

If $y_1(x)$ and $y_2(x)$ are two independent solutions of the equation, then

$$y(x) = c_1y_1(x) + c_2y_2(x)$$

is the general solution of the equation Equation. We have the following theorem

Theorem: Suppose $y_1(x)$ and $y_2(x)$ are two independent solutions of the equation Equation, then the coefficient functions $P(x)$ and $Q(x)$ are uniquely determined by $y_1(x)$ and $y_2(x)$.

Proof: By hypothesis, $y_1(x)$ and $y_2(x)$ are two independent solutions of the equation. We have

$$1. \begin{cases} y''_1 + P(x)y'_1(x) + Q(x)y_1(x) = 0 \\ y''_2(x) + P(x)y'_2(x) + Q(x)y_2(x) = 0 \end{cases}$$

The equations can be treated as *a* linear homogeneous equations with unknown $P(x)$ and $Q(x)$. Since $y_1(x)$ and $y_2(x)$ are independent, their Wronskian

2. $$W(x) = \begin{vmatrix} y_1 & y_2 \\ y'_1 & y'_2 \end{vmatrix} = y_1(x)y'_2(x) - y'_1(x)y_2(x)$$

is nonzero. Then the linear equations has unique solutions for $P(x)$ and $Q(x)$ which are given by

3. $$P(x) = -\frac{y_1(x)y''_2(x) - y''_1(x)y_2(x)}{y_1(x)y'_2(x) - y'_1(x)y_2(x)} = -\frac{W'(x)}{W(x)}$$

and

4. $$Q(x) = -\frac{y'_1(x)y''_2(x) - y''_1(x)y'_2(x)}{W(x)}$$

Theorem: Suppose $u(x)$ is *a* non-zero solution of the equation Equation, then

$$v(x) = u(x)\int_{x_0}^{x} \exp\left(-\int_{x_0}^{\xi} P(t)dt\right)\frac{d\xi}{[u(\xi)]^2}$$

is *a* particular solution to the equation.

Proof: By hypothesis, $u(x)$ and $v(x)$ are solutions of the equation. Then we have

1. $u''(x) + P(x) + Q(x)u(x) = 0$

   and

2. $v''(x) + P(x) + Q(x)v(x) = 0$

   From (1) × $v(x)$ – (2) × $u(x)$, we get

   $(uv'' - u''v) + P(x)(uv' - u'v) = 0$

   that is

3. $(uv' - u'v)' + P(x)(uv' - u'v) = 0$

   which is *a* linear first-order equation in $(uv' - u'v)$, and we can obtain easily about its Solution:

$$uv' - u'v = M\exp\left(-\int_{x_0}^{x} P(t)dt\right)$$

4. $$\frac{uv' - u'v}{u^2} = \frac{1}{u^2(x)}\exp\left(-\int_{x_0}^{x} P(t)dt\right)$$

The above equation can be reduced into

$$\left(\frac{v}{u}\right)' = \frac{1}{u^2(x)}\exp\left(-\int_{x_0}^{x} P(t)dt\right)$$

5. $$\frac{v}{u} = \int_{x_0}^{x} \frac{1}{u^2(\xi)}\exp\left(-\int_{x_0}^{x} P(t)dt\right) d\xi$$

Finally we obtain

6. $$v(x) = u(x)\int_{x_0}^{x} \exp\left(-\int_{x_0}^{\xi} P(t)dt\right)\frac{d\xi}{[u(\xi)]^2}$$

Example: Assume that $y(x) = e^x$ is *a* particular solution of the linear differential equation

$$\frac{d^2y}{dx^2} + Q(x)y = 0$$

Find the general solution and the coefficient function $Q(x)$.

Solution: In the above ODE, $P(x) = 0$, from Equation, we have the general solution

$$v(x) = \int_{x_0}^{x} \exp\left(-\int_{x_0}^{\xi} P(t)dt\right)\frac{d\xi}{e^{e\xi}} = e^x \int_{x_0}^{\xi} \exp(-2\xi)d\xi$$

$$= e^x\left(\frac{\exp(-2x) - \exp(-2x_0)}{-2}\right) = -\frac{1}{2}e^{-x} + Ce^x$$

where C is the integration constant.

Substituting the general solution back to the original ODE, we have

$$\frac{d^2}{dx^2}\left(-\frac{1}{2}e^{-x} + Ce^x\right) + Q(x)\left(-\frac{1}{2}e^{-x} + Ce^x\right) = 0$$

$$\Rightarrow \left(-\frac{1}{2}e^{-x} + Ce^x\right) + Q(x)\left(-\frac{1}{2}e^{-x} + Ce^x\right) = 0 \Rightarrow Q(x) = -1$$

Hence the general solution of the equation is

$$y(x) = c_1e^x + c_2e^{-x}$$

SECOND-ORDER HOMOGENEOUS ODE WITH CONSTANT COEFFICIENTS

THE CHARACTERISTIC EQUATION

The second-order differential equation with constant coefficients has the general form

$$\frac{d^2y}{dx^2} + a_1\frac{dy}{dx} + a_0y = f(x)$$

in which a_1 and a_2 are real constants. When $f(x)$ is non-trivial (i.e. $f(x) \neq 0$) the equation is nonhomogeneous second-order differential $f(x) = 0$ equation with constant coefficients. Otherwise, if the equation becomes

$$\frac{d^2y}{dx^2} + a_1\frac{dy}{dx} + a_0y = 0$$

which is homogeneous second-order differential equation with constant coefficients. In this section, we only consider the second-order homogeneous ODE with constant coefficients.

If we assume that is $y(x) = A \exp(\lambda x)$ a particular solutions of the above homogeneous equation, then we can substitute it back to the equation

$$\lambda^2 \exp(\lambda x) + a_1\lambda A\exp(\lambda x) + a_0\lambda A \exp(\lambda x) = 0$$

and it is reduced to an algebraic equation

$$\lambda^2 + a_1\lambda\, a_0 = 0$$

Equation is called the characteristic equation of the second-order ODE.

THE GENERAL SOLUTION

Usually, the characteristic equation Equation has two complex roots λ_1 and λ_2, we have the following cases:

Case 1. λ_1 and λ_2 are both Real and Distinct

In this case, we have the two linear independent solutions $\exp(\lambda_1 x)$ and $\exp(\lambda_2 x)$ and the general solution is

$$y(x) = c_1\exp(\lambda_1 x) + c_2 \exp(\lambda_2 x)$$

In the special case $\lambda_1 = -\lambda_2$, the solution Equation can be rewritten as

$$y(x) = k_1\sinh \lambda_1 x + k_2 \cosh \lambda_2 x$$

Case 2. $\lambda_1 = a + ib$, *a* Complex Number

In this case, since a_1 and a_0 are assumed to be real, the roots of Equation must appear in conjugate pairs; thus the other root is $\lambda_2 = a - ib$. Two linear independent solutions are $\exp[(a + ib)\,x]$ and $\exp[(a - ib)\,x]$ and the general solution is

$$y(x) = c_1 \exp[(a + ib)\,x] + c_2 \exp[(a - ib)\,x]$$

which is algebraic equivalent to

$$y(x) = c_1e^{ax} \cos bx + c_2e^{ax} \sin bx$$

Case (3). $\lambda_1 = \lambda_2$

In this case, we have two linear independent solutions exp $(\lambda_1 x)$ and x exp$(\lambda_1 x)$ and the general solution is

$$y(x) = c_1 \exp(\lambda_1 x) + c_2 \exp(\lambda_2 x)$$

Example: Solve the following differential equations:

1. $\frac{d^2y}{dx^2} + 10\frac{dy}{dx} + 21y = 0$

2. $\frac{d^2y}{dx^2} - 6\frac{dy}{dx} + 25y = 0$

Solution:

1. The characteristic equation is $\lambda^2 + 10\lambda + 21 = 0$

   The roots of the above equation are 3, 7; hence the solution is

$$y = c_1 e^{3x} + c_2 e^{7x}$$

2. The characteristic equation is $\lambda^2 - 6\lambda + 25 = 0$

The roots of the above equation are $3 \pm 4i$; hence the solution is

$$y = e^{3x}(c_1 \cos 4x + c_2 \sin 4x)$$

Chapter 9

Non-homogeneous Differential Equations

It's now time to start thinking about how to solve nonhomogeneous differential equations. A second order, linear nonhomogeneous differential equation is

$$y''+p(t)y'+q(t)y=g(t)$$

where $g(t)$ is a non-zero function. Note that we didn't go with constant coefficients here because everything that we're going to do in this section doesn't require it. Also, we're using a coefficient of 1 on the second derivative just to make some of the work a little easier to write down. It is not required to be a 1.

Before talking about how to solve one of these we need to get some basics out of the way, which is the point of this section. First, we will call

$$y''+p(t)y'+q(t)y=0$$

the associated homogeneous differential.

Now, let's take a look at the following theorem.

Theorem: Suppose that $Y_1(t)$ and $Y_2(t)$ are two solutions and that $y_1(t)$ and $y_2(t)$ are a fundamental set of solutions to the associated homogeneous differential equation then,

$$Y_1(t)-Y_2(t)$$

is a solution and it can be written as

$$Y_1(t)-Y_2(t)=c_1y_1(t)+c_2y_2(t).$$

Note the notation used here. Capital letters referred to solutions to (1) while lower case letters referred to solutions. This is a fairly common convention when dealing with nonhomogeneous differential equations. This theorem is easy enough to prove so let's do that. To prove that $Y_1(t)-Y_2(t)$ is a solution all we need to do is plug this into the differential equation and check it.

$$(Y_1-Y_2)''+p(t)(Y_1-Y_2)'+q(t)(Y_1-Y_2)=0$$

$$Y_1''+p(t)Y_1'+q(t)Y_1-(Y_2''+P(t)Y_2'+q(t)Y_2)=0$$

$$g(t) - g(t) = 0$$

$$0 = 0.$$

We used the fact that $Y_1(t)$ and $Y_2(t)$ are two solutions in the third step. Because they are solutions we know that

$$Y_1'' + p(t)Y_1' + q(t)Y_1 = g(t)$$

$$Y_1'' + p(t)Y_2' + q(t)Y_2 = g(t).$$

So, we were able to prove that the difference of the two solutions.

Proving that

$$Y_1(t) - Y_2(t) = c_1 y_1(t) + c_2 y_2(t)$$

is even easier. Since $y_1(t)$ and $y_2(t)$ are a fundamental set of solutions we know that they form a general solution and so any solution can be written in the form

$$y(t) = c_1 y_1(t) + c_2 y_2(t).$$

Well, $Y_1(t)$ - $Y_2(t)$ is a solution, as we've shown above, therefore it can be written as-

$$Y_1(t) - Y_2(t) = c_1 y_1(t) + c_2 y_2(t).$$

So, what does this theorem do for us? We can use this theorem to write down the form of the general solution. Let's suppose that $y(t)$ is the general solution and that $Y_P(t)$ is any solution that we can get our hands on. Then using the second part of our theorem we know that

$$y(t) - Y_p(t) = c_1 y_1(t) = c_1 y_1(t) + c_2 y_2(t)$$

where $y_1(t)$ and $y_2(t)$ are a fundamental set of solutions. Solving for $y(t)$ gives,

$$y(t) = c_1 y_1(t) + c_2 y_2(t) + Y_P(t).$$

We will call

$$y_c(t) = c_1 y_1(t) + c_2 y_2(t)$$

the complimentary solution and $Y_P(t)$ a particular solution. The general solution to a differential equation can then be written as.

$$y(t) = y_c(t) + Y_P(t).$$

So, to solve a nonhomogeneous differential equation, we will need to solve the homogeneous differential equation, which for constant coefficient differential equations is pretty easy to do, and we'll need a solution.

This seems to be a circular argument. In order to write down a solution we need a solution. However, this isn't the problem that it seems to be. There are ways to find a solution. They just won't, in general, be the general solution. In fact, the next two sections are devoted to exactly that, finding a particular solution to a nonhomogeneous differential equation.

There are two common methods for finding particular solutions: Undetermined Coefficients and Variation of Parameters. Both have their advantages and disadvantages as you will see in the next couple of sections.

UNDETERMINED COEFFICIENTS

In this section we will take a look at the first method that can be used to find a particular solution to a nonhomogeneous differential equation.

$$y''+p(t)y'+q(t)y=g(t).$$

One of the main advantages of this method is that it reduces the problem down to an algebra problem. The algebra can get messy on occasion, but for most of the problems it will not be terribly difficult. Another nice thing about this method is that the complimentary solution will not be explicitly required, although as we will see knowledge of the complimentary solution will be needed in some cases and so we'll generally find that as well.

There are two disadvantages to this method. First, it will only work for a fairly small class of $g(t)$'s. The class of $g(t)$'s for which the method works, does include some of the more common functions, however, there are many functions out there for which undetermined coefficients simply won't work. Second, it is generally only useful for constant coefficient differential equations.

The method is quite simple. All that we need to do is look at $g(t)$ and make a guess as to the form of $Y_p(t)$ leaving the coefficient(s) undetermined (and hence the name of the method). Plug the guess into the differential equation and see if we can determine values of the coefficients. If we can determine values for the coefficients then we guessed correctly, if we can't find values for the coefficients then we guessed incorrectly.

It's usually easier to see this method in action rather than to try and describe it, so let's jump into some examples.

Example: Determine a particular solution to

$$y''-4y'-12y=3e^{5t}.$$

Solution: The point here is to find a particular solution, however the first thing that we're going to do is find the complimentary solution to this differential equation. Recall that the complimentary solution comes from solving,

$$y''-4y'-12y=0.$$

The characteristic equation for this differential equation and its roots are.

$$r^2-4r-12=(r-6)(r+2)=0 \quad \Rightarrow \quad r_1=-2,\ r_2=6.$$

The complimentary solution is then,

$$y_c(t) = c_1 e^{-2t} + c_2 e^{6t}.$$

At this point the reason for doing this first will not be apparent, however we want you in the habit of finding it before we start the work to find a particular solution. Eventually, as we'll see, having the complimentary solution in hand will be helpful and so it's best to be in the habit of finding it first prior to doing the work for undetermined coefficients.

Now, let's proceed with finding a particular solution. As mentioned prior to the start of this example we need to make a guess as to the form of a particular solution to this differential equation. Since *g(t)* is an exponential and we know that exponentials never just appear or disappear in the differentiation process it seems that a likely form of the particular solution would be

$$Y_P(t) = Ae^{5t}.$$

Now, all that we need to do is do a couple of derivatives, plug this into the differential equation and see if we can determine what *A* needs to be. Plugging into the differential equation gives

$$25Ae^{5t} - 4(5Ae^{5t}) - 12(Ae^{5t}) = 3e^{5t}$$

$$-7Ae^{5t} = 3e^{5t}.$$

So, in order for our guess to be a solution we will need to choose *A* so that the coefficients of the exponentials on either side of the equal sign are the same. In other words we need to choose *A* so that,

$$-7A = 3 \Rightarrow A = -\frac{3}{7}.$$

Okay, we found a value for the coefficient. This means that we guessed correctly. A particular solution to the differential equation is then,

$$Y_p(t) = -\frac{3}{7}e^{5t}.$$

Before proceeding any further let's again note that we started off the solution above by finding the complimentary solution. This is not technically part the method of Undetermined Coefficients however, as we'll eventually see, having this in had before we make our guess for the particular solution can save us a lot of work and/or headache. Finding the complimentary solution first is simply a good habit to have so we'll try to get you in the habit over the course of the next few examples. At this point do not worry about why it is a good habit. We'll eventually see why it is a good habit.

Now, back to the work at hand. Notice in the last example that we kept saying "a" particular solution, not "the" particular solution. This is because there are other possibilities out there for the particular solution we've just managed to find one of them. Any of them will work when it comes to writing down the general solution to the differential equation.

Speaking of which... This section is devoted to finding particular solutions and most of the examples will be finding only the particular solution. However, we should do at least one full blown IVP to make sure that we can say that we've done one.

Example: Solve the following IVP

$$y''-4y'-12y=3e^{5t} \quad y(0)=\frac{18}{7} \quad y'(0)=-\frac{1}{7}.$$

Solution: We know that the general solution will be of the form,

$$y(t)=y_c(y)+Y_P(t)$$

and we already have both the complimentary and particular solution from the first example so we don't really need to do any extra work for this problem.

One of the more common mistakes in these problems is to find the complimentary solution and then, because we're probably in the habit of doing it, apply the initial conditions to the complimentary solution to find the constants. This however, is incorrect. The complimentary solution is only the solution to the homogeneous differential equation and we are after a solution to the nonhomogeneous differential equation and the initial conditions must satisfy that solution instead of the complimentary solution.

So, we need the general solution to the nonhomogeneous differential equation. Taking the complimentary solution and the particular solution that we found in the previous example we get the following for a general solution and its derivative.

$$y(t)=c_1e^{-2t}+c_2e^{6t}-\frac{3}{7}e^{5t}$$

$$y'(t)=-2c_1e^{-2t}+6c_2e^{6t}-\frac{15}{7}e^{5t}.$$

Now, apply the initial conditions to these.

$$\frac{18}{7}=y(0)=c_1+c_2-\frac{3}{7}$$

$$-\frac{1}{7}=y'(0)=-2c_1+6c_2-\frac{15}{7}.$$

Solving this system gives $c_1 = 2$ and $c_2 = 1$. The actual solution is then.

$$y(t)=2e^{-2t}+e^{6t}-\frac{3}{7}e^{5t}.$$

This will be the only IVP in this section so don't forget how these are done for nonhomogeneous differential equations!

Let's take a look at another example that will give the second type of *g(t)* for which undetermined coefficients will work.

Example: Find a particular solution for the following differential equation.

$$y''-4y'-12y=\sin(2t).$$

Solution:

1. Again, let's note that we should probably find the complimentary solution before we proceed onto the guess for a particular solution. However, because the homogeneous differential equation for this example is the same as that for the first example we won't bother with that here.
2. Now, let's take our experience from the first example and apply that here. The first example had an exponential function in the *g(t)* and our guess was an exponential. This differential equation has a sine so let's try the following guess for the particular solution.

$$Y_P(t)=A\sin(2t).$$

Differentiating and plugging into the differential equation gives,

$$-4A\sin(2t)-4(2A\cos(2t))-12(A\sin(2t))=\sin(2t).$$

Collecting like terms yields

$$-16A\sin(2t)-8A\cos(2t)=\sin(2t).$$

We need to pick A so that we get the same function on both sides of the equal sign. This means that the coefficients of the sines and cosines must be equal. Or,

$$\cos(2t): \quad -8A=0 \quad \Rightarrow \quad A=0$$

$$\sin(2t): \quad -16A=0 \quad \Rightarrow \quad A=-\frac{1}{16}.$$

Notice two things. First, since there is no cosine on the right hand side this means that the coefficient must be zero on that side.

More importantly we have a serious problem here. In order for the cosine to drop out, as it must in order for the guess to satisfy the differential equation, we need to set $A = 0$, but if $A = 0$, the sine will also drop out and that can't happen. Likewise, choosing A to keep the sine around will also keep the cosine around.

What this means is that our initial guess was wrong. If we get multiple values of the same constant or are unable to find the value of a constant then we have guessed wrong.

One of the nicer aspects of this method is that when we guess wrong our work will often suggest a fix. In this case the problem was the cosine that cropped up. So, to counter this let's add a cosine to our guess. Our new guess is

$$Y_P(t)=A\cos(2t)+B\sin(2t).$$

Plugging this into the differential equation and collecting like terms gives,

$$-4A\cos(2t) - 4B\sin(2t) - 4(-2A\sin(2t)) + 2B\cos(2t)) - 12(A\cos(2t)$$
$$+B\sin(2t)) = \sin(2t)$$
$$(-4A - 8B - 12A)\cos(2t) + (-4B + 8A - 12B)\sin(2t) = \sin(2t)$$
$$(-16A - 8B)\cos(2t) + (8A - 16B)\sin(2t) = \sin(2t).$$

Now, set the coefficients equal

$$\cos(2t): \quad -16\text{A} - 8B = 0$$
$$\sin(2t): \quad 8\text{A} - 16B = 1.$$

Solving this system gives us

$$A = \frac{1}{40} \qquad B = -\frac{1}{40}.$$

We found constants and this time we guessed correctly. A particular solution to the differential equation is then,

$$Y_P(t) = \frac{1}{40}\cos(2t) - \frac{1}{20}\sin(2r).$$

Notice that if we had had a cosine instead of a sine in the last example then our guess would have been the same. In fact, if both a sine and a cosine had shown up we will see that the same guess will also work.

Let's take a look at the third and final type of basic *g(t)* that we can have. There are other types of *g(t)* that we can have, but as we will see they will all come back to two types that we've already done as well as the next one.

Example: Find a particular solution for the following differential equation.

$$y'' - 4y' - 12y = 2t^3 - t + 3.$$

Solution: Once, again we will generally want the complimentary solution in hand first, but again we're working with the same homogeneous differential equation so we'll again just refer to the first example. For this example *g(t)* is a cubic polynomial. For this we will need the following guess for the particular solution.

$$Y_P(t) = At^3 + Bt^3 + Ct + D.$$

Notice that even though *g(t)* doesn't have a t^2 in it our guess will still need one! So, differentiate and plug into the differential equation.

$$6At + 2B - 4(3At^2 + 2Bt + C) - 12(At^3 + Bt^2 + Ct + D) = 2t^3 - t + 3$$
$$-12At^3 + (-12A - 12B)t^2 + (6A - 8B - 12C)t + 2B - 4C - 12D = 2t^3 - t + 3.$$

Now, as we've done in the previous examples we will need the coefficients of the terms on both sides of the equal sign to be the same so set coefficients equal and solve.

$$t^3: \quad -12A = 2 \quad \Rightarrow \quad A = -\frac{1}{6}$$

$$t^2: \quad -12A - 12B = 0 \quad \Rightarrow \quad B = -\frac{1}{6}$$

$$t^1: \quad 6A - 8B - 12C = -1 \quad \Rightarrow \quad C = -\frac{1}{9}$$

$$t^0: \quad 2B - 4C - 12D = 3 \quad \Rightarrow \quad D = -\frac{5}{27}.$$

Notice that in this case it was very easy to solve for the constants. The first equation gave *A*. Then once we knew *A* the second equation gave *B*, *etc*. A particular solution for this differential equation is then

$$Y_P(t) = -\frac{1}{6}t^3 + \frac{1}{6}t^2 - \frac{1}{9}t - \frac{5}{27}.$$

Now that we've gone over the three basic kinds of functions that we can use undetermined coefficients on let's summarize.

g(t)	$Y_P(t)$ guess
$ae^{\beta t}$	$Ae^{\beta t}$
$a\cos(\beta t)$	$A\cos(\beta t) + B\sin(\beta t)$
$b\sin(\beta t)$	$A\cos(\beta t) + B\sin(\beta t)$
$a\cos(\beta t) + b\sin(\beta t)$	$A\cos(\beta t) + B\sin(\beta t)$
n^{th} degree polynomial	$A_n t^n + A_{n-1}t^{n-1} + \cdots A_1 t + A_0.$

Notice that there are really only three kinds of functions given above. If you think about it the single cosine and single sine functions are really special cases of the case where both the sine and cosine are present. Also, we have not yet justified the guess for the case where both a sine and a cosine show up. We will justify this later.

We now need move on to some more complicated functions. The more complicated functions arise by taking products and sums of the basic kinds of functions. Let's first look at products.

Example: Find a particular solution for the following differential equation.

$$y'' - 4y' - 12y = te^{4t}.$$

Solution: You're probably getting tired of the opening comment, but again find the complimentary solution first really a good idea but again we've already done the work in the first example so we won't do it again here. We promise that eventually you'll see why we keep using the same homogeneous problem and why we say it's a good idea to have the complimentary solution in hand first. At this point all we're trying to do is reinforce the habit of finding the complimentary solution first. Okay, let's start off by writing down the guesses for the individual pieces of the function. The guess for the t would be

$$At + B$$

while the guess for the exponential would be

$$Ce^{4t}.$$

Now, since we've got a product of two functions it seems like taking a product of the guesses for the individual pieces might work. Doing this would give

$$Ce^{4t}(At + B).$$

However, we will have problems with this. As we will see, when we plug our guess into the differential equation we will only get two equations out of this. The problem is that with this guess we've got three unknown constants. With only two equations we won't be able to solve for all the constants. This is easy to fix however. Let's notice that we could do the following

$$Ce^{4t}(At + B) = e^{4t}(ACt + BC).$$

If we multiply the C through, we can see that the guess can be written in such a way that there are really only two constants. So, we will use the following for our guess.

$$Y_P(t) = e^{4t}(At + B).$$

Notice that this is nothing more than the guess for the t with an exponential tacked on for good measure. Now that we've got our guess, let's differentiate, plug into the differential equation and collect like terms.

$$e^{4t}(16At + 16B + 8A) - 4(e^{4t}(4At + 4B + A)) - 12(e^{4t}(At + B)) = te^{4t}$$

$$(16A - 16A - 12A)te^{4t} + (16B + 8A - 16B - 4A - 12B)e^{4t} = te^{4t}$$

$$-12Ate^{4t} + (4A - 12B)e^{4t} = te^{4t}.$$

Note that when we're collecting like terms we want the coefficient of each term to have only constants in it. Following this rule we will get two terms when we collect like terms. Now, set coefficients equal.

$$te^{4t}: \quad -12A = 1 \quad \Rightarrow \quad A = -\frac{1}{12}$$

$$e^{4t}: \quad 4A - 12B = 0 \quad \Rightarrow \quad B = -\frac{1}{36}.$$

A particular solution for this differential equation is then

$$Y_P(t) = e^{4t}\left(-\frac{t}{12} - \frac{1}{36}\right) = -\frac{1}{36}(3t+1)\,e^{4t}.$$

This last example illustrated the general rule that we will follow when products involve an exponential. When a product involves an exponential we will first strip out the exponential and write down the guess for tne portion of the function without the exponential, then we will go back and tack on the exponential without any leading coefficient.

Let's take a look at some more products. In the interest of brevity we will just write down the guess for a particular solution and not go through all the details of finding the constants. Also, because we aren't going to give an actual differential equation we can't deal with finding the complimentary solution first.

Example: Write down the form of the particular solution to

$$y'' + p(t)y' + q(t)y = g(t)$$

for the following *g(t)*'s.

1. $g(t) = 16e^{7t}\sin(10t)$
2. $g(t) = (9t^2 - 103t)\cos t$
3. $g(t) = -e^{-2t}(3-5t)\cos(9t).$

Solution:

1. $g(t) = 16e^{-7t}\sin(10t).$

So, we have an exponential in the function. Remember the rule. We will ignore the exponential and write down a guess for 16 sin(10*t*) then put the exponential back in.

The guess for the sine is

$$A\cos(10t) + B\sin(10t).$$

Now, for the actual guess for the particular solution we'll take the above guess and tack an exponential onto it. This gives,

$$Y_P(t) = e^{7t}(A\cos(10t) + B\sin(10t)).$$

One final note before we move onto the next part. The 16 in front of the function has absolutely no bearing on our guess. Any constants multiplying the whole function are ignored.

2. $g(t) = (9t^2 - 103t)\cos t.$

We will start this one the same way that we initially started the previous example. The guess for the polynomial is

$$At^2 + Bt + C$$

and the guess for the cosine is

$$D\cos t + E\sin t.$$

If we multiply the two guesses we get.

$$(At^2 + Bt + C)(D\cos t + E\sin t).$$

Let's simplify things up a little. First multiply the polynomial through as follows.

$$(At^2 + Bt + C)(D\cos t) + (At^2 + Bt + C)(E\sin t)$$

$$(ADt^2 + BDt + CD)\cos t + (AEt^2 + BEt + CE)\sin t.$$

Notice that everywhere one of the unknown constants occurs it is in a product of unknown constants. This means that if we went through and used this as our guess the system of equations that we would need to solve for the unknown constants would have products of the unknowns in them. These types of systems are generally very difficult to solve.

So, to avoid this we will do the same thing that we did in the previous example. Everywhere we see a product of constants we will rename it and call it a single constant. The guess that we'll use for this function will be.

$$Y_P(t) = (At^2 + Bt + C)\cos t + (Dt^2 + Et + F)\sin t.$$

This is a general rule that we will use when faced with a product of a polynomial and a trig function. We write down the guess for the polynomial and then multiply that by a cosine. We then write down the guess for the polynomial again, using different coefficients, and multiply this by a sine.

3. $g(t) = -e^{-2t}(3 - 5t)\cos(9t)$.

This final part has all three parts to it. First we will ignore the exponential and write down a guess for.

$$-(3 - 5t)\cos(9t).$$

The minus sign can also be ignored. The guess for this is

$$(At + B)\cos(9t) + (Ct + D)\sin(9t).$$

Now, tack an exponential back on and we're done.

$$Y_P(t) = e^{-2t}(At + B)\cos(9t) + e^{-2t}(Ct + D)\sin(9t).$$

Notice that we put the exponential on both terms.

There a couple of general rules that you need to remember for products.

If $g(t)$ contains an exponential, ignore it and write down the guess for the remainder. Then tack the exponential back on without any leading coefficient.

1. For products of polynomials and trig functions you first write down the guess for just the polynomial and multiply that by the appropriate cosine. Then add on a new guess for the polynomial with different coefficients and multiply that by the appropriate sine.

If you can remember these two rules you can't go wrong with products. Writing down the guesses for products is usually not that difficult. The difficulty arises when you need to actually find the constants. Now, let's take a look at sums of the basic components and/or products of the basic components. To do this we'll need the following fact.

FACT

If $Y_{P1}(t)$ is a particular solution for

$$y''+p(t)y'+q(t)y=g_1(t)$$

and if $Y_{P2}(t)$ is a particular solution for

$$y''+p(t)y'+q(t)y=g_2(t)$$

then $Y_{P1}(t)+Y_{P2}(t)$ is a particular solution for

$$y''+p(t)y'+q(t)y=g_1(t)+g_2(t).$$

This fact can be used to both find particular solutions to differential equations that have sums in then and to write down guess for functions that have sums in them.

Example: Find a particular solution for the following differential equation.

$$y''-4y'-12y=3e^{5t}+\sin(2t)+te^{4t}.$$

Solution: This example is the reason that we've been using the same homogeneous differential equation for all the previous examples. There is nothing to do with this problem. All that we need to do it go back to the appropriate examples above and get the particular solution from that example and add them all together.

Doing this gives

$$Y_P(t)=-\frac{3}{7}e^{5t}+\frac{1}{40}\cos(2t)-\frac{1}{20}\sin(2t)-\frac{1}{36}(3t+1)\,e^{4t}.$$

Let's take a look at a couple of other examples. As with the products we'll just get guesses here and not worry about actually finding the coefficients.

Example: Write down the form of the particular solution to

$$y''+p(t)y'+q(t)y=g(t)$$

for the following $g(t)$'s.

1. $g(t) = 4\cos(6t) - 9\sin(6t)$
2. $g(t) = -2\cos t + \sin(14t) - 5\sin(14t)$
3. $g(t) = e^{7t} + 6$
4. $g(t) = 6t^2 - 7\sin(3t) + 9$
5. $g(t) = 10e^t - 5te^{-8t} + 2e^{-8t}$
6. $g(t) = t^2 \cos t - 5t\sin t$
7. $g(t) = 5e^{-3t} + e^{-3t}\cos(6t) - \sin(6t)$.

Solution:

1. $g(t) = 4\cos(6t) - 9\sin(6t)$.

This first one we've actually already told you how to do. This is in the table of the basic functions. However we wanted to justify the guess that we put down there. Using the fact on sums of function we would be tempted to write down a guess for the cosine and a guess for the sine. This would give.

$$\underbrace{A\cos(6t) + B\sin(6t)}_{\text{guess for the cosine}} + \underbrace{C\cos(6t) + D\sin(6t)}_{\text{guess for the sine}}.$$

So, we would get a cosine from each guess and a sine from each guess. The problem with this as a guess is that we are only going to get two equations to solve after plugging into the differential equation and yet we have 4 unknowns. We will never be able to solve for each of the constants.

To fix this notice that we can combine some terms as follows.

$$(A + C)\cos(6t) + (B + D)\sin(6t).$$

Upon doing this we can see that we've really got a single cosine with a coefficient and a single sine with a coefficient and so we may as well just use

$$Y_P(t) = A\cos(6t) + B\sin(6t).$$

The general rule of thumb for writing down guesses for functions that involve sums is to always combine like terms into single terms with single coefficients. This will greatly simplify the work required to find the coefficients.

2. $g(t) = -2\sin t + \sin(14t) - 5\cos(14t)$.

For this one we will get two sets of sines and cosines. This will arise because we have two different arguments in them. We will get on set for the sine with just a t as its argument and we'll get another set for the sine and cosine with the $14t$ as their arguments.

The guess for this function is

$$Y_P(t) = A\cos t + B\sin + C\cos(14t) + D\sin(14t)$$

3. $g(t) = e^{7t} + 6.$

The main point of this problem is dealing with the constant. But that isn't too bad. We just wanted to make sure that an example of that is somewhere in the notes. If you recall that a constant is nothing more than a zeroth degree polynomial the guess becomes clear.

The guess for this function is

$$Y_P(t) = Ae^{7t} + B.$$

4. $g(t) = 6t^2 - 7\sin(3t) + 9.$

This one can be a little tricky if you aren't paying attention. Let's first rewrite the function

$$g(t) = 6t^2 - 7\sin(3t) + 9 \text{ as}$$

$$g(t) = 6t^2 + 9 - 7\sin(3t).$$

All we did was move the 9. However upon doing that we see that the function is really a sum of a quadratic polynomial and a sine. The guess for this is then

$$Y_P = At^2 + Bt + C + D\cos(3t) + E\sin(3t).$$

If we don't do this and treat the function as the sum of three terms we would get

$$At^2 + Bt + C + D\cos(3t) + E\sin(3t) + G.$$

and as with the first part in this example we would end up with two terms that are essentially the same (the C and the G) and so would need to be combined. An added step that isn't really necessary if we first rewrite the function.

Look for problems where rearranging the function can simplify the initial guess.

5. $g(t) = 10e^t - 5te^{-8t} + 2e^{-8t}.$

So, this look like we've got a sum of three terms here. Let's write down a guess for that.

$$Ae^t + (Bt + C)e^{-8t} + De^{-8t}.$$

Notice however that if we were to multiply the exponential in the second term through we would end up with two terms that are essentially the same and would need to be combined. This is a case where the guess for one term is completely contained in the guess for a different term. When this happens we just drop the guess that's already included in the other term.

So, the guess here is actually.

$$Y_P(t) = Ae^t + (Bt + C)e^{-8t}.$$

Notice that this arose because we had two terms in our *g(t)* whose only difference was the polynomial that sat in front of them. When this happens we look at the term that contains the largest degree polynomial, write down the guess for that and don't bother writing down the guess for the other term as that guess will be completely contained in the first guess.

6. $g(t) = t^2 \cos t - 5t \sin t.$

In this case we've got two terms whose guess without the polynomials in front of them would be the same. Therefore, we will take the one with the largest degree polynomial in front of it and write down the guess for that one and ignore the other term. So, the guess for the function is

$$Y_P(t) = (Ae^2 + Bt + C)\cos t + (Dt^2 + Et + F)\sin t.$$

7. $g(t) = 5e^{-3t} + e^{-3t} \cos(6t) - \sin(6t).$

This last part is designed to make sure you understand the general rule that we used in the last two parts. This time there really are three terms and we will need a guess for each term. The guess here is

$$Y_P(t) = Ae^{-3t} + e^{-3t}(B\cos(6t) + C\sin(6t)) + D\cos(6t) + E\sin(6t).$$

We can only combine guesses if they are identical up to the constant. So we can't combine the first exponential with the second because the second is really multiplied by a cosine and a sine and so the two exponentials are in fact different functions. Likewise, the last sine and cosine can't be combined with those in the middle term because the sine and cosine in the middle term are in fact multiplied by an exponential and so are different.

So, when dealing with sums of functions make sure that you look for identical guesses that may or may not be contained in other guesses and combine them. This will simplify your work later on.

We have one last topic in this section that needs to be dealt with. In the first few examples we were constantly harping on the usefulness of having the complimentary solution in hand before making the guess for a particular solution. We never gave any reason for this other that "trust us". It is now time to see why having the complimentary solution in hand first it useful. This is best shown with an example so let's jump into one.

Example: Find a particular solution for the following differential equation.

$$y'' - 4y' - 12y = e^{6t}.$$

Solution: This problem seems almost too simple to be given this late in the section.

This is especially true given the ease of finding a particular solution for $g(t)$'s that are just exponential functions. Also, because the point of this example is to illustrate why it is generally a good idea to have the complimentary solution in hand first we'll let's go ahead and recall the complimentary solution first. Here it is,

$$y_c(t) = c_1 e^{-2t} + c_2 e^{6t}.$$

Now, without worrying about the complimentary solution for a couple more seconds let's go ahead and get to work on the particular solution. There is not much to the guess here. From our previous work we know that the guess for the particular solution should be,

$$Y_P(y) = Ae^{6t}.$$

Plugging this into the differential equation gives,

$$36Ae^{6t} - 24Ae^{6t} - 12Ae^{6t} = e^{6t}$$

$$0 = e^{6t}.$$

Clearly an exponential can't be zero. So, what went wrong? We finally need the complimentary solution. Notice that the second term in the complimentary solution (listed above) is exactly our guess for the form of the particular solution and now recall that both portions of the complimentary solution are solutions to the homogeneous differential equation,

$$y'' - 4y' - 12y = 0.$$

In other words, we had better have gotten zero by plugging our guess into the differential equation, it is a solution to the homogeneous differential equation!

So, how do we fix this? The way that we fix this is to add a t to our guess as follows.

$$Y_P(t) = Ate^{6t}.$$

Plugging this into our differential equation gives,

$$(12Ae^{6t} + 36Ate^{6t}) - 4(Ae^{6t} + 6Ate^{6t}) - 12Ate^{6t} = e^{6t}$$

$$(36A - 24A - 12A)te^{6t} + (12A - 4A)e^{6t} = e^{6t}$$

$$8Ae^{6t} = e^{6t}.$$

Now, we can set coefficients equal.

$$8A = 1 \quad \Rightarrow \quad A = \frac{1}{8}.$$

So, the particular solution in this case is,

$$Y_P(t) = \frac{t}{8} e^{6t}.$$

So, what did we learn from this last example. While technically we don't need the complimentary solution to do undetermined coefficients, you can go through a lot of work only to figure out at the end that you needed to add in a t to the guess because it appeared in the complimentary solution. This work is avoidable if we first find the complimentary solution and comparing our guess to the complimentary solution and seeing if any portion of your guess shows up in the complimentary solution.

If a portion of your guess does show up in the complimentary solution then we'll need to modify that portion of the guess by adding in a t to the portion of the guess that is causing the problems. We do need to be a little careful and make sure that we add the t in the correct place however. The following set of examples will show you how to do this.

Example: Write down the guess for the particular solution to the given differential equation. Do not find the coefficients.

1. $y''+3y'-28y=7t+e^{-7t}-1$

4. $y''-100y=9t^2e^{10t}+\cos t-t\sin t$

3. $4y''-100y=e^{-2t}\sin\left(\frac{t}{2}\right)+6t\cos\left(\frac{t}{2}\right)$

4. $4y''+16y'+17y=e^{-2t}\sin\left(\frac{t}{2}\right)+6t\cos\left(\frac{t}{2}\right)$

5. $y''+8y'+16y=e^{-4t}+(t^2+5)e^{-4t}$.

Solution: In these solutions we'll leave the details of checking the complimentary solution to you.

1. $y''+3y'-28y=7t+e^{-7t}-1$.

The complimentary solution is

$$y_c(t)=c_1e^{4t}+c_2e^{-7t}.$$

Remembering to put the "-1" with the $7t$ gives a first guess for the particular solution.-

$$Y_P(t)=At+B+Ce^{-7t}.$$

Notice that the last term in the guess is the last term in the complimentary solution. The first two terms however aren't a problem and don't appear in the complimentary solution. Therefore, we will only add a t onto the last term. The correct guess for the form of the particular solution is.

$$Y_P(t)=At+B+Cte^{-7t}$$

(b) $y''-100y=9t^2e^{10t}+\cos t-t\sin t.$

The complimentary solution is

$$y_c(t)=c_1e^{10t}+c_2e^{-10t}.$$

A first guess for the particular solution is

$$Y_P(t)=(At^2+Bt+C)e^{10t}+(Et+F)\cos t+(Gt+H)\sin t.$$

Notice that if we multiplied the exponential term through the parenthesis that we would end up getting part of the complimentary solution showing up. Since the problem part arises from the first term the *whole* first term will get multiplied by *t*. The second and third terms are okay as they are.

The correct guess for the form of the particular solution in this case is.

$$Y_P(t)=t(At^2+Bt+C)\,e^{10t}+(Et+F)\cos t+(Gt+H)\sin t.$$

So, in general, if you were to multiply out a guess and any part of a that shows up in the complimentary solution, then the whole term will get a *t* not just the problem portion of the term.

3. $4y''+y=e^{-2t}\sin\left(\frac{t}{2}\right)+6t\cos\left(\frac{t}{2}\right).$

The complimentary solution is

$$y_c(t)=c_1\cos\left(\frac{t}{2}\right)+c_2\sin\left(\frac{t}{2}\right).$$

A first guess for the particular solution is

$$Y_P(t)=e^{-2t}\left(A\cos\left(\frac{t}{2}\right)+B\sin\left(\frac{t}{2}\right)\right)+(Cr+D)\cos\left(\frac{t}{2}\right)+(Et+F)\sin\left(\frac{t}{2}\right).$$

In this case both the second and third terms contain portions of the complimentary solution. The first term doesn't however, since upon multiplying out, both the sine and the cosine would have an exponential with them and that isn't part of the complimentary solution. We only need to worry about terms showing up in the complimentary solution if the only difference between the complimentary solution term and the particular guess term is the constant in front of them.

So, in this case the second and third terms will get a *t* while the first won't

The correct guess for the form of the particular solution is.

$$Y_P(t)=e^{-2t}\left(A\cos\left(\frac{t}{2}\right)+B\sin\left(\frac{t}{2}\right)\right)+t(Ct+D)\cos\left(\frac{t}{2}\right)+t(Et+F)\sin\left(\frac{t}{2}\right).$$

4. $4y''+16y'+17y = e^{-2t}\sin\left(\frac{t}{2}\right)+6t\cos\left(\frac{t}{2}\right)$.

To get this problem we changed the differential equation from the last example and left the *g(t)* alone. The complimentary solution this time is

$$y_c(t) = c_1 e^{-2t}\cos\left(\frac{t}{2}\right)+c_2 e^{-2t}\sin\left(\frac{t}{2}\right).$$

As with the last part, a first guess for the particular solution is

$$Y_P(t) = e^{-2t}\left(A\cos\left(\frac{t}{2}\right)+B\sin\left(\frac{t}{2}\right)\right)+t(Ct+D)\cos\left(\frac{t}{2}\right)+(Et+F)\sin\left(\frac{t}{2}\right).$$

This time however it is the first term that causes problems and not the second or third. In fact, the first term is exactly the complimentary solution and so it will need a *t*. Recall that we will only have a problem with a term in our guess if it only differs from the complimentary solution by a constant. The second and third terms in our guess don't have the exponential in them and so they don't differ from the complimentary solution by only a constant.

The correct guess for the form of the particular solution is.

$$Y_P(t) = e^{-2t}\left(A\cos\left(\frac{t}{2}\right)+B\sin\left(\frac{t}{2}\right)\right)+(Ct+D)\cos\left(\frac{t}{2}\right)+(Et+F)\sin\left(\frac{t}{2}\right).$$

5. $y''+8y'+16y = e^{-4t}+(t^2+5)e^{-4t}$.

The complimentary solution is

$$y_c(t) = c_1 e^{-4t}+c_2 t e^{-4t}.$$

The two terms in *g(t)* are identical with the exception of a polynomial in front of them. So this means that we only need to look at the term with the highest degree polynomial in front of it. A first guess for the particular solution is

$$Y_P(t) = (Ar^2+Bt+C)e^{-4t}.$$

Notice that if we multiplied the exponential term through the parenthesis the last two terms would be the complimentary solution. Therefore, we will need to multiply this whole thing by a *t*.

The next guess for the particular solution is then.

$$Y_P(t) = t(Ar^2+Bt+C)e^{-4t}.$$

This still causes problems however. If we multiplied the *t* and the exponential through, the last term will still be in the complimentary solution. In this case, unlike the previous ones, a *t* wasn't sufficient to fix the problem. So, we will add in another *t* to our guess.

The correct guess for the form of the particular solution is.

$$Y_P(t) = t^2(Ar^2 + Bt + C)e^{-4t}.$$

Upon multiplying this out none of the terms are in the complimentary solution and so it will be okay.

As this last set of examples has shown, we really should have the complimentary solution in hand before even writing down the first guess for the particular solution. By doing this we can compare our guess to the complimentary solution and if any of the terms from your particular solution show up we will know that we'll have problems. Once the problem is identified we can add a t to the problem term(s) and compare our new guess to the complimentary solution. If there are no problems we can proceed with the problem, if there are problems add in another t and compare again.

Can you see a general rule as to when a t will be needed and when a t^2 will be needed for second order differential equations?

VARIATION OF PARAMETERS

In the last section we looked at the method of undetermined coefficients for finding a particular solution to

$$p(t)y'' + q(t)y' + r(t)y = g(t)$$

and we saw that while it reduced things down to just an algebra problem, the algebra could become quite messy. On top of that undetermined coefficients will only work for a fairly small class of functions.

The method of Variation of Parameters is a much more general method that can be used in many more cases. However, there are two disadvantages to the method. First, the complimentary solution is absolutely required to do the problem. This is in contrast to the method of undetermined coefficients where it was advisable to have the complimentary solution on hand, but was not required. Second, as we will see, in order to complete the method we will be doing a couple of integrals and there is no guarantee that we will be able to do the integrals. So, while it will always be possible to write down a formula to get the particular solution, we may not be able to actually find it if the integrals are too difficult or if we are unable to find the complimentary solution.

We're going to derive the formula for variation of parameters. We'll start off by acknowledging that the complimentary solution to the equation is

$$y_c(t) = c_1 y_1(t) + c_2 y_2(t).$$

Remember as well that this is the general solution to the homogeneous differential equation.

$$p(t)y'' + q(t)y' + r(t)y = 0.$$

Also recall that in order to write down the complimentary solution we know that $y_1(t)$ and $y_2(t)$ are a fundamental set of solutions. What we're going to do is see if we can find a pair of functions, $u_1(t)$ and $u_2(t)$ so that

$$Y_P(t) = u_1(t)y_1(t) + u_2(t)y_2(t)$$

will be a solution. We have two unknowns here and so we'll need two equations eventually. One equation is easy. Our proposed solution must satisfy the differential equation, so we'll get the first equation by plugging our proposed solution. The second equation can come from a variety of places. We are going to get our second equation simply by making an assumption that will make our work easier. We'll say more about this shortly.

So, let's start. If we're going to plug our proposed solution into the differential equation we're going to need some derivatives so let's get those. The first derivative is

$$Y_P'(t) = u_1'y_1 + u_1y_1' + u_2'y_2 + u_2y_2'.$$

Here's the assumption. Simply to make the first derivative easier to deal with we are going to assume that whatever $u_1(t)$ and $u_2(t)$ are they will satisfy the following.

$$u_1'y_1 + u_2'y_2 = 0.$$

Now, there is no reason ahead of time to believe that this can be done. However, we will see that this will work out. We simply make this assumption on the hope that it won't cause problems down the road and to make the first derivative easier so don't get excited about it. With this assumption the first derivative becomes.

$$Y_P'(t) = u_1y_1' + u_2y_2'.$$

The second derivative is then,

$$Y_p''(t) = u_1'y_1' + u_1y_1' + u_1y_1'' + u_2'y_2' + u_2y_2''.$$

Plug the solution and its derivatives into (1).

$$p(t)(u_1'y_1' + u_1y_1'' + u_2'y_2' + u_2y_1'') + q(t)(u_1y_1' + u_2y_2') + r(t)(u_1y_1 + u_2y_2) = g(t).$$

Rearranging a little gives the following.

$$p(t)(u_1'y_1' + u_2'y_2') + u_1(t)(p(t)y_1'' + q(t)y_1' + r(t)y_1) + u_2(t)(p(t)y_2''$$
$$+r(t)y_2) = g(t).$$

Now, both $y_1(t)$ and $y_2(t)$ are solutions and so the second and third terms are zero. Acknowledging this and rearranging a little gives us,

$$p(t)(u_1'y_1' + u_2'y_2') + u_1(t)(0) + u_2(t)(0) = g(t)$$
$$u_1'y_1' + u_2'y_2' = \frac{g(t)}{p(t)}.$$

We've almost got the two equations that we need. Before proceeding we're going to go back and make a further assumption. The last equation, is actually the one that we want, however, in order to make things simpler for us we are going to assume that the function $p(t) = 1$. In other words, we are going to go back and start working with the differential equation,

$$y''+q(t)y'+r(t)y=g(t).$$

If the coefficient of the second derivative isn't one divide it out so that it becomes a one. The formula that we're going to be getting will assume this! Upon doing this the two equations that we want so solve for the unknown functions are

$$u_1'y_1+u_2'y_2=0$$

$$u_1'y_1'+u_2'y_2'=g(t).$$

Note that in this system we know the two solutions and so the only two unknowns here are u_1' and u_1'. Solving this system is actually quite simple. First, solve for u_1' u_1' u_1' and plug this and do some simplification.

$$u_1'=-\frac{u_2'y_2}{y_1}$$

$$\left(-\frac{u_2'y_2}{y_1}\right)y_1'+u_2'y_2'=g(t)$$

$$u_2'\left(y_2'-\frac{y_2y_1'}{y_1}\right)=g(t)$$

$$u_2'\left(\frac{y_1y_2'-y_2y_1'}{y_1}\right)=g(t)$$

$$u_2'=\frac{y_1g(t)}{y_1y_2'-y_2y_1'}.$$

So, we now have an expression for u_2'. Plugging this will give us an expression for u_1'.

$$u_1'=-\frac{y_2g(t)}{y_1y_2'-y_2y_1'}.$$

Next, let's notice that

$$W(y_1,y_2)=y_1y_2'-y_2y_1'\neq 0.$$

Recall that $y_1(t)$ and $y_2(t)$ are a fundamental set of solutions and so we know that the

Wronskian won't be zero! Finally, all that we need to do is integrate in order to determine what $u_1(t)$ and $u_2(t)$ are. Doing this gives,

$$u_1(t) = -\int \frac{y_2 g(t)}{W(y_1, y_2)} dt \qquad u_2(t) = \int \frac{y_1 g(t)}{W(y_1, y_2)} dt.$$

So, provided we can do these integrals, a particular solution to the differential equation is

$$\begin{aligned} Y_P(t) &= y_1 u_1 + y_2 u_2 \\ &= -y_1 \int \frac{y_2 g(t)}{W(y_1, y_2)} dt + y_2 \int \frac{y_1 g(t)}{W(y_1, y_2)} dt. \end{aligned}$$

So, let's summarize up what we've determined here.

VARIATION OF PARAMETERS

Consider the differential equation,

$$y'' + q(t)y' + r(t)y = g(t).$$

Assume that $y_1(t)$ and $y_2(t)$ are a fundamental set of solutions for

$$y' + q(t)y' + r(t)y = 0.$$

Then a particular solution to the nonhomogeneous differential equation is,

$$Y_P(t) = -y_1 \int \frac{y_2 g(t)}{W(y_1, y_2)} dt + y_2 \int \frac{y_1 g(t)}{W(y_1, y_2)} dt.$$

Depending on the person and the problem, some will find the formula easier to memorize and use, while others will find the process used to get the formula easier. The examples in this section will be done using the formula.

Before proceeding with a couple of examples let's first address the issues involving the constants of integration that will arise out of the integrals. Putting in the constants of integration will give the following.

$$\begin{aligned} Y_P(t) &= -y_1 \left(\int \frac{y_2 g(t)}{W(y_1, y_2)} dt + c \right) + y_2 \left(\int \frac{y_1 g(t)}{W(y_1, y_2)} dt + k \right) \\ &= -y_1 \int \frac{y_2 g(t)}{W(y_1, y_2)} dt + y_2 \int \frac{y_1 g(t)}{W(y_1, y_2)} dt + (-c y_1 + k y_2). \end{aligned}$$

The final quantity in the parenthesis is nothing more than the complimentary solution with $c_1 = -c$ and $c_2 = k$ and we know that if we plug this into the differential equation it will simplify out to zero since it is the solution to the homogeneous differential equation. In other words, these terms add nothing to the particular solution and so we will go ahead and assume that $c = 0$ and $k = 0$ in all the examples. One final note before we proceed with examples. Do not worry about which of your two solutions in the complimentary solution is $y_1(t)$ and which one is $y_2(t)$. It doesn't matter. You will get the same answer no matter which one you choose to be $y_1(t)$ and which one you choose to be $y_2(t)$.

Let's work a couple of examples now.

Example: Find a general solution to the following differential equation.

$$2y'+18y = 6\tan(3t).$$

Solution: First, since the formula for variation of parameters requires a coefficient of a one in front of the second derivative let's take care of that before we forget. The differential equation that we'll actually be solving is

$$y''+9y = 3\tan(3t).$$

We'll leave it to you to verify that the complimentary solution for this differential equation is

$$y_c(t) = c_1\cos(3t) + c_2\sin(3t).$$

So, we have

$$y_1(t) = \cos(3t) \qquad y_2(t) = \sin(3t).$$

The Wronskian of these two functions is

$$W\begin{vmatrix} \cos(3t) & \sin(3t) \\ -3\sin(3t) & 3\cos(3t) \end{vmatrix} = 3\cos^2(3t) + 3\sin^2(3t) = 3.$$

The particular solution is then,

$$\begin{aligned} Y_P(t) &= -\cos(3t)\int\frac{3\sin(3t)\tan(3t)}{3}\,dt + \sin(3t)\int\frac{3\cos(3t)\tan(3t)}{3}\,dt \\ &= -\cos(3t)\int\frac{3\sin^2(3t)}{\cos(3t)}\,dt + \sin(3t)\int\sin(3t)\,dt \\ &= -\cos(3t)\int\frac{1-\cos^2(3t)}{\cos(3t)}\,dt + \sin(3t)\int\sin(3t)\,dt \\ &= -\cos(3t)\int\sec(3t) - \cos(3t)\,dt + \sin(3t)\int\sin(3t)\,dt \\ &= -\frac{\cos(3t)}{3}(\ln|\sec(3t)+\tan(3t)| - \sin(3t)) + \frac{\sin(3t)}{3}(-\cos(3t)) \\ &= -\frac{\cos(3t)}{3}\ln|\sec(3t)+\tan(3t)|. \end{aligned}$$

The general solution is,

$$y(t) = c_1 \cos(3t) + c_2 \sin(3t) - \frac{\cos(3t)}{3} \ln|\sec(3t) + \tan(3t)|.$$

Example: Find a general solution to the following differential equation.

$$y'' - 2y' + y = \frac{e^t}{t^2 + 1}.$$

Solution: We first need the complimentary solution for this differential equation. We'll leave it to you to verify that the complimentary solution is,

$$y_c(t) = c_1 e^t + c_2 t e^t.$$

So, we have

$$y_1(t) = e^t \qquad y_2(t) = te^t.$$

The Wronskian of these two functions is

$$W = \begin{vmatrix} e^t & te^t \\ e^t & e^t + te^t \end{vmatrix} = e^t(e^t + te^t) - e^t(te^t) = e^{2t}.$$

The particular solution is then,

$$Y_P = -e^t \int \frac{te^t e^t}{e^{2t}(t^2+1)} dt + te^t \int \frac{e^t e^t}{e^{2t}(t^2+1)} dt$$

$$= e^t \int \frac{t}{e^2 + 1} dt + te^t \int \frac{1}{t^2+1} dt$$

$$= \frac{1}{2} e^t \ln(1 + t^2) + te^t \tan^{-1}(t).$$

The general solution is,

$$y(t) = c_1 e^t + c_2 te^t - \frac{1}{2} e^t \ln(1 + t^2) + te^t \tan^{-1}(t).$$

This method can also be used on non-constant coefficient differential equations, provided we know a fundamental set of solutions for the associated homogeneous differential equation.

Example: Find the general solution to

$$ty'' - (t+1)y'' + y = t^2$$

given that

$$y_1(t) = e^t \qquad y_2(t) = t + 1$$

form a fundamental set of solutions for the homogeneous differential equation.

Solution: As with the first example, we first need to divide out by a *t*.

$$y''-\left(1+\frac{1}{t}\right)y'+\frac{1}{t}y=t.$$

The Wronskian for the fundamental set of solutions is

$$W=\begin{vmatrix} e^t & t+1 \\ e^t & 1 \end{vmatrix}=e^t-e^t(t+1)=-te^t.$$

The particular solution is.

$$\begin{aligned} Y_P &= -e^t\int\frac{(t+1)t}{-te^t}dt+(t+1)\int\frac{e^t(t)}{-te^t}dt \\ &= e^t\int(t+1)e^{-t}dt-(t+1)\int dt \\ &= e^t\left(-e^{-t}(t+2)\right)-(t+1)t \\ &= -t^2-2t-2. \end{aligned}$$

The general solution for this differential equation is.

$$y(t)=c_1e^t+c_2(t+1)-t^2-2t-2.$$

We need to address one more topic about the solution to the previous example. The solution can be simplified down somewhat if we do the following.

$$\begin{aligned} y(t) &= c_1e^t+c_2(t+1)-t^2-2t-2 \\ &= c_1e^t+c_2(t+1)-t^2-2(t+1) \\ &= c_1e^t+(c_2+2)(t+1)-t^2. \end{aligned}$$

Now, since is an unknown constant subtracting 2 from it won't change that fact. So we can just write the c_2-2as $c_2$and be done with it. Here is a simplified version of the solution for this example.

$$y(t)=c_1e^t+c_2(t+1)-t^2.$$

MECHANICAL VIBRATIONS

It's now time to take a look at an application of second order differential equations. We're going to take a look at mechanical vibrations. In particular we are going to look at a mass that is hanging from a spring.

Vibrations can occur in pretty much all branches of engineering and so what we're going to be doing here can be easily adapted to other situations, usually with just a change in notation.

Let's get the situation setup. We are going to start with a spring of length l, called the natural length, and we're going to hook an object with mass m up to it. When the object is attached to the spring the spring will stretch a length of L. We will call the equilibrium position the position of the center of gravity for the object as it hangs on the spring with no movement.

Below is sketch of the spring with and without the object attached to it.

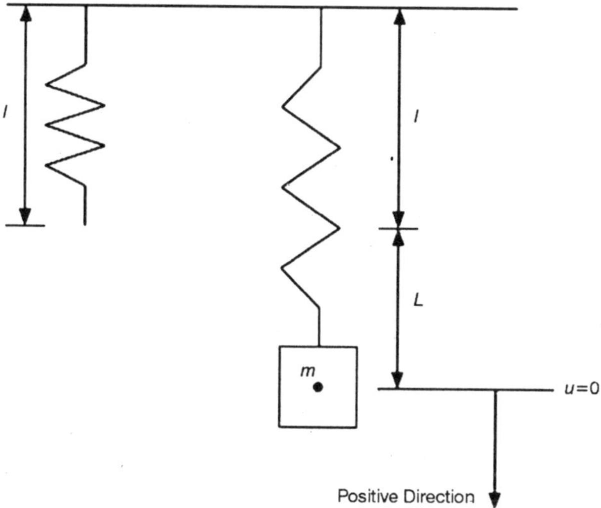

As denoted in the sketch we are going to assume that all forces, velocities, and displacements in the downward direction will be positive. All forces, velocities, and displacements in the upward direction will be negative.

Also, as shown in the sketch above, we will measure all displacement of the mass from its equilibrium position. Therefore, the $u = 0$ position will correspond to the center of gravity for the mass as it hangs on the spring and is at rest (*i.e.* no movement).

Now, we need to develop a differential equation that will give the displacement of the object at any time t. First, recall Newton's Second Law of Motion.

$$ma = F.$$

In this case we will use the second derivative of the displacement, u, for the acceleration and so Newton's Second Law becomes,

$$mu'' = F(t, u, u').$$

We now need to determine all the forces that will act upon the object. There are four forces that we will assume act upon the object. Two that will always act on the object and two that may or may not act upon the object.

Here is a list of the forces that will act upon the object.

Grasvity, F_g

The force due to gravity will always act upon the object of course. This force is

$$F_g = mg.$$

Spring, F_s

We are going to assume that Hooke's Law will govern the force that the spring exerts on the object. This force will always be present as well and is

$$F_s = -k(L+u).$$

Hooke's Law tells us that the force exerted by a spring will be the spring constant, $k > 0$, times the displacement of the spring from its natural length. For our set up the displacement from the springs natural length is $L + u$ and the minus sign is in there to make sure that the force always has the correct direction.

Let's make sure that this force does what we expect it to. If the object is at rest in its equilibrium position the displacement is L and the force is simply $F_s = kL$ which will act in the upward position as it should since the spring has been stretched from its natural length.

If the spring has been stretched further down from the equilibrium position then will be positive and F_s will be negative acting to pull the object back up as it should be.

Next, if the object has been moved up past it's equilibrium point, but not yet to it's natural length then u will be negative, but still less than L and so $L + u$ will be positive and once again F_s will be negative acting to pull the object up.

Finally, if the object has been moved upwards so that the spring is now compressed, then u will be negative and greater than L. Therefore, $L + u$ will be negative and now F_s will be positive acting to push the object down.

So, it looks like this force will act as we expect that it should.

Dampin'g, F_d

The next force that we need to consider is damping. This force may or may not be present for any given problem

Dampers work to counteract any movement. There are several ways to define a damping force. The one that we'll use is the following.

$$F_d = -\gamma u'$$

where, *ã> 0* is the damping coefficient. Let's think for a minute about how this force will act. If the object is moving downward, then the velocity (u' $u'u'$) will be positive and so F_d will be negative and acting to pull the object back up. Likewise, if the object is moving upward, the velocity (u' $u'u'$) will be negative and so F_d will be positive and acting to push the object back down.

In other words, the damping force as we've defined it will always act to counter the current motion of the object and so will act to damp out any motion in the object.

External Forces, *F(t)*

This is the catch all force. If there are any other forces that we decide we want to act on our object we lump them in here and call it good. We typically call *F(t)* the forcing function. Putting all of these together gives us the following for Newton's Second Law.

$$mu'' = mg - k(L+u) - \gamma u' + F(t).$$

Or, upon rewriting, we get,

$$mu'' + \gamma u' + ku = mg - kL + F(t).$$

Now, when the object is at rest in its equilibrium position there are exactly two forces acting on the object, the force due to gravity and the force due to the spring. Also, since the object is at rest (*i.e.* not moving) these two forces must be canceling each other out. This means that we must have,

$$mg = kL.$$

Using this in Newton's Second Law gives us the final version of the differential equation that we'll work with.

$$mu'' + \gamma u' + ku = F(t).$$

Along with this differential equation we will have the following initial conditions.

$u'(0) = u_0$ Initial displacement from the equilibrium position

$u'(0) = u_0'$ Initial velocity.

Note that we'll determine the spring constant, *k*.

FREE, UNDAMPED VIBRATIONS

This is the simplest case that we can consider. Free or unforced vibrations means that *F(t)* = *0* and undamped vibrations means that *ã*= *0*. In this case the differential equation becomes,

$$mu'' + ku = 0.$$

This is easy enough to solve in general. The characteristic equation has the roots,

$$r = \pm i\sqrt{\frac{k}{m}}.$$

This is usually reduced to,

$$r = \pm\, \omega_0\, i$$

where,

$$\omega_0 = \sqrt{\frac{k}{m}}.$$

and ù0 is called the natural frequency. Recall as well that $m > 0$ and $k > 0$ and so we can guarantee that this quantity will be complex. The solution in this case is then

$$u(t) = c_1 \cos(\omega_0 t) + c_2 \sin(\omega_0 t).$$

We can write (4) in the following form,

$$u(t) = R\cos(\omega_0 t - \delta)$$

where R is the amplitude of the displacement and äis the phase shift or phase angle of the displacement.

So, assuming that we have c_1 and c_2 how do we determine R and ä? Let's start with equation and use a trig identity to write it as

$$u(t) = R\cos(\delta)\cos(\omega_0 t) + R\sin(\delta)\sin(\omega_0 t).$$

Now, R and äare constants and so if we compare we can see that

$$c_1 = R\cos\delta \qquad c_2 = R\cos\delta.$$

We can find R in the following way.

$$c_1^2 + c_2^2 = R^2\cos^2\delta + R^2\sin^2\delta = R^2.$$

Taking the square root of both sides and assuming that R is positive will give

$$R = \sqrt{c_1^2 + c_2^2}.$$

Finding äis just as easy. We'll start with

$$\frac{c_2}{c_1} = \frac{R\sin\delta}{R\cos\delta} = \tan\delta.$$

Taking the inverse tangent of both sides gives,

$$\delta = \tan^{-1}\left(\frac{c_2}{c_1}\right).$$

Before we work any examples let's talk a little bit about units of mass and the British vs. metric system differences.

Recall that the weight of the object is given by

$$W = mg$$

where m is the mass of the object and g is the gravitational acceleration. For the examples in this problem we'll be using the following values for g.

$$\text{British: } g = 32 \text{ ft/s}^2$$

$$\text{Metric: } g = 9.8 \text{ m/s}^2$$

This is not the standard 32.2 ft/s^2 or 9.81 m/s^2, but using these will make some of the numbers come out a little nicer.

In the metric system the mass of objects is given in kilograms (kg) and there is nothing for us to do. However, in the British system we tend to be given the weight of an object in pounds (yes, pounds are the units of weight not mass...) and so we'll need to compute the mass for these problems. At this point we should probably work an example of all this to see how this stuff works.

Example: A 16 lb object stretches a spring $\frac{8}{9}$ ft by itself. There is no damping and no external forces acting on the system. The spring is initially displaced 6 inches upwards from its equilibrium position and given an initial velocity of 1 ft/sec downward. Find the displacement at any time t, $u(t)$.

Solution: We first need to set up the IVP for the problem. This requires us to get our hands on m and k.

This is the British system so we'll need to compute the mass.

$$m = \frac{W}{g} = \frac{16}{32} = \frac{1}{2}.$$

Now, let's get k. We can use the fact that $mg = kL$ to find k. Don't forget that we'll need all of our length units the same. We'll use feet for the unit of measurement for this problem.

$$k = \frac{mg}{L} = \frac{16}{8/9} = 18.$$

We can now set up the IVP.

$$\frac{1}{2}u'' + 18u = 0 \qquad u(0) = -\frac{1}{2} \qquad u'(0) = 1.$$

For the initial conditions recall that upward displacement/motion is negative while downward displacement/motion is positive. Also, since we decided to do everything in feet we had to convert the initial displacement to feet.

Now, to solve this we can either go through the characteristic equation or we can just jump straight to the formula that we derived above. We'll do it that way. First, we need the natural frequency,

$$\omega_0 = \sqrt{\frac{18}{1/2}} = \sqrt{36} = 6.$$

The general solution, along with its derivative, is then,

$$u(t) = c_1 \cos(6t) + c_2 \sin(6t)$$

$$u'(t) = -c_1 6c_1 \sin(6t) + 6c_2 \cos(6t).$$

Applying the initial conditions gives

$$-\frac{1}{2} = u(0) = c_1 \qquad c_1 = -\frac{1}{2}$$

$$1 = u'(0) = 6c_2 \cos(6t) \quad c_1 = -\frac{1}{6}.$$

The displacement at any time t is then

$$u(t) = -\frac{1}{2}\cos(6t) + \frac{1}{6}\sin(6t).$$

Now, let's convert this to a single cosine. First let's get the amplitude, R.

$$R = \sqrt{\left(-\frac{1}{2}\right)^2 + \left(\frac{1}{6}\right)^2} = \frac{\sqrt{10}}{6} = 0.52705.$$

You can use either the exact value here or a decimal approximation. Often the decimal approximation will be easier. Now let's get the phase shift.

$$\delta = \tan^{-1}\left(\frac{1/6}{-1/2}\right) = -0.32175.$$

We need to be careful with this part. The phase angle found above is in Quadrant IV, but there is also an angle in Quadrant II that would work as well. We get this second angleπ ππby adding onto the first angle. So, we actually have two angles. They are

$$\delta_1 = -0.32175$$

$$\delta_2 = \delta_1 + \pi = 2.81984.$$

We need to decide which of these phase shifts is correct, because only one will be correct. To do this recall that

$$c_1 = R\cos\delta$$

$$c_2 = R\sin\delta.$$

Now, since we are assuming that R is positive this means that the sign of cosä will be the same as the sign of c_1 and the sign of sinä will be the same as the sign of c_2. So, for this particular case we must have cos*ä* < *0* and sin*ä* > *0*. This means that the phase shift must be in Quadrant II and so the second angle is the one that we need.

So, after all of this the displacement at any time t is.

$$u(t) = 0.52705\cos(6t - 2.81984).$$

Here is a sketch of the displacement for the first 5 seconds.

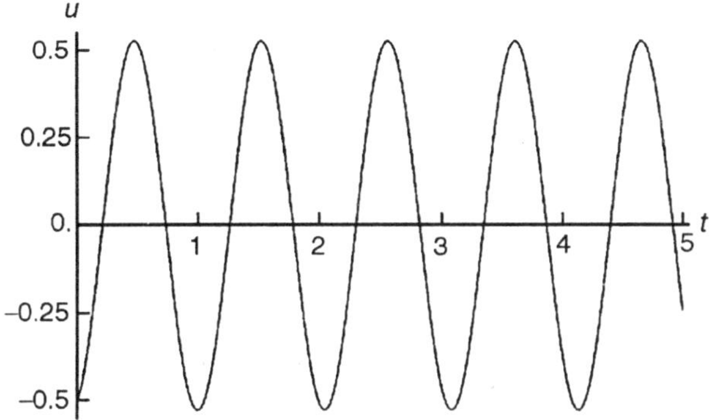

Now, let's take a look at a slightly more realistic situation. No vibration will go on forever. So let's add in a damper and see what happens now.

FREE, DAMPED VIBRATIONS

He are still going to assume that there will be no external forces acting on the system, with the exception of damping of course. In this case the differential equation will be.

$$mu'' + \gamma u' + ku = 0$$

where m, ä, and k are all positive constants. Upon solving for the roots of the characteristic equation we get the following.

$$r_{1,2} = \frac{-\gamma \pm \sqrt{\gamma^2 - 4mk}}{2m}.$$

We will have three cases here.

1. $\gamma^2 - 4mk = 0.$

In this case we will get a double root out of the characteristic equation and the displacement at any time t will be.

$$u(t) = c_1 e^{\frac{\gamma t}{2m}} + c_1 t e^{\frac{\gamma t}{2m}}.$$

Notice that as $t \to \infty$the displacement will approach zero and so the damping in this

This case is called critical damping and will happen when the damping coefficient is,

$$.\gamma^2 - 4mk = 0$$

$$\gamma^2 = 4mk$$

$$\gamma = 2\sqrt{mk} = \gamma_{CR}.$$

The value of the damping coefficient that gives critical damping is called the critical damping coefficient and denoted by $ã_{CR}$.

2. $\gamma^2 - 4mk > 0$.

In this case let's rewrite the roots a little.

$$r_{1,2} = \frac{-\gamma \pm \sqrt{\gamma^2 - 4mk}}{2m}$$

$$= \frac{-\gamma \pm \gamma\sqrt{1 - \frac{4mk}{\gamma^2}}}{2m}$$

$$= -\frac{\gamma}{2m}\left(1 \pm \sqrt{1 - \frac{4mk}{\gamma^2}}\right).$$

Also notice that from our initial assumption that we have,

$$\gamma^2 > 4mk$$

$$1 > \frac{4mk}{\gamma^2}.$$

Using this we can see that the fraction under the square root above is less than one. Then if the quantity under the square root is less than one, this means that the square root of this quantity is also going to be less than one. In other words,

$$\sqrt{1 - \frac{4mk}{\gamma^2}} < 1.$$

Why is this important? Well, the quantity in the parenthesis is now one plus/minus a number that is less than one. This means that the quantity in the parenthesis is guaranteed to be positive and so the two roots in this case are guaranteed to be negative. Therefore the displacement at any time t is,

$$u(t) = c_1 e^{k_1 t} + c_2 e^{r_2 t}$$

and will approach zero as $t \to \infty$. So, once again the damper does what it is supposed to do.

This case will occur when

$$\gamma^2 > 4mk$$

$$\gamma > 2\sqrt{mk}$$

$$\gamma > \gamma_{CR}$$

and is called over damping.

3. $\gamma^2 - 4mk < 0.$

In this case we will get complex roots out of the characteristic equation.

$$r_{1,2} = \frac{-\gamma}{2m} \pm \frac{\sqrt{\gamma^2 - 4mk}}{2m} = \lambda \pm \mu i$$

where the real part is guaranteed to be negative and so the displacement is

$$\begin{aligned} u(t) &= c_1 e^{\lambda t} \cos(\mu t) + c_2 e^{\lambda t} \sin(\mu t) \\ &= e^{\lambda t}(c_1 \cos(\mu t) + c_2 \sin(\mu t)) \\ &= Re^{\lambda t} \cos(\mu t - \delta). \end{aligned}$$

Notice that we reduced the sine and cosine down to a single cosine in this case as we did in the undamped case. Also, since *ë< 0* the displacement will approach zero as $t \rightarrow \infty$and the damper will also work as it's supposed to in this case.

We will get this case will occur when

$$\gamma^2 > 4mk$$

$$\gamma > 2\sqrt{mk}$$

$$\gamma > \gamma_{CR}$$

and is called under damping.

Let's take a look at a couple of examples here with damping.

Example: Take the spring and mass system from the first example and attach a damper to it that will exert a force of 12 lbs when the velocity is 2 ft/s. Find the displacement at any time *t*, *u(t)*.

Solution: The mass and spring constant were already found in the first example so we won't do the work here. We do need to find the damping coefficient however. To do this we will use the formula for the damping force given above with one modification. The original damping force formula is,

$$F_d = -\gamma u'.$$

However, remember that the force and the velocity are always acting in opposite directions. So, if the velocity is upward (*i.e.* negative) the force will be downward (*i.e.* positive) and so the minus in the formula will cancel against the minus in the velocity. Likewise, if the velocity is downward (*i.e.* positive) the force will be upwards (*i.e.* negative) and in this case the minus sign in the formula will cancel against the minus in the force. In other words, we can drop the minus sign in the formula and use

$$F_d = -\gamma u'$$

and then just ignore any signs for the force and velocity.

Doing this gives us the following for the damping coefficient

$$12 = \gamma(2) \quad \Rightarrow \quad \gamma = 6.$$

The IVP for this example is then,

$$\frac{1}{2}u'' + 6u' + 18u = 0 \qquad u(0) = -\frac{1}{2} \qquad u'(0) = 1.$$

Before solving let's check to see what kind of damping we've got. To do this all we need is the critical damping coefficient.

$$\gamma_{CR} = 2\sqrt{km} = 2\sqrt{(18)\left(\frac{1}{2}\right)} = 2\sqrt{9} = 6.$$

So, it looks like we've got critical damping. Note that this means that when we go to solve the differential equation we should get a double root.

Speaking of solving, let's do that. I'll leave the details to you to check that the displacement at any time t is.

$$u(t) = -\frac{1}{2}e^{-6t} - 2te^{-6t}.$$

Here is a sketch of the displacement during the first 3 seconds.

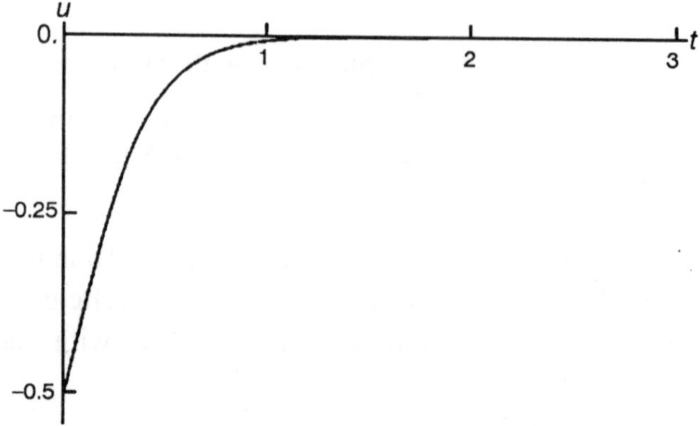

Notice that the "vibration" in the system is not really a true vibration as we tend to think of them. In the critical damping case there isn't going to be a real oscillation about the equilibrium point that we tend to associate with vibrations. The damping in this system is strong enough to force the "vibration" to die out before it ever really gets a chance to do much in the way of oscillation.

Example: Take the spring and mass system from the first example and this time let's attach a damper to it that will exert a force of 17 lbs when the velocity is 2 ft/s. Find the displacement at any time *t*, *u(t)*.

Solution: So, the only difference between this example and the previous example is damping force. So let's find the damping coefficient

$$17 = \gamma(2) \quad \Rightarrow \quad \gamma = \frac{17}{2} = 8.5 > \gamma_{CR}.$$

So it looks like we've got over damping this time around so we should expect to get two real distinct roots from the characteristic equation and they should both be negative. The IVP for this example is,

$$\frac{1}{2}u'' + \frac{17}{2}u' + 18u = 0 \qquad \text{u(0)=} -\frac{1}{2} \qquad u'(0) = 1.$$

This one's a little messier than the previous example so we'll do a couple of the steps, leaving it to you to fill in the blanks. The roots of the characteristic equation are

$$r_{1,2} = \frac{-17 \pm \sqrt{145}}{2} = -2.472, -14,4208.$$

In this case it will be easier to just convert to decimals and go that route. Note that, as predicted we got two real, distinct and negative roots. The general and actual solution for this example are then,

$$u(t) = c_1 e^{-2.4792t} + c_2 e^{-145208t}$$

$$u(t) = -0.5198e^{-2.4792t} + 0.0199e^{-145208t}.$$

Here's a sketch of the displacement for this example.

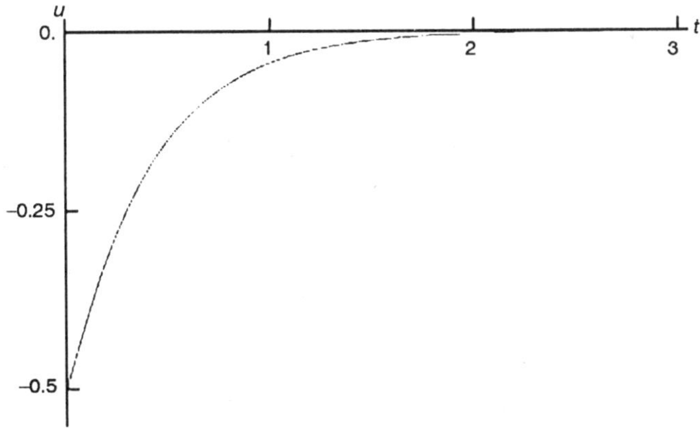

Notice an interesting thing here about the displacement here. Even though we are "over" damped in this case, it actually takes longer for the vibration to die out than in the critical damping case. Sometimes this happens, although it will not always be the case that over damping will allow the vibration to continue longer than the critical damping case.

Also notice that, as with the critical damping case, we don't get a vibration in the sense that we usually think of them. Again, the damping is strong enough to force the vibration do die out quick enough so that we don't see much, if any, of the oscillation that we typically associate with vibrations.

Let's take a look at one more example before moving on the next type of vibrations.

Example: Take the spring and mass system from the first example and for this example let's attach a damper to it that will exert a force of 5 lbs when the velocity is 2 ft/s. Find the displacement at any time t, $u(t)$.

Solution: So, let's get the damping coefficient.

$$s = \gamma(2) \quad \Rightarrow \quad \gamma = \frac{5}{2} = 2.5 < \gamma_{CR}.$$

So it's under damping this time. That shouldn't be too surprising given the first two examples. The IVP for this example is,

$$\frac{1}{2}u'' + \frac{5}{2}u' + 18u = 0 \qquad u(0) = -\frac{1}{2} \quad u'(0) = 1.$$

In this case the roots of the characteristic equation are

$$r_{1,2} = \frac{-5 \pm \sqrt{119}i}{2}.$$

They are complex as we expected to get since we are in the under damped case. The general solution and actual solution are

$$u(t) = e^{-\frac{5t}{2}}\left(c_1 \cos\left(\frac{\sqrt{119}}{2}t\right) + c_2 \sin\left(\frac{\sqrt{119}}{2}t\right)\right)$$

$$u(t) = e^{-\frac{5t}{2}}\left(-0.5\cos\left(\frac{\sqrt{119}}{2}t\right) - 0.04583\sin\left(\frac{\sqrt{119}}{2}t\right)\right).$$

Let's convert this to a single cosine as we did in the undamped case.

$$R = \sqrt{(-0.5)^2 + (-0.04583)^2} = 0.502096$$

$$\delta_1 = \tan^{-1}\left(\frac{-0.04583}{-0.5}\right) = 0.09051 \qquad \text{OR} \quad \delta_2 = \delta_1 + \pi = 3.2321.$$

As with the undamped case we can use the coefficients of the cosine and the sine to determine which phase shift that we should use. The coefficient of the cosine (c_1) is negative and so cosä must also be negative. Likewise, the coefficient of the sine (c_2) is also negative and so sinä must also be negative. This means that ämust be in the Quadrant III and so the second angle is the one that we want.

The displacement is then

$$u(t) = 0.502096e^{-\frac{5t}{2}}\cos\left(\frac{\sqrt{119}}{2}t - 3.2321\right).$$

Here is a sketch of this displacement.

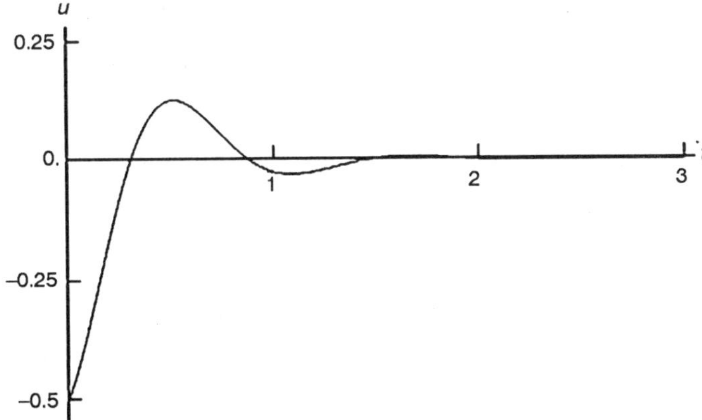

In this case we finally got want we usually consider to be a true vibration. In fact that is the point of critical damping. As we increase the damping coefficient, the critical damping coefficient will be the first one in which a true oscillation in the displacement will not occur. For all values of the damping coefficient larger than this (*i.e.* over damping) we will also not see a true oscillation in the displacement.

From a physical standpoint critical (and over) damping is usually preferred to under damping. Think of the shock absorbers in your car. When you hit a bump you don't want to spend the next few minutes bouncing up and down while the vibration set up by the bump die out. You would like there to be as little movement as possible. In other words, you will want to set up the shock absorbers in your car so get at the least critical damping so that you can avoid the oscillations that will arise from an under damped case.

It's now time to look at systems in which we allow other external forces to act on the object in the system.

UNDAMPED, FORCED VIBRATIONS

We will first take a look at the undamped case. The differential equation in this case is

$$mu'' + ku = F(t).$$

This is just a nonhomogeneous differential equation and we know how to solve these. The general solution will be

$$u(t) = u_c(t) + U_P(t)$$

where the complimentary solution is the solution to the free, undamped vibration case. To get the particular solution we can use either undetermined coefficients or variation of parameters depending on which we find easier for a given forcing function.

There is a particular type of forcing function that we should take a look at since it leads to some interesting results. Let's suppose that the forcing function is a simple periodic function of the form

$$F(t) = F_0 \cos(\omega t) \quad \text{OR} \quad F(t) = F_0 \sin(\omega t).$$

For the purposes of this discussion we'll use the first one. Using this, the IVP becomes,

$$mu'' + ku = F_0 \cos(\omega t).$$

The complimentary solution, as pointed out above, is just

$$u_c(t) = c_1 \cos(\omega_0 t) + c_2 \sin(\omega_0 t)$$

where ω_0 is the natural frequency.

We will need to be careful in finding a particular solution. The reason for this will be clear if we use undetermined coefficients. With undetermined coefficients our guess for the form of the particular solution would be,

$$U_P(t) = A\cos(\omega t) + B\sin(\omega t).$$

Now, this guess will be problems if $\omega_0 = \omega$. If this were to happen the guess for the particular solution is exactly the complimentary solution and so we'd need to add in a t. Of course if we don't have $\omega_0 = \omega$then there will be nothing wrong with the guess.

So, we will need to look at this in two cases.

2. $\omega_0 \neq \omega$

In this case our initial guess is okay since it won't be the complimentary solution. Upon differentiating the guess and plugging it into the differential equation and simplifying we get,

$$(-m\omega^2 A + kA)\cos(wt) + (-m\omega^2 B + kB)\sin(wt) = F_0 \cos(wt).$$

Setting coefficients equal gives us,

$$\cos(\omega t): \quad (-m\omega^2 + k)A = F_0 \quad \Rightarrow \quad A = \frac{F_0}{k - m\omega^2}$$

$$\sin(\omega t): \quad (-m\omega^2 + k)B = F_0 \quad \Rightarrow \quad B = 0$$

The particular solution is then

$$U_P(t) = \frac{F_0}{k - m\omega^2}\cos(\omega t)$$

$$= \frac{F_0}{m\left(\frac{k}{m} - \omega^2\right)}\cos(\omega t)$$

$$= \frac{F_0}{m\left(\omega_0^2 - \omega^2\right)}\cos(\omega t).$$

Note that we rearranged things a little. Depending on the form that you'd like the displacement to be in we can have either of the following.

$$u(t) = c_1\cos(\omega_0 t) + c_2\sin(\omega_0 t) + \frac{F_0}{m\left(\omega_0^2 - \omega^2\right)}\cos(\omega t)$$

$$u(t) = R\cos(\omega_0 t - \delta) + \frac{F_0}{m\left(\omega_0^2 - \omega^2\right)}\cos(\omega t).$$

If we used the sine form of the forcing function we could get a similar formula.

1. $\omega_0 = \omega$.

In this case we will need to add in a t to the guess for the particular solution.

$$U_P(t) = At\cos(\omega_0 t) + Bt\sin(\omega_0 t).$$

Note that we went ahead and acknowledge that $\omega_0 = \omega$in our guess. Acknowledging this will help with some simplification that we'll need to do later on. Differentiating our guess, plugging it into the differential equation and simplifying gives us the following.

$$(-m\omega_0^2 + k)At\cos(wt) + (-mw_0^2 + k)Bt\sin(wt) +$$
$$2m\omega_0 B\cos(\omega t) - 2m\omega_0 A\sin(wt) = F_0\cos(wt).$$

Before setting coefficients equal, let's remember the definition of the natural frequency and note that

$$-m\omega_0^2 + k = -m\left(\sqrt{\frac{k}{m}}\right)^2 + k = -m\left(\frac{k}{m}\right) + k = 0.$$

So, the first two terms actually drop out (which is a very good thing…) and this gives us,

$$2m\omega_0 B\cos(\omega t) - 2m\omega_0 A\sin(wt) = F_0\cos(wt).$$

Now let's set coefficient equal.

$$\cos(\omega t): \quad 2m\omega_0 B = F_0 \quad \Rightarrow \quad B = \frac{F}{2m\omega_0}$$

$$\sin(\omega t): \quad 2m\omega_0 A = F_0 \quad \Rightarrow \quad A = 0.$$

In this case the particular will be,

$$U_P(t) = \frac{F_0}{2m\omega_0} t \sin(\omega_0 t).$$

The displacement for this case is then

$$u(t) = c_1 \cos(\omega_0 t) + c_0 \sin(\omega_0 t) + \frac{F_0}{2m\omega_0} t \sin(\omega_0 t)$$

$$u(t) = R\cos(\omega_0 t - \delta) + \frac{F_0}{2m\omega_0} t \sin(\omega_0 t)$$

depending on the form that you prefer for the displacement.

So, what was the point of the two cases here? Well in the first case, $\omega_0 \neq \omega$our displacement function consists of two cosines and is nice and well behaved for all time.

In contrast, the second case, $\omega_0 = \omega$will have some serious issues at t increases. The addition of the t in the particular solution will mean that we are going to see an oscillation that grows in amplitude as t increases. This case is called resonance and we would generally like to avoid this at all costs.

In this case resonance arose by assuming that the forcing function was,

$$F(t) = F_0 \cos(\omega t).$$

We would also have the possibility of resonance if we assumed a forcing function of the form.

$$F(t) = F_0 \sin(\omega t).$$

We should also take care to not assume that a forcing function will be in one of these two forms. Forcing functions can come in a wide variety of forms. If we do run into a forcing function different from the one that used here you will have to go through undetermined coefficients or variation of parameters to determine the particular solution.

Example: A 3 kg object is attached to spring and will stretch the spring 392 mm by itself. There is no damping in the system and a forcing function of the form

$$F(t) = 10\cos(\omega t)$$

is attached to the object and the system will experience resonance. If the object is initially displaced 20 cm downward from its equilibrium position and given a velocity of 10 cm/sec upward find the displacement at any time t.

Solution: Since we are in the metric system we won't need to find mass as it's been given to us. Also, for all calculations we'll be converting all lengths over to meters.

The first thing we need to do is find k.

$$k = \frac{mg}{L} = \frac{(3)(9.8)}{0.392} = 75.$$

Now, we are told that the system experiences resonance so let's go ahead and get the natural frequency so we can completely set up the IVP.

$$\omega_0 = \sqrt{\frac{k}{m}} = \sqrt{\frac{75}{3}} = 5.$$

The IVP for this is then

$$3u'' + 75u = 10\cos(5t) \qquad u(0) = 0.2 \qquad u'(0) = -0.1.$$

Solution: wise there isn't a whole lot to do here. The complimentary solution is the free undamped solution which is easy to get and for the particular solution we can just use the formula that we derived above.

The general solution is then,

$$u(t) = c_1\cos(5t) + c_2\sin(5t) + \frac{10}{2(3)(5)}t\sin(5t)$$

$$u(t) = c_1\cos(5t) + c_2\sin(5t) + \frac{1}{3}t\sin(5t).$$

Applying the initial conditions gives the displacement at any time t. We'll leave the details to you to check.

$$u(t) = \frac{1}{5}\cos(5t) - \frac{1}{50}\sin(5t) + \frac{1}{3}t\sin(5t).$$

The last thing that we'll do is combine the first two terms into a single cosine.

$$R = \sqrt{\left(\frac{1}{5}\right)^2 + \left(-\frac{1}{50}\right)^2} = 0.200998$$

$$\delta_1 = \tan^{-1}\left(\frac{-1/50}{1/5}\right) = -0.099669 \qquad \delta_2 = \delta_1 + \pi = 3.041924.$$

In this case the coefficient of the cosine is positive and the coefficient of the sine is negative. This forces cosä to be positive and sinä to be negative.

This means that the phase shift needs to be in Quadrant IV and so the first one is the correct phase shift this time.

The displacement then becomes,

$$u(t) = 0.200998\cos(5t + 0.09969) + \frac{1}{3}t\sin(5t).$$

Here is a sketch of the displacement for this example.

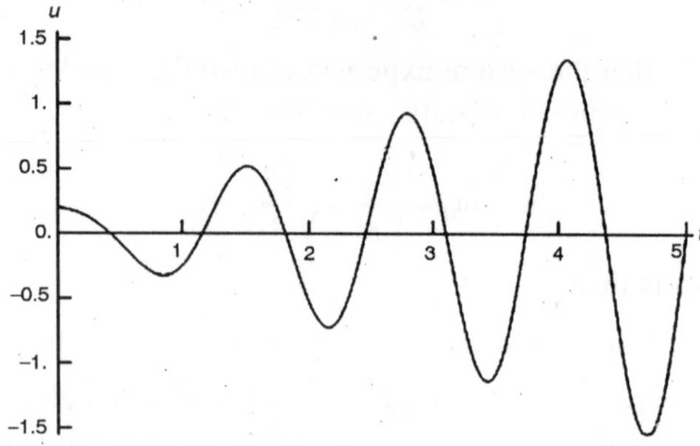

It's now time to look at the final vibration case.

FORCED, DAMPED VIBRATIONS

This is the full blown case where we consider every last possible force that can act upon the system. The differential equation for this case is,

$$mu'' + \gamma u' + ku = F(t).$$

The displacement function this time will be,

$$u(t) = u_c(t) + U_P(t)$$

where the complimentary solution will be the solution to the free, damped case and the particular solution will be found using undetermined coefficients or variation of parameter, whichever is most convenient to use.

There are a couple of things to note here about this case. First, from our work back in the free, damped case we know that the complimentary solution will approach zero as t increases. Because of this the complimentary solution is often called the transient solution in this case.

Also, because of this behavior the displacement will start too look more and more like the particular solution as t increases and so the particular solution is often called the steady state solution or forced response.

Let's work one final example before leaving this section. As with the previous examples, we're going to leave most of the details out for you to check.

Example: Take the system from the last example and add in a damper that will exert a force of 45 Newtons when then velocity is 50 cm/sec.

Solution: So, all we need to do is compute the damping coefficient for this problem then pull everything else down from the previous problem. The damping coefficient is

$$F_d = \gamma u'$$
$$45 = \gamma(0.5)$$
$$\gamma = 90.$$

The IVP for this problem is.

$$3u'' + 90u' + 75u = 10\cos(5t) \qquad u(0) = 0.2 \qquad u'(0) = -0.1.$$

The complimentary solution for this example is

$$u_c(t) = c_1 e^{(-15+10\sqrt{2})t} + c_2 e^{(-15+10\sqrt{2})t}$$

$$u_c(t) = c_1 e^{-0.8579t} + c_2 e^{-29.1421t}.$$

For the particular solution we the form will be,

$$U_P(t) = A\cos(5t) + B\sin(5t).$$

Plugging this into the differential equation and simplifying gives us,

$$450B\cos(5t) - 450A\sin(5t) = 10\cos(5t).$$

Setting coefficient equal gives,

$$U_P(t) = \frac{1}{45}\sin(5t).$$

The general solution is then

$$u(t) = c_1 e^{-0.8579t} + c_2 e^{-29.1421t} + \frac{1}{45}\sin(5t).$$

Applying the initial condition gives

$$u(t) = 0.1986e^{-0.8579t} + 0.00139e^{-29.1421t} + \frac{1}{45}\sin(5t).$$

Here is a sketch of the displacement for this example.

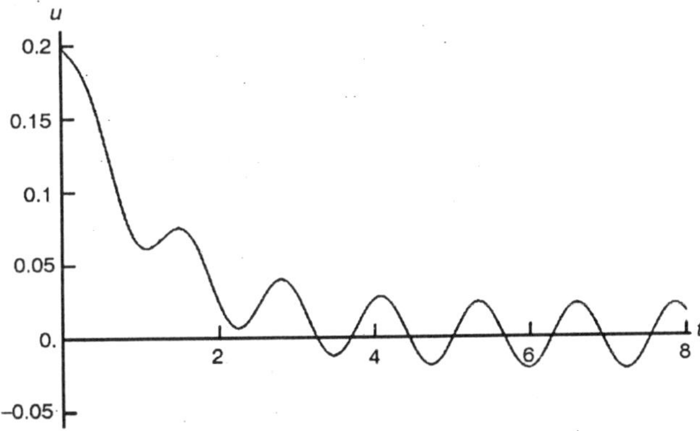

Chapter 10

Differential Equations in Space

Until now we have studied partial differential equations in one space dimension, x. Now we are ready to consider some problems in two or three space dimensions. Fortunately, the technique of separation of variables in more dimensions presents only a few new conceptual issues, and the technical complications are quite manageable.

In three space dimensions the *wave equation* has the form:

$$\frac{\partial^2 u}{\partial t^2} = c^2 \nabla.\nabla u := c^2 \left(\frac{\partial^2 u}{\partial x^2} + \frac{\partial^2 u}{\partial y^2} + \frac{\partial^2 u}{\partial z^2} \right).$$

It describes many sorts of waves that move through space, such as acoustic waves in air or another fluid medium, where c is the speed of sound in that medium. It also describes any component of an electromagnetic field in free space if c is the speed of light. The expression on the right is c^2 times the *Laplacean* of u, which is abbreviated ∇^2 u (some texts prefer the notation Δ u). It is sometimes convenient to think of ∇^2 u as the Laplace operator ∇^2 applied to u.

The heat equation in more than one space dimension likewise involves the Laplace operator:

$$u_t = k \nabla^2 u.$$

Both these equations describe how a quantity will change in time - but it might be static. If an electromagnetic field or a temperature is at equilibrium, the time derivatives will be 0, and we are led to the potential equation of Laplace,

$$\nabla^2 u: = \frac{\partial^2 u}{\partial x^2} + \frac{\partial^2 u}{\partial y^2} = 0$$

or, in three dimensions,

$$\nabla^2 u: = \frac{\partial^2 u}{\partial x^2} + \frac{\partial^2 u}{\partial y^2} + \frac{\partial^2 u}{\partial z^2}$$

Laplace's equation arises in many other applications as well. It is the equation which describes the

1. Displacement of an elastic membrane
2. Electrostatic potential function
3. Equilibrium concentration of a suspensate in a still fluid and still further physical quantities.

Solutions of Laplace's equation are called *harmonic functions*.

Let's begin with the two-dimensional Laplace equation on a rectangle

$$0 < x < a$$

$$0 < y < b.$$

This could arise from the wave equation for a long, rectangular wave-guide, assuming that the electric field is independent of z and static, that is, independent of t; or it might also arise from heat flow in a rectangular solid at equilibrium. In either case we are left with an equation in only two independent variables, x and y. It should thus be no more complicated than the earlier equations in the variables t and x.

Laplace's equation is the prototype of an elliptic equation, with different qualitative properties from either hyperbolic equations or parabolic equations.

The auxiliary conditions that are imposed are quite different from those for the wave equation. Specifically, we are normally given boundary conditions on the entire boundary, and are not specially concerned with an "initial" value of either variable x or y. Dirichlet boundary conditions for this problem are typically nonhomogeneous, of the form:

$$u(0,y) = f_1(y)$$

$$u(a,y) = f_2(y)$$

$$u(x,0) = f_3(x)$$

$$u(x,b) = f_4(x)$$

Where the four boundary functions are supposed given. In electromagnetic theory, for example, they are determined if the charge is measured on the boundary. In thermodynamics, they are the temperature determined by the temperature at the edge of the rectangle.

Separation of variables works much as it did for the heat equation and the wave equation. As with our earlier equations, we begin with the ansatz that u is a product solution

$$u(x,y) = X(x)\ Y(y)$$

and substitute into Laplace's equation; it is again convenient to divide the result by u. We use first information from the partial differential equation, then the information from homogeneous boundary conditions, and lastly information from inhomogeneous boundary conditions, which are treated much like the initial conditions for the earlier PDE's.

Rather than tackling four non-homogeneous boundary conditions all at once, we begin by setting three of the four boundary functions to 0. For definiteness, we also make a specific choice of the fourth while developing the ideas:

Example: Let us solve Laplace's equation with boundary conditions which are homogeneous on three of the four sides of a rectangle:

$$u(0,y) = 0$$

$$u(1,y) = 1$$

$$u(x,0) = 0$$

$$u(x,2) = 0.$$

Solution. Separation begins much as before, but note that boundary condition (9.8) is not homogeneous, so it is not consistent with the superposition principle. If we add two functions satisfying (9.8), for example, we will get a function with the boundary value 2, not 1, when x = a. As in our earlier solutions by separating variables, let us guess that the solution is a product, u(x,y) = X(x) Y(y), and subsitute into Laplace's equation. Dividing through by X(x) Y(y), we find that

$$\frac{X''(x)}{X(x)} + \frac{Y''(x)}{Y(x)} = 0$$

Evidently, if the y-independent quantity X″/X equals the x-independent quantity – Y″/Y, both must be a constant, and we have the familiar ordinary differential equations,

$$X'' = \mu X$$

$$-Y'' = \mu Y$$

(notice the difference of sign).

A good rule of thumb in this subject is to deal with the most homogeneous boundaries first. It is the variable y which has two homogeneous boundary conditions in this case, and it satisfies essentially the same eigenvalue problem as we have seen in previous chapters:

$$-Y'' = \mu Y$$

$$Y(0) = 0,\ Y(2) = 0$$

At this stage we can probably recognize the eigenfunctions and eigenvalues as

$$Y_n(y) = \sin\left(\frac{n\pi}{2}y\right), \mu_n = \left(\frac{n\pi}{2}\right)^2.$$

This same constant μ_n enters in to the X equation, but with the other sign. The general solution for X is a linear combination of exp(n π x/2) and exp(– n π x/2), rather than sines and cosines. There is a better choice for the basis of this solution space, though, namely

$$X_n(x) = C_1 \cosh(n \pi x/2) + C_2 \sinh(n \pi x/2).$$

One homogeneous boundary condition applies to this function, forcing

$$X_n(0) = 0.$$

The Dirichlet boundary condition thus eliminates the hyperbolic cosine, allowing us to normalize the function X_n as

$$X_n(x) = \sinh(n \pi x/2).$$

For the general solution we get:

$$u(x, y) = \sum_{k=1}^{\infty} c_k X_k(x) Y_k(y) = \sum_{k=1}^{\infty} c_n \sin(n \pi y/2) \sinh(n x/2).$$

There is only one boundary condition left, the nonhomogeneous at x = 1. Before incorporating it, we make a general linear combination of all the product solutions which are consistent with what we know so far:

$$u(x, y) = \sum_{k=1}^{\infty} c_n \sin(n \pi y/2) \sinh(n \pi x/2)$$

This is a Fourier sine series in the variable y, with the complication that the coefficients have an extra factor,

$$b_n = c_n \sinh(n \pi/2).$$

The series must be matched to the series for the function $f_2(y) = 1$, the coefficients in which can be calculated with the familiar formula. If

$$\sum_{n=1}^{\infty} b_n \sin(n \pi y/2) = 1,$$

then the Fourier coefficients are easily calculated as:

$$b_n = 4/(n \pi) \text{ if n is odd, and otherwise } 0.$$

To find c_n, now divide by sinh(n π/2). The solution becomes

$$u(x, y) = \sum_{\substack{n=1 \\ n\,odd}}^{\infty} \frac{4}{n\pi} \sin(n \pi y/2) \frac{\sinh(n \pi x/2)}{\sinh(n \pi/2)}.$$

It wasn't critical that the function at x = 1 was just the constant function; given any other function of y at x =1 as a Fourier sine series,

$$u(1, y) = f_2(y) = \sum_{n=1}^{\infty} b_{2n} \sin(n\pi y/2),$$

the solution we get would still be of the form:

$$u(x, y) = \sum_{n=1}^{\infty} b_{2n} \sin(n\pi y/2) \frac{\sinh(n\pi x/2)}{\sinh(n\pi/2)}.$$

Now let us return to the full problem with boundary conditions. Don't start over and rederive everything! Use the solution we already obtained as a springboard, as follows.

Let us rename the solution we just obtained for the simplified boundary conditions

$$u(0,y) = 0$$
$$u(a,y) = f_2(y)$$
$$u(x,0) = 0$$
$$u(x,b) = 0,$$

with only the function f_2 different from 0; for future bookkeeping u_2 will be better than u:

$$u_2(x, y) = \sum_{n=1}^{\infty} b_{2n} \sin(n\pi y/2) \frac{\sinh(n\pi x/2)}{\sinh(n a\pi/2)}.$$

(Notice the a in the denominator, which was fixed as 1 in the model problem.)

Next, suppose that we had the boundary conditions:

$$u(0,y) = 0$$
$$u(a,y) = 0$$
$$u(x,0) = 0$$
$$u(x,b) = f_4.$$

The only differences here are that f_2 becomes f_4, a becomes b, and the variables x and y are switched. Thus:

$$u_4(x, y) = \sum_{n=1}^{\infty} b_{4n} \sin(n\pi y/2) \frac{\sinh(n\pi x/2)}{\sinh(n b\pi/2)}.$$

As the next piece of the puzzle, suppose that

$$u(0,y) = f_1(y)$$
$$u(a,y) = 0$$
$$u(x,0) = 0$$
$$u(x,b) = 0.$$

We can get these boundary conditions from our original ones by interchanging x and a – x. Notice that this does not affect the potential equation, because under this change of variable there are two compensating changes of sign in u_{xx}; each time you use the chain rule there is a factor of –1. The answer has to be:

$$\mu = \frac{X''(x)}{X(x)} + \frac{Y''(y)}{Y(y)} + \frac{Z''(z)}{Z(z)}.$$

Similarly, if

$$u(0,y) = 0$$
$$u(a,y) = 0$$
$$u(x,0) = f_3(x)$$
$$u(x,b) = 0,$$

then

$$\mu = \frac{\nabla^2 Q(x)}{Q(x)}.$$

Because of the principle of superposition, the entire solution of the BVP (boundary-value problem) with boundary conditions is the sum:

$$u(x,y) = u_1(x,y) + u_2(x,y) + u_3(x,y) + u_4(x,y).$$

Let us turn our attention now to a fully multidimensional problem. There will be few new concepts, though as we shall see the additional dimensions require some extra book-keeping with several indices to label the pieces of the solution correlating with the various dimensions. We illustrate the topic in a specific example.

Example: How a mathematician cooks a cube steak. Consider how to cook a steak which is one meter on a side (it is a whale steak) under the following conditions.

boundary conditions: The steak will be put into a preheated oven at temperature 200 C at time t = 0.

IC: At time t = 0, the steak is pulled directly from the freezer (u(0, x,y,z) = 0) and put into the oven for four hours.

We wish to find the temperature throughout the interior of the steak at that time. Since the mathematician shops at a terrible meat market (but doesn't really notice), we may as well assume that the steak has the thermal properties of wood, so that in the heat equation, k = 2500 cm^2/hr.

Solution. Instead of directly solving for the temperature, let u(t,x,y,z) be the temperature minus 200 C, in order to have homogeneous Dirichlet boundary conditions. With this change, for all t > 0 the conditions on the six faces of the cube are

$$u(t,0,y,z) = 0$$

$$u(t,x,0,z) = 0$$

$$u(t,x,y,0) = 0$$

$$u(t,100,y,z) = 0$$

$$u(t,x,100,z) = 0$$

$$u(t,x,y,100) = 0,$$

$$u(t,x,y,100) = 0,$$

$$u(t,x,y,100) = 0,$$

while the PDE is the usual heat equation (HE),

$$u_t := \frac{\partial u}{\partial t} = k\nabla^2 u,$$

and the initial condition becomes

$$u(0,x,y,z) = -200$$

Solution.

1. Construct a general solution by separation of variables.

Suppose as usual that a solution of the heat equation is a product of the form

$$u(t,x) = T(t)\, Q(x);$$

A new feature here is the vector variable x. Plugging into the heat equation and dividing by T Q, we find:

$$\frac{T'(t)}{T(t)} = k\frac{\nabla^2 Q(x)}{Q(x)}.$$

We have a multidimensional eigenvalue problem to solve, *viz.*,

$$-\nabla^2 Q(x) = \mu\, Q(x),$$

Subject to the Dirichlet boundary conditions. I inserted the minus sign, as before, because experience teaches me to expect Mu to be positive this way. (It is a matter of book-keeping and not essential.)

How do we solve this equation? Why, by separating variables again, of course! Let

$$Q(x) = X(x)\ Y(y)\ Z(z)$$

and evaluate

$$\mu = \frac{\nabla^2 Q(x)}{Q(x)}.$$

The result, after canceling common factors is

$$\mu = \frac{X''(x)}{X(x)} + \frac{y''(y)}{Y(y)} + \frac{Z''(z)}{Z(z)}.$$

Which is a very curious equation, since any one of the three terms on the right could be isolated on one side of the equation. For instance, if we solve for X''/X, we find

$$X''(x)/X(x) = (\text{something independent of } x).$$

Similarly for Y''/Y and Z''/Z.

The conclusion is that we have *three* one-dimensional eigenvalue problems, all precisely like the familiar eigenvalue problem from Chapter 6:

$$-X'' = {}_1 X$$

$$-Y'' = {}_2 Y$$

$$-Z'' = {}_3 Z$$

If we compare, we see that

$$= {}_1 + {}_2 + {}_3,$$

and by this stage we can immediately see that ${}_1$ and X are of the form

$${}_1 = (m_1/100)^2,\ X(x) = \sin(m_1\ x/100),\ m_1 = 1, 2, \ldots.$$

Likewise,

$${}_2 = (m_2/100)^2,\ Y(y) = \sin(m_2\ y/100),\ m_2 = 1, 2, \ldots,$$

and

$${}_3 = (m_3/100)^2,\ Z(z) = \sin(m_3\ z/100),\ m_3 = 1, 2, \ldots.$$

The product solutions thus look like:

$$u_{m_1 m_2 m_3}(t,x,y,z) = \exp\left(-\frac{\pi^2}{10000}(m_1^2 + m_2^2 + m_3^2)kt\right)\sin\left(\frac{m_2 \pi x}{100}\right)\sin\left(\frac{m_2 \pi y}{100}\right)\sin\left(\frac{m_3 \pi z}{100}\right).$$

The general solution is a triple sum:

$$u(t,xy,z) = \sum_{m_1=1}^{\infty} \sum_{m_2=1}^{\infty} \sum_{m_3=1}^{\infty} c_{m_1 m_2 m_3} u_{m_1 m_2 m_3}(t,x,y,z),$$

where the multi-indexed constants $c_{m1,m2,m3}$ can have arbitrary values, subject only to convergence of the series.

2. The constants will be determined by the initial conditions when we set t = 0, obtaining a triple Fourier sine series. In this model problem the coefficients need to satisfy

$$u(0,x,y,z) = -200 = \sum_{m_1=1}^{\infty} \sum_{m_2=1}^{\infty} \sum_{m_3=1}^{\infty} c_{m_1 m_2 m_3} u_{m_1 m_2 m_3}(0,x,y,z),$$

$$= \sum_{m_1} \sum_{m_2} \sum_{m_3} c_{m_1 m_2 m_3} \sin\left(\frac{m_1 \pi x}{100}\right) \sin\left(\frac{m_2 \pi y}{100}\right) \sin\left(\frac{m_3 \pi z}{100}\right).$$

There are two ways to proceed here. It turns out that the products of three sines compose a complete orthogonal set for the cube, so we could directly evaluate the coefficients by the use of projections in function space. In this case the inner product uses a thre-dimensional integral over the cube.

You may find it more congenial, however, to rely on the usual, one-variable Fourier sine series, as follows. Imagine for the moment that y and z are fixed. What remains is a function of x, and we could use the orthogonality of the functions $\sin(m_1 \pi x/L)$ to remove the sum over m_1:

$$\int_0^{100} u(0,x,y,z) \sin\left(\frac{n_1 \pi x}{100}\right) dx = \sum_{m_1} \sum_{m_2} \sum_{m_3} c_{m_1 m_2 m_3} \int_0^{100} \sin\left(\frac{n_1 \pi x}{100}\right) \sin\left(\frac{m_1 \pi x}{100}\right) dx$$

$$\times \sin\left(\frac{m_2 \pi y}{100}\right) \sin\left(\frac{m_3 \pi z}{100}\right)$$

$$= \sum_{m_1} \sum_{m_2} \sum_{m_3} c_{m_1 m_2 m_3} 50 \delta_{n_1 m} \sin\left(\frac{m_2 \pi y}{100}\right) \sin\left(\frac{m_3 \pi z}{100}\right)$$

$$= \sum_{m_2} \sum_{m_3} c_{m_1 m_2 m_3} 50 \sin\left(\frac{m_2 \pi y}{100}\right) \sin\left(\frac{m_3 \pi z}{100}\right)$$

The point here is that $\sin(m_1\ \pi\ x/100)$ and $\sin(n_1\ \pi x/100)$ are orthogonal unless $m_1 = n_1$, so only one term survives from the sum over m_1.

If we now multiply by $\sin(n_2 \pi y/100)$ and integrate from 0 to 100, which will eliminate all but one term in the sum over m_2:

$$\int_0^{100}\int_0^{100} u(0,x,y,z)\sin\left(\frac{n_1\pi x}{100}\right)\sin\left(\frac{n_2\pi y}{100}\right)dx\,dy = 50^2\sum_{m_3} c_{m_1m_2m_3}\sin\left(\frac{m_3\pi z}{100}\right).$$

Finally, let us multiply the result by $\sin(n_3 \pi z/100)$ and integrate over z:

$$\int_0^{100}\int_0^{100}\int_0^{100} u(0,x,y,z)\sin\left(\frac{n_1\pi x}{100}\right)\sin\left(\frac{n_2\pi y}{100}\right)\sin\left(\frac{n_3\pi z}{100}\right)dx\,dy\,dz = 50^3 c_{n_1n_2n_3}.$$

DIFFERENTIAL EQUATIONS ON A DISK [MH]

Let us look again at the potential equation of Laplace, but this time in polar coordinates. This equation would apply to the equilibrium temperature distribution in a thin disk insulated on its two flat faces, or to a long cylinder along the z-axis, provided that the temperature does not depend on z. A temperature could be prescribed on the outside edge as a function of the angular variable producing a boundary-value problem of the form

$$\nabla.\nabla = u\frac{\partial^2 u}{\partial r^2} + \frac{1\partial u}{r\partial\theta^2} = 0.$$

with

$$u(a, \theta) = f(\theta)$$

The PDE (10.1) is what you get if you change variables using the 2-variable chain rule. I shall leave this to the exercises, but notice at least that is dimensionally consistent: r has units of length and partial/ partial r of length^{-1}, whereas θ is dimensionless. Hence each term has dimensions length^{-2}.

If we try to find a product solution for of the form

$u(r, \theta) = R(r)\, Q(\theta)$, we readily find

$$0 = \frac{\nabla.\nabla u}{u} = \frac{R''(r)}{R(r)} + \frac{1}{r}\frac{R'(r)}{R(r)} + \frac{1}{r^2}\frac{Q''(\theta)}{Q(\theta)}$$

The equation for Q is the familiar one that gives sines and cosines, but what are the boundary conditions?

Answer: *periodic BC*. Since there is no physical difference between θ and $\theta + 2\pi$, we must have:

$$Q(\theta + 2\pi) = Q(\theta) \text{ (PBC)}$$

In order to arrange periodicity, we take solutions in the form

$$Q(\theta,m) = a(m)\cos(m\,\theta) + b(m)\sin(m\,\theta),$$

Where m is an integer, or, equivalently, $Q(\theta,m) = c(m) \exp(i\ m\ \theta) + c(-m) \exp(-i\ m\ \theta)$. The function R satisfies the *equidimensional*, or Euler, equation:

$$R'' + (1/r) R' - (m^2/r^2) R = 0 \quad (10.5)$$

Equidimensional equations can be solved by making the guess $R = r^a$ and figuring out what the constant a has to be. In this case, $a = \pm m$, forced by the periodicity of the solutions (10.3), so the general solution is of the form:

$$a_0 + \sum_{m=1}^{\infty} r^m \left(a_m \cos(m\theta) + b_m \sin(m\theta)\right)$$

$$+\sum_{m=1}^{\infty} r^{-m} \left(a_{-m} \cos(m\theta) + b_{-m} \sin(m\theta)\right).$$

If we are studying the equilibrium temperature on a disk, the solutions containing r^{-m}, $m > 0$, are unphysical, so in this case the general solution will be of the form:

$$a_0 + \sum_{m=1}^{\infty} r^m \left(a_m \cos(m\theta) + b_m \sin(m\theta)\right)$$

Where the sum now runs from m =1 to infinity. I have written a_0 separately for the same reasons as with Fourier series.

Example: Solve with the boundary condition $u(1, \theta) = (\cos(\theta))^2$.

Solution: To find a solution in the interior, we need to determine the coefficients. This is done by substituting r = 1 and comparing:

$$a_0 + \mathrm{Sum}(a_m \cos(m\ \theta) + b_m \sin(m\ \theta) = (\cos(\theta))^2$$

In other words, the coefficients a[m] and b[m] are none other than the usual Fourier coefficients for the function f, here $(\cos(\theta))^2$. Rather than doing integrals here, the easiest way to find these Fourier coefficients is to use the double-angle formula in the form:

$$\cos(2\ \theta) = 2 \cos^2(\theta) - 1$$

Remember if you can find Fourier coefficients by some trick, the answer is perfectly correct, since the Fourier series is uniquely determined. This is great thing about uniqueness theorems in mathematics - they allow you find an answer by the easiest method available, without worrying that it might be a different answer from the one you would get from a harder but more standard procedure. We conclude that

$$(\cos(\theta))^2 = \frac{1}{2} + \frac{1}{2} \cos(2\theta),$$

or in other words that $a_0 = 1/2$, $a_2 = 1/2$, and all other coefficients are 0.

That is the solution on the boundary of the disk (r = 1); inside the disk the temperature is:

$$u(r, \theta) = \frac{1}{2} + \frac{1}{2} r^2 \cos (2, \theta)$$

Next, let us look again at a time-dependent problem, such as the heat equation on a disk, in polar coordinates.

Setting k = 1 for convenience, the heat equation in takes on the form:

$$u_t = \nabla^2 u = \frac{\partial^2 u}{\partial r^2} + \frac{1}{r}\frac{\partial u}{\partial r} + \frac{1}{r^2}\frac{\partial^2 u}{\partial \theta^2}.$$

As before, we begin by searching for product solutions. It is convenient to separate vector variables one at a time in the product solution. If

$$u = T(t)\ V(r, \theta),$$

then we find that

$$T'/T = (\text{Grad}^2\ V)/V$$

and we can make the usual argument that the left side is independent of space while the right side is independent of time, hence both equal some constant, called $-\lambda$. The solutions for T are of the form

$$T(t) = \text{const. } \text{Exp}(-\lambda t),$$

where λ should be determined from the boundary conditions in space.

Example: Suppose that a thin disk of radius A is insulated on its faces while the round edge is held at temperature 0. Find the eigenfunctions and eigenvalues for the spatial part of the separated equation.

Solution: On the edge we have Dirichlet boundary conditions of the form $u(t,A, \theta) = 0$ This two-dimensional problem, resembles except that in the eigenvalue equation,

$$-\nabla\ v(r,\theta) = \lambda\ v(r,\theta),$$

The eigenvalue need not be zero. The constant may be zero, in which case the disk is at equilibrium and because of the boundary conditions at r = A, the equilibrium temperature is 0 throughout.

This trivial solution is not considered an eigenfunction, since it is not useful in building a series for a general solution.

Separating variables again by writing

$$V = R(r)\ Q(\theta),$$

leads to the same eigenvalue equation for $Q(\theta)$ as before. The boundary conditions are also the same, periodic, so the solutions are the same sines and cosines, indexed by the integer m. Something new occurs for the other part of the solution, R. Evaluating

$$\lambda = -\frac{\nabla^2 v}{v}$$

with the knowledge that $Q(\theta) = \sin(2\ m\ \theta)$ or $\cos(2\ m\ \theta)$, we get an ordinary differential equation for R. In place of the Euler, which was solved with power functions, we get Bessel's equation,

$$R'' + (1/r)\ R' + (\lambda - m^2/r^2)\ R = 0,$$

which no longer has elementary solutions. Just as with the equidimensional equation, near $r = 0$ the solutions can behave like either r^m or r^{-m}, but the latter are unphysical because the temperature can't diverge at $r = 0$. The regular solutions, behaving roughly like r^m near $r = 0$, are called *regular Bessel functions* and, with a standard choice of the overall constant, denoted $J_m(\lambda^{1/2}\ r)$. Mathematica uses the notation BesselJ. They are normalized so that $J_m x) \sim x^m/(2\ m!)$ as $x -> 0$:

In: = Plot[BesselJ[0,x],{x,0,25}]

Out =

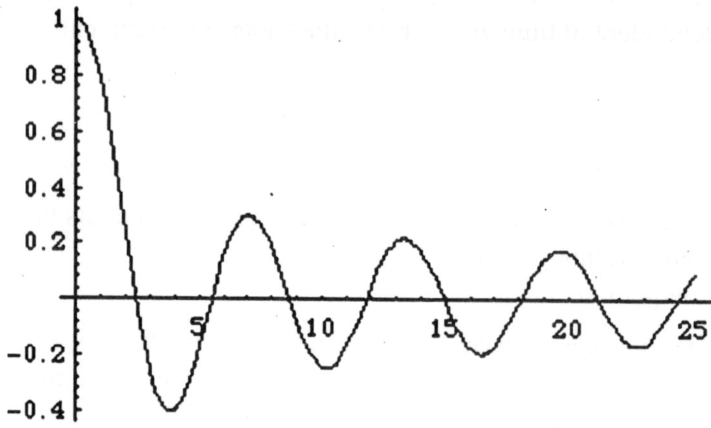

The oscillations are damped, and not quite as evenly spaced as those of the sines and cosines, but, as you might expect, there are many similarities between these functions and the sines and cosines.

With the aid of a computer and software, Bessel functions should not be considered much more exotic or hard to use than sines and cosines.

In order to satisfy the boundary condition at A, we must fix so that one of these nodes happens at r = A. It turns out that the nodes of the Bessel functions are tabulated functions, and they are traditionally denoted

$j_{m,n}$ = the n-th positive value of x for which $J_m[x] = 0$.

```
In = Plot[BesselJ[1,x],{x,0,25}]
Out =
```

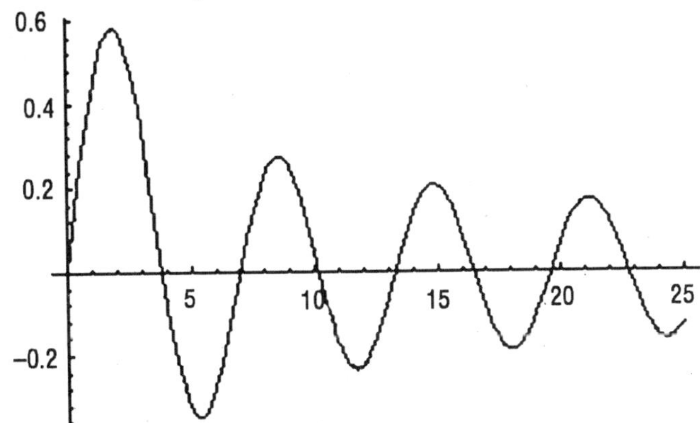

They are also quite easy for Mathematica to compute. There is a Mathematica package which tabulates them, called BesselZeros.m, but let's proceed on our own:. Looking at the graph, we choose reasonable guesses for these numbers, 2,5,9,15, etc., and let the FindRoot command locate the nearest actual root to our guess:

```
In: = FindRoot[{BesselJ[0,j[0,1]] = = 0,BesselJ[0,j[0,2]] = = 0,\
BesselJ[0,j[0,3]] = 0,BesselJ[0,j[0,4]] = = 0},\
{j[0,1],2}, {j[0,2],5}, {j[0,3],9},{j[0,4],15} ]
Out = {j[0, 1] → 2.40483, j[0, 2] → 5.52008, j[0, 3] → 8.65373,
j[0, 4] → 14.9309}
In = FindRoot[{BesselJ[1,j[1,1]] = = 0,BesselJ[1,j[1,2]] = = 0,\
BesselJ[1,j[1,3]] = 0,BesselJ[1,j[1,4]] = = 0},\
{j[1,1],4}, {j[1,2],10}, {j[1,3],10},{j[1,4],20} ]
Out = { j[1, 1] → 3.83171, j[1, 2] → 10.1735, j[1, 3] → 10.1735,
j[1, 4] → 19.6159}
```

And so forth. The important point is that today Bessel functions and their zeroes j_{mn} should be regarded as known, easily accessible numbers, nearly as accessible as and values of the sine.

The eigenfunctions and eigenvalues for Model Problem X.2 are (as usual, up to a constant, nonzero multiple):

$$J_m\left(\frac{i_{mn} r}{A}\right) \text{ and } \lambda_{mn} = \left(\frac{i_{mn}}{A}\right)^2, m = 0,1,\ldots;=1,2,\ldots.$$

Before continuing with the heat equation on the disk, a few more remarks about Bessel functions are in order. One of the ways in which Bessel functions resemble sines and cosines is that they are the source of some complete sets, with which we can define Fourier-Bessel series. This is not an accident, of course; the eigenfunctions of any eigenvalue problem which satisfies the technical condition of self-adjointness form a complete set by the spectral theorem of J. von Neumann. Moreover, the eigenfunctions are orthogonal. (If the eigenvalues are different this is automatically true, and even if there are several eigenfunctions with the same eigenvalue, they can be recombined as an orthogonal set.)

Here is how it works for Bessel functions: Just as we have different Fourier series, which use sines, cosines, or both, there are different Fourier-Bessel series. Each series uses the J's with a fixed index (like m) in place of the expressions n π x/L or 2n π x/L which appear in Fourier series, there is an expression involving $j_{m,n}$.

ORTHOGONALITY

$$\int_0^A J_m\left(\frac{i_{mn} r}{A}\right) J_m\left(\frac{i_{mp} r}{A}\right) r\, dr = \frac{A^2}{2}(J_{m+1}(i_{mn}))^2 \delta_{np}.$$

You may at first think that the factor of r in this integral is peculiar, but it is really inherited from the two-dimensional area element in polar coordinates, which is

$$d\text{ Area} = r\, dr\, d.$$

The alert reader will perhaps recall that when inner products were introduced we learned that the integrals could incorporate weights like the extra r here. The product solutions for $V(r, \theta)$ found above can be written either with sines and cosines, or with exponential functions, and I shall make the latter choice:

$$\left\{ J_{|m|}\left(\frac{i_{mn} r}{A}\right) \exp(i m \theta) \right\}_{-\infty < m < \infty\ n=1,2,\ldots}$$

are orthogonal in the sense of the two-dimensional inner product

$$(f, g) := \int_0^A \int_0^A f(r,\theta)\overline{g(r,\theta)}\, r\, dr\, d\theta.$$

Here, m can be positive, negative, or 0. With the normalization, the orthogonality relationship is:

$$\int\int_1^1 J_{|m|}\left(\frac{i_{|m|} r}{A}\right) \exp(im\theta) J_{|q|}\left(\frac{i_{|q|p} r}{A}\right) \exp(-iq\theta) r\, dr\, d\theta = \pi A^2 (J_{m+1}(i_{mn}))^2 \delta_{mq} \delta_{np}.$$

Note that if we were to consider the wave equation rather than the heat equation, with Dirichlet boundary conditions on a circle, the spatial dependence of the normal modes would be the same as the eigenfunctions of the Laplace operator as found in Model Problem. Several of these normal modes may be viewed in a Mathematica notebook.

The following Fourier-Bessel series are useful, and result from identities involving the derivatives of Bessel functions:

$$r^p = 2A^p \sum_{n=1}^{\infty} \frac{1}{i_{pn} J_{p+1}(i_{pn})} J_p\left(\frac{i_{pn} r}{A}\right)$$

(valid for each fixed m, 0 <= r < A).

$$1-\left(\frac{r}{A}\right)^2 = \sum_{n=1}^{\infty} \frac{8}{i_{0n}{}^3 J_1(i_{0n})} J_0\left(\frac{i_{0n} r}{A}\right)$$

(valid for 0 <= r < A).

Including the time-dependence for the product solutions solving the heat equation yields:

$$u_{mn}(t, r, \theta) = J_{|m|}\left(\frac{i_{mn} r}{A}\right) \exp(i m \theta) \exp\left(-\left(\frac{i_{mn}}{A}\right)^2_t\right).$$

Let us work a specific initial-boundary-value problem.

Example: Suppose that the ends of a cylinder of length L and radius A are insulated and the curved side is held at temperature 0. Set up the one-dimensional eigenvalue problems which make up the spatial part of the separated equation.

Solution: On the ends we have Neumann boundary conditions of the form

$$\frac{\partial u(t,r,\theta,0)}{\partial z} = \frac{\partial u(t,r,\theta,L)}{\partial z} = 0, \text{ (NBC)}$$

While on the side we have Dirichlet boundary conditions of the form

$$u(t,A,\theta,z) = 0 \qquad \text{(DBC)}$$

The z part also separates off easily; if $U = V(r, \theta)\, Z(z)$, then we get:

$$\lambda - = (\nabla^2 V)/V + Z''/Z,$$

or,

$$-Z''/Z = \lambda + \nabla^2 V)/V = \text{constant} =: \mu.$$

The eigenvalue problem for Z gives us familiar eigenvalues and eigenfunctions:

$$Z(z,k) = \cos(k \text{ pi } z/L),$$

$$\text{mu}(k) = (k \text{ pi}/L)^2.$$

The remaining part V is just like the V of Model Problem same set of Bessel functions:

$$J_m\left(\frac{i_{mn} r}{A}\right), \; m=0,1,\ldots; \; n = 1,2,\ldots.$$

The Z-dependence contributes to the eigenvalue, however, so that

$$\lambda_{mnk} = (j_{mn}/A)^2 + (k\pi/L)^2.$$

Chapter 11

Great Balls of Differential Equation

In this chapter we solve some partial differential equations with spherical symmetry in three dimensions. Conceptually, this chapter will be much like the previous one, but the geometry will now be three-dimensional. If there is a radial symmetry in a three-dimensional physical problem, the best coordinate system to use is probably the spherical system, with variables

$$\begin{aligned} r &= \text{distance from the origin} \\ \theta &= \text{colatitude} \\ \phi &= \text{longitude.} \end{aligned}$$

The colatitude is the angle measured down from the pole rather than up from the equator, which would be the usual geographic coordinate of latitude. Since we use radians for all angles, the relationship is

$$\text{colatitude} = \pi/2 - \text{latitude},$$

and the colatitude runs from 0 to π. The longitude runs from 0 to 2π. You may wish to review the spherical coordinate system at this stage in your vector calculus textbook. Let us again begin with the potential equation of Laplace. A temperature could be prescribed on the outside of a ball as a function of the angular variables θ, ϕ producing a boundary-value problem of the form

$$\nabla^2 f = \frac{\partial^2 f}{\partial r^2} + \frac{2}{r}\frac{\partial f}{\partial r} + \frac{1}{r^2}\frac{\partial^2 f}{\partial \theta^2} + \frac{\cot(\theta)}{r^2}\frac{\partial f}{\partial \theta} + \frac{1}{r^2 \sin^2(\theta)}\frac{\partial^2 f}{\partial \varphi^2} = 0.$$

with $u(a, \theta, \phi) = f(\theta, \phi)$

To determine the temperature in the interior from measurements on the boundary, we must solve for u. If we try to find a product solution for of the form

$$u(r, \theta) = R(r)\, Q(\theta, \phi),$$

we readily come up with

$$0 = \frac{\nabla.\nabla u}{u} = \frac{R''(r)}{R(r)} + \frac{2}{r^2}\frac{R'(r)}{R(r)}\frac{1}{r^2}\frac{\nabla^2 Q(\theta,\varphi)}{Q(\theta,\varphi)}.$$

This equation has some features of the multidimensional problem and some of those of the problems on disks. To separate variables, we look at the last term. If we solve for it, we see that

$$\frac{\nabla^2 Q(\theta,\varphi)}{Q(\theta,\varphi)}.$$

equals a function of r alone; since it does not depend on r, it must be a constant. Thus we have a two-dimensional eigenvalue problem of the form

$$\frac{\nabla^2 Q(\theta,\varphi)}{Q(\theta,\varphi)} = -\mu$$

Equivalently, after multiplying through by $-r^2Q$,

$$\frac{\partial^2 Q}{\partial\theta^2} + \cot(\theta)\frac{\partial Q}{\partial\theta} + \frac{1}{\sin^2(\theta)}\frac{\partial^2 Q}{\partial\varphi^2} = -\mu Q(\theta,\varphi).$$

Notice that the variable ϕ explicitly appears only in the term

$$\frac{\partial^2 Q}{\partial\varphi^2}$$

If $Q = \Theta\,(\theta)\,\Theta\,(\phi)$, we can further separate variables. Making this substitution and dividing by Q leads to

$$\frac{(\Theta'' + \cot(\theta)\Theta)}{\Theta} + \frac{1}{\sin^2(\theta)}\left(\frac{\Phi''}{\Phi}\right) = -\mu$$

and all the dependence on ϕ is isolated in the term Φ''/Φ Since the rest of the equation does not involve ϕ, we conclude that this term is a constant, which leads us to our most familiar eigenvalue problem,

$$-\Phi'' = \lambda\Phi$$

What about boundary conditions? The variable ϕ is periodic - if you increase your longitude by 2π radians, you will come completely around the earth to your starting point. This problem is so familiar that we can write down the solution from memory:

$$\Phi_m(\varphi) = \exp(i\, m\, \varphi), -\infty\, m < \infty,$$

That is, m can be a positive or negative integer or zero. (Sines and cosines would serve just as well, of course, but then $m >= 0$.) The eigenvalue associated with Φ_m is m^2. Looking back and substituting $\Phi''/\Phi = -m^2$, we get an ordinary differential equation for θ, which a step or two of algebra puts into the form

$$\Phi'' + \cot(\theta)\Phi' + \left(\mu - \frac{m^2}{\sin^2(\theta)}\right)\Phi = 0$$

This is probably not striking the reader as the friendliest equation ever seen, and even hours of substitution cannot reduce it to a familiar equation, although some simplification is achieved by substituting $z = \cos(\theta)$, $P(z) = \Phi(\theta)$. The result is known as *Legendre's equation,*

$$\frac{d}{dz}\left((1-z^2)\frac{dP}{dz}\right) + \left(\mu - \frac{m^2}{1-z^2}\right)P = 0,$$

And for general values of m and u its solutions are new transcendental functions, just as the Bessel functions were new functions defined as the solutions of a new equation in the previous chapter. The variable z runs from – 1 to +1, and Legendre's equation becomes singular at those points, which means that some solutions are bounded at the end points while others diverge. In effect we have a boundary condition at two end points, and the boundary conditions determine the eigenvalues u. The simplest case is when $m = 0$, and in this case you may be astonished to know that the regular solutions of Legendre's equation are the Legendre polynomials Pl, $l = 0, 1, \ldots$! The eigenvalues are of the form $u = l(l+1)$. When m is not 0, the eigenvalues are still of the form

$u = l(l+1)$, with the restriction that $l >= |m|$, and the bounded solutions are the *associated Legendre polynomials*, defined for $m > 0$ by

$$P_l^m(z) = (-1)^m (1-z^2)^{m/2} \frac{d^m}{dx^m} P_l(z).$$

The usual Legendre polynomials result from setting $m = 0$, and when $m < 0$,

$$P_l^m(z) = (-1)^{|m|} \frac{(l+m)!}{(l-m)!} P_l^{|m|}(z).$$

Fortunately, these functions are all known to Mathematica, where the notation is LegendreP[l,x]

LegendreP[l,m,x]

While Mathematica can take care of doing calculations with special functions like the Legendre polynomials, you may find it helpful to read some further discussion of the Legendre polynomials when doing the exercises. The term *spherical harmonic* refers to an eigenfunction of the Laplace operator, which is a normalized product of Legendre polynomials with exp(i m ϕ):

$$Y_l^m = \sqrt{\frac{(2l+1)(l-m)!}{4\pi(l+m)!}}\, P_l^m(\cos(\theta))\exp(im\varphi);$$

Some useful formulae are:

$$-\left(\cot(\theta)\frac{\partial}{\partial\theta}+\frac{1}{\sin^2(\theta)}\frac{\partial^2}{\partial\varphi^2}\right)Y_l^m = -r^2\,\tilde{N}^2\,Y_l^m = l(l+1)\,Y_l^m;$$

$$\int_0^l\int_0^l Y_l^m\overline{Y_l^m}\sin(\theta)d\theta d\varphi = \delta_{mm}.\delta_{ll},$$

and

$$\frac{(2l+1)(l-m)!}{2(l+m)!}\int_{-1}^{1}P_l^m(z)P_l^m(z)dz = \delta_{ll}.$$

The one remaining variable to consider is r, and when Q is a spherical harmonic, can be simplified to

$$r^2R''(r)+2rR'(r)-l(l+1)R(r) = 0$$

This is again an equidimensional equation, so it has solutions of the form $R = r^a$, where a is a root of the polynomial

$$0 = a(a-1) + 2a - l\,(l+1),$$

That is $a = l$ or $-l-1$. The general solution for R is a linear combination of r^l and r^{-l-1}. The latter solution is singular at $r = 0$, so it can be excluded if the differential equation is operative at the origin. The general solution can then be written as

$$\sum_{l=1}^{\infty}\sum_{m=-l}^{l} c_{im}r^lY_l^m(\theta,\varphi).$$

The coefficients cl m in the linear combination will be determined by the boundary conditions.

Example: Solve on the interior of the ball of radius 1 centered at the origin, with the boundary condition

$$u(1,\theta,\phi) = (\cos(\theta))^2$$

on its surface.

Solution. The boundary data are azimuthally symmetric (independent of ϕ), so the solution will also be independent of ϕ. Equivalently, all terms with $m! = 0$ can be dropped from the general solution. This means that the spherical harmonics are just multiples of our old friends, the Legendre polynomials, with the variable $z = \cos(\theta)$. The general solution with this simplification is

$$\sum_{l=1}^{\infty}c_{l0}r^ly_l^0(\theta,\varphi) = \sum_{l=1}^{\infty}b_lr^ly^lP_l(z),$$

where the coefficients b_l differ from the coefficients c_{l0} by a normalization factor. That factor is unimportant, since at this stage we are just as ignorant about the values of the c's as of the b's, but in case it is reassuring,

$$b_l = \sqrt{\frac{(2l+1)}{2}}$$

The boundary corresponds to r =1, and in terms of the variable z, the function u on the boundary is z^2. In order to determine the b's we need to represent z^2 as a Legendre series. The coefficients are obtained by projecting z^2 onto the various Legendre polynomials. Alternatively, because of the uniqueness theorem for Legendre series, we can be confident that if we find the coefficients with a shortcut, we have the correct answer. Recalling that $P_0(z) = 1$ and

$$P_2(z) = \frac{3}{2}z^2 - \frac{1}{2},$$

we can easily determine that

$$z^2 = (2/3)\, P_2(z) + (1/3)\, P_0(z).$$

In other words, $b_0 = 1/3$ and $b_2 = 2/3$, while all other coefficients are 0. The solution to the boundary-value problem is:

$$u = \frac{2}{3}r^2 P_2(z) + \frac{1}{3}r^0 P_0(z)$$

$$= \frac{2}{3}r^2\left(\frac{3}{2}z^2 - \frac{1}{2}\right) + \frac{1}{3}$$

$$= r^2\cos^2(\theta) - \frac{1}{3}r^2 + \frac{1}{3}$$

Example: A classic problem is electrostatics is to find the potential inside a sphere when one hemisphere is kept at potential +1 and the other at 0. This can be solved in two different-looking ways, depending on the orientation of the hemispheres. Here I will solve the problem in one orientation, leaving the other to the exercises (yours is the easier way). The electrostatic potential solves Laplace's equation, and we may take as our boundary conditions:

$$u(1, \theta, \phi) = 1 \text{ for } 0 <= \phi < \pi, -1 \text{ for } \pi <= \phi < 2\pi.$$

Solution. The general solution and we compare with the boundary conditions at r =1. We need to find the coefficients in the series

$$\sum_{l=1}^{\infty}\sum_{n-l}^{l} c_{lm} y_l^m(\theta, \varphi) = F(\varphi).$$

The function on the right side equals 1 or –1, depending on the value of ϕ. To determine the coefficients we multiply both sides of this expression by

$$\overline{Y_l^m} = \sqrt{\frac{(2l+1)(l-m)!}{4\pi(l+m)!}} P_i^m(\cos(\theta))\exp(-im\varphi)$$

and integrate by $\sin(\theta)d\theta\, d\phi$. Because of the orthonormality of the spherical harmonics, the left side will reduce to the one term, cl' m. Let us do the ϕ integration first, since it is exactly like the integrals we did earlier when finding the Fourier series for the square pulse. We find that

$$\int F(\varphi)\exp(-im\varphi)d\varphi = \frac{4}{im}$$

when m' is odd, and 0 when m' is even. For the integrations it is somewhat more efficient to use the variable z = cos(θ). we shall use Mathematica to calculate several of the coefficients:

```
In: = c[l_,m_]: = (4/(I m)) Sqrt[(2 l + 1) (l-m)!/(4 Pi (l + m)!)]
   Integrate[LegendreP[l,m,z], {z,-1,1}]
In: = c[1, 1]
```

$$\text{Out} = \text{I Sqrt}\left[\frac{3\text{Pi}}{2}\right]$$

In: = c[3,1]

$$\text{Out} = \frac{3\text{I}}{16}\text{ Sqrt}\left[\frac{7\text{Pi}}{3}\right]$$

In: = c[3,3]

$$\text{Out} = \frac{5\text{I}}{16}\text{ Sqrt}\left[\frac{7\text{Pi}}{5}\right]$$

In: = c[5,1]

$$\text{Out} = \frac{15\text{I}}{64}\text{ Sqrt}\left[\frac{11\text{Pi}}{30}\right]$$

In: = c[5,3]

$$\text{Out} = \frac{35\text{I}}{128}\text{ Sqrt}\left[\frac{11\text{Pi}}{35}\right]$$

In: = c [5,5]

$$\text{Out} = \frac{21I}{128}\ \text{Sqrt}\left[\frac{11\text{Pi}}{7}\right]$$

In: = c[7,1]

$$\text{Out} = \left[\frac{175I}{2048}\right]\ \text{Sqrt}\left[\frac{15\text{Pi}}{14}\right]$$

And so on.

Next, let us look again at a time-dependent problem, such as the heat equation,

$$u_t\,(t,x) = k\,\nabla^2\,u(t,x).$$

As usual, we begin by looking for the normal modes, product solutions with the time-dependence separated: u(t,x) = T(t) U(x). Upon substitution and dividing by u we get

$$\frac{T'(t)}{T(t)} = k\frac{\nabla^2 U(x)}{U(x)},$$

so if

$$-\,\nabla^2\,U(x) = \mu\,U(x),$$

then we have T'(t) = – u k T(t). This is an elementary equation with solution

$$T(t) = T(0)\,\exp(-\,u\,k\,t).$$

The values of u have to come from, which would be the Laplace's equation if u = 0. When u > 0, we can still rely on our analysis of Laplace's equation up to a point, especially if the equation is defined on a domain with spherical symmetry.

The values of u will be determined as eigenvalues, once we have imposed boundary conditions.

Example: A mathematician has recently received tenure and can now afford a better cut of meat, the round steak. Having too little spare time to enroll in Cordon Bleu classes, he still cooks by the unsophisticated method of pulling the round steak - an exact sphere 10 cm. in radius - directly out of the freezer, at 0 C, and putting it into a preheated oven, at 200C.

We wish to find the temperature at the center of the steak after, say, an hour. Since the steak is of higher quality, let's assume that the thermal constant k = 25 cm^2/hr.

Solution. The boundary condition,

$$u(t,10,\,\theta,\phi) = 200$$

Is nonhomogeneous, so we modify the problem by redefining v = u – 200. The initial condition for v will be v(0,x) = – 200, and v will still solve the wave equation. We shall solve for v, and then add 200 at the end. If we seek product solutions v = T(t) U(x), equation for U will become a well-posed eigenvalue problem when we supplement it with the boundary condition (now homogeneous):

$$-\nabla^2 U(x) = \mu\, U(x),$$
$$U(10, \theta, \phi) = 0$$

Instead of going through the entire separation procedure again, let's speed matters up by noting that the angular part of this equation is identical. We can surmise from this that the normal modes will be of the form

$$U(x) = R(r)\, Y_l^m(\theta, \varphi)$$

Indeed, if we calculate the effect of the Laplace operator on this, we find

$$\nabla^2 R(r) Y_l^m(\theta,\varphi) = \left(R''(r) + \frac{2}{1} R'(r) - \frac{l(l+1)}{r^2} R(r) \right) Y_l^m(\theta,\varphi).$$

Because the spherical harmonics are the eigenfunctions of the angular part of the Laplace operator, they are still a common factor when the product solution is acted upon by the operator. The eigenvalue equation for u boils down to an ordinary differential equation for the function R:

$$r^2 R''(r) + 2rR'(r) - (l(l+1) + \mu r^2)R(r) = 0$$

Equation should remind you of Bessel's equation; the only difference is the factor of 2 in the second term. As with Bessel's equation, the term with u prevents the equation from being equidimensional, and thus from being solved by a simple r^a. It is in fact a form of Bessel's equation, except that when the extra factor of 2 is scaled out, the index gets messed up. According to Mathematica:

In: = DSolve[r^2 R''[r] + 2 r R'[r] – (l (l+1) – mu r^2) R[r] = = 0, R[r],r]

Out = {{R[r] – >

> (BesselJ[Sqrt[$\frac{1}{4}$ + l + l^2], Sqrt[mu] r] C [1] +

> BesselY[Sqrt[$\frac{1}{4}$ + l + l^2], Sqrt[mu] r] C[2])/ Sqrt[r]}}

The subscript of the Bessel function simplifies to $l + 1/2$. As in the previous chapter, we reject the Y Bessel functions because they are singular at r = 0. You may now be worrying that we are about to encounter still new complications when investigating Bessel functions with fractional indices, but take heart - when the index is half an odd integer, the Bessel functions reduce to more familiar functions:

$$J_{1/2}(z) = \left(\frac{2}{\pi 2}\right)^{1/2} \sin(z),$$

$$J_{3/2}(z) = \left(\frac{2}{\pi 2}\right)^{1/2} \left(\frac{\sin(z)}{2} - \cos(z)\right),$$

and, in general,

$$J_{l+1/2}(z) = (-1)^l \left(\frac{2}{\pi}\right)^{1/2} z^{l+1/2} \left(\frac{1}{z}\frac{d}{dz}\right)^l \frac{\sin(z)}{z}$$

Here are some plots for $J_{1/2}$, $J_{3/2}$, and $J_{5/2}$.

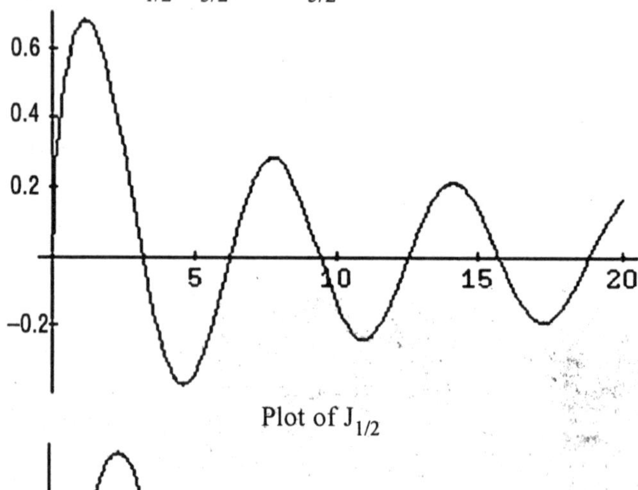

Plot of $J_{1/2}$

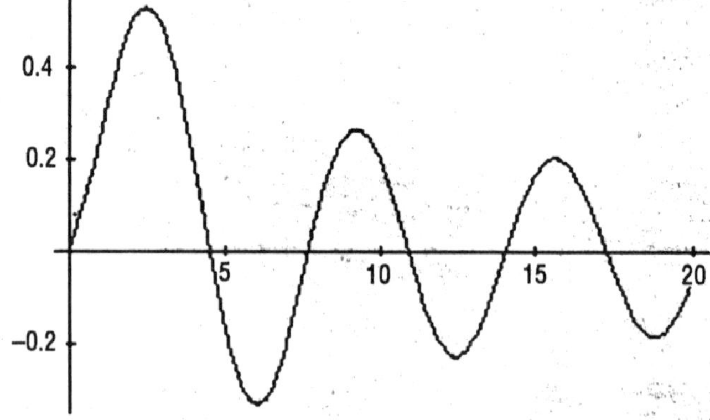

Plot of $J_{3/2}$

As with the Bessel functions of integer index, these functions oscillate and damp down, and we can determine the eigenvalues, at least numerically. If $l = 0$, then it is clear that $J_{1/2}(z) = 0$ when $z = n$, but otherwise it is best to find them numerically.

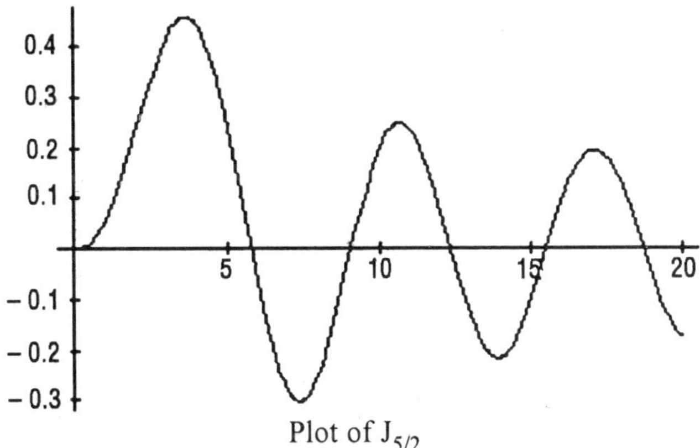

Plot of $J_{5/2}$

We use Mathematica and begin the numerical calculations by estimating the positions of the zeroes from the graphs of the Bessel functions:

```
In: =
FindRoot[{BesselJ[3/2,j[3/2,1]] = = 0,BesselJ[3/2,j[3/2,2]] = = 0, \
BesselJ[3/2,j[3/2,3]] = = 0,BesselJ[3/2,j[3/2,4]] = = 0}, \
{j[3/2,1],4}, {j[3/2,2],7}, {j[3/2,3],11},{j[3/2,4],15} ]
Out =
{j[3/2,1] → 4.493409457909064175, j[3/2,2] → 7.725251836937706892,
j[3/2,3] – > 10.90412165942889983, j[3/2,4] → 14.06619391277779873}
FindRoot[{BesselJ[5/2,j[5/2,1]] = = 0,BesselJ[5/2,j[5/2,2]] = = 0,.\
BesselJ[5/2,j[5/2,3]] = = 0,BesselJ[5/2,j[5/2,4]] = = 0}, \
{j[5/2,1],6}, {j[5/2,2],9}, {j[5/2,3],12},{j[5/2,4],16} ]
{j[5/2,1] → 5.763459196891101 09, j[5/2,2] → 9.095011330476355156,
j[5/2,3] → 12.32294097056643961, j[5/2,4] → 15.51460301083520084}
```

In order to satisfy the boundary condition at r = 10, the eigenvalue u must be such that

$$\sqrt{\mu_{ln}10} = j_{l+1/2}, n,$$

where $j_{l+1/2}$ is the n-th zero of a Bessel function, which was called j[l + 1/2,n] in the Mathematica code. Solving for the eigenvalues,

$$\mu_{ln} = \frac{j^2 l + 1/2, n}{100}.$$

Although the eigenvalue could in principle depend on the index m as well, it does not happen to in this problem. The eigenfunctions of the Laplace operator are thus:

$$U_{lnn}(x) = \frac{1}{lr} J_{l+1/2}\left(j_{i+1/2,n} r/10\right) Y_l^m(\theta,\varphi),$$

and the general soiution to the heat equation in the ball is

$$v(t,x) = \sum_{lmn=1}^{\infty} c_{lmn} U_{lmn}(x) \exp(-25\mu_{in} t)$$

$$= \sum_{lmn=1}^{\infty} c_{lmn} \frac{1}{lr} J_{1/2}\left(j_{l+1/2,n} r/10\right) Y_l^m(\theta,\varphi) \exp\left(-j^2_{l+1/2,n} t/4\right).$$

The coefficients are determined by comparison with the initial condition, which is:

$$-200 = \sum_{lmn=1}^{\infty} c_{lmn} \frac{1}{lr} J_{l+1/2}\left(j_{l+1/2,n} r/10\right) Y_l^m(\theta,\varphi).$$

The standard way to proceed would be to use the orthogonality properties of the spherical harmonics and the Bessel functions. The spherical harmonics are chosen to be orthonormal with respect to the angular measure sin(θ) dθ dϕ, as we know. Since the eigenfunctions are orthogonal when integrated over the volume of the ball, by r^2 sin(θ) dθ dϕ dr, the orthogonality relationship of these Bessel functions will incorporate the extra r^2:

$$\int_0^{10} \frac{1}{lr} J_{l+1/2}\left(j_{l+1/2,n} r/10\right) \frac{1}{lr} J_{l+1/2}\left(j_{l+1/2,n} r/10\right) r^2 \, dr = 0$$

unless n = n'

For this problem there is an easier way. If we note that there is no dependence on θ or ϕ, we can tell that the initial condition is orthogonal to all the spherical harmonics except for

$$Y_0^0(\theta,\varphi) = \frac{1}{\sqrt{4\pi}};$$

Since this spherical harmonics is a constant function, the best approximation to the constant initial condition is proportional to it. The effect of this is that the indices *l* and m are set to 0. The same argument does not apply to the r variable, since none of the Bessel functions is a constant. The Fourier Bessel series we need is:

$$-200 = \sum_{lmn=1}^{\infty} c_{00m} \frac{1}{14\pi} \frac{1}{lr} {}_{1/2}\left(j_{l+1/2,n} r/10\right)$$

$$= \sum_{lmn=1}^{\infty} c_{00m} \frac{1}{2\pi r} \sqrt{\frac{20}{j_{1/2,n}}} \sin(n\pi r/10)$$

If we multiply this by r, it becomes a Fourier sine series, of the form

$$-200r = \sum_{lmn=1}^{\infty} b_n \sin(n\pi \cdot r/10)\ !$$

We can find these coefficients by using the standard formula for Fourier coefficients,

In: = b[n_] = (2/10) Integrate[(– 200 r) Sin[n Pi r/10], {r,0,10}]

$$\text{Out} = \frac{4000(-1)^n}{nPi}$$

(It is better to work directly with the coefficients b_n rather than the c_{00n} along with its additional coefficients, which just complicate matters.) When we bring back the t dependence, we have

$$v(t,x) = \sum_{n=1}^{\infty} \frac{(-1)^n 4000}{n\pi r} \sin(n\pi\, r/10)\exp(-n^2\pi^2 t/4),$$

so

$$u(t,x) = 200 + \sum_{n=1}^{\infty} \frac{(-1)^n 4000}{n\pi r} \sin(n\pi\, r/10)\exp(-n^2\pi^2 t/4).$$

To calculate the temperature the center of the steak, we encounter a 0/0 form, so it is useful to recall the limit sin(x)/x – > 1 as x – > 0. The numerical value from the first 7 terms in the series is thus:

In: = N[200 + Sum[(–1)^n 400 Exp[– n^2 Pi^2/4], {n, 1, 7}]]

Out = 166.099

GEOMETRY AND INTEGRAL OPERATORS

The following four problems illustrate, in a simple way, the primary concerns of the next several chapters. The first is a problem about matrices and vectors, and it will be our guide to solving integral equations and differential equations.

Example: Find the *inverse* of a matrix. Example: Let u and v be vectors in R^2, and

$$B = \begin{pmatrix} 3 & -2 \\ -1 & 1 \end{pmatrix}$$

Then

1. $\begin{pmatrix} 1 & 2 \\ 1 & 3 \end{pmatrix} u = v$

If and only if u = B v.

The equivalence of these two matrix-vector equations is easy to establish, and you have no doubt learned long ago how to construct the *inverse* matrix B such that statement (b) is equivalent to statement (a).

Example: Find the inverse of an integral operator. Example: Let u and v be functions of x in the interval [0,1], and let K(x,t):= 1 + x t. The function u is a solution to

1. $u(x) = \int_0^1 K(x,t)u(t)\,dt + x^2$ for x in [0,1]

   if and only if

2. $u(x) = x2 - \int_0^1 K(x,t)u(t)\,dt + x^2$ for x in [0,1].

If one supposes u is given by the formula (2), then the integral calculus will show that u satisfies (1). On the other hand, the task of deriving a formula for u from the relationship in (1) involves unfamiliar techniques, which we shall discuss in this course.

Example: Find the integral solution operator for an ordinary differential equation. Example: Let f and g be continuous functions of x in the interval [0,1], and

$$K(x, t) = \begin{cases} x(1-t) \text{ if } x \le t \\ t(1-x) \text{ if } t \le x. \end{cases}$$

The function g is a solution for

1. $g'' = -f$ and $g(0) = g(1) = 0$

   if and only if

2. $g(x) = \int_0^1 K(x,t)f(t)\,dt,$

Once again, the task of deriving this connection is probably somewhat mysterious, but the result can be checked with the calculus:

VERIFICATION OF MODEL PROBLEM

(a) = > (b) Suppose that f is continuous on [0,1] and $g'' = -f$ with $g(0) = g(1) = 0$. Suppose also that K is as given by sample problem (3). Then

$$\int_0^1 K(x,t)f(t)\,dt, = -\int_0^1 K(x,t)g''(t)\,dt,$$

$$= -(1-x)\int_0^x tg''(t)\,dt$$

Using integration by parts this last line can be rewritten as

$$-(1-x)\left([xg'(x)-0g'(0)]-\int_0^x g'(t)\,dt\right)$$

$$-x\left([(1-1)g'(1)-(1-x)g'(x)]+\int_x^1 g'(t)\,dt\right)$$

= – (1– x)[x g'(x) – (g(x) – g(0)}

– x[– (1– x) g'(x) + (g(1) – g(x))]

= (1– x) g(x) + x g(x) = g(x).

To get the last line we used the assumption that g(1) = g(0) = 0.

(b) = > (a) Again, suppose that f is continuous and, now, suppose that

$$g(x)=\int_0^1 K(x,t)f(t)dt.$$

Then
$$g(0)=\int_0^1 K(0,t)f(t)dt=0,$$

$$g(1)=\int_0^1 K(1,t)f(t)dt=0,$$

and
$$g(x)=(1-x)\int_0^x tf(t)dt+\int_x^1 (1-t)f(t)dt,$$

$$g'(x)=-\int_0^x tf(t)dt+\int_x^1 (1-t)f(t)dt,$$

$$g''(x)=-xf(x)-(1-x)f(x)=-f(x).$$

As you can see, it is not hard to show that these two statements are equivalent. In the next few chapters you will learn how, given statement (a), you can construct K such that statement (b) is equivalent to statement (a). Perhaps you can do this already.

Example: Find the solution to a boundary-value problem.

Example: $u_{xx}+u_{yy}=0$, for $y>0$ and all x, with boundary data that u(x,0) = sin(x), and the "condition at infinity" that u(x,y) remains finite as y – >.

The solution, as can be verified by elementary calculus, is

$$u(x,y) = e^{-y} \sin(x).$$

We shall later learn how to obtain such a solution by means of integral transforms of the boundary data. Most often, we shall take the interval on which our functions are defined to be [0,1]. Of course, we do not work in the class of *all* functions on [0,1]; rather, in the spirit of Chapters I-II, we ask that the linear space should consist of functions f for which

$$\int_0^1 | f(x) |^2 \, dx < \infty.$$

Then we have the inner product space we called $L^2([0,1])$. We recall that the inner product of two functions is given by

$$< f, g > = \int_0^1 f(x)\, g(x)\, dx$$

and the norm of f is defined as the square-root of the inner product of a function with itself:

$$\|f\|2 = \int_0^1 | f(x) |^2 \, dx.$$

(Compare with the norm in R^n.)

It does not seem appropriate to study in detail the nature of $L^2[0,1]$ at this time. Rather, suffice it to say that the space is large enough to contain all continuous functions - even functions which are continuous except at a finite number of places. The interested student can learn more about $L^2[0,1]$ by looking in standard books on real analysis or Hilbert space.

Suppose $\{f_p\}$ is a sequence of functions in $L^2([0,1])$. It is valuable to consider the possible meanings for the statement that $\lim_p f_p(x) = g(x)$. There are three useful interpretations for our purposes:

1. The sequence $\{f_p\}$ converges *pointwise* to g at each x in [0,1] provided that for each x in [0,1], $\lim_p f_p(x) = g(x)$.

   Sometimes we modify this to convergence *pointwise a.e.* when the set on which the convergence fails is a null set.
2. The sequence converges to g *uniformly* on [0,1] provided that

   $\lim_p \sup_x |f_p(x) - g(x)| = 0.$
3. The sequence converges to g *in norm* if

   $\lim_p \| f_p - g \| = 0.$

Convergence in norm is the same as r.m.s. convergence, as described in chapter II. Uniform convergence on a finite interval implies both of the other two notions of convergence, but examples show that other implications among these notions are not generally valid. (Compare with the notions of convergence for sequences of vectors in R^n.)

In this section we study one of the most common types of integral equation. As an example, given a function called the *kernel*

K: [0,1]x[0,1] – > R

and a function f: [0,1] – > R, we seek a function y such that for each x in [0,1],

$$y(x) = \int_0^1 K(x,t)y(t)dt + f(x).$$

Such equations are called Fredholm equations of the second kind. An equation of the form

$$0 = \int_0^1 K(x,t)y(t)dt + f(x).$$

is a Fredholm equation of the first kind.

The requirements in this section on K and f will be that

$$\int_0^1\int_0^1 |K(x,t)|^2\, dx\, dt < \infty \text{ and } \int_0^1 |f(x)|^2\, dx < \infty$$

These requirements are met if K and f are continuous.

For simplicity, we denote by K the linear function given by

$$K(y)(x) = \int_0^1 K(x,t)y(t)dt.$$

Note that K has a domain large enough to contain all functions y which are continuous on [0,1]. Also, if y is continuous then K(y) is a function and its value at x is denoted K(y)(x). In spoken conversation, it is not so easy to distinguish the number valued function K and the function valued K. The bold character will be used in these notes to denoted the latter.

It is well to note the resemblance of this function K to the multiplication of a vector u by a matrix A:

$$A(u)(p) = \sum_{q=1}^{\infty} A(p,q)u(q).$$

This formula has the same form as that for K given above.

It is a historical accident that differential equations were understood before integral equations. Often an integral equation can be converted into a differential equation or vice versa, so many of the laws of nature which we think of as differential equations might just as well have been developed as integral equations initially. In some instances it is easier to differentiate than to integrate, but at other times integral operators are more tractable.

In this course integral operators will be called upon to solve differential equations, and this is one of their main uses.

They have many other uses as well, most notably in the theory of filtering and signal processing. In most of these applications the integral and differential operators are linear transformations. The analogy between linear transformations and matrices is deep and useful.

Just as a matrix has an adjoint, the integral operator K has an adjoint, denoted K*. The adjoint plays an important role in the theory and use of integral equations.

In order to understand K*, one must consider $< K(f), g >$ and seek K* such that $< Kf, g > = < f, K^*g >$.

$$< K(f), g > = \int_0^1 K(f)(x)g(x)dx$$

$$= \int_0^1\int_0^1 K(x,t)f(t)g(x)dt\,dx.$$

An examination of these last equations leads one to guess that K* is given by

$$K^*(g)(t) = \int_0^1 K(x,t)g(x)dx.$$

or, keeping t as the variable of integration,

$$K^*(g)(x) = \int_0^1 K(x,t)g(t)dt.$$

Those last equations verified that

$$< K(f), g > = < f, K^*(g) >.$$

Care has to be taken to watch whether the "variable of integration" is t or x in the integrals involved.

In summary, if K is the kernel associated with the linear operator K, then the kernel associated with K* is given by K*(x,y) [[equivalence]] K(y,x). It is of value to compare how to get K* from K with the process of how to get A* from A:

$$A^*_{p,q} = A_{q,p}.$$

Consistent with the rather standard notation we have adopted above, it is clear that a briefer representation of the equation

$$y(x) = \int_0^1 K(x,t)y(t)dt + f(x).$$

is the concise equation y = K(y) + f, or (1 – K) y = f.

Example: Suppose that

$$K(x, t) = \begin{cases} (x-t)^2 & \text{if } 0 < x < t < 1 \\ 0 & \text{if } 0 < t < x < 1 \end{cases}$$

To get K*, let's use other letters for the argument of K* and K to avoid confusion. Suppose that $0 < u < v < 1$. Then, K*(u,v) = K(v,u) = 0. In a similar manner, K*(u,v) = $(u-v)^2$ if $0 < v < u < 1$. Note that K* is not K.

$$K^*(x,t) = \begin{cases} 0 & \text{if } 0 < x < t < 1 \\ (x-t)^2 & \text{if } 0 < t < x < 1 \end{cases}.$$

The discussion of this example has been algebraic to this point. Consider this geometric notion that is suggested by the alternate name for "self-adjoint", namely, some call K "symmetric" if K(x,t) = K(t,x).

The geometric name suggests a picture and the picture is the graph of K. The K of this example is not symmetric in x and t. Its graph is not symmetric about the line x = t. The function K is different from the function K*.

$$\frac{1}{r}\frac{\left(r\frac{\partial u}{\partial r}\right)}{r} + \frac{1}{r^2}\frac{\partial^2 u}{\partial \theta^2} = 0$$

THE FREDHOLM ALTERNATIVE THEOREMS

A first understanding of the problem of solving an integral equation

$$y = Ky + f$$

can be gotten by referring to the Fredholm Alternative Theorems in this context.

1. Exactly one of the following holds:

   (a) (First Alternative) if f is in $L^2\{0,1\}$, then

$$y(x) = \int_0^1 K(x,t)y(t)dt + f(x)$$

Has one and only one solution.

(b) (Second Alternative)

$$y(x) = \int_0^1 K(x,t)y(t)dt$$

Has a nontrivial solution.

(a) If the first alternative holds for the equation

$$y(x) = \int_0^1 K(x,t)y(t)dt + f(x)$$

Then it also holds for the equation

$$z(x) = \int_0^1 K(t,x)z(t)dt + g(x).$$

(b) In either alternative, the equation

$$y(x) = \int_0^1 K(x,t)y(t)dt$$

and its adjoint equation

$$z(x) = \int_0^1 K(x,t)z(t)dt$$

Have the same number of linearly independent solutions.

Suppose the second alternative holds. Then

$$y(x) = \int_0^1 K(x,t)y(t)dt + f(x)$$

Has a solution if and only if

$$\int_0^1 f(t)z(t)dt = 0$$

For each solution z of the adjoint equation

$$z(x) = \int_0^1 K(x,t)z(t)dt$$

Comparing this context for the Fredholm Alternative Theorems with an understanding of matrix examples seems irresistible. Since these ideas will re-occur in each section, the student should pause to make these comparisons.

Example: Suppose that E is the linear space of continuous functions on the interval [–1,1]. with

$$< f, g > = \int_{-1}^1 f(x)g(x)dx$$

and that

$$K(x,y) = \frac{1}{2} + x(3t^2 - 1).$$

Then $$K^*(x,t) = \frac{1}{2} + x(3x^2 - 1)$$

for $$< K(f), g > = \int_{-1}^1 K(f)(t)g(t)dt$$

$$= \int_{-1}^1 \left(\int_{-1}^1 \left[\frac{1}{2} + t(3s^2 - 1) \right] f(s)ds \right) g(t)dt$$

$$= \int_{-1}^1 \int_{-1}^1 \left[\frac{1}{2} + t(3s^2 - 1) \right] f(s)g(t)ds\, dt$$

$$= \int_{-1}^1 \int_{-1}^1 \left[\frac{1}{2} + s(3t^2 - 1) \right] f(t)g(s)dt\, ds$$

$$= \int_{-1}^{1} f(t) \int_{-1}^{1} \left[\frac{1}{2} + s(3t^2 - 1) \right] g(s) ds\, dt$$

$$= \int_{-1}^{1} f(t) K^*(g)(t) dt.$$

The equation y = K(y) has a non-trivial solution: the constant function 1. To see this, one computes

$$K[1](x) = \int_{-1}^{1} \left[\frac{1}{2} + x(3t^2 - 1) \right] 1 dt = 1 + 0x = 1.$$

One implication of these computations is that the problem y = Ky + f is a second alternative problem. It may be verified that y(x) = 1 is also a nontrivial solution for y = K*y. It follows from the third of the Fredholm alternative theorems that a necessary condition for y = Ky + f to have a solution is that

$$0 = \langle f, 1 \rangle = \int_{-1}^{1} f(x) 1\, dx.$$

Note that one such f is $f(x) = x + x^3$.

SOLVING Y = KY + F

In this chapter we shall learn how to solve integral equations in three situations:

1. K has a separable kernel,
2. K has norm less than one,
3. K is approximated by K's with separable kernels.

These terms will be explained as they are encountered. First we discuss the separable case.

Definition: Suppose that there are an integer n and functions

$$\{a_p(x)\}_{p=1}^{n} \text{ and } \{b_p(x)\}_{p=1}^{n}$$

Such that, for each p, a_p and b_p are in $L^2[0,1]$. Then K has a *separable* kernel if its kernel is given by

$$K(x, t) = \sum_{p=1}^{n} a_p(x) b_p(t)$$

Another term for operators K of this type is *finite-rank*, and we shall see that they can be considered as matrices of finite rank.

With the supposition that K is separable, it is not hard to find y such that y = Ky + f, for this equation can be re-written as

$$y(x) = \sum_{p=1}^{n} a_p(x) \int_0^1 b_p(t)y(t)dt + f(x)$$

or, using the notation of inner products,

$$y(x) = \sum_{p=1}^{n} a_p(x) < b_p, y > f(x).$$

We can see that if the sequence

$$\{a_p(x)\}_{p=1}^{n}$$

Of functions on [0,1] is a linearly independent sequence, then y will have the following special form:

There is a sequence $\{c_p\}$ of numbers such that

$$y(x) = \sum_{p=1}^{n} c_p a_p(x) f(x).$$

Why is this? The definite integrals over t are just numbers. Even though we do not know their values yet, we can call them c_p and procede to determine their values with a bit of algebraic labor. Suppose

$$y(x) = \sum_{p=1}^{n} c_p a_p(x) + f(x)$$

Substitute this in the equation to be solved:

$$\sum_{p=1}^{n} c_p a_p(x) + f(x) = \sum_{p=1}^{n} a_p(x) < b_p, \sum_{p=1}^{n} c_p a_p + f > + f(x)$$

and we see that

$$c_p = \sum_{p=1}^{n} < b_p, a_p > c_p + < b_p, f >.$$

This now reduces to a matrix problem:

$$\begin{pmatrix} c_1 \\ c_2 \\ \cdot \\ \cdot \\ c_n \end{pmatrix} = \begin{pmatrix} < b_1, a_1 > & .. & < b_1, a_n > \\ < b_2, a_1 > & .. & < b_2; a_n > \\ < b_n; a_1 > & .. & . \\ < b_n; a_1 > & .. & < b; a_n > \end{pmatrix} \begin{pmatrix} c_1 \\ c_2 \\ \cdot \\ \cdot \\ c_n \end{pmatrix} + \begin{pmatrix} < b_1; f > \\ < b_2; f > \\ \cdot \\ \cdot \\ < b_n; f > \end{pmatrix}$$

Define K and f to be the matrix and vector so defined that the last equation is rewritten as

$$c = Kc + f$$

We now employ ideas from linear algebra. The equation c = K c + f has exactly one solution provided

$$\det(1 - K) \neq 0.$$

Once the sequence

$$\{c_p\}_{p=1}^{n}$$

Is found, we have a formula for y(x).

Theorem: If K satisfies the condition that

$$\max_x \int_0^1 |K(x,t)|\,dt < 1,$$

Then $\lim_p \phi_p(x)$ exists and the convergence is uniform on [0,1] – in the sense that if $u = \lim_p \phi_p$ then

$$\lim_p \max_x |u(x) - \phi_p(x)| = 0.$$

Proof. Note that

$$|\phi_1(x) - \phi_0(x)| = |\int_0^1 K(s,t)f(t)dt| \leq \int_0^1 |K(x,t)dt \max_x |f(x)|.$$

Furthermore, if p is a positive integer, the distance between successive iterates can be computed:

$$|\phi_{p+1}(x) - \phi_p(x)| = |\int_0^1 |K(x,t)[\phi_p(t) - \phi_{p-1}(t)]dt|$$

$$\leq \int_0^1 |K(x,t)dt \max_x |\phi_p(x) - \phi_{p-1}(x)|$$

Inductively, this does not exceed

$$\left[\max_x \int_0^1 |K(x,t)|\,dt\right]^{p+1} \max_x |f(x)|.$$

Thus, if

$$r = \max_x \int_0^1 |K(x,t)|\,dt$$

and n > m then

$$|\phi_n(x) - \phi_m(x)| \le \frac{r^{m+1}}{1-r} \max_x |f(x)|.$$

Hence, the sequence $\{\phi_p\}$ of functions converges uniformly on [0,1] to a limit function and this limit provides a solution to the equation

$$u(x) = \int_0^1 K(x,t)u(t)dt + f(x).$$

Corollary: If

$$r = \max_x \int_0^1 K(x,t)dt$$

and

$$u = \lim_p \phi_p$$

Then

$$\max_x |u(x) - \phi_m(x)| \le \frac{r^{m+1}}{1-r} \max_x |f(x)|.$$

Sometimes it is convenient to express the iteration as an infinite series, called the Neumann series, i.e., the sum of $\psi_n = \phi_n\phi_{n-1}$. We reason this way in the next example.

Model: Consider the integral equation

$$u(x) = \int_0^1 \exp(t - x)u(t)dt + g(x).$$

Where g(x) is given. We wish to solve for u(x), and we try the method of iteration.

We begin with the guess $\psi_0 = g(x)$, and calculate the next couple of iterates:

$$\psi_1(x) = K[g](x) = \int_0^\lambda \exp(t - x)g(t)dt.$$

$$\psi_2(x) = K^\lambda\,[g](x) = \int_0^\lambda \exp(t_2 - x) \int_0^\lambda \exp(t - t_2)g(t)dtdt_1.$$

This integral can be simplified by reversing the order of integration. Setting the limits takes a moment of reflection, and may be helped by the following diagram:

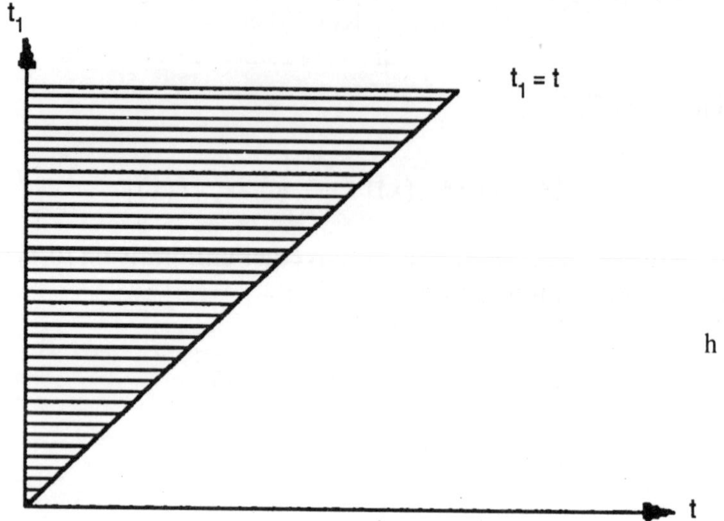

The relationship of the variables is $0 < t < t_1 < x$. If the first (inside) integral is in the variable t, then it runs from 0 to t_1, and then the second integral in the variable t_1 runs from 0 to x. If we reverse the order, the first integral, in the variable t_1, runs from t to x, and the second integral runs from 0 to x. We find that $\psi_2(x)$ is:

$$\int_0^{\lambda} \exp(t_2 - x) \int_0^{\lambda} \exp(t - t_2) g(t) dt dt_1$$

$$= \int_0^{\lambda} g(t) \int_0^{\lambda} \exp(t_1 - x) \exp(t - t_2) dt_1 dt \int_1^{\lambda} g(t)(x - t) \exp(t - x) dt.$$

If we now calculate the further iterates, we find inductively that

$$\psi_\pi(x) = \int_0^{\lambda} g(t) \frac{(x-t)^{\pi-1}}{(n-1)!} \exp(t - x) dt.$$

The special feature about this example is that the iterates can be summed up, and when we recall that

$$\sum_{\pi=1}^{\infty} \frac{(x-t)^{\pi-1}}{(n-1)!} \exp(x - t).$$

we get:

$$u(x) = \sum_{\pi=1}^{\infty} \psi_\pi(x) = \psi_0(x) + \sum_{\pi=1}^{\infty} \psi_0(x) - g(t) dt.$$

We leave it as an exercise to check this solution.

It is a miracle when the series for K sums in closed form like this, but that is not important in applications, since the convergence of the Neumann series implies that we can calculate the answer to any desired accuracy.

There is different, independent, way in which K can be considered small, which leads to convergence of the iteration process in the norm of $L^2[0,1]$. This hypothesis asks that

$$\int_0^1\int_0^1 | K(x,t)|^2 \, dt \, dx < 1.$$

Theorem: If K satisfies the Hilbert- Schmidt condition, then $\lim_p {}_p(x)$ exists and the convergence is in the r.m.s. sense, that is:

$$\lim_p \| u(x) - f_p(x) \| = 0.$$

Indication of Proof. The analysis of the nature of the convergence will go like this:

$$\|\phi_1 - \phi_2\|^2$$

is defined to be

$$\int_0^1 | \phi_1(x) - \phi_0(x)|^2 \, dx$$

$$= \int_0^1 | \int_0^1 K(x,t)f(t)dt \,|^2 \, dx$$

$$\leq \int_0^1\int_0^1 | K(x,t)|^2 \, dx \, dt \int_0^1 | f(t)|^2 \, dt.$$

As a consequence, the sequence ϕ_n is Cauchy convergent:

$$\|\phi_{\pi} - \phi_{\pi}\| \leq \frac{r^{m+1}}{1-r} \| f \|.$$

Let's state the conclusion in a careful way:

Corollary: If $r := \sqrt{\int_0^1\int_0 | K(x,t)|^2 \, dt \, dx}$,

Then

$$u = \lim_p \phi_p$$

Converges and

$$||u - \phi_\pi|| \pounds \leq \frac{r^{m+1}}{1-r} || f ||.$$

Definition: The *resolvent* of an operator K is the inverse operator $(\lambda - K)^{-1}$. This is same as the *solution operator* for the equation

$$y = Ky + f$$

Often the number is taken as 1, and unless stated otherwise we shall always do this.

Before addressing the final case - where

1. K does not have a separable kernel,
2. The smallness condition fails, and
3. The Hilbert-Schmidt smallness condition fails, we generate resolvents for the integral equations.

Re-examining the iteration process:

$$\phi_0(x) = f(x),$$

$$\phi_1(x) = K\phi_0(x) + f(x)$$

$$\phi_2(x) = K(K(\phi_0))x + K(f)(x) + f(x)$$

One writes $\phi_0 = f$, $\phi_1 = Kf + f$, $\phi_2 = K[Kf + f] + f = K^2 f + Kf + f, ...$

In fact, with

$$Kf(x) = \int_0^1 K(x,t)f(t)dt$$

$$K^2f(x) = \int_0^1 K(x,t)[K(f)](t)dt$$

$$= \int_0^1 K(x,t)[\int_0^1 K(t,s)f(s)ds]dt$$

$$= \int_0^1[\int_0^1 K(x,t)K(t,s)dt]f(s)ds$$

Hence, the kernel K_2 associated with K^2 is

$$K_2(x, t) = \int_0^1 K(x,s)K(s,t)ds.$$

Inductively,

$$K_n f(x) = \int_0^1 K(x,t)[K^{n-1}f](t)dt\ K_\pi(x,t)$$

$$= \int_0^1 K(x,s)K_{n-1}(s,t)\,ds$$

and $$\phi_\pi(x) = f(x) + \sum_{p=1}^{\infty} K^p f(x).$$

We have, in this section, conditions which imply that

$$\Sigma_{p=1} K^p f$$

Converges and that its limit y satisfies y = Ky + f. We have thus defined the resolvent for K,

$$(1 - K)^{-1} = 1 + R, \text{ where}$$

$$R = \Sigma_{p=1} K^p.$$

Note that R operates on elements of $L^2[0,1]$, and, subject to convergence, it has an integral kernel. The solution of

$$y = Ky + f$$

Has a very similar form to the equation itself.

$$y(x) = (1 + R)\,[f] = f(x) + \int_0^1 R(x,t)f(t)dt$$

Indeed,

$$f = -Rf + y.$$

Algebraically, we can identifying (1 + R) with

$$(1 - K)^{-1} = 1 + K(1 - K)^{-1},$$

So that

$$R = K\,(1 - K)^{-1}.$$

In case K neither has a separable kernel nor is small, then the next resort is to approximate K with an operator which has a separable kernel.

Theorem. If $\int_0^1\int_0^1 |K(x,t)|^2\,dx\,dt < \infty$

Then there are kernels K_n and G such that

1. $K = K_n + G$,
2. K_n has a separable kernel, and
3. $\int_0^1\int_0^1 |G(x,t)|^2\, dx\, dt < 1.$

In the succeeding pages, we show how to compute K_n and G. However, we first illustrate that the problem is - in theory - solved if we have such a resolution of K into K_n and G. We seek y such that

$$y = Ky + f = K_n y + Gy + f$$

or

$$y - Gy = K_n y + f.$$

Use the resolvent for G:

$$(1 - G)^{-1} = 1 + RG,$$

to get that

$$y = K_n y + RG(K_n y + f) + f$$

$$= [K_n + RGK_n]y + (RGf + f).$$

Define z to be RGf + f, or, what is the same, solve the equation

$$z = Gz + f.$$

We can solve this equation because G is small. Now, we seek y such that

$$y = (K_n + RGK_n)y + z.$$

Re-writing this as an integral equation, we seek y such that

$$y(x) = \int_0^1 H(x,t)y(t)dt + z(x)$$

Where

$$H(x,t) = K_n(x,t) + \int_0^1 R_G(x,s)K_n(s,t)ds.$$

What is astonishing is that this last integral equation is separable! To see this, suppose

$$K_n(x,t) = \sum_{p=1}^{n} a_p(x)b_p(t)$$

Then

$$\int_0^1 R_G(x,s)K_n(s,t)ds$$

$$= \int_0^1 R_G(x,s) \sum_{p=1}^n a_p(s) b_p(t) ds$$

$$= \sum_{p=1}^n a_p \int_0^1 R_G(x,s) a_p(s) ds \; b_p(t)$$

So, here is the conclusion. If K is K_n + G as in the above Theorem, in order to solve y = Ky + f, use the fact that

$$\int_0^1 \int_0^1 | R_G(x,t) |^2 \, dx \, dt < 1$$

Tzo form the resolvent for RG; then find z such that z = (1+R G)f. Finally, solve the separable equation y = (K_n + RGK_n)y + z

We now must address the question of how to achieve the decomposition of K into K_n + G. The ideas are familiar to us from earlier chapter on Fourier series. In summary of those ideas, recall that if p and q are integers, then

$$\int_0^1 \sin(p\pi x) \sin(q\pi x) dx = 0 \text{ if } p \neq q$$

$$= \frac{1}{2} \text{if } p = q.$$

We seek A_{pq} such that

$$K(x,y) = \sum_{p=1}^{\infty} \sum_{qa=1}^{\infty} A_{pq} \sin(p\pi x) \sin(q\pi y).$$

In fact, by integrating both sides of this last equation after multiplying by sin(m π x) sin(n π y), we have

$$\int_0^1 \int_0^1 K(x,y) \sin(m\pi x) \sin(n\pi y) dx dy = \frac{A_{mn}}{4}$$

From the theory of Fourier series,

$$\text{lin}_n \sum_{p=1}^n \sum_{q=1}^n A_{pq} \sin(p\pi x) \sin(q\pi y) = K(x,y)$$

In the sense that

$$\left| \int_0^1 \int_0^1 [K(x,y) - \sum_{p=1}^n \sum_{q=1}^n A_{pq} \sin(p\pi x) \sin(q\pi y)]^2 dx \, dy \right| \to 0$$

as n – > *. Let n be an integer such that

$$|\int_0^1\int_0^1[K(s,y)-\sum_{p=1}^{n}\sum_{p=1}^{n}A_{pq}\sin(p\pi x)\sin(q\pi y)]^2 dx\ dy|<1.$$

Define K_n and G by

$$K_n(x, y) = \sum_{p=1}^{n}\sum_{q=1}^{n}A_{pq}\sin(p\pi x)\sin(q\pi y)$$

and

$$G = K - K_n.$$

Then these three requirements are met:

1. $K = K_n + G$,
2. K_n is separable, and
3. $|\int_0^1\int_0^1 | G(x,y)|^2\ dx\ dy < 1$

Thus, we have an analysis of an integral equation y = Ky + f where

$$\int_0^1\int_0^1 | K(x,y)|^2\ dx\ dy < \infty$$

The engineer will want to know about approximations. Here are two appropriate questions:

(a) Suppose one hopes to solve y = Ky + f and that K_n is separable and approximates K. How well does the solution u for u = K_n u + f approximate y?

(b) Suppose K = K_n + G and

$$s = \sum_{q=1}^{n} G^p$$

G^p approximates RG. How well does the solution u for

$$u = [K_n + SK_n]\ u + [1+S\ ]f$$

Approximate y?

Chapter 12

Step Functions

Before proceeding we should take a look at one more function. Without Laplace transforms it would be much more difficult to solve differential equations that involve this function in $g(t)$.

The function is the Heaviside function and is defined as,

$$u_c(t) = \begin{cases} 0 & \text{if } t < c \\ 1 & \text{if } t \geq c \end{cases}.$$

Here is a graph of the Heaviside function.

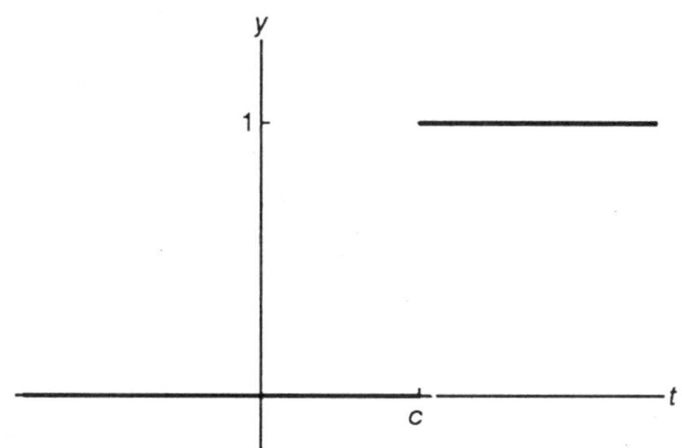

Heaviside functions are often called step functions. Here is some alternate notation for Heaviside functions.

$$u_c(t) = u(t-c) = H(t-c).$$

We can think of the Heaviside function as a switch that is off until $t = c$ at which point it turns on and takes a value of 1. So what if we want a switch that will turn on and takes some other value, say 4, or --7?

Heaviside functions can only take values of 0 or 1, but we can use them to get other kinds of switches. For instance $4u_c(t)$ is a switch that is off until $t = c$ and then turns on and takes a value of 4. Likewise, $-7u_c(t)$ will be a switch that will take a value of -7 when it turns on.

Now, suppose that we want a switch that is on (with a value of 1) and then turns off at $t = c$. We can use Heaviside functions to represent this as well. The following function will exhibit this kind of behavior.

$$1 - u_c(t) = \begin{cases} 1-0=1 & \text{if } t < c \\ 1-1=0 & \text{if } t \geq c \end{cases}.$$

Prior to $t = c$ the Heaviside is off and so has a value of zero. The function as whole then for $t < c$ has a value of 1. When we hit $t = c$ the Heaviside function wi!l turn on and the function will now take a value of 0.

We can also modify this so that it has values other than 1 when it is one. For instance,

$$3 - 3u_c(t)$$

will be a switch that has a value of 3 until it turns off at $t = c$.

We can also use Heaviside functions to represent much more complicated switches.

Example: Write the following function (or switch) in terms of Heaviside functions.

$$f(t) = \begin{cases} -4 & \text{if } t < 6 \\ 25 & \text{if } 6 \leq t < 6 \\ 16 & \text{if } 8 \leq t < 30 \\ 10 & \text{if } t \geq 30 \end{cases}.$$

Solution: There are three sudden shifts in this function and so (hopefully) it's clear that we're going to need three Heaviside functions here, one for each shift in the function. Here's the function in terms of Heaviside functions.

$$f(t) = -4 + 29u_6(t) - 9u_8(t) - 6u_{30}(t).$$

It's fairly easy to verify this.

In the first interval, $t < 6$ all three Heaviside functions are off and the function has the value

$$f(t) = -4.$$

Notice that when we know that Heaviside functions are on or off we tend to not write them at all as we did in this case.

In the next interval, $6 \leq t < 8$the first Heaviside function is now on while the remaining two are still off. So, in this case the function has the value.

$$f(t) = -4. + 29 = 25.$$

In the third interval, $8 \leq t < 30$the first two Heaviside functions are one while the last remains off. Here the function has the value.

$$f(t) = -4 + 29 - 9 = 16.$$

In the last interval, $t \geq 30$all three Heaviside function are one and the function has the value.

$$f(t) = -4 + 29 - 9 - 6 = 10.$$

So, the function has the correct value in all the intervals.

All of this is fine, but if we continue the idea of using Heaviside function to represent switches, we really need to acknowledge that most switches will not turn on and take constant values. Most switches will turn on and vary continually with the value of *t*.

So, let's consider the following function.

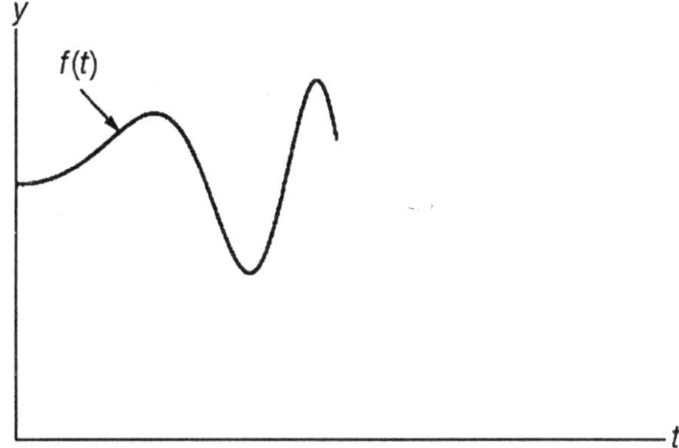

We would like a switch that is off until $t = c$ and then turns on and takes the values above. By this we mean that when $t = c$ we want the switch to turn on and take the value of $f(0)$ and when $t = c + 4$ we want the switch to turn on and take the value of $f(4)$, *etc.* In other words, we want the switch to look like the following,

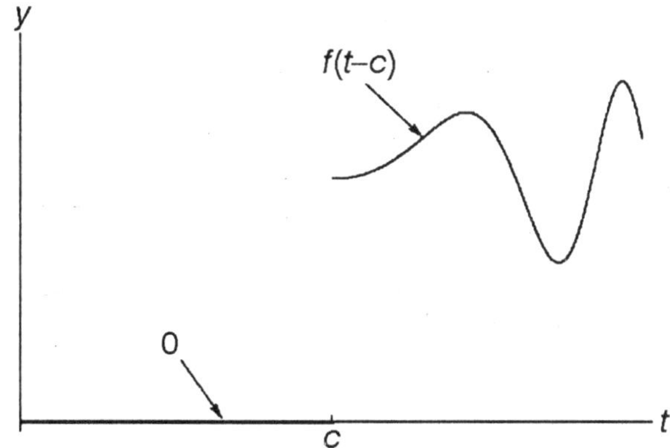

Notice that in order to take the values that we want the switch to take it needs to turn on and take the values of $f(t - c)$! We can use Heaviside functions to help us represent this switch as well. Using Heaviside functions this switch can be wrote as

$$g(t) = u_c(t) f(t - c).$$

Okay, we've talked a lot about Heaviside functions to this point, but we haven't even touched on Laplace transforms yet. So, let's start thinking about that. Let's determine the Laplace transform. This is actually easy enough to derive so let's do that. Plugging into the definition of the Laplace transform gives,

$$\mathfrak{L}\{u_c(t)f(t-c)\} = \int_0^\infty e^{-st}u_c(t)f(t-c)dt$$
$$= \int_0^\infty e^{-st}f(t-c)dt.$$

Notice that we took advantage of the fact that the Heaviside function will be zero if $t < c$ and 1 otherwise. This means that we can drop the Heaviside function and start the integral at c instead of 0. Now use the substitution $u = t - c$ and the integral becomes,

$$\mathfrak{L}\{u_c(t)f(t-c)\} = \int_0^\infty e^{-s(u+c)}f(u)du$$
$$= \int_0^\infty e^{-su}e^{-cs}(u)du.$$

The second exponential has no u's in it and so it can be factored out of the integral. Note as well that in the substitution process the lower limit of integration went back to 0.

$$\mathfrak{L}\{u_c(t)f(t-c)\} = e^{-cs}\int_0^\infty e^{-su}f(u)du.$$

Now, the integral left is nothing more than the integral that we would need to compute if we were going to find the Laplace transform of *f(t)*. Therefore, we get the following formula

$$\mathfrak{L}\{u_c(t)f(t-c)\} = e^{-cs}F(s).$$

In order to use the function *f(t)* must be shifted buy c, the same value that is used in the Heaviside function. Also note that we only take the transform of *f(t)* and not *f(t-c)*! We can also turn this around to get a useful formula for inverse Laplace transforms.-

$$\mathfrak{L}^{-1}\{e^{-cs}F(s)\} = u_c(t)f(t-c).$$

We can use to get the Laplace transform of a Heaviside function by itself. To do this we will consider the function to by $f(t) = 1$. Doing this gives us

$$\mathfrak{L}\{u_c(t)\} = \mathfrak{L}\{u_c(t)\bullet 1\} = e^{-cs}\mathfrak{L}\{1\} = \frac{1}{s}e^{-cs} = \frac{e^{-cs}}{s}.$$

Putting all of this together leads to the following two formulas.

$$\mathfrak{L}\{u_c(t)\} = \frac{e^{-cs}}{s} \qquad \mathfrak{L}^{-1}\left\{\frac{e^{-cs}}{s}\right\} = u_c(t).$$

Let' do some examples.

Example: Find the Laplace transform of each of the following.

1. $g(t) = 10u_{12}(t) + 2(t-6)^3 u_6(t) - (7 - e^{12-3t})u_4(t)$
2. $f(t) = -t^2 u_3(t) + \cos(t))u_5(t)$
3. $h(t) = \begin{cases} t^4 & \text{if } t < 5 \\ t^4 + 3\sin\left(\dfrac{t}{10} - \dfrac{1}{2}\right) & \text{if } t \geq 5 \end{cases}$
4. $f(t) = \begin{cases} t & \text{if } t < 6 \\ -8 + (t-6)^2 & \text{if } t \geq 6 \end{cases}$.

Solution: In all of these problems remember that the function MUST be in the form

$$u_c(t) f(t-c)$$

before we start taking transforms. If it isn't in that form we will have to put it into that form!

1. $g(t) = 10u_{12}(t) + 2(t-6)^3 u_6(t) - \left(7 - e^{12-3t}\right)u_4(t)$.

So there are three terms in this function. The first is simply a Heaviside function and so we can use on this term. The second and third terms however have functions with them and we need to identify the functions that are shifted for each of these. In the second term it the following function is shifted,

$$f(t) = 2t^3 \quad \Rightarrow \quad f(t-6) = 2(t-6)^3$$

and this has been shifted by the correct amount.

The third function uses,

$$f(t) = 7 - e^{-3t} \quad \Rightarrow \quad f(t-4) = 7 - e^{-3(t-4)} = 7 - e^{12-3t}$$

which has also been shifted by the correct amount.

With these functions identified we can now take the transform of the function.

$$G(s) = \frac{10e^{-12s}}{s} + e^{-6s}\frac{2(3!)}{s^{3+1}} - \left(\frac{7}{s} - \frac{1}{s+3}\right)e^{-4s}$$

$$= \frac{10e^{-12s}}{s} + \frac{12e^{-6s}}{s^{3+1}} - \left(\frac{7}{s} - \frac{1}{s+3}\right)e^{-4s}$$

2. $f(t) = -t^2 u_3(t) + \cos(t)\, u_5(t)$.

This part is going to cause some problems. There are two terms and neither has been shifted by the proper amount. The first term needs to be shifted by 3 and the second needs to be shifted by 5. So, since they haven't been shifted, we will need to force the issue. We will need to add in the shifts, and then take them back out of course. Here they are.

$$f(t) = -(t-3+3)^2 u_3(t) + \cos(t-5+5)\, u_5(t).$$

Now we still have some potential problems here. The first function is still not really sifted correctly, so we'll need to use

$$(a+b)^2 a^2 + 2ab + b^2$$

to get this shifted correctly.

The second term can be dealt with in one of two ways. The first would be to use the formula

$$\cos(a+b) = \cos(a)\cos(b) - \sin(a)\sin(b)$$

to break it up into cosines and sines with arguments of t-5 which will be shifted as we expect. There is an easier way to do this one however. From out table of Laplace transforms we can see that if

$$g(t) = \cos(t+5) \quad \Rightarrow \quad g(t-5) = \cos(t-5+5).$$

This will make our life a little easier so we'll do it this way. Now, breaking up the first term and leaving the second term along gives us,

$$f(t) - ((t-3)^2 + 6(t-3) + 9)\, u_3(t) + \cos(t-5+5)\, u_5(t).$$

Okay, so it looks like the two functions that have been shifted here are

$$g(t) = t^2 + 6t + 9$$
$$g(t) = \cos(t+5).$$

Taking the transform then gives,

$$F(s) = -\left(\frac{2}{s^3} + \frac{6}{s^2} + \frac{9}{s}\right) e^{-3s} + \left(\frac{s\cos(5) - \sin(5)}{s^2+1}\right) e^{-5s}.$$

It's messy, especially the second term, but there it is. Also, do not get excited about the cos (5) and sin (5). They are just numbers.

3. $h(t) = \begin{cases} t^4 & \text{if } t < 5 \\ t^4 + 3\sin\left(\dfrac{t}{10} - \dfrac{1}{2}\right) & \text{if } t \ge 5 \end{cases}$

This one isn't as bad as it might look on the surface. The first thing that we need to do is write it in terms of Heaviside functions.

$$h(t) = t^4 + 3u_5(t)\sin\left(\frac{t}{10} - \frac{1}{2}\right)$$

$$= t^4 + 3u_5(t)\sin\left(\frac{t}{10}(t-5)\right).$$

Since the t^4 is in both terms there isn't anything to do when we add in the Heaviside function. The only thing that gets added in is the sine term. Notice as well that the sine has been shifted by the proper amount.

All we need to do now is to take the transform.

$$H(s) = \frac{4!}{s^5} + \frac{3\left(\frac{1}{10}\right)e^{-5s}}{s^2 + \left(\frac{1}{10}\right)^2} = \frac{24}{s^5} + \frac{\frac{3}{10}e^{-5s}}{s^2 + \frac{1}{100}}$$

4. $f(t) = \begin{cases} t & \text{if } t < 6 \\ -8 + (t-6)^2 & \text{if } t \geq 6 \end{cases}$.

Again, the first thing that we need to do is write the function in terms of Heaviside functions.

$$f(t) = t + \left(-8 - t + (t-6)^2\right)u_6(t).$$

We had to add in a "-8" in the second term since that appears in the second part and we also had to subtract a t in the second term since the t in the first portion is no longer there. This subtraction of the t adds a problem because the second function is no longer correctly shifted. This is easier to fix than the previous example however.-

Here is the corrected function.

$$f(t) = t + \left(-8 - t + (t-6+6) + (t-6)^2\right)u_6(t)$$

$$= t + \left(-8 - (t-6) - 6 + (t-6)^2\right)u_6(t)$$

$$= t + \left(-14 - (t-6) + (t-6)^2\right)u_6(t)$$

So, in the second term it looks like we are shifting

$$g(t) = t^2 - t - 14.$$

The transform is then,

$$F(s) = \frac{1}{s^2} + \left(\frac{2}{s^3} - \frac{1}{s^2} - \frac{14}{s}\right)e^{-6s}.$$

Without the Heaviside function taking Laplace transforms is not a terribly difficult process provided we have our trusty table of transforms. However, with the advent of Heaviside functions, taking transforms can become a fairly messy process on occasion.

So, let's do some inverse Laplace transforms to see how they are done.

Example: Find the inverse Laplace transform of each of the following.

1. $H(s) = \dfrac{se^{-4s}}{(3s+2)(s-2)}$

2. $G(s) = \dfrac{5e^{-6s} - 3e^{-11s}}{(s+2)(s^2+9)}$

3. $F(s) = \dfrac{4s + e^{-s}}{(s-1)(s+2)}$

4. $G(s) = \dfrac{3s + 8e^{-20s} - 2se^{-3s} + 6e^{-7s}}{s^2(s+3)}$

Solution: Notice that in order to use this formula the exponential doesn't really enter into the mix until the very end. The vast majority of the process is finding the inverse transform of the stuff without the exponential.

In these problems we are not going to go into detail on many of the inverse transforms. If you need a refresher on some of the basics of inverse transforms go back and take a look at the previous section.

1. $H(s) = \dfrac{se^{-4s}}{(3s+2)(s-2)}$

In light of the comments above let's first rewrite the transform in the following way.

$$H(s) = e^{-4s} \frac{s}{(3s+2)(s-2)} = e^{-4s} F(s).$$

Now, this problem really comes down to needing *f(t)*. So, let's do that. We'll need to partial fraction *F(s)* up. Here's the partial fraction decomposition.

$$F(s) = \frac{A}{3s+2} + \frac{B}{s-2}.$$

Setting numerators equal gives,

$$s = A(s-2) + B(3s+2).$$

We'll find the constants here by selecting values of s. Doing this gives,

$$s = 2 \qquad 2 = 8B \quad \Rightarrow \quad B = \frac{1}{4}$$

$$s = -\frac{2}{3} \qquad -\frac{2}{3} = -\frac{8}{3}A \quad \Rightarrow \quad B = \frac{1}{4}.$$

So, the partial fraction decomposition becomes,

$$F(s) = \frac{\frac{1}{4}}{3\left(s + \frac{2}{3}\right)} + \frac{\frac{1}{4}}{s - 2}.$$

Notice that we factored a 3 out of the denominator in order to actually do the inverse transform. The inverse transform of this is then,

$$f(t) = \frac{1}{12}e^{\frac{2t}{3}} + \frac{1}{4}e^{2t}.$$

Now, let's go back and do the actual problem. The original transform was,

$$H(s) = e^{-4s}F(s).$$

Note that we didn't bother to plug in *F(s)*. There really isn't a reason to plug it back in. Let's write down the inverse transform in terms of symbols. The inverse transform is,

$$h(t) = u_4(t)f(t-4)$$

where, *f(t)* is,

$$f(t) = \frac{1}{12}e^{-\frac{2t}{3}} + \frac{1}{4}e^{2t}.$$

This is all the farther that we'll go with the answer. There really isn't any reason to plug in *f(t)* at this point. It would make the function longer and definitely messier. We will give almost all of our answers to these types of inverse transforms in this form.

2. $G(s) = \dfrac{5e^{-6s}}{(s+2)(s^2+9)}$.

This problem is not as difficult as it might at first appear to be. Because there are two exponentials we will need to deal with them separately eventually. Now, this might lead us to conclude that the best way to deal with this function is to split it up as follows,

$$G(s) = e^{-6s}\frac{5}{(s+2)(s^2+9)} - e^{-11s}\frac{3}{(s+2)(s^2+9)}.$$

Notice that we factored out the exponential, as we did in the last example, since we would need to do that eventually anyway. This is where a fairly common complication arises. Many people will call the first function *F(s)* and the second function *H(s)* and the partial fraction both of them.

However, if instead of just factoring out the exponential we would also factor out the coefficient we would get,

$$G(s) = 5e^{-6s}\frac{1}{(s+2)(s^2+9)} - 3e^{-11s}\frac{1}{(s+2)(s^2+9)}.$$

Upon doing this we can see that the two functions are in fact the same function. The only difference is the constant that was in the numerator. So, the way that we'll do these problems is to first notice that both of the exponentials have only constants as coefficients. Instead of breaking things up then, we will simply factor out the whole numerator and get,

$$G(s) = \left(5e^{-6s} - 3e^{-11s}\right)\frac{1}{(s+2)(s^2+9)}\left(5e^{-6s} - 3e^{-11s}\right)F(s)$$

and now we will just partial fraction *F(s)*.

Here is the partial fraction decomposition.

$$F(s) = \frac{A}{s+2} + \frac{Bs+C}{s^2+9}.$$

Setting numerators equal and combining gives us,

$$\begin{aligned} 1 &= A\left(s^2+9\right) + (s+2)(Bs+C) \\ &= (A+B)s^2 + (2B+C)s + 9A + 2C. \end{aligned}$$

Setting coefficient equal and solving gives,

$$\left.\begin{aligned} s^2 &: \quad A+B=0 \\ s^1 &: \; 2B+C=0 \\ s^0 &: 9A+2C=1 \end{aligned}\right\} \Rightarrow \quad A = \frac{1}{13}, \quad B = -\frac{1}{13}, C = \frac{2}{13}.$$

Substituting back into the transform gives and fixing up the numerators as needed gives,

$$\begin{aligned} F(s) &= \frac{1}{13}\left(\frac{1}{s+2} + \frac{-s+2}{s^2+9}\right) \\ &= \frac{1}{13}\left(\frac{1}{s+2} - \frac{s}{s^2+9} + \frac{2\frac{3}{3}}{s^2+9}\right). \end{aligned}$$

As we did in the previous section we factored out the common denominator to make our work a little simpler. Taking the inverse transform then gives,

$$f(t) = \frac{1}{13}\left(e^{-2t} - \cos(3t) + \frac{2}{3}\sin(3t)\right).$$

At this point we can go back and start thinking about the original problem.

$$G(s) = \left(5e^{-6s} - 3e^{-11s}\right)F(s)$$
$$= 5e^{-6s}F(s) - 3e^{-11s}F(s).$$

We'll also need to distribute the *F(s)* through as well in order to get the correct inverse transform. Recall that in order to take the inverse transform you must have a single exponential times a single transform. This means that we must multiply the *F(s)* through the parenthesis. We can now take the inverse transform,

$$g(t) = 5u_6(t)f(t-6) - 3u_u(t)f(t-11)$$

where,
$$f(t) = \frac{1}{13}\left(e^{-2t} - \cos(3t) + \frac{2}{3}\sin(3t)\right)$$

3. $F(s) = \dfrac{4s + e^{-s}}{(s-1)(s+2)}$.

In this case, unlike the previous part, we will need to break up the transform since one term has a constant in it and the other has an *s*. Note as well that we don't consider the exponential in this, only its coefficient. Breaking up the transform gives,

$$F(s) = \frac{4s}{(s-1)(s+2)} + e^{-s}\frac{1}{(s-1)(s+2)} = G(s) + e^{-s}H(s).$$

We will need to partial fraction both of these terms up. We'll start with *G(s)*.

$$G(s) = \frac{A}{s-1} + \frac{B}{s+2}.$$

Setting numerators equal gives,

$$4s = A(s+2) + B(s-1).$$

Now, pick values of *s* to find the constants.

$$s = -2 \qquad -8 = -3B \quad \Rightarrow \quad B = \frac{8}{3}$$
$$s = 1 \qquad -4 = 3A \quad \Rightarrow \quad A = \frac{4}{3}$$

So *G(s)* and its inverse transform is,

$$G(s) = \frac{\frac{4}{3}}{s-1} + \frac{\frac{8}{3}}{s+2}$$

$$g(t) = \frac{4}{3}e^{t} + \frac{8}{3}e^{-2t}.$$

Now, repeat the process for *H(s)*.

$$H(s) = \frac{A}{s-1} + \frac{B}{s+2}.$$

Setting numerators equal gives,

$$1 = A(s+2) + B(s-1).$$

Now, pick values of *s* to find the constants.

$$s = -2 \qquad 1 = -3B \qquad \Rightarrow \qquad B = \frac{1}{3}.$$

$$s = 1 \qquad 1 = 3A \qquad \Rightarrow \qquad A = \frac{1}{3}.$$

So *H(s)* and its inverse transform is,

$$H(s) = \frac{\frac{1}{3}}{s-1} - \frac{\frac{1}{3}}{s+2}$$

$$h(t) = \frac{1}{3}e' - \frac{1}{3}e^{-2t}.$$

Putting all of this together gives the following,

$$F(s) = G(s) + e^{-s}H(s)$$

$$f(t) = g(t) + u_1(t)h\,(t-1)$$

where,

$$g(t) = \frac{4}{3}e' + \frac{8}{3}e^{-2t} \quad \text{and} \quad h(t) = \frac{1}{3}e' - \frac{1}{3}e^{-2t}$$

4. $G(s) = \dfrac{3s + 8e^{-20s} - 2e^{-3s} + 6e^{-7s}}{s^2(s+3)}$

This one looks messier than it actually is. Let's first rearrange the numerator a little.

$$G(s) = \frac{s\left(3 - 2e^{-3s}\right) + \left(8e^{20s} + 6e^{-7s}\right)}{s^2(s+3)}.$$

In this form it looks like we can break this up into two pieces that will require partial fractions. When we break these up we should always try and break things up into as few pieces as possible for the partial fractioning. Doing this can save you a great deal of unnecessary work. Breaking up the transform as suggested above gives,

$$G(s) = \left(3 - 2e^{-3s}\right)\frac{1}{s(s+3)} + \left(8e^{-20s} + 6e^{-7s}\right)\frac{1}{s^2(s+3)}$$

$$= \left(3 - 2e^{-3s}\right)F(s) + \left(8e^{-20s} + 6e^{-7s}\right)H(s).$$

Note that we canceled an s in *F(s)*. You should always simplify as much a possible before doing the partial fractions.

Let's partial fraction up *F(s)* first.

$$F(s) = \frac{A}{s} + \frac{B}{s+3}.$$

Setting numerators equal gives, $1 = A(s+3) + Bs.$

Now, pick values of s to find the constants.

$$s = -3 \qquad 1 = -3B \qquad \Rightarrow \qquad B = \frac{1}{3}$$

$$s = 0 \qquad 1 = -3A \qquad \Rightarrow \qquad A = \frac{1}{3}.$$

So *F(s)* and its inverse transform is,

$$F(s) = \frac{\frac{1}{3}}{s} - \frac{\frac{1}{3}}{s+3}$$

$$f(t) = \frac{1}{s} - \frac{1}{3}e^{-3t}.$$

Now partial fraction *H(s)*.

$$H(s) = \frac{A}{s} - \frac{B}{s^2} + \frac{C}{s+3}.$$

Setting numerators equal gives,

$$1 = As(s+3) + B(s+3) + Cs^2.$$

Pick values of s to find the constants.

$$s = -3 \qquad 1 = 9C \qquad \Rightarrow \qquad C = \frac{1}{9}$$

$$s = 0 \qquad 1 = -3B \qquad \Rightarrow \qquad B = \frac{1}{3}$$

$$s = 1 \qquad 1 = 4A + 4B + C = 4A + \frac{13}{3} \qquad \Rightarrow \qquad B = \frac{1}{9}.$$

So *H(s)* and its inverse transform is,

$$H(s) = -\frac{\frac{1}{9}}{s} + \frac{\frac{1}{3}}{s^2} + \frac{\frac{1}{9}}{s+3}$$

$$h(t) = -\frac{1}{9} + \frac{1}{3}t + \frac{1}{9}e^{-3t}.$$

Now, let's go back to the original problem, remembering to multiply the transform through the parenthesis.

$$G(t) = 3F(s) - 2e^{-3s}F(s) + 8e^{-20s}H(s) + 6e^{-7s}H(s).$$

Taking the inverse transform gives,

$$g(t) = 3f(t) - 2u_3(t)f(t-3) + 8u_{20}(t)h(t-20) + 6u_7(t)h(t-7).$$

So, as this example has shown, these can be a somewhat messy. However, the mess is really only that of notation and amount of work. The actual partial fraction work was identical to the previous sections work. The main difference in this section is we had to do more of it. As far as the inverse transform process goes. Again, the vast majority of that was identical to the previous section as well.

So, don't let the apparent messiness of these problems get you to decide that you can't do them. Generally they aren't as bad as they seem initially.

SOLVING IVP'S WITH LAPLACE TRANSFORMS

It's now time to get back to differential equations. We've spent the last three sections learning how to take Laplace transforms and how to take inverse Laplace transforms. These are going to be invaluable skills for the next couple of sections so don't forget what we learned there. Before proceeding into differential equations we will need one more formula. We will need to know how to take the Laplace transform of a derivative. First recall that $f^{(n)}$ denotes the nth derivative of the function f. We now have the following fact.

FACT

Suppose that f, f', f'',...$f^{(n-1)}$ are all continuous functions and $f^{(n)}$ is a piecewise continuous function. Then,-

$$\mathfrak{L}\left\{f^{(n)}\right\} = s^n F(s) - s^{n-1}(0) - s^{n-2}f'(0) - \cdots - sf^{(n-2)}(0) - f^{(n-1)}(0).$$

Since we are going to be dealing with second order differential equations it will be convenient to have the Laplace transform of the first two derivatives.

Notice that the two function evaluations that appear in these formulas, $y(0)$and $y'(0)$, are often what we've been using for initial condition in out IVP's. So, this means that if we are to use these formulas to solve an IVP we will need initial conditions at $t = 0$.

While Laplace transforms are particularly useful for nonhomogeneous differential equations which have Heaviside functions in the forcing function we'll start off with a couple of fairly simple problems to illustrate how the process works.

Example: Solve the following IVP.

$$y' - 10y + 9y = 5t, \qquad y(0) = -1 \quad y'(0) = 2.$$

Solution: The first step in using Laplace transforms to solve an IVP is to take the transform of every term in the differential equation.

$$\mathcal{L}\{y''\}-10\mathcal{L}\{y'\}+9\mathcal{L}\{y\}=\mathcal{L}\{5t\}.$$

Using the appropriate formulas from our table of Laplace transforms gives us the following.

$$5^2Y(s)-sy(0)-y'(0)-10(sY(s)-y(0))+9Y(s)=\frac{5}{s^2}.$$

Plug in the initial conditions and collect all the terms that have a *Y(s)* in them.

$$\left(s^2-10s+9\right)Y(s)+s-12=\frac{5}{s^2}.$$

Solve for *Y(s)*.

$$Y(s)=\frac{5}{s^2(s-9)(s-1)}+\frac{12-s}{(s-9)(s-1)}.$$

At this point it's convenient to recall just what we're trying to do. We are trying to find the solution, *y(t)*, to an IVP. What we've managed to find at this point is not the solution, but its Laplace transform. So, in order to find the solution all that we need to do is to take the inverse transform.

Before doing that let's notice that in its present form we will have to do partial fractions twice. However, if we combine the two terms up we will only be doing partial fractions once. Not only that, but the denominator for the combined term will be identical to the denominator of the first term. This means that we are going to partial fraction up a term with that denominator no matter what so we might as well make the numerator slightly messier and then just partial fraction once.

This is one of those things where we are apparently making the problem messier, but in the process we are going to save ourselves a fair amount of work!

Combining the two terms gives,

$$Y(s)=\frac{5+12s^2-s^2}{s^2(s-9)(s-1)}.$$

The partial fraction decomposition for this transform is,

$$Y(s)=\frac{A}{s}+\frac{B}{s^2}+\frac{C}{s-9}+\frac{D}{s-1}.$$

Setting numerators equal gives,

$$5+12s^2-s^3=As(s-9)(s-1)+B(s-9)(s-1)+Cs^2(s-1)+Ds^2(s-9).$$

Picking appropriate values of s and solving for the constants gives,

$$s = 0 \qquad 5 = 9B \qquad \Rightarrow \qquad B = \frac{5}{9}$$

$$s = 1 \qquad 16 = -8D \qquad \Rightarrow \qquad D = -2$$

$$s = 9 \qquad 248 = 648C \qquad \Rightarrow \qquad C = \frac{31}{81}$$

$$s = 2 \qquad 45 = -14A + \frac{4345}{81} \qquad \Rightarrow \qquad B = \frac{50}{81}$$

Plugging in the constants gives,

$$Y(s)\frac{\frac{50}{81}}{s} + \frac{\frac{5}{9}}{s^2} + \frac{\frac{31}{81}}{s-9} - \frac{2}{s-1}.$$

Finally taking the inverse transform gives us the solution to the IVP.

$$y(t)\frac{50}{81} + \frac{5}{9}t + \frac{31}{81}e^{5t} - 2e^t.$$

That was a fair amount of work for a problem that probably could have been solved much quicker using the techniques for the previous chapter. The point of this problem however, was to show how we would use Laplace transforms to solve an IVP.

There are a couple of things to note here about using Laplace transforms to solve an IVP. First, using Laplace transforms reduces a differential equation down to an algebra problem. In the case of the last example the algebra was probably more complicated than the straight forward approach from the last chapter. However, in later problems this will be reversed. The algebra, while still very messy, will often be easier than a straight forward approach.

Second, unlike the approach in the last chapter, we did not need to first find a general solution, differentiate this, plug in the initial conditions and then solve for the constants to get the solution. With Laplace transforms, the initial conditions are applied during the first step and at the end we get the actual solution instead of a general solution. In many of the later problems Laplace transforms will make the problems significantly easier to work than if we had done the straight forward approach of the last chapter. Also, as we will see, there are some differential equations that simply can't be done using the techniques from the last chapter and so, in those cases, Laplace transforms will be our only solution.

Let's take a look at another fairly simple problem.

Example: Solve the following IVP.

$$2y' + 3y' - 2y = te^{-2t}, \qquad y(0) = 0 \quad y'(0) = -2.$$

Solution: As with the first example, let's first take the Laplace transform of all the terms in the differential equation. We'll the plug in the initial conditions to get,

$$2(s^2Y(s) - sy(0) - y'(0)) + 3(sY(s) - y(0)) - 2Y(s) = \frac{1}{(s+2)^2}$$

$$(2s^2 + 3s - 2)Y(s) + 4 = \frac{1}{(s+2)^2}$$

Now solve for $Y(s)$.

$$Y(s) = \frac{1}{(2s-1)(s+2)^3} - \frac{4}{(2s-1)(s+2)}.$$

Now, as we did in the last example we'll go ahead and combine the two terms together as we will have to partial fraction up the first denominator anyway, so we may as well make the numerator a little more complex and just do a single partial fraction. This will give,

$$Y(s) = \frac{1 - 4(s+2)^2}{(2s-1)(s+2)}$$

$$= \frac{-4s^2 - 16s - 15}{(2s-1)(s+2)^3}.$$

The partial fraction decomposition is then,

$$Y(s) = \frac{A}{2s-1} + \frac{B}{s+2} + \frac{C}{(s+2)^2} + \frac{D}{(s+2)}.$$

Setting numerator equal gives,

$$-4s^2 - 16s - 15 = A(s+2)^3 + B(2s-1)(s+2)^2 + C(2s-1)(s+2) + D(2s-1)$$

$$= (A+2B)s^3 + (6A+7B+2C)s^2 + (12A+4B+2D)s$$

$$+ 8A - 4B - 2C - D.$$

In this case it's probably easier to just set coefficients equal and solve the resulting system of equation rather than pick values of s. So, here is the system and its solution.

$$\left.\begin{array}{ll} s^3: & A + 2B = 0 \\ s^2: & 6A + 7B + 2C = -4 \\ s^1: & 12A + 4B + 3C + 2D = -16 \\ s^0: & 8A - 4B - 2C - D = -15 \end{array}\right\} \Rightarrow \begin{array}{ll} A = -\frac{192}{125} & B = \frac{96}{125} \\ C = -\frac{2}{25} & B = \frac{1}{5} \end{array}$$

We will get a common denominator of 125 on all these coefficients and factor that out when we go to plug them back into the transform. Doing this gives,

$$Y(s) = \frac{1}{125}\left(\frac{-192}{s\left(s-\frac{1}{2}\right)} + \frac{96}{s+2} - \frac{10}{s+2} - \frac{10}{(s+2)^2} - \frac{25\frac{2!}{2!}}{(s+2)^3} \right).$$

Notice that w[illegible] had to factor a 2 out of the denominator of the first term and fix up the numerator of the last term in order to get them to match up to the correct entries in our table of transforms.

Taking the inverse transform then gives,

$$Y(t) = \frac{1}{125}\left(-96e^{\frac{1}{2}} + 96e^{-2t} - 10te^{-2t} \frac{25}{2} t^2 e^{-2t} \right).$$

Example: Solve the following IVP.

$$y' - 6y' + 15y = 2\sin(3t), \qquad y(0) = -1 \quad y'(0) = -4.$$

Solution: Take the Laplace transform of everything and plug in the initial conditions.

$$s^2 y(s) - sy(0) - y'(0) - 6(sY(s) - y(0)) + 15y(s) = 2\frac{3}{s^2+9}$$

$$\left(s^2 - 6s + 15\right)Y(s) + s - 2 = \frac{6}{s^2+9}.$$

Now solve for *Y(s)* and combine into a single term as we did in the previous two examples.

$$Y(s) = \frac{-s^3 + 2s^2 - 9s + 24}{(s^2+9)(s^2-6s+15)}.$$

Now, do the partial fractions on this. First let's get the partial fraction decomposition.

$$Y(s) = \frac{As+B}{s^2+9} + \frac{Cs+D}{s^2-6s+15}.$$

Now, setting numerators equal gives,

$$\begin{aligned} -s^3 + 2s^2 - 9s + 24 &= (As+B)(s^2-6s+15) + (Cs+D)(s^2+9) \\ &= (A+C)s^2 + (-6A+B+D)s^2 + (15A-6B+9C)s + 15B + 9D. \end{aligned}$$

Setting coefficients equal and solving for the constants gives,

$$\left.\begin{aligned} s^3 &: & A + C &= -1 \\ s^2 &: & -6A + B + D &= 2 \\ s^1 &: & 15A - 6B + 9c &= -9 \\ s^0 &: & 15B + 9D &= 24 \end{aligned}\right\} \Rightarrow \begin{aligned} A &= \frac{1}{10} & B &= \frac{1}{10} \\ C &= \frac{11}{10} & D &= \frac{5}{2} \end{aligned}.$$

Now, plug these into the decomposition, complete the square on the denominator of the second term and then fix up the numerators for the inverse transform process.

$$Y(s) = \frac{1}{10}\left(\frac{s+1}{s^2+9} + \frac{-11s+25}{s^2-6s+15}\right)$$

$$= \frac{1}{10}\left(\frac{s+1}{s^2+9} + \frac{-11(s-3+3)+25}{(s-3)^2+6}\right)$$

$$= \frac{1}{10}\left(\frac{s}{s^2+9} + \frac{1\frac{2}{3}}{s^2+9} - \frac{11(s-3)}{(s\quad 3)^2+6} - \frac{8\frac{\sqrt{6}}{\sqrt{6}}}{(s-3)^2+6}\right).$$

Finally, take the inverse transform.

$$y(t) = \frac{1}{10}\left(\cos(3t)\frac{1}{3} + \frac{1}{3}\sin(3t) - 11e^{3t}\cos\left(\sqrt{6t}\right) - \frac{8}{\sqrt{6}}e^{3t}\sin\left(\sqrt{6t}\right)\right).$$

To this point we've only looked at IVP's in which the initial values were at $t = 0$. This is because we need the initial values to be at this point in order to take the Laplace transform of the derivatives. The problem with all of this is that there are IVP's out there in the world that have initial values at places other than $t = 0$. Laplace transforms would not be as useful as it is if we couldn't use it on these types of IVP's. So, we need to take a look at an example in which the initial conditions are not at $t = 0$ in order to see how to handle these kinds of problems.

Example: Solve the following IVP.

$$y' + 4y' = \cos(t-3) + 4t, \qquad y(3) = 0 \quad y'(3) = 7.$$

Solution: The first thing that we will need to do here it to take care of the fact that initial conditions are not at $t = 0$. The only way that we can take the Laplace transform of the derivatives is to have the initial conditions at $t = 0$.

This means that we will need to formulate the IVP in such a way that the initial conditions are at $t = 0$. This is actually fairly simple to do, however we will need to do a change of variable to make it work. We are going to define

$$\eta = t - 3 \qquad \Rightarrow \qquad t = \eta + 3.$$

Let's start with the original differential equation.

$$y'(t) + 4y'(t) = \cos(t-3) + 4t.$$

Notice that we put in the *(t)* part on the derivatives to make sure that we get things correct here. We will next substitute in for t.

$$y'(\eta+3) + 4y'(\eta+3) = \cos(\eta) + 4(\eta+3).$$

Now, to simplify life a little let's define,

$$u(\eta) = y(\eta + 3).$$

Then, by the chain rule, we get that the following for the first derivative.

$$u'(\eta) = \frac{du}{d\eta} = \frac{dy}{dt}\frac{dt}{d\eta} = y'(\eta + 3).$$

By a similar argument we get the following for the second derivative.

$$u'(\eta) = y''(\eta + 3).$$

The initial conditions for *u(ç)* are,

$$u(0) = y(0+3) = y(3) = 0$$

$$u'(0) = y'(0+3) = y'(3) = 7.$$

The IVP under these new variables is then,

$$u'' + 4u' = \cos(\eta) + 4\eta + 12, \quad u(0) = 0 \quad u'(0) = 7.$$

This is an IVP that we can use Laplace transforms on provided we replace all the *t*'s with ç's. So, taking the Laplace transform of this new differential equation and plugging in the new initial conditions gives,

$$\begin{aligned} s^2U(s) - su(0) - u'(0) + 4(sU(s) - u(0)) &= \frac{s}{s^2+1} + \frac{4}{s^2} + \frac{12}{s} \\ (s^2 + 4s)U(s) - 7 &= \frac{s}{s^2+1} + \frac{4+12s}{s^2}. \end{aligned}$$

Solving for *U(s)* gives,

$$\begin{aligned} (s^2 + 4s)\,U(s) &= \frac{s}{s^2+1} + \frac{4+12s+7s^2}{s^2} \\ U(s) &= \frac{1}{(s+4)(s^2+1)} + \frac{4+12s+7s^2}{s^2(s+4)}. \end{aligned}$$

Note that unlike the previous examples we did not completely combine all the terms this time. In all the previous examples we did this because the denominator of one of the terms was the common denominator for all the terms. Therefore, upon combining, all we did was make the numerator a little messier, and reduced the number of partial fractions required down from two to one. Note that all the terms in this transform that had only powers of *s* in denominator where combined for exactly this reason.

In this transform however, if we combined both of the remaining terms into a single term we would be left with a fairly involved partial fraction problem. Therefore, in this case, it would probably be easier to just do partial fractions twice.

We've done several partial fractions problems in this section and many partial fraction problems in the previous couple of sections so we're going to leave the details of the partial fractioning to you to check. Partial fractioning each of the terms in our transform gives us the following.

$$\frac{1}{(s+4)\left(s^2+1\right)}=\frac{\frac{1}{17}}{s+4}+\frac{1}{17}\left(\frac{-s+4}{s^2+1}\right)$$

$$\frac{4+12s+7s^2}{s^3(s+4)}=\frac{1}{s^3}+\frac{\frac{11}{4}}{s^3}-\frac{\frac{17}{16}}{s+4}.$$

Plugging these into our transform and combining like terms gives us

$$\begin{aligned} U(s) &= \frac{1}{s^3}+\frac{\frac{11}{4}}{s^2}+\frac{\frac{17}{16}}{s}-\frac{\frac{273}{272}}{s+4}+\frac{1}{17}\left(\frac{-s+4}{s^2+1}\right) \\ &= \frac{1\frac{2!}{2!}}{s^3}+\frac{\frac{11}{4}}{s^2}+\frac{\frac{17}{16}}{s}-\frac{\frac{273}{272}}{s+4}+\frac{1}{17}\left(\frac{-s}{s^2+1}+\frac{4}{s^2+1}\right). \end{aligned}$$

Now, taking the inverse transform will give the solution to our new IVP. Don't forget to use ç's instead of *t*'s!

$$u(\eta)=\frac{1}{2}\eta^2+\frac{11}{2}\eta+\frac{17}{16}-\frac{273}{272}e^{-4\eta}+\frac{1}{17}(4\sin(\eta)-\cos(\eta)).$$

This is not the solution that we are after of course. We are after *y(t)*. However, we can get this by noticing that

$$y(t)=y(\eta+3)=u(\eta)=u(t-3).$$

So the solution to the original IVP is,

$$y(t)=\frac{1}{2}(t-3)^2+\frac{11}{4}(t-3)+\frac{17}{16}-\frac{273}{272}e^{-4(t-3)}+\frac{1}{17}(4\sin(t-3)-\cos(t-3)).$$

$$y(t)=\frac{1}{2}t^2-\frac{1}{4}t-\frac{43}{16}-\frac{273}{272}e^{-4(t-3)}+\frac{1}{17}(4\sin(t-3)-\cos(t-3)).$$

So, we can now do IVP's that don't have initial conditions that are at $t = 0$. We also saw in the last example that it isn't always the best to combine all the terms into a single partial fraction problem as we have been doing prior to this example.The examples worked in this section would have just as easy, if not easier, if we had used techniques from the previous chapter. They were worked here using Laplace transforms to illustrate the technique and method.

NONCONSTANT COEFFICIENT IVP'S

In this section we are going to see how Laplace transforms can be used to solve some differential equations that do not have constant coefficients. This is not always an easy thing to do. However, there are some simple cases that can be done.

To do this we will need a quick fact.

FACT

If *f(t)* is a piecewise continuous function on (0, ∞) of exponential order then,

$$\lim_{s \to \infty} F(s) = 0.$$

A function *f(t)* is said to be of exponential order áif there is exist positive constants *T* and *M* such that

$$|f(t)| \leq Me^{at} \quad \text{for all } t \geq T.$$

Put in other words, a function that is of exponential order will grow no faster than

$$Me^{at}$$

for some *M* and áand all sufficiently large *t*. One way to check whether a function is of exponential order or not is to compute the following limit.

$$\lim_{t \to \infty} \frac{|f(t)|}{e^{at}}.$$

If this limit is finite for some áthen the function will be of exponential order á. Likewise, if the limit is infinite for every áthen the function is not of exponential order. Almost all of the functions that you are liable to deal with in a first course in differential equations are of exponential order. A good example of a function that is not of exponential order is

$$f(t) = e^{t^3}.$$

We can check this by computing the above limit.

$$\lim_{t \to \infty} \frac{e^{t^3}}{e^{at}} = \lim_{t \to \infty} e^{t^3 - at} = \lim_{t \to \infty} e^{e(t^2 - a)} = \infty.$$

This is true for any value of áand so the function is not of exponential order.

Do not worry too much about this exponential order stuff. This fact is occasionally needed in using Laplace transforms with non constant coefficients.

So, let's take a look at an example.

Example: Solve the following IVP.

$$y'' + 3ty' - 6y = 2, \qquad y(0) = 0 \quad y'(0) = 0.$$

Solution: Laplace transforms tells us that,

$$\begin{aligned} \mathfrak{L}\{ty'\} &= -\frac{d}{ds}(\mathfrak{L}\{y'\}) \\ &= \frac{d}{ds}(sY(s) - y(0)) \\ &= -sY'(s) - Y(s). \end{aligned}$$

So, upon taking the Laplace transforms of everything and plugging in the initial conditions we get,

$$s^2Y(s) - sy(0) - y'(0) + 3(-sy'(s) - Y(s)) - 6Y(s) = \frac{2}{s}$$

$$-3sY'(s) + (s^2 - 9)\,Y(s) = \frac{2}{s}$$

$$Y'(s) + \left(\frac{3}{s} - \frac{s}{3}\right)Y(s) = \frac{2}{3s^2}.$$

Unlike the examples in the previous section where we ended up with a transform for the solution, here we get a linear first order differential equation that must be solved in order to get a transform for the solution.

The integrating factor for this differential equation is,

$$\mu(t) = e^{\int\left(\frac{3}{s}-\frac{s}{3}\right)dt} = e^{\ln(s^3)-\frac{s^2}{6}=s^2e^{\frac{s^2}{6}}}.$$

Multiplying through, integrating and solving for *Y(s)* gives,

$$\begin{aligned} \int\left(s^3e^{\frac{s^2}{s}Y}(s)\right)'ds &= \int -\frac{2}{3}se^{-\frac{s^2}{6}ds} \\ s^3e^{-\frac{s^2}{6}}Y(s) &= 2e^{-\frac{s^2}{6}} + c \\ Y(s) &= \frac{2}{s^3} + c\frac{e^{\frac{s^2}{6}}}{s^3}. \end{aligned}$$

Now, we have a transform for the solution. However that second term looks unlike anything we've seen to this point. This is where the fact about the transforms of exponential order functions comes into play. We are going to assume that whatever our solution is, it is of exponential order. This means that

$$\lim_{s\to\infty}\left(\frac{2}{s^3} + \frac{ce^{\frac{s^2}{6}}}{s^3}\right) = 0.$$

The first term does go to zero in the limit. The second term however, will only go to zero if $c = 0$. Therefore, we must have $c = 0$ in order for this to be the transform of our solution.

So, the transform of our solution, as well as the solution is,

$$Y(s) = \frac{2}{s^3} \qquad y(t) = t^2.$$

I'll leave it to you to verify that this is in fact a solution if you'd like to.

Now, not all nonconstant differential equations need to use. So, let's take a look at one more example.

Example: Solve the following IVP.

$$ty' - ty' + y = 2, \qquad \mathrm{y}(0) = 2 \quad y'(0) = -4.$$

Solution: From the first example we have,

$$\mathcal{L}\{ty'\} = -sY'(s) - Y(s).$$

We'll also need,

$$\begin{aligned} \mathcal{L}\{ty''\} &= -\frac{d}{ds}(\mathcal{L}\{ty''\}) \\ &= -\frac{d}{ds}(s^2Y(s) - sy(0) - y'(0)) \\ &= -s^2Y'(s) - 2sY(s) + y(0). \end{aligned}$$

Taking the Laplace transform of everything and plugging in the initial conditions gives,

$$-s^2Y'(s) - 2sY(s) + y(0) - (-sY')(s) - Y(s)) + Y(s) = \frac{2}{s}$$

$$(s - s^2)Y'(s) + (2 - 2s)Y(s) + 2 = \frac{2}{s}$$

$$s(1-s)Y'(s) + 2(1-s)Y(s) = \frac{2(1-s)}{s^2}$$

$$Y'(s) + \frac{2}{s}Y(s) = \frac{2}{s^2}.$$

Once again we have a linear first order differential equation that we must solve in order to get a transform for the solution. Notice as well that we never used the second initial condition in this work.

That is okay, we will use it eventually.

Since this linear differential equation is much easier to solve compared to the first one, we'll leave the details to you. Upon solving the differential equation we get,

$$Y'(s) = \frac{2}{s} + = \frac{c}{s^2}.$$

Now, this transform goes to zero for all values of c and we can take the inverse transform of the second term. Therefore, we won't need to get rid of the second term as did in the previous example.

Taking the inverse transform gives,

$$y(t) = 2 + ct.$$

Now, is where we will use the second initial condition. Upon differentiating and plugging in the second initial condition we can see that $c = -4$.

So, the solution to this IVP is,

$$y(t) = 2 - 4t.$$

So, we've seen how to use Laplace transforms to solve some nonconstant coefficient differential equations. Notice however that all we did was add in an occasional t to the coefficients. We couldn't get too complicated with the coefficients. If we had we would not have been able to easily use Laplace transforms to solve them.

Sometimes Laplace transforms can be used to solve nonconstant differential equations, however, in general, nonconstant differential equations are still very difficult to solve.

IVP'S WITH STEP FUNCTIONS

In this section we will use Laplace transforms to solve IVP's which contain Heaviside functions in the forcing function. This is where Laplace transform really starts to come into its own as a solution method.

To work these problems we'll just need to remember the following two formulas,

$$\mathfrak{L}\{u_c(t)f(t-c)\} = e^{-cs}F(s) \qquad \text{where} \quad F(s) = \mathfrak{L}\{f(t)\}$$

$$\mathfrak{L}\{e^{-cs}F(s)\} = u_c(t)f(t-c) \qquad \text{where} \quad f(s) = \mathfrak{L}^{-1}\{F(s)\}$$

In other words, we will always need to remember that in order to take the transform of a function that involves a Heaviside we've got to make sure the function has been properly shifted.

Let's work an example.

Example: Solve the following IVP.

$$y'' - y' + 5y = 4 + u_2(t)\, e^{4-2t}, \qquad y(0) = 2 \quad y'(0) = -1.$$

Solution: First let's rewrite the forcing function to make sure that it's being shifted correctly and to identify the function that is actually being shifted.

$$y''-y'+5y=4+u_2(t)\,e^{-2(t-2)}.$$

So, it is being shifted correctly and the function that is being shifted is e^{-2t}. Taking the Laplace transform of everything and plugging in the initial conditions gives,

$$s^2Y(s)-sy(0)-y'(0)-(sY(s)-y(0))+5Y(s)=\frac{4}{s}+\frac{e^{-2s}}{s+2}$$

$$(s^2-s+5)Y(s)-2s+3=\frac{4}{s}+\frac{e^{-2s}}{s+2}.$$

Now solve for *Y(s)*.

$$(s^2-s+5)Y(s) = \frac{4}{s}+\frac{e^{-2s}}{s+2}+2s-3$$

$$(s^2-s+5)Y(s) = \frac{2s^2-3s+4}{s}+\frac{e^{-2s}}{s+2}$$

$$Y(s) = \frac{2s^2-3s+4}{s(s^2-s+5)}+e^{-2s}\frac{1}{(s+2)(s^2-s+5)}$$

$$Y(s) = F(s)+e^{-2s}G(s).$$

Notice that we combined a couple of terms to simplify things a little. Now we need to partial fraction *F(s)* and *G(s)*. We'll leave it to you to check the details of the partial fractions.

$$F(s)=\frac{2s^2-3s+4}{s(s^2-s+5)}=\frac{1}{5}\left(\frac{4}{s}+\frac{6s-11}{s^2-s+5}\right)$$

$$G(s)=\frac{1}{(s+2)(s^2-s+5)}=\frac{1}{11}\left(\frac{1}{s+5}-\frac{s-3}{s+s+5}\right).$$

We now need to do the inverse transforms on each of these. We'll start with *F(s)*.

$$F(s) = \frac{1}{5}\left(\frac{4}{s}+\frac{6\left(s-\frac{1}{2}+\frac{1}{2}\right)-11}{\left(s-\frac{1}{2}\right)+\frac{19}{4}}\right)$$

$$= \frac{1}{5}\left(\frac{4}{s}+\frac{6\left(s-\frac{1}{2}\right)}{\left(s-\frac{1}{2}\right)^2+\frac{19}{4}}-\frac{8\frac{\sqrt{2}}{2}-\frac{2}{\sqrt{19}}}{\left(s-\frac{1}{2}\right)^2+\frac{19}{4}}\right)$$

$$f(t) = \frac{1}{5}\left(4 + 6e^{\frac{t}{2}}\cos\left(\frac{\sqrt{19}}{2}t\right) - \frac{16}{\sqrt{19}}e^{\frac{t}{2}} + \left(\frac{\sqrt{19}}{2}t\right)\right).$$

Now $G(s)$.

$$G(s) = \frac{1}{11}\left(\frac{1}{s+2} + \frac{s - \frac{1}{2} + \frac{1}{2} - 3}{\left(s - \frac{1}{2}\right)^2 + \frac{19}{4}}\right)$$

$$= \frac{1}{11}\left(\frac{1}{s+2} - \frac{s - \frac{1}{2}}{\left(s - \frac{1}{2}\right)^2 + \frac{19}{4}} + \frac{\frac{5}{2}\frac{\sqrt{19}}{\sqrt{19}}}{\left(s - \frac{1}{2}\right)^2 + \frac{19}{4}}\right)$$

$$g(t) = \frac{1}{11}\left(e^{-2t} - e^{\frac{t}{2}}\cos\left(\frac{\sqrt{19}}{2}t\right) + \frac{5}{\sqrt{19}}e^{\frac{t}{2}}\sin\left(\frac{\sqrt{19}}{2}t\right)\right).$$

Okay, we can now get the solution to the differential equation. Starting with the transform we get,

$$y(s) = F(s) + e^{-2s}G(s)$$

$$y(t) = f(t) + u_2(t)g(t-2)$$

where $f(t)$ and $g(t)$ are the functions shown above.

There is can be a fair amount of work involved in solving differential equations that involve Heaviside functions.

Let's take a look at another example or two.

Example: Solve the following IVP.

$$y'' - y' = \cos(2t) + \cos(2t - 12)u_6(t).\ y(0) = -4,\ y'(0) = 0.$$

Solution: Let's rewrite the differential equation so we can identify the function that is actually being shifted.

$$y'' - y' = \cos(2t) + \cos(2(t-6))u_6(t).$$

So, the function that is being shifted is and it is being shifted correctly. Taking the Laplace transform of everything and plugging in the initial conditions gives,

$$s^2Y(s) - sy(0) - y'(0) - (sY(s) - y(0)) = \frac{s}{s^2+4} + \frac{se^{-6s}}{s^2+4}$$

$$(s^2 - s)Y(s) + 4s - 4 = \frac{s}{s^2+4} + \frac{se^{-6s}}{s^2+4}$$

Now solve for $Y(s)$.

$$(s^2 - s)Y(s) = \frac{s + se^{-6s}}{s^2 + 4} = 4s + 4$$

$$Y(s) = \frac{s\left(1 + e^{-6s}\right)}{(s-1)\left(s^2+4\right)} - 4\frac{s-1}{s(s-1)}$$

$$= \frac{1 + e^{-6s}}{(s-1)\left(s^2+4\right)} - \frac{4}{s}$$

$$Y(s) = \left(1 + e^{-6s}\right)F(s) + \frac{4}{s}.$$

Notice that we combined the first two terms to simplify things a little. Also there was some canceling going on in this one. Do not expect that to happen on a regular basis. We now need to partial fraction *F(s)*. We'll leave the details to you to check.

$$F(s) = \frac{1}{(s-1)\left(s^2+4\right)} = \frac{1}{5}\left(\frac{1}{s-1} - \frac{s+1}{s^2+1}\right)$$

$$f(t) = \frac{1}{5}\left(e^t - \cos(2t) - \frac{1}{2}\sin(2t)\right).$$

Okay, we can now get the solution to the differential equation. Starting with the transform we get,

$$Y(s) = F(s) + F(s)e^{-6s} + \frac{4}{s}$$

$$y(t) = f(t) + u_6(t)f(t-6) + 4.$$

where *f(t)* is given above.

Example: Solve the following IVP.

$$y'' - 5y' - 14y = 9 - u_3(t) + 4(t-1)u_1(t) \quad y(0) = 0,\ y'(0) = 10.$$

Solution: Let's take the Laplace transform of everything and note that in the third term we are shifting $4t$.

$$s^2Y(s) - sy(0) - y'(0) - 5(sY(s) - y(0)) - 14Y(s) = \frac{9}{s} + \frac{e^{-3s}}{s} + 4\frac{e^{-s}}{s^2}$$

$$(s^2 - 5s - 14)Y(s) - 10 = \frac{9 + e^{-3s}}{s} - 4\frac{e^{-s}}{s^2}.$$

Now solve for *Y(s)*.

$$\left(s^2 - 5s - 14\right)Y(s) - 10 = \frac{9 + e^{-3s}}{s} - 4\frac{e^{-s}}{s^2}$$

$$Y(s) = \frac{9+e^{-3s}}{s(s-7)(s+2)} + \frac{4e^{-s}}{s^2(s-7)(s+2)} + \frac{10}{(s-7)(s+2)}$$

$$Y(s) = \left(9+e^{-3s}\right)F(s) + 4e^{-s}G(s) + H(s)$$

So, we have three functions that we'll need to partial fraction for this problem. I'll leave it to you to check the details.

$$F(s) = \frac{1}{s(s-7)(s+2)} = -\frac{1}{14}\frac{1}{s} + \frac{1}{63}\frac{1}{s-7} + \frac{1}{18}\frac{1}{s+2}$$

$$f(t) = \frac{1}{14} + \frac{1}{63}e^{7t} + \frac{1}{18}e^{-2t}$$

$$G(s) = \frac{1}{s^2(s-7)(s+2)} = \frac{5}{196}\frac{1}{s} + \frac{1}{14}\frac{1}{s^2} + \frac{1}{441}\frac{1}{s-7} - \frac{1}{36}\frac{1}{s+2}$$

$$g(t) = \frac{5}{196}\frac{1}{14}t + \frac{1}{441}e^{7t} - \frac{1}{36}e^{-2t}$$

$$H(s) = \frac{10}{(s-7)(s+2)} = \frac{10}{9}\frac{1}{s-7} - \frac{10}{9}\frac{1}{s+2}$$

$$h(t) = \frac{10}{9}e^{7t} - \frac{10}{9}e^{-2t}.$$

Okay, we can now get the solution to the differential equation. Starting with the transform we get,

$$Y(s) = 9F(s) + e^{-3s}F(s) + 4e^{-s}G(s) + H(s)$$

$$y(t) = 9f(t) + u_3(t)f(t-3) + 4u_1(t)g(t-1) + h(t).$$

where $f(t)$, $g(t)$ and $h(t)$ are given above.

Let's work one more example.

Example: Solve the following IVP.

$$y''(t) = 3y' + 2y = (t), \qquad y(0) = 0 \quad y'(0) = -2$$

where,

$$g(t) = \begin{cases} 2 & t < 6 \\ t & 6 \le t < 10. \\ 4 & t \ge 10 \end{cases}$$

Solution: The first step is to get *g(t)* written in terms of Heaviside functions so that we can take the transform.

$$g(t) = 2 + (t-2)u_6(t) + (4-t)u_{10}(t).$$

Now, while this is *g(t)* written in terms of Heaviside functions it is not yet in proper form for us to take the transform. Remember that each function must be shifted by a proper amount. So, getting things set up for the proper shifts gives us,

$$g(t) = 2 + (t-6+6-2)u_6(t) + (4-(t-10+10))u_{10}(t)$$

$$g(t) = 2 + (t-6+4)u_6(t) + (4-(t-10))u_{10}(t).$$

So, for the first Heaviside it looks like $f(t) = t + 4$is the function that is being shifted and for the second Heaviside it looks like $f(t) = -6 - t$is being shifted.

Now take the Laplace transform of everything and plug in the initial conditions.

$$s^2Y(s) - sy(0) - y'(0) + 3(sY(s) - y(0)) + 2Y(s) = \frac{2}{s} + e^{-6s}\left(\frac{1}{s^2} + \frac{4}{s}\right) - e^{-6s}\left(\frac{1}{s^2} + \frac{6}{s}\right)$$

$$(s^2 + 3s + 2)Y(s) + 2 = \frac{2}{s} + e^{-6s}\left(\frac{1}{s^2} + \frac{4}{s}\right) - e^{-6s}\left(\frac{1}{s^2} + \frac{6}{s}\right)$$

Solve for *Y(s)*.

$$\left(s^2 + 3s + 2\right)Y(s) = \frac{2}{s} + e^{-6s}\left(\frac{1}{s^2} + \frac{4}{s}\right) - e^{-10s}\left(\frac{1}{s^2} + \frac{6}{s}\right) - 2$$

$$\left(s^2 + 3s + 2\right)Y(s) = \frac{2 + 4e^{-6s} - 4e^{-10s}}{s} + \frac{e^{-6s} - e^{-10s}}{s^4} - 2$$

$$Y(s) = \frac{2 + 4e^{-6s} - 6e^{-10s}}{s(s+1)(s+2)} + \frac{e^{-6s} - e^{-10s}}{s^2(s+1)(s+2)} - \frac{2}{(s+1)(s+2)}$$

$$Y(s) = \left(2 + 4e^{-6s} - 6e^{-10s}\right)F(s) + \left(e^{-6s} - e^{-10s}\right)G(s) - H(s).$$

Now, in the solving process we simplified things into as few terms as possible. Even doing this, it looks like we'll still need to do three partial fractions.

.I'll leave the details of the partial fractioning to you to verify. The partial fraction form and inverse transform of each of these are.

$$F(s) = \frac{1}{s(s+1)(s+2)} = \frac{\frac{1}{2}}{s} - \frac{1}{s+1} + \frac{\frac{1}{2}}{s+2}$$

$$f(t) = \frac{1}{2} - e^{-1} + \frac{1}{2}e^{-2t}.$$

$$G(s) = \frac{1}{s^2(s+1)(s+2)} = -\frac{\frac{3}{4}}{s} + \frac{\frac{1}{2}}{s^2} + \frac{\frac{1}{2}}{s^2} + \frac{\frac{1}{4}}{s+2}$$

$$g(t) = -\frac{3}{4} + \frac{1}{2}t + e^{-t}\frac{1}{4}e^{-2t}$$

$$H(s) = \frac{2}{(s+1)(s+2)} = \frac{2}{s+1} - \frac{2}{s+2}$$

$$h(t) = 2e^{-t} - 2e^{-2t}.$$

Putting this all back together is going to be a little messy. First rewrite the transform a little to make the inverse transform process possible.

$$Y(s) = 2F(s) + e^{-6s}(4F(s) + G(s)) - e^{-10s}(6F(s) + G(s)) - H(s).$$

Now, taking the inverse transform of all the pieces gives us the final solution to the IVP.

$$y(t) = 2f(t) - h(t) + u_6(t)(4f(t-6) + g(t-6)) - u_{10}(t)(6f(t-10) + g(t-10))$$

where *f(t)*, *g(t)*, and *h(t)* are defined above.

So, the answer to this example is a little messy to write down, but overall the work here wasn't too terribly bad.

Before proceeding with the next section let's see how we would have had to solve this IVP if we hadn't had Laplace transforms. To solve this IVP we would have had to solve three separate IVP's. One for each portion of *g(t)*. Here is a list of the IVP's that we would have had to solve.

1. $0 < t < 6$

$$y'' + 3y + 2y = 2, \qquad y(0) = 0 \quad y'(0) = -2.$$

The solution to this IVP, with some work, can be made to look like,

$$y_1(t) = 2f(t) - h(t)$$

2. $6 \le t < 10$

$$y'' + 3y' + 2 = t, \qquad y(6) = y_1(6) \qquad y'(6) = y_1(6)$$

where, $y_1(t)$ is the solution to the first IVP. The solution to this IVP, with some work, can be made to look like,

$$y_2(t) = 2f(t) - h(t) + 4f(t-6) + g(t-6).$$

3. $t \ge 10$

$$y'' + 3y' + 2y = 4, \qquad y(10) = y_2(10) \qquad y'(10) = y_2(10)$$

where, $y_2(t)$ is the solution to the second IVP. The solution to this IVP, with some work, can be made to look like,

$$y_3(t) = 2f(t) - h(t) + 4f(t-6) + g(t-6) - f(t-10) - g(t-10).$$

There is a considerable amount of work required to solve all three of these and in each of these the forcing function is not that complicated. Using Laplace transforms saved us a fair amount of work.

DIRAC DELTA FUNCTION

When we first introduced Heaviside functions we noted that we could think of them as switches changing the forcing function, $g(t)$, at specified times. However, Heaviside functions are really not suited to forcing functions that exert a "large" force over a "small" time frame.

Examples of this kind of forcing function would be a hammer striking an object or a short in an electrical system. In both of these cases a large force (or voltage) would be exerted on the system over a very short time frame. The Dirac Delta function is used to deal with these kinds of forcing function.

DIRAC DELTA FUNCTION

There are many ways to actually define the Dirac Delta function. There three main properties of the Dirac Delta function that we need to be aware of. These are,

1. $\delta(t - a) = 0,\ t \neq a.$
2. $\int_{a=\varepsilon}^{a+c} \delta(t-a)dt = 1, \quad \varepsilon > 0.$
3. $\int_{a=\varepsilon}^{a+c} f(t)\,\delta\,(t=1)dt = f(a), \quad \varepsilon > 0.$

At the Dirac Delta function is sometimes thought of has having an "infinite" value. So, the Dirac Delta function is a function that is zero everywhere except one point and at that point it can be thought of as either undefined or as having an "infinite" value.

Note that the integrals in the second and third property are actually true for any interval containing $t = a$, provided it's not one of the endpoints. The limits given here are needed to prove the properties and so they are also given in the properties. We will however use the fact that they are true provided we are integrating over an interval containing $t = a$.

This is a very strange function. It is zero everywhere except one point and yet the integral of any interval containing that one point has a value of 1. The Dirac Delta function is not a real function as we think of them. It is instead an example of something called a generalized function or distribution.

Despite the strangeness of this "function" it does a very nice job of modeling sudden shocks or large forces to a system. Before solving an IVP we will need the transform of the Dirac Delta function. We can use the third property above to get this.

$$\mathfrak{L}\{\delta(t-a)\} = \int_0^\infty e^{-st}\delta(t-a)dt = e^{-as} \qquad \text{provided } a > 0.$$

Note that often the second and third properties are given with limits of infinity and negative infinity, but they are valid for any interval in which $t = a$is in the interior of the interval.

With this we can now solve an IVP that involves a Dirac Delta function.

Example: Solve the following IVP.

$$y'' + 2y' - 15y = 6\delta\,(t-9), \qquad y(0) = -5 \quad y'(0) = 7.$$

Solution:As with all previous problems we'll first take the Laplace transform of everything in the differential equation and apply the initial conditions.

$$s^2Y(s) - sy(0) - y'(0) + 2(sY(s) - y(0)) - 15Y(s) = 6e^{-9t}$$

$$(s^2 + 2s - 15)Y(s) + 5s + 3 = 6e^{-9s}.$$

Now solve for *Y(s)*.

$$\begin{aligned} Y(s) &= \frac{6e^{-9s}}{(s+5)(s-3)} - \frac{5s+3}{(s+5)(s-3)} \\ &= 6e^{-9s}F(s) - G(s). \end{aligned}$$

We'll leave it to you to verify the partial fractions and their inverse transforms are,

$$F(s) = \frac{1}{(s+5)(s-3)} = \frac{\frac{1}{8}}{s-3} - \frac{\frac{1}{8}}{s+5}$$

$$f(1) = \frac{1}{8}e^{3t} - \frac{1}{8}e^{-5t}$$

$$G(s) = \frac{5s+3}{(s+5)(s-3)} = \frac{\frac{9}{4}}{s-3} + \frac{\frac{11}{4}}{s+5}$$

$$g(t) = \frac{9}{4}e^{3t} + \frac{11}{4}e^{-5t}.$$

The solution is then,

$$Y(s) = 6e^{-9s}F(s) - G(s)$$

$$y(t) = 6u_9(t)f(t-9) - g(t).$$

where, *f(t)* and *g(t)* are defined above.

Example: Solve the following IVP.

$$2y''+10y=3u_{12}(t)-5\delta(t-4), \qquad y(0)=-1 \quad y'(0)=-2.$$

Solution: Take the Laplace transform of everything in the differential equation and apply the initial conditions.

$$2(s^2Y(s)-sy(0)-y'(0))+10Y(s)=\frac{3e^{-12s}}{s}-5e^{-4s}$$

$$(2(s^2+10)Y(s)+2s+4=\frac{3e^{-12s}}{s}-5e^{-4s}.$$

Now solve for $Y(s)$.

$$Y(s)=\frac{3e^{-12s}}{s(2s^2+10)}-\frac{5e^{-4s}}{2s^2+10}-\frac{2s+4}{2s^2+10}$$

$$=3e^{-12s}F(s)-5e^{-4s}G(s)-H(s).$$

We'll need to partial fraction the first function. The remaining two will just need a little work and they'll be ready. I'll leave the details to you to check.

$$F(s)=\frac{1}{s(2s^2+10)}=\frac{1}{10}\frac{1}{s}-\frac{1}{10}\frac{s}{s^2+5}$$

$$f(t)=\frac{1}{10}-\frac{1}{10}\cos\left(\sqrt{5}t\right)$$

$$g(t)=\frac{1}{2\sqrt{5}}\sin\left(\sqrt{5}t\right)$$

$$h(t)=\cos\left(\sqrt{5}t\right)+\frac{2}{\sqrt{5}}\sin\left(\sqrt{5}t\right).$$

The solution is then,

$$Y(s)=3e^{-12s}F(s)-5e^{-4s}G(s)-H(s)$$

$$y(t)=3u_{12}(t)f(t-12)-5u_4(t)g(t-4)-h(t).$$

where, $f(t)$, $g(t)$ and $h(t)$ are defined above.

So, with the exception of the new function these work the same way that all the problems that we've seen to this point work. Note as well that the exponential was introduced into the transform by the Dirac Delta function, but once in the transform it doesn't matter where it came from.

In other words, when we went to the inverse transforms it came back out as a Heaviside function.

Before proceeding to the next section let's take a quick side trip and note that we can relate the Heaviside function and the Dirac Delta function. Start with the following integral.

$$\int_{-\infty}^{t} \delta(u-a)du = \begin{cases} 0 & \text{if } t < a \\ 1 & \text{if } t > a \end{cases}.$$

However, this is precisely the definition of the Heaviside function. So,

$$\int_{-\infty}^{t} \delta(u-a)du = u_a(t).$$

Now, recalling the Fundamental Theorem of Calculus, we get,

$$u'_a(t)\left(\int_{-\infty}^{t} \delta(u \quad a)du\right) = \delta(t-a).$$

CONVOLUTION INTEGRALS

On occasion we will run across transforms of the form,

$$H(s) = F(s)\,G(s)$$

that can't be dealt with easily using partial fractions. We would like a way to take the inverse transform of such a transform. We can use a convolution integral to do this.

CONVOLUTION INTEGRAL

If $f(t)$ and $g(t)$ are piecewise continuous function on $(0, \infty)$then the convolution integral of $f(t)$ and $g(t)$ is,

$$(f * g)(t) = \int_0^t f(t-\tau)g(\tau)d\tau.$$

A nice property of convolution integrals is.

$$(f * g)(t) = (g * f)(t).$$

Or,

$$\int_0^t f(t-\tau)g(\tau)d\tau = \int_0^t f(\tau)g(t-\tau)d\tau.$$

The following fact will allow us to take the inverse transforms of a product of transforms.

FACT

$$\mathfrak{L}\{f * g\} = F(s)G(s) \qquad \mathfrak{L}^{-1}\{F(s)\,G(s)\} = (f * g)(t).$$

Let's work a quick example to see how this can be used.

Example: Use a convolution integral to find the inverse transform of the following transform.

$$H(s) = \frac{1}{(s^2 + a^2)^2}.$$

Solution: Now, since we are going to use a convolution integral here we will need to write it as a product whose terms are easy to find the inverse transforms of. This is easy to do in this case.

$$H(s)=\left(\frac{1}{s^2+a^2}\right)\left(\frac{1}{s^2+a^2}\right).$$

So, in this case we have,

$$F(s)=G(s)=\frac{1}{s^2+a^2} \quad \Rightarrow \quad f(t)=g(t)=\frac{1}{a}\sin(at).$$

Using a convolution integral *h(t)* is,

$$\begin{aligned} h(t) &= (f * g)(t) \\ &= \frac{1}{a^2}\int_0^t \sin(at-a\tau)\sin(a\tau)d\tau \\ &= \frac{1}{2a^3}\left(\sin(at)-at\cos(at)\right). \end{aligned}$$

This is exactly what we would have gotten by using #11 from the table.

Convolution integrals are very useful in the following kinds of problems.

Example: Solve the following IVP

$$4y''+y=h(t), \qquad y(0)=3 \quad y'(0)=-7.$$

Solution: First, notice that the forcing function in this case has not been specified. Prior to this section we would not have been able to get a solution to this IVP. With convolution integrals we will be able to get a solution to this kind of IVP. The solution will be in terms of *g(t)* but it will be a solution.

Take the Laplace transform of all the terms and plug in the initial conditions.

$$4(s^2Y(s)-sy(0)-y'(0))+Y(s)=G(s)$$

$$(4s^2+1)Y(s)-12s+28=G(s).$$

Notice here that all we could do for the forcing function was to write down *G(s)* for its transform. Now, solve for *Y(s)*.

$$(4s^2+1)Y(s) = G(s)+12s+28$$

$$Y(s) = \frac{12s}{4\left(s^2+\frac{1}{4}\right)}+\frac{G(s)}{4\left(s^2+\frac{1}{4}\right)}.$$

We factored out a 4 from the denominator in preparation for the inverse transform process. To take inverse transforms we'll need to split up the first term and we'll also rewrite the second term a little.

$$Y(s) = \frac{12s-28}{4\left(s^2+\frac{1}{4}\right)} + \frac{G(s)}{4\left(s^2+\frac{1}{4}\right)}$$

$$= \frac{3s}{s^2+\frac{1}{4}} + \frac{7\frac{2}{2}}{s^2+\frac{1}{4}} + \frac{1}{4}G(s)\frac{\frac{2}{2}}{s^2+\frac{1}{4}}.$$

Now, the first two terms are easy to inverse transform. We'll need to use a convolution integral on the last term. The two functions that we will be using are,

$$g(t) \quad f(t) = 2\sin\left(\frac{t}{2}\right).$$

We can shift either of the two functions in the convolution integral. We'll shift *g(t)* in our solution. Taking the inverse transform gives us,

$$y(t) = 3\cos\left(\frac{t}{2}\right) + 14\sin\left(\frac{t}{2}\right) + \frac{1}{2}\int_0^t \sin\left(\frac{\tau}{2}\right) g(t-\tau)\,d\tau.$$

So, once we decide on a *g(t)* all we need to do is to an integral and we'll have the solution.

As this last example has shown, using convolution integrals will allow us to solve IVP's with general forcing functions. This could be very convenient in cases where we have a variety of possible forcing functions and don't which one we're going to use. With a convolution integral all that we need to do in these cases is solve the IVP once then go back and evaluate an integral for each possible *g(t)*. This will save us the work of having to solve the IVP for each and every *g(t)*.

Chapter 13

High Order Ordinary Differential Equations

THE CHARACTERISTIC EQUATION

Equations of the order higher than one, in general, are not solvable in terms of known functions. However, there are *a* number of important linear differential equations which can be solved by elementary method, and higher order ODEs with constant coefficients are one of such *a* family. The equation has *a* general form

$$a_n y^{(n)} + a_{n-1} y^{(n-1)} + \cdots + a_1 y' + a_0 y = f(x)$$

when $f(x)$ is non-trivial (i.e. $f(x) \neq 0$) the equation is nonhomogeneous nth-order differential equation with constant coefficients. Otherwise, if $f(x) = 0$ the equation becomes

$$a_n y^{(n)} + a_{n-1} y^{(n-1)} + \cdots + a_1 y' + a_0 y = 0$$

which is homogeneous nth-order differential equation with constant coefficients.

If $y_1, y_2, \ldots y_m$ are m particular solutions of the above homogeneous equation, then for all real values of the parameters $c_1, c_2, \ldots c_m$ is also *a* solution

$$y = c_1 y_2 + c_2 y_2 + \ldots + c_m y_m = \sum_{k=1}^{m} c_k y_k$$

To solve the homogeneous equation Equation substitute which leads to *a* characteristic equation

$$a_n \lambda^n + a_{n-1} \lambda^{n-1} + \cdots + a_1 \lambda + a_0 = 0$$

whose roots (real or complex) are $r_1, r_2, \ldots, r_n$. The equation is known as the characteristic equation of the differential equation.

THE METHOD OF UNDETERMINED COEFFICIENTS

For each real root of multiplicity k determines k linearly independent solutions e^{rx},

xe^{rx}, ... $x^{k-1}e^{rx}$, for $k \geq 1$. Since the constant coefficient a_i are real, complex roots occurs in conjugate pair, i.e. if $a + ib$ is the root, $a - ib$ is also the root of Equation. Each complex root $a + ib$ of multiplicity k thus y ields $2k$ linearly independent real solutions

$$e^{ex}\cos bx \; xe^{ex}\cos bx \cdots x^{k-1}e^{ex}\cos bx$$

$$e^{ex}\sin bx \; xe^{ex}\sin bx \cdots x^{k-1}e^{ex}\sin bx$$

The general solution is then *a* linear combination of the solutions obtained above. To solve the nonhomogeneous equation, (i.e. $f(x) \neq 0$) we ought to be able to find one particular solution of the equation g(x). Then the complete solution is the sum of the complementary function G(x) obtained by putting $f(x) = 0$ in equation and $g(x)$ i.e.

$$y = G(x) + g(x)$$

If $f(x)$ is *a* function for which repeated differentiation y ields only *a* finite number of linearly independent functions, appearing possibly in linear combination, then *a* particular integral Y for the nonhomogeneous linear equation can be found through the procedures:

$$a_n y^{(n)} + a_{n-1}y^{(n-1)} + \cdots + a_1 y' + a_0 y = f(x)$$

1. Assume Y to be an arbitrary linear combination of all the linearly independent functions which arise from f(x) by repeated differentiation.
2. Substitute Y into the given differential equation.
3. Determine the arbitrary constants in Y.

The class of function f(x) possessing only *a* finite number of linearly independent derivatives consists of the simple functions:

k

x^n(n is *a* positive integer)

e^{kx}

sin kx

cos kx

Example: Solve the following differential equations:

1. $\dfrac{d^3y}{dx^3} - 6\dfrac{d^2y}{dx^2} + 11\dfrac{dy}{dx} - 6y = 0$
2. $\dfrac{d^4y}{dx^4} - 2\dfrac{d^2y}{dx^2} + y = 0$

Solution: 1. The characteristic equation is $\lambda^3 - 6\lambda^2 + 11\lambda - 6 = 0$

The roots of the above equation are 1, 2, 3; hence the solution is

$$y = c_1e^x + c_2e^{2x} + c_3e^{3x}$$

2. The characteristic equation is $\lambda^4 - 2\lambda^2 + 1 = 0$

The roots of the above equation are 1, –1 (repeated roots); hence the solution is

$$y = (c_1 + c_2x)e^x + (c_3 + c_4x)e^{-x}$$

Example: Use the method of undetermined coefficients to find the complete solution of the following differential equations:

1. $\dfrac{d^2y}{dx^2} - \dfrac{dy}{dx} - 2y = e^{3x}$

2. $\dfrac{d^2y}{dx^2} - \dfrac{dy}{dx} - 2y = \sin 2x$

Solution:

1. The characteristic equation is $\lambda^2 - \lambda - 2 = 0$

The roots of the above equation are – 1, 2; hence the general solution is

$$y = c_1e^{-x} + c_2e^{2x}$$

To find the particular solution, we assume that

$$y_p = Ae^{3x}$$

Thus, $y'_p = 3Ae^{3x}$ and $y''_p = 9Ae^{3x}$. Substituting these results into the differential equation, we have

$$9Ae^{3x} - 3Ae^{3x} - 2Ae^{3x} = e^{3x} \Rightarrow 4Ae^{3x} = e^{3x}$$

It follows that $A = \dfrac{1}{4}$ so that $y_p = \dfrac{1}{4}e^{3x}$ and the complete solution is

$$y = c_1e^{-x} + c_2e^{2x} \frac{1}{4}e^{3x}$$

2. From the results of (a), we have the general solution

$$y = c_1e^{-x} + c_2e^{2x}$$

To find the particular solution, we assume that

$$y_p = A\sin 2x + B\cos 2x$$

Thus, $y_p = 2A\cos 2x - 2B\sin^2 x$ and $y''_p = -4A\sin 2x - 4B\cos 2x$. Substituting these results into the differential equation, we have

$$(-4A\sin 2x - 4B\cos 2x) -- (2A\cos 2x - 2B\sin 2x) - 2(A\sin 2x - 4B\cos 2x) = \sin 2x$$

or equivalently,

$$(-6A + 2B)\sin 2x + (-6B - 2A)\cos 2x = (1)\sin 2x + -(0)\sin 2x + (0)\cos 2x$$

Equating coefficients of like terms, we obtain

$$-6A + 2B = 1 \; -2A - 6B = 0$$

Solving the above system, we find that $A = -\frac{3}{20}$ and B $= \frac{1}{20}$. Then the complete solution is

$$y = c_1 e^{-x} c_2 e^{2x} - \frac{3}{20}\sin 2x + \frac{1}{20}\cos 2x$$

VARIATION OF PARAMETERS

In general, if $f(x)$ in Equation is not one of the types of functions considered above, or if the differential equation does not have constant coefficients, then we cannot apply the method of undetermined coefficients and the variation of parameters is preferred.

Variation of parameters is another method for finding *a* particular solution of the nth-order differential equation.

$$a_n(x)y^{(x)} + a_{n-1}(x)y^{(n-1)} + \cdots + a_1(x)y' + a_0(x)y = f(x)$$

Once the solution of the associated homogeneous equation

$$a_n(x)y^{(n)} + a_{n-1}(x)y^{(n-1)} + \cdots + a_1(x)y' + a_0(x)y = 0$$

is known. If $y_1, y_2, \ldots, y_m$ are m particular solutions of the above homogeneous equation, then for all real values of the parameters v_1, v_2, ..., v_m is also *a* general solution

$$y = v_1 y_1 + v_2 y_2 + \cdots + v_m y_m = \sum_{k-1}^{m} v_k y_k$$

The Method

A particular solution of Equation has the form

$$yp\;(x) = v_1(x)y_1(x) = v_2(x)y_2(x) + \cdots + v_m(x)y_m(x)$$

where $y_1(x), y_2(x)\cdots, y_m(x)$ are given in Equation and $v_1(x), v_2(x)\cdots v_m(x)$ are unknown functions of x which still must be determined.

To find $v_1(x), v_2(x)\cdots v_m(x)$, first solve the following linear equations simultaneously for $v_1(x), v_2(x)\cdots v_m(x)$:

$$\begin{cases} v_1'y_1v_2'y_2+\cdots+v_m'y_m=0 \\ v_1'y_1v_2'y_2'+\cdots+v_m'y_m'=0 \\ \vdots \\ v_1'y_1{}^{(n-2)}v_2'y_2{}^{(n-2)}+\cdots+v_m'y_m{}^{(n-2)}=0 \\ v_1'y_1{}^{(n-1)}v_2'y_2{}^{(n-1)}+\cdots+v_m'y_m{}^{(n-1)}=f(x) \end{cases}$$

Then integrate each to obtain v'_i disregarding all constants of integration. This is permissible because we are seeking only one particular solution. The above mentioned method is known as the method of variation of parameters.

Proof: We are now going to prove that Equation will give us *a* particular solution of the differential equation for second order.

When $n = 2$, assume *a* particular solution of the form

$$yp(x) = v_1(x)y_1(x)+v_2(x)y_2(x)$$

and Equation is reduced into

$$\begin{cases} v_1'(x)y_1(x)+v_2'(x)y_2(x)=0 \\ v_1'(x)y_1'(x)+v_2'(x)y_2'(x)=f(x) \end{cases}$$

Then

$$\begin{aligned} y'p(x) &= v_1(x)y_1'(x)+v_2(x)y_2'(x)+v_1'(x)y_1(x)+v_2'(x)y_2(x) \\ &= v_1(x)y_1'(x)+v_2(x)y_2'(x) \end{aligned}$$

and

$$\begin{aligned} y''_p(x) &= v_1(x)y_1''(x)+v_2(x)y_2''(x)+v_1'(x)y_1'(x)+v_2'(x)y_2'(x) \\ &= v_1(x)y_1''(x)+v_2(x)y_2''(x)+f(x) \end{aligned}$$

Substituting the above values of $v_1(x)y_1''(x), y_2'(x)$, and $y_p(x)$ back to the Equation, we obtain

$$y_p''+a_1y_p'+a_0y_p = (v_1y_1''+v_2y_2''+f)+a_1(v_1y_1'+v_2y_2')+a_0(v_1y_1+v_2y_2)$$

$$= v_1(y_1''a_1y_2' + a_0y_1) + v_2(y_2'' + a_1y_2' + a_0y_2 + f(x)$$

$$= f(x)$$

Example: Use the method of variation of parameters to find the complete solution of the differential equations $\frac{d^2y}{dx^2} + 4y\sin^2 2x$.

Solution: The general solution is $y = c_1 \cos 2x + c_2 \sin 3x$. We can assume that the particular solution takes the form

$$y_p = v_1 \cos 2x + v_2 \sin 3x$$

Then from Eqution, we have

$$\begin{cases} v_1' \cos 2x + v_2' \sin 2x = 0 \\ v_1'(-2\sin 2x) + v_2'(2\cos 2x)\sin^2 = 2x \end{cases}$$

The solutions of this set of equations is

$$\begin{cases} v_1' = -\frac{1}{2}\sin^3 2x \\ v_1' = \frac{1}{2}\sin^2 2x \cos 2x \end{cases}$$

Thus,

$$\begin{cases} v_1' = -\frac{1}{2}\int \sin^3 2x\, dx = \frac{1}{4}\cos 2x - \frac{1}{2}\cos^3 2x \\ v_2 = -\frac{1}{2}\int \sin^2 2x \cos 2x\, dx = \frac{1}{12}\sin^3 2x \end{cases}$$

Hence the particular solution is

$$y_p = \left[\frac{1}{4}\cos 2x - \frac{1}{12}\cos^3 2x\right]\cos 2x + \left[\frac{1}{12}\sin^3 2x\right]\sin 2x$$

$$= \frac{1}{6}\cos^3 2x + \frac{1}{12}\sin^3 2x$$

The general solution is thus

$$y = c_1 \cos 2x + c_2 \sin 2x + \frac{1}{6}\cos^3 2x + \frac{1}{12}\sin^3 2x$$

LINEAR ALGEBRA

MATRIX ALGEBRA

A rectangular array A of mn complex numbers arranged in m rows and n columns is called a matrix of size $m \times n$. A $m + n$ matrix is usually represented in the form:-

$$A = \begin{bmatrix} a_{11} & a_{12} & \cdots & a_{1n} \\ a_{21} & a_{22} & \cdots & a_{2n} \\ \vdots & \vdots & \vdots & \vdots \\ a_{m1} & a_{m2} & \cdots & a_{mn} \end{bmatrix}$$

OR more compactly by $A = [a_{ij}]_{m \times n}$.

The entries a_{ij} in the matrix is called element of the matrix. With i and j given, the element a_{ij} in the ith row and the jth column is said to have row index i and column index j. A $n \times 1$ matrix a (i.e. a matrix having one column only) is called a column vector. An $\times$ n matrix b (i.e. a matrix having one row only) is called a row vector.

$$a = \begin{bmatrix} a_1 \\ a_2 \\ \vdots \\ a_n \end{bmatrix} b = \begin{bmatrix} b_1 & b_2 & \cdots & b_n \end{bmatrix}$$

The norm of a column vector a and a row vector b is defined as:

$$|a| = \sqrt{|a_1|^2 + |a_2|^2 + \cdots + |a_n|^2} \text{ or}$$

$$|b| = \sqrt{|b_1|^2 + |b_2|^2 + \cdots + |b_n|^2}$$

and they are said to be normalized if or $|a| = 1$ or $|b| = 1$.

Two m $\times$ n matrices $A = [a_{ij}]$ $B = [b_{ij}]$ and are equal if and only if their corresponding elements are equal, i.e. $a_{ij} = b_{ij}$, for all $i = 1, 2, ..., m; j = 1, 2, ..., n$. Matrices of different sizes cannot be equal. A $n \times n$ matrix is called a square matrix of order n. A matrix, whether square or not, in which every element is z ero is called a z ero matrix and

$$O = \begin{bmatrix} 0 & 0 & \cdots & 0 \\ 0 & 0 & \cdots & 0 \\ \vdots & \vdots & \vdots & \vdots \\ 0 & 0 & \cdots & 0 \end{bmatrix}$$

is denoted by the symbol O, i.e..

THE ALGEBRA OF MATRICES

1. The sum (or difference) of two $m \times n$ matrices $A = [a_{ij}]$ and $B = [b_{ij}]$ is the $m \times n$ matrix $C = [c_{ij}]$, where

   $c_{ij} = a_{ij} + b_{ij}$ (or $c_{ij} = a_{ij} - b_{ij}$), for all $i = 1, 2, \ldots m; j = 1, 2, \ldots n$.

   N.B. We usually write $C = A + B$ (or $C = A - B$).

2. The scalar multiplication of *a* m *n* matrix $A = [a_{ij}]$ by a scalar k (k is *a* real or complex number) is *a* m *n* matrix $D = [d_{ij}]$, where $d_{ij} = ka_{ij}$, for all.

   and is written as $D = kA$.

   The difference between two matrices A and B can be written as $A - B = A + (-1)B$.

   For all positive integers m and n, for all scalars k and l, and for all $m \times n$ matrices $A = [a_{ij}]$, and $B = [b_{ij}]$ and $C = [c_{ij}]$, we have the following algebraic laws:

   (a) $(A + B) + C = A + (B + C)$

   (b) $A + B = B + A$

   (c) $A + O = O + A$

   (d) $A + (-A) = O$

   (e) $k(A \pm B) = kA \pm kB$

   (f) $(k \pm l)A = kA \pm lA$

   (g) $(k\,l)A = k(l\,A)$

   (h) $1\,A = A$

3. The product of a $m \times p$ matrix $A = [a_{ij}]$ and a $p \times n$ matrix $B = [b_{ij}]$ is a $m \times n$ matrix $C = [c_{ij}]$, where

   $$c_{ij} = \sum_{k-1}^{p} a_{ik} b_{kj}\text{ , for all } i = 1, 2, \ldots m; j = 1, 2, \ldots n.$$

   And is written as $C = AB$. In general, AB ¹ BA.

   For all scalar c, and for all $m \times p$ matrix A, $p \times n$ matrices B and C, $m \times n$ matrix D, we have the following algebraic laws:

   (a) $c(AB) = (cA)B = A(cB)$

   (b) $A(B + C) = AB + AC$

   (c) $(B + C)D = BD + CD$

   (e) $A(BC) = (AB)C$

4. For any two column vectors $u = \begin{bmatrix} u_1 \\ u_2 \\ \vdots \\ u_n \end{bmatrix}$ and $v = \begin{bmatrix} v_1 \\ v_2 \\ \vdots \\ v_n \end{bmatrix}$ of order n, the dot product between u and v is given by the products,

$$u.v = \bar{u}^T v = \bar{u}_1 v_1 + \bar{u}_2 v_2 + \cdots + \bar{u}_n v_n$$

Two vectors u and v are said to be orthogonal if and only if.

5. In any square matrix A, the sum of the diagonal elements is called trace and it is denoted by TrA, i.e. if $\begin{bmatrix} a_{11} & a_{12} & \cdots & a_{1n} \\ a_{21} & a_{22} & \cdots & a_{2n} \\ \vdots & \vdots & \vdots & \vdots \\ a_{n1} & a_{n2} & \cdots & a_{nn} \end{bmatrix}$, then

$$TrA = a_{11} + a_{22} + \cdots + a_{nn} = \sum_{j-1}^{n} a_{jj}$$

The trace of matrices has the following properties:

(a) $Tr(AB) = \text{Tr(BA)}$

(b) $Tr(A^n) = (TrA)^n$

SPECIAL MATRICES

1. The transpose of a m n matrix $A = [a_{ij}]$ is the $n \times m$ matrix $A^T = [b_{ij}]$, where $b_{ij} = a_{ji}$. For all matrices A, B, and any scalar c, the transposition of matrices has the following properties:-

   (a) $(A^T)^T = A$

   (ii)$(cA)^T = cA^T\,(A \pm B)T = A^T \pm B^T$

   (b) If is defined

   (c) If AB is defined $(AB)^T = B^T A^T$

2. A square matrix $A = [a_{ij}]$ of order n is called an identity matrix of order n if and only if $a_{ij} = \begin{cases} 1 & \text{if } i = j \\ 0 & \text{if } i \neq j \end{cases}$ and is denoted by I_n.

For instance: $I_2 = \begin{bmatrix} 1 & 0 \\ 0 & 1 \end{bmatrix}, I_3 = \begin{bmatrix} 1 & 0 & 0 \\ 0 & 1 & 0 \\ 0 & 0 & 1 \end{bmatrix}.$

It can be shown that for any square matrix A and identity matrix I, both are of the same order, $AI = IA = A$.

3. A square matrix $D = [d_{ij}]$ of order n is said to be diagonal if $d_{ij} = 0$ for $i \neq j$ i.e.

$$\begin{bmatrix} d_{11} & 0 & \cdots & 0 \\ 0 & d_{22} & \cdots & 0 \\ \vdots & \vdots & \vdots & \vdots \\ 0 & 0 & \cdots & d_{nn} \end{bmatrix}$$

N.B. Obviously $D^N = \begin{bmatrix} d_{11}^N & 0 & \cdots & 0 \\ 0 & d_{22}^N & \cdots & 0 \\ \vdots & \vdots & \vdots & \vdots \\ 0 & 0 & \cdots & d_{nn}^N \end{bmatrix}$ for any positive integer N.

4. A square matrix $A = [a_{ij}]$ of order n is said to be
   (a) Symmetric if and only if $A^T = A$.
   (b) Skew-symmetric if and only if $A^T = -A$.
5. The conjugate of *a* matrix A is the matrix $\overline{A}$ whose elements are, respectively, the conjugates of the elements of A. The matrix defined by

$$A^* = (\overline{A})^T = \overline{(A^T)}$$

is called the hermitian conjugate of A.

6. (a) The matrix A is said to be hermitian if and only if $A^* = A$.
   (b) A square matrix A is said to be unitary if

$$A^*A = AA^* = I$$

MULTIPLICATIVE INVERSE OF *a* SQUARE MATRIX

Let

$$A = \begin{bmatrix} a_{11} & a_{12} & \cdots & a_{1n} \\ a_{21} & a_{22} & \cdots & a_{2n} \\ \vdots & \vdots & \vdots & \vdots \\ a_{n1} & a_{n2} & \cdots & a_{nn} \end{bmatrix}$$

be *a* square matrix of order n.

1. The determinant of A, denoted by detA or |A|, is defined as the determinant:

$$\det A = \begin{bmatrix} a_{11} & a_{12} & \cdots & a_{1n} \\ a_{21} & a_{22} & \cdots & a_{2n} \\ \vdots & \vdots & \vdots & \vdots \\ a_{n1} & a_{n2} & \cdots & a_{nn} \end{bmatrix}$$

matrices Direct from the definitions and the properties of determinants, for any square matrices A, B of the same order any scalar c, we have

(a) $\det(cA) = c^n \, detA.$

(b) $\det(AB) = \det A \det B$

(c) $\det A^T = \det A.$

(d) $\det \overline{A} = \overline{\det A}$

(e) $|\det A| = 1$ if A is unitary

(f) $\det I = 1$

Proof:

1. If A is unitary, from definition

$$A^*A = \overline{A}A = I$$

Consider the determinant of both sides, we obtain

$$\det\left(\overline{A}^T\right)\det A = \det 1 \Rightarrow \overline{\det A} \det A = 1$$

$$\therefore |\det A|^2 = 1 \Rightarrow |\det A| = 1 |\because| \det A| > 0)$$

2. Let $A = [a_{ij}]$ be a square matrix of order n. The cofactor matrix of A, denoted by cof A, is defined by cof $A = [A_{ij}]_{n\times n}$, where A_{ij} is the cofactor of a_{ij}, for every $i, j = 1, 2, ..., n$.

3. Let $A = [a_{ij}]$ be a square matrix of order n. The transpose of the cofactor of A, i.e. $(\text{cof } A)^T$, is called the adjoint of A, denoted by *adj A*.

4. *A* square matrix *A* of order *n* is said to be non-singular or invertible if and only if there exists *a* square matrix B such that

   $AB = BA = I$ where I is an identity matrix of order n, and the matrix *B* is called the multiplicative inverse or simply inverse of A, which is denoted by A^{-1}, i.e.

$$AA^{-1} = A^{-1}A = I.$$

N.B. The inverse of *a* non-singular matrix is unique.

Theorem: For any square matrix A of order n,

$$A(adj\ A) = (adj\ A)A = (det\ A)I$$

where I is an identity matrix of order n.

Proof: Let $A = (a_{ij})$, then adj $A = (A_{ij})^T$, where A_{ij} is the cofactor of a_{ij}.

$$A\ (adj\ A) = \begin{bmatrix} a_{11} & a_{12} & \cdots & a_{1n} \\ a_{21} & a_{22} & \cdots & a_{2n} \\ \vdots & \vdots & \vdots & \vdots \\ a_{n1} & a_{n2} & \cdots & a_{nn} \end{bmatrix} \begin{bmatrix} A_{11} & A_{12} & \cdots & A_{1n} \\ A_{21} & A_{22} & \cdots & A_{2n} \\ \vdots & \vdots & \vdots & \vdots \\ A_{n1} & A_{n2} & \cdots & A_{nn} \end{bmatrix}$$

$$= \begin{bmatrix} \det A & 0 & \cdots & \cdots & 0 \\ 0 & \det A & \cdots & \cdots & 0 \\ 0 & 0 & \det A & & 0 \\ \vdots & \vdots & \vdots & \cdots & \vdots \\ 0 & 0 & 0 & \cdots & \det A \end{bmatrix} = (\det A)I$$

Similarly, it can also be shown that $(adj\ A)A = (det\ A)I$.

1 A square matrix A is non-singular if and only if $\det A \neq 0$.

2. Let A be a square matrix, if $\det A \neq 0$, then A is non-singular and

$$A^{-1} = \frac{1}{\det A} \operatorname{adj} A .$$

N.B. For any non-singular matrix A of order 2 $A = \begin{bmatrix} a & b \\ c & d \end{bmatrix}$, its multiplicative inverse is $A^{-1} = \dfrac{1}{\det A}\begin{bmatrix} d & -b \\ -c & d \end{bmatrix}$

3. A square matrix A is said to be singular or not invertible if and only if the inverse A^{-1} of A does not exist.

N.B. A square matrix A is singular if and only if $\det A = 0$.

4. Let A, B are non-singular square matrix of order n and any scalar c, we have
   (a) A^{-1} is non-singular and $(A^{-1})^{-1} = A$,
   (b) AB is non-singular and $(AB)^{-1} = B^{-1}A^{-1}$
   (c) A^n is non-singular for any positive integer n and $(A^n)^{-1} = (A^{-1})^n$
   (d) λA is non-singular for any non-zero scalar c and $(cA)^{-1} = \frac{1}{c}A^{-1}$

(e) A^T is non-singular and $(A^T)^{-1} = (A^{-1})^T$

5. If $U^{-1}AU = B$, then

   (a) $TrA = TrB$

   (b) $\det A = \det B$

Example: Given *a* square complex unitary matrix Q of order 2 such that

$$Q^*Q = QQ^* = I$$

1. Show that detQ = 1.
2. Hence, show that the general form of Q is and

$$Q = \begin{bmatrix} \alpha & \beta \\ -\bar{\beta} & \bar{\alpha} \end{bmatrix} \text{ and } \alpha\bar{\alpha} + \beta\bar{\beta} = 1$$

where α and β are arbitrary complex numbers and $\bar{\alpha}$ and $\bar{\beta}$ are their respective complex conjugates.

Solution:

1. $\det(Q^*Q) = \det 1 \Rightarrow \overline{\det Q}\det Q = 1 \Rightarrow |\det Q|^2 = 1$

   $\therefore$ $\det Q = 1$

2. Let $Q = \begin{bmatrix} \alpha & \beta \\ \gamma & \delta \end{bmatrix}$ where $\alpha, \beta, \gamma, \delta$ are arbitrary complex numbers. Using the results in (a), we have $\overline{\det Q}\det Q = 1$. If we expand the given equality, we obtain

$$\begin{cases} \alpha\bar{\alpha} + \gamma\bar{\gamma} = 1 \\ \beta\bar{\beta} + \delta\bar{\delta} = 1 \\ \bar{\alpha}\beta + \bar{\gamma}\delta = 0 \end{cases}$$

From det Q = 1, we further obtain

$$\alpha\delta - \beta\gamma = 1$$

From Equation, we get $\delta = -\bar{\alpha}\dfrac{\beta}{\bar{\gamma}}$ and put it back to Equation

$$-\frac{\beta}{\bar{\gamma}}(\alpha\bar{\alpha} + \gamma\bar{\gamma}) = 1 \;\Rightarrow\; \gamma = -\bar{\beta}$$

and $\delta = \bar{\alpha}$. Hence we have

$$Q = \begin{bmatrix} \alpha & \beta \\ -\bar{\beta} & \bar{\alpha} \end{bmatrix} \text{ where } \alpha\bar{\alpha} + \beta\bar{\beta} = 1$$

Example: The special theory of relativity shows that the transformation between *a* moving frame (x', ict') of velocity v with respect to *a* stationary observer frame (x, ict) is given by the Lorentz transformation matrix

$$\begin{bmatrix} x' \\ ict' \end{bmatrix} = \Lambda \begin{bmatrix} x \\ ict \end{bmatrix}$$

where $\Lambda = \begin{bmatrix} \cosh\theta & i\sinh\theta \\ -i\sinh\theta & \cosh\theta \end{bmatrix}$ and $\tanh\theta = \frac{v}{c}$ (c is the velocity of light).

1. Find Λ^{-1}.

$$\begin{cases} x = \dfrac{x' + vt'}{\sqrt{1 - \dfrac{v^2}{c^2}}} \\ t = \dfrac{t' + \dfrac{vx'}{c^2}}{\sqrt{1 - \dfrac{v^2}{c^2}}} \end{cases}$$

Hence show that

2. Given that

$$\Lambda_1 = \begin{bmatrix} \cosh\theta_1 & i\sinh\theta_1 \\ -i\sinh\theta_1 & \cosh\theta_1 \end{bmatrix} \tanh\theta_1 = \frac{v_1}{c} \text{ and}$$

$$\Lambda_2 = \begin{bmatrix} \cosh\theta_2 & i\sinh\theta_2 \\ -i\sinh\theta_2 & \cosh\theta_2 \end{bmatrix} \tanh\theta_2 = \frac{v_2}{c}$$

Show that if $\Lambda = \Lambda_1\,\Lambda_2$ where

$$\Lambda = \begin{bmatrix} \cosh\theta & i\sinh\theta \\ -i\sinh\theta & \cosh\theta \end{bmatrix} \tanh\theta = \frac{v}{c}$$

then $v = \dfrac{v_1 + v_2}{1 + \dfrac{v_1 v_2}{c^2}}$.

Solution:

1. $\det \Lambda = \begin{bmatrix} \cosh\theta & i\sinh\theta \\ -i\sinh\theta & \cosh\theta \end{bmatrix} = \cosh^2\theta - \sinh^2\theta = 1$

$$\therefore \quad \Lambda^{-1} = \frac{1}{\det \Lambda}\begin{bmatrix} \cosh\theta & i\sinh\theta \\ -i\sinh\theta & \cosh\theta \end{bmatrix} = \begin{bmatrix} \cosh\theta & -i\sinh\theta \\ i\sinh\theta & \cosh\theta \end{bmatrix}$$

$$\begin{bmatrix} x \\ ict \end{bmatrix} = \Lambda^{-1}\begin{bmatrix} x' \\ ict'' \end{bmatrix} = \begin{bmatrix} \cosh\theta & -i\sinh\theta \\ i\sinh\theta & \cosh\theta \end{bmatrix} = \begin{bmatrix} x'\cosh\theta + ct'i\sinh\theta \\ ix'\sinh\theta + ict'\cosh\theta \end{bmatrix}$$

$$\therefore \quad \begin{cases} x = x'\cosh\theta + ct'i\sinh\theta \\ t = \dfrac{x'}{c}\sinh\theta + t'\cosh\theta \end{cases}$$

Since and $\tanh\theta = \dfrac{v}{c}$ and $1 - \tanh^2\theta = \dfrac{1}{\cosh^2\theta}$

$$\therefore \quad \cosh\theta = \frac{1}{\sqrt{1 - \dfrac{v^2}{c^2}}} \text{ and } \sinh\theta = \tanh\theta\cosh\theta = \frac{v/c}{\sqrt{1 - \dfrac{v^2}{c^2}}}$$

Hence we have

$$\therefore \quad \begin{cases} x = x'\cosh\theta + ct'i\sinh\theta \\ t = \dfrac{x'}{c}\sinh\theta + t'\cosh\theta \end{cases} \Rightarrow \begin{cases} x = \dfrac{x' + vt'}{\sqrt{1 - \dfrac{v^2}{c^2}}} \\ t = \dfrac{t' + \dfrac{vx'}{c^2}}{\sqrt{1 - \dfrac{v^2}{c^2}}} \end{cases}$$

2. $\Lambda = \Lambda_1\Lambda_2 = \begin{bmatrix} \cosh\theta_1 & i\sinh\theta_1 \\ -i\sinh\theta_1 & \cosh\theta_1 \end{bmatrix}\begin{bmatrix} \cosh\theta_2 & i\sinh\theta_2 \\ -i\sinh\theta_2 & \cosh\theta_2 \end{bmatrix}$

$$= \begin{bmatrix} \cosh\theta_1\cosh\theta_2 + \sinh\theta_1 i\sinh\theta_2 & i(i\sinh\theta_1\cosh\theta_2 + \cosh\theta_1\sinh\theta_2 \\ -i(i\sinh\theta_1\cosh_2\cosh\theta_1 + \sinh\theta_2) & \cosh\theta_1\cosh\theta_2 + \sinh\theta_1 i\sinh\theta_2 \end{bmatrix}$$

Hence we have

$$\begin{cases} \cosh\theta = \cosh\theta_1 + \cosh\theta_1 + i\sinh\theta_2\sinh\theta_1\sinh\theta_2 \\ \sinh\theta = \sinh\theta_1 + \cosh\theta_2 + \cosh\theta_2\cosh\theta_1\sinh\theta_2 \end{cases}$$

i.e. $\dfrac{v}{c}$ $\tan\theta = \dfrac{\sinh\theta}{\cosh\theta} = \dfrac{\sinh\theta_1 + \cosh\theta_2 + \cosh\theta_1\sinh\theta_2}{\cosh\theta_1 + \cosh\theta_2 + \sinh\theta_1\sinh\theta_2} = \dfrac{\tanh\theta_1 + \tanh\theta_2}{1 + \tanh\theta_1\tanh\theta_2}$

$$\therefore \quad v = \frac{v_1 + v_2}{1 + \dfrac{v_1 v_2}{c^2}}$$

$$\therefore \quad \Lambda^{-1} = \frac{1}{\det \Lambda}\begin{bmatrix} \cosh\theta & i\sinh\theta \\ -i\sinh\theta & \cosh\theta \end{bmatrix} = \begin{bmatrix} \cosh\theta & -i\sinh\theta \\ i\sinh\theta & \cosh\theta \end{bmatrix}$$

$$\begin{bmatrix} x \\ ict \end{bmatrix} = \Lambda^{-1}\begin{bmatrix} x' \\ ict'' \end{bmatrix} = \begin{bmatrix} \cosh\theta & -i\sinh\theta \\ i\sinh\theta & \cosh\theta \end{bmatrix} = \begin{bmatrix} x'\cosh\theta + ct'i\sinh\theta \\ ix'\sinh\theta + ict'\cosh\theta \end{bmatrix}$$

$$\therefore \quad \begin{cases} x = x'\cosh\theta + ct'i\sinh\theta \\ t = \dfrac{x'}{c}\sinh\theta + t'\cosh\theta \end{cases}$$

Since and $\tanh\theta = \frac{v}{c}$ and $1 - \tanh^2\theta = \frac{1}{\cosh^2\theta}$

$$\therefore \quad \cosh\theta = \frac{1}{\sqrt{1-\dfrac{v^2}{c^2}}} \text{ and } \sinh\theta = \tanh\theta\cosh\theta = \frac{v/c}{\sqrt{1-\dfrac{v^2}{c^2}}}$$

Hence we have

$$\therefore \quad \begin{cases} x = x'\cosh\theta + ct'i\sinh\theta \\ t = \dfrac{x'}{c}\sinh\theta + t'\cosh\theta \end{cases} \Rightarrow \begin{cases} x = \dfrac{x' + vt'}{\sqrt{1-\dfrac{v^2}{c^2}}} \\ t = \dfrac{t' + \dfrac{vx'}{c^2}}{\sqrt{1-\dfrac{v^2}{c^2}}} \end{cases}$$

2. $$\Lambda = \Lambda_1\Lambda_2 = \begin{bmatrix} \cosh\theta_1 & i\sinh\theta_1 \\ -i\sinh\theta_1 & \cosh\theta_1 \end{bmatrix}\begin{bmatrix} \cosh\theta_2 & i\sinh\theta_2 \\ -i\sinh\theta_2 & \cosh\theta_2 \end{bmatrix}$$

$$= \begin{bmatrix} \cosh\theta_1\cosh\theta_2 + \sinh\theta_1 i\sinh\theta_2 & i(i\sinh\theta_1\cosh\theta_2 + \cosh\theta_1\sinh\theta_2 \\ -i(i\sinh\theta_1\cosh_2\cosh\theta_1 + \sinh\theta_2) & \cosh\theta_1\cosh\theta_2 + \sinh\theta_1 i\sinh\theta_2 \end{bmatrix}$$

Hence we have

$$\begin{cases} \cosh\theta = \cosh\theta_1 + \cosh\theta_1 + i\sinh\theta_2\sinh\theta_1\sinh\theta_2 \\ \sinh\theta = \sinh\theta_1 + \cosh\theta_2 + \cosh\theta_2\cosh\theta_1\sinh\theta_2 \end{cases}$$

i.e. $$\frac{v}{c}\tan\theta = \frac{\sinh\theta}{\cosh\theta} = \frac{\sinh\theta_1 + \cosh\theta_2 + \cosh\theta_1\sinh\theta_2}{\cosh\theta_1 + \cosh\theta_2 + \sinh\theta_1\sinh\theta_2} = \frac{\tanh\theta_1 + \tanh\theta_2}{1 + \tanh\theta_1\tanh\theta_2}$$

$$\therefore \quad v = \frac{v_1 + v_2}{1 + \dfrac{v_1 v_2}{c^2}}$$

Example: The special theory of relativity shows that the transformation between *a* moving frame (*x*', *ict*') of velocity v with respect to *a* stationary observer frame (x, ict) is given by the Lorentz transformation matrix

$$\begin{bmatrix} x' \\ ict' \end{bmatrix} = \Lambda \begin{bmatrix} x \\ ict \end{bmatrix}$$

where $\Lambda = \begin{bmatrix} \cosh\theta & i\sinh\theta \\ -i\sinh\theta & \cosh\theta \end{bmatrix}$ and $\tanh\theta = \frac{v}{c}$ (c is the velocity of light).

1. Find Λ^{-1}.

$$\begin{cases} x = \dfrac{x' + vt'}{\sqrt{1 - \dfrac{v^2}{c^2}}} \\ t = \dfrac{t' + \dfrac{vx'}{c^2}}{\sqrt{1 - \dfrac{v^2}{c^2}}} \end{cases}$$

   Hence show that

2. Given that

$$\Lambda_1 = \begin{bmatrix} \cosh\theta_1 & i\sinh\theta_1 \\ -i\sinh\theta_1 & \cosh\theta_1 \end{bmatrix} \tanh\theta_1 = \frac{v_1}{c} \text{ and}$$

$$\Lambda_2 = \begin{bmatrix} \cosh\theta_2 & i\sinh\theta_2 \\ -i\sinh\theta_2 & \cosh\theta_2 \end{bmatrix} \tanh\theta_2 = \frac{v_2}{c}$$

   Show that if $\Lambda = \Lambda_1 \Lambda_2$ where

$$\Lambda = \begin{bmatrix} \cosh\theta & i\sinh\theta \\ -i\sinh\theta & \cosh\theta \end{bmatrix} \tanh\theta = \frac{v}{c}$$

   then $v = \dfrac{v_1 + v_2}{1 + \dfrac{v_1 v_2}{c^2}}$.

Solution:

1. $\det \Lambda = \begin{vmatrix} \cosh\theta & i\sinh\theta \\ -i\sinh\theta & \cosh\theta \end{vmatrix} = \cosh^2\theta - \sinh^2\theta = 1$

Encyclopaedia

of

ORDINARY DIFFERENTIAL EQUATION

Encyclopaedia of ORDINARY DIFFERENTIAL EQUATION

Volume 2

Dr. Rakesh Kumar Pandey

ANMOL PUBLICATIONS PVT. LTD.
NEW DELHI - 110 002 (INDIA)

ANMOL PUBLICATIONS PVT. LTD.
Regd. Office: 4360/4, Ansari Road, Daryaganj,
New Delhi-110 002 (India)
Ph.: 23278000, 23261597
Branch Office: No. 1015, Ist Main Road, BSK IIIrd Stage
IIIrd Phase, IIIrd Block,
Bangalore-560 085 (India)
Tel.: 080-41723429
Visit us at: www.anmolpublications.com

Encyclopaedia of Ordinary Differential Equation

First Edition, 2009
ISBN 978-81-261-4106-7 (Set)

PRINTED IN INDIA

Printed at Mehra Offset Press, Delhi.

Contents

Preface

The Encyclopaedia of Ordinary Differential Equation is intended both as an introductory text and as a reference book for those interested in studying Ordinary Differential Equation. This encyclopaedia gives an up-to-date account of the theories for the *Equations and their applications*. All the chapters are beautifully illuminated towards the approach found in these volumes. The Encyclopaedia starts with Introduction to First Order Equation making it suitable for those with little background in mathematics. Both the volumes are good size covering all the essential topics is in adequate depth. It covers topics similar to other books targeted at the same audience.

Written in a clear, precise and readable manner, this encyclopaedia is designed to provide undergraduate mathematics students with a sound and inspiring introduction to the main themes of Ordinary Differential Equations. The encyclopaedia explains the origins of various types of differential equations. Gives in detail suitable method of solution to different situation. This encyclopaedia is designed to cover the basics thoroughly and then move on. Each page of these books is packed with direction, examples, and a world of understanding. This encyclopaedia is an excellent support and resource for the teacher. This is an outstanding resource for anyone interested on knowing more about the art of teaching.

This Encyclopaedia is well written and easy to follow chapters on many different areas. The layout and style are cleverly designed to keep students working through the chapters. As a reminder of useful techniques, these volumes are definitely valuable. The scope of this encyclopaedia is limited to linear differential equations of the first order, linear differential equation of higher order, differential equations of higher degree and special methods of solution of differential equations of second order, keeping in view the requirement of undergraduate students.

Author

List of Symbols

$\pm$	$\not\subset$	$\subset$	$\supset$	$\subseteq$	$\in$	$\notin$	∂
$\neq$	$\equiv$	$\approx$	$\angle$	∇	Σ	Π	Ω
ϕ	α	β	χ	ε	γ	η	λ
μ	ν	$\perp$	π	θ	ρ	σ	τ
$\Re$	ω	ξ	ψ	ζ	Υ	$\exists$	$\int$
$\iint$	$\iiint$	$\oint$	$\times$	$\sqrt{}$	$\therefore$	$\because$	Δ
$\geq$	$\leq$	$\Rightarrow$	$\Downarrow$	$\Leftrightarrow$	∞	$\propto$	$\mathbb{R}$
$\mathbb{Z}$	$\mathbb{N}$	$\mathbb{C}$	$\mathbb{Q}$				

Chapter 14

Line Surface and Volume Integerals

LINE INTEGRALS

A curve *C* in two-dimensional space is defined by

$$C = \{(x, y) | \ x = x(t), y = y(t), a \leq t \leq \beta\}$$

where α and β specify initial and final points *A* and *B* of the curve, respectively. Suppose *a* function *f*(*x*, *y*) is defined along a curve *C* joining *A* to *B*.

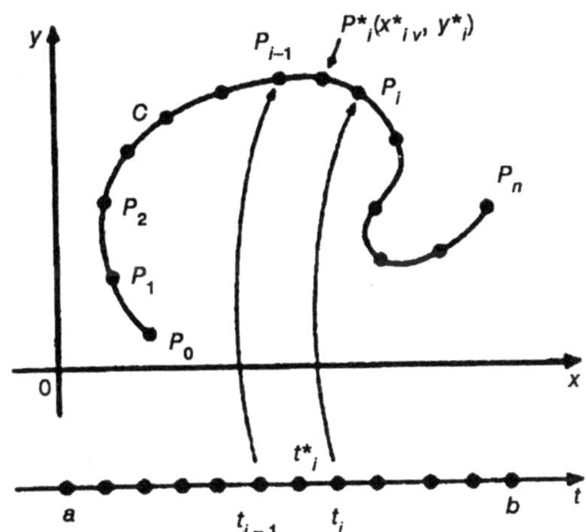

Fig. A Curve in Two-dimensional Space

We call the limit the line integral of *f*(*x*, *y*) along the curve *C*, and denote it by

$$\int_C f(x,y)\,dS = \lim_{\|P\|\to 0} \sum_{i=1}^{n} f(x_i^*, y_i^*)\Delta S_i$$

Then the line integral of *f*(*x*, *y*) along *C* can be evaluated by means of the following definite integral:

$$\int_C f(x,y)\,dS = \int_{\alpha}^{\beta} f[x(t), y(t)]\sqrt{\left(\frac{dx}{dt}\right)^2 + \left(\frac{dy}{dt}\right)^2}\,dt$$

Note: $$dS = \sqrt{(dx)^2 + (dy)^2}$$

$$= \sqrt{\left[\left(\frac{dx}{dt}\right)^2 + \left(\frac{dy}{dt}\right)^2 (dt)^2\right]} = \sqrt{\left(\frac{dx}{dt}\right)^2 + \left(\frac{dy}{dt}\right)^2}\, dt$$

The generalization to three-dimensional space can be made easily from Equation. *A* curve *C* in space is defined by three functions

$$C = \{(x, y, z)|\ x = x(t), y = y(t)\ z = z(t), \alpha \le t \le \beta\}$$

where α and β specify initial and final points *A* and B of the curve, respectively. Suppose a function $f(x, y, z)$ is defined along *a* curve *C* joining *A* to B. The line integral in 3-spaces is given by the following expression

$$\int_C f(x,y,z)\,dS = \int_\alpha^\beta f[x(t), y(t), z(t)] \sqrt{\left(\frac{dx}{dt}\right)^2 + \left(\frac{dy}{dt}\right)^2 + \left(\frac{dz}{dt}\right)^2}\, dt$$

If equations for *C* are given in the form *C*:

$$C = \{(x, \dot{y}, z)|\ y = y(x), z = z(x), x_A \le x \le x_B\}$$

then

$$\int_C f(x,y,z)\,dS = \int_{x_A}^{x_=} f[x, y(x), z(x)] \sqrt{1 + \left(\frac{dy}{dx}\right)^2 + \left(\frac{dz}{dx}\right)^2}\, dx$$

Suppose *a* function $f(x, y)$ is positive along *a* curve *C* in the *x* y-plane. If at each point of *C* we draw a vertical line of height $z = f(x, y)$, then dS is an elemental piece of length along C, we interpret of line integral $\int_C f(x,y)\,dS$ as the total area of the vertical wall. It is obvious that the line integral $\int_C dS$ represents the length of the curve *C*. When *C* is *a* closed curve in the *xy*-plane that does not cross itself, we write $\oint_{C_1} f(x,y)\,dS$ and $\oint_{C_2} f(x,y)\,dS$

LINE INTEGRALS INVOLVING VECTOR FUNCTIONS

If S is *a* measure of length along *a* curve *C* from *A* to *B*, and if S is chosen equal to z ero at *A*, then *a* unit tangent vector pointing in the direction of motion along *C* $T = \frac{dr}{ds}$ is:. Consequently, if $f(x, y, z)$ is the tangential component of $F(x, y, z)$ along *C*, then $f(x, y, z) =$

$F.\ T = F.\dfrac{dr}{ds}$. If the components of the vector field $F(x, y, z)$ are $F(x, y, z) = P(x, y, z)i + Q(x, y, z)j + R(x, y, z)k$ then

$$\int_C F \cdot dr = \int_C (Pi + Qj + Rk) \cdot (dxi + dyj + dzk)$$

$$= \int_C (Pdx + Qdy + Rdz)$$

So
$$\int_C f(x, y, z)\, dS = \int_C F \cdot dr = \int_C (Pdx + Qdy + Rdz)$$

If *a* curve C has initial and final points A and B, then the curve that traces the same points but has initial and final points B and A is denoted by $-C$. Then

$$\int_C F \cdot dr = -\int_{-C} F \cdot dr$$

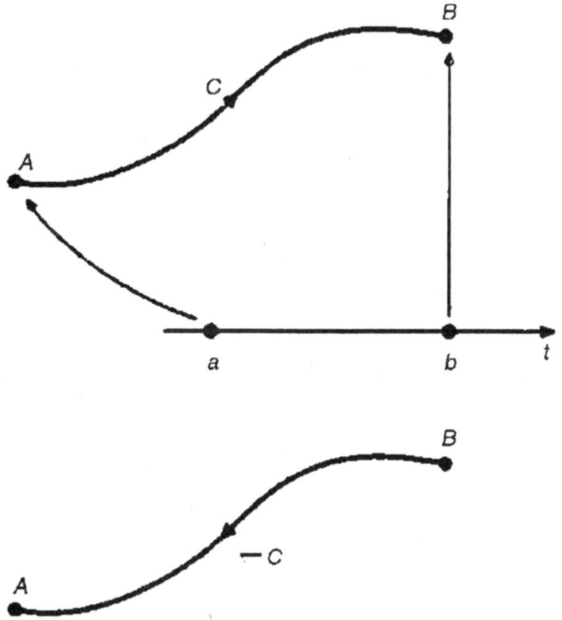

Fig. Line Integrals Involving Vector Functions

Definition: *A* line integral $\int_C F \cdot dr$ is said to be independent of path in *a* domain D if for each pair of points A and B in D, the value of the line integral $\int_C F \cdot dr$ is the same for all piece-smooth paths C in D from A to B.

Example: If $A = (3x^2 + 6y)i - 14yzj + 20xz^2k$, evaluate $\int_C A \cdot dr$ from (0, 0, 0) to (1, 1, 1) along the following paths C:

1. $C = \{(x, y, z) | x = t, y = t^2, z = t^3\}$.
2. The straight lines from (0, 0, 0) to (1, 0, 0), then to (1, 1, 0), and then to (1, 1, 1).

Solution:

1. If $x = t, y = t^2, z = t^3$, points (0, 0, 0) and (1, 1, 1) correspond to $t = 0$ and $t = 1$ respectively. Then

$$\int_C A \cdot dr = \int_C (3x^2 + 6y)\,d(t) - 14yz\,d(t^2) + 20xz^2 d(t^3)$$

$$= \int_0^1 (3t^2 + 6t^2)dt - 28(t^2)(t^3)d(t^2) + 20(t)(t^3)^2 d(t^3)$$

$$= \int_0^1 9t^2 dt - 28t^2 dt + 60t^2 dt$$

$$= \int_0^1 (9t^2 - 28t^6 + 60t^9)\,dt = 5$$

2. Along the straight line from (0, 0, 0) to (1, 0, 0) $y = 0$, $z = 0$, $dy = 0$, $dz = 0$ while x varies from 0 to 1. Then the integral over this part of the path is

$$\int_0^1 (3x^2 + 6(0))\,dx - 14(0)(0)(0) + 20x(0)^2 = \int_0^1 3x^2 dx = 1$$

Along the straight line from (1, 0, 0) to (1, 1, 0) $x = 1$, $z = 0$, dx = 0, dz = 0 while y varies from 0 to 1. Then the integral over this part of the path is

$$\int_0^1 (3(1)^2 + 6y)0 - 14y(0)\,dy + (0)(0) + 20(1)(0)^2(0) = 0$$

Along the straight line from (1, 1, 0) to (1, 1, 1) $x = 1$, $y = 1$, $dx = 0$, dy = 0 while z varies from 0 to 1. Then the integral over this part of the path is

$$\int_0^1 (3(1)^2 + 6(1)0 - 14z(0) + 20(1)z^2 dz = \int_0^1 20z^2 dz = \frac{20}{3}$$

(c) The straight line joining (0, 0, 0) to (1, 1, 1) is given in parametric form $x = t$, $y = t$, $z = t$. Then

$$\int_C A \cdot dr = \int_0^1 (3t^2 + 6t^2)\,dt - 14(t)(t)\,dt + 20(t)(t)^2\,dt$$

$$= \int_0^1 9t^2\,dt - 28t^3\,dt + 60t^2 dt = \int_0^1 (6t - 11t^2 + 20t^3)\,dt = \frac{13}{3}.$$

Suppose $P(x, y, z)$, $Q(x, y, z)$, $R(x, y, z)$ are continuous functions in some domain D. The line integral $\int_C F \cdot dr = \int_C (Pdx = Qdy + Rdz)$ is independent of path in D if and only if these exists a function $\phi(x, y, z)$ defined in D such that

$$\nabla\phi(x, y, z) = F(x, y, z) = P(x, y, z)i + Q(x, y, z)j + R(x, y, z)k$$

Theorem: When *a* line integral is independent of path in *a* domain D, and A and B are points in D, then $\int_C F \cdot dr = \phi(x_B, y_B, z_B) - \phi(x_A, y_A, z_A)$ where $\nabla\phi = F$ for every piecewise smooth curve C in D from A to B.

Proof: Let $F(x, y, z) = P(x, y, z)i + Q(x, y, z)j + R(x, y, z)k$. By hypothesis, $\int_C F \cdot dr$ is independent of the path C joining any two points A and B $A(x_A, y_A, z_A)$ and $B(x_B, y_B, z_B)$.

Then define *a* scalar function *f* with respect *a* reference point $O(x_0, y_0, z_0)$ by using *a* line integral joining the points $P(x, y, z)$ and $O(x_0, y_0, z_0)$ along *a* path C_o:

$$\phi(x, y, z) = \int_{(x_0, y_0, z_0)}^{(x,y,z)} F \cdot dr = \int_{C_p} F \cdot dr$$

Direct from the definition, we have

$$\int_C F \cdot dr = \int_{C_B} F \cdot dr - \int_{C_A} F \cdot dr = \phi(x_B, y_B, z_B) - \phi(x_A, y_A, z_A)$$

On the other hand, let us consider

$$\phi(x+\Delta x, y, z) - \phi(x, y, z) = \int_{(x,y,z)}^{(x+\Delta x, y, z)} F \cdot dr = \int_{(x,y,z)}^{(x+\Delta x, y, z)} (Pdx + Qdy + Rdz)$$

Since the last integral must be independent of the path joining (x, y, z) and $(x + Dx, y, z)$, we must choose the path to be *a* straight line joining these points so that dy and dz are *z* ero. Then

$$\frac{\phi(x+\Delta x, y, z) - \phi(x, y, z)}{\Delta x} = \frac{1}{\Delta x}\int_{(x,y,z)}^{(x+\Delta x, y, z)} Pdx$$

Taking the limit of both sides as $\Delta x \to 0$, we have $P = \frac{\partial\phi}{\partial x}$.

Similarly, we can show that $Q = \frac{\partial\phi}{\partial x}$ and $R = \frac{\partial\phi}{\partial z}$.

Therefore we have $F = \nabla\phi$.

Corollary: The line integral $\int_C F \cdot dr$ is independent of path in *a* domain *D* if and only if $\oint_C F \cdot dr = 0$ for every closed path in *D*.

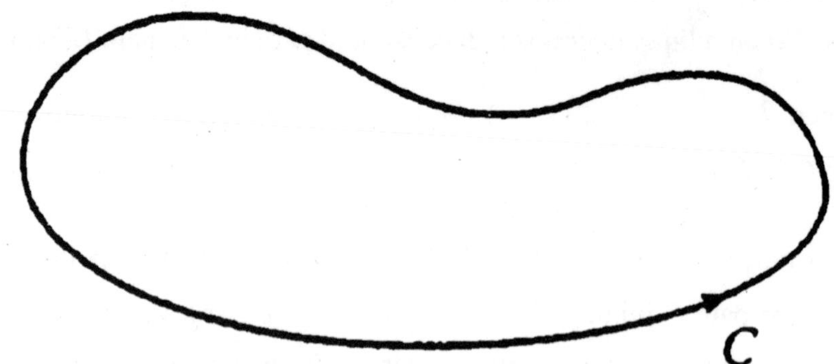

Fig. Line Integral is Independent of Path in *a* Domain

Theorem: Let D be a domain in which $P(x, y, z)$, $Q(x, y, z)$ and $R(x, y, z)$ have continuous first derivatives.

If the line integral $\int_C F \cdot dr = \int_C (Pdx = Qdy + Rdz)$ is independent of path in *D*, then $\nabla \times F = 0$ in *D*. Conversely, if $\nabla \times F = 0$ in *D*, then the line integral is independent of path in *D*.

Proof: By theorem, there exists *a* scalar function ϕ such that $F = \nabla\phi$. Hence

$$\nabla \times F = \nabla \times (\nabla\phi) = 0$$

Note: For line integrals in the *xy*-plane, $\nabla \times F = 0$ can be stated more simply as

$$\frac{\partial P}{\partial x} = \frac{\partial P}{\partial y}$$

SURFACE INTEGRALS INVOLVING VECTOR FUNCTIONS

Surface integral appears in the same form as line integrals, the element of area also being a vector, dS.

Often this area element is written as $dS = ndS$ in which *n* is *a* unit (normal) vector to indicate the positive direction.

First, if the surface is closed, we agree to take the outward normal as positive.

Second, if the surface is opened, the positive normal is determined by the right-hand grip rule as shown in the following figure

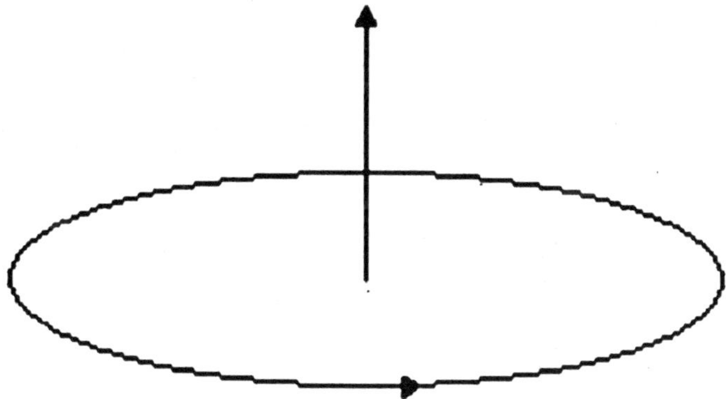

Analogous to the line integrals, surface integral of the vector field (see Fig. 9.4) $F(x, y, z) = P(x, y, z)i + Q(x, y, z)j + R(x, y, z)k$ appear in the form

$$\iint_S F \cdot dS = \iint_S F \cdot n dS$$

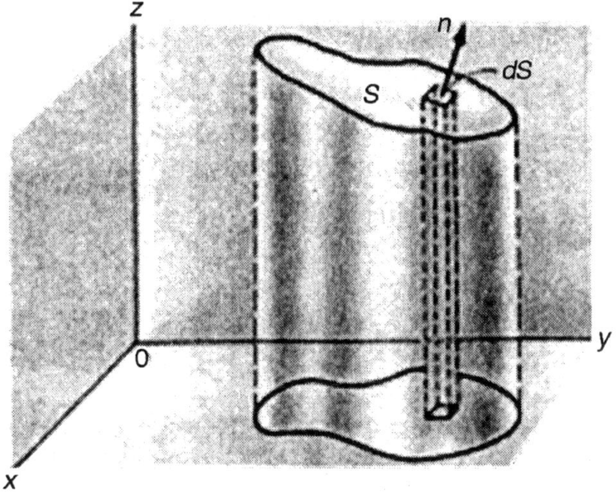

Fig. Surface Integrals Involving Vector Functions

Example: If $F = 4xzi - y^2j + yzk$, evaluate $\iint_S F \cdot n dS$ where S is the surface of the cube bounded by $x = 0, x = 1, y = 0, y = 1, z = 0, z = 1$.

Solution: Face DEFG: $n = i, x = 1$. Then

$$\iint_{DEFG} F \cdot n dS = \int_0^1 \int_0^1 (4zi - y^2 j + yzk) \cdot i dy dz = \int_0^1 \int_0^1 4z dy dz = 2$$

Face $ABCO$: $n = -i, x = 0$. Then

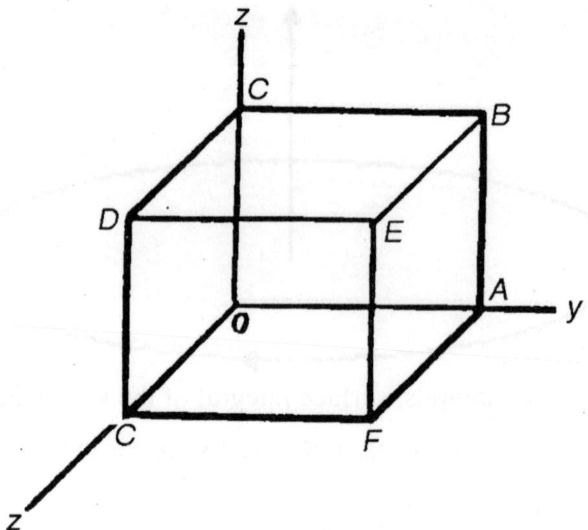

Fig. The Surface of the Cube

$$\iint_{ABCO} F \cdot ndS = \int_0^1\int_0^1 (-y^2 j + yzk)\cdot(-i)dydz = 0$$

Face $ABEF$: $n = j$, $y = 1$. Then

$$\iint_{ABEF} F \cdot ndS = \int_0^1\int_0^1 (4xzi - j + zk)\cdot jdxdz = \int_0^1\int_0^1 -dxdz = -1$$

Face $OGDC$: $n = -j$, $y = 0$. Then

$$\iint_{OGDC} F \cdot ndS = \int_0^1\int_0^1 (4xzi).(-j)dxdz = 0$$

Face $BCDE$: $n = k$, $z = 1$. Then

$$\iint_{BCDE} F \cdot ndS = \int_0^1\int_0^1 (4xi - y^2 j + tk)\cdot kdxdz = \int_0^1\int_0^1 yd\,xdy = \frac{1}{2}$$

Face $AFGO$: $n = -k$, $z = 0$. Then

$$\iint_{AFGO} F \cdot ndS = \int_0^1\int_0^1 (-y^2 j).(-k)dxdy = 0$$

Adding, $\iint_{DEFG} F \cdot ndS = 2 + 0 + (-1) + 0 + \frac{1}{2} + 0 = \frac{3}{2}$

Remark: The surface integral $\iint_S F \cdot dS$ is sometimes interpreted as *a* flux through the given surface.

VOLUME INTEGRALS INVOLVING VECTOR FUNCTIONS

Volume integrals are somewhat simpler, for the volume element dV is *a* scalar quantity. If the components of the vector field $F(x, y, z)$ are

$$F(x, y, z) = P(x, y, z)i + Q(x, y, z)j + R(x, y, z)k$$

then we have

$$\iiint_V F \cdot dV = \iiint_V (Pi + Qj + Rk)\,dV = i\iiint_V P\,dV + j\iiint_V Q\,dV + k\iiint_V R\,dV$$

Example: If $F = 2xzi - xj + y^2k$, evaluate where V is the region bounded by the surfaces $x = 0, y = 0, y = 6, z = x^2, z = 4$.

Solution:

$$\iiint_V F\,dV = \int_{x-0}^{2}\int_{y-0}^{6}\int_{z-x^2}^{4}(2xzi - xj + y^2k)\,dz\,dy\,dx$$

$$= i\int_{x-0}^{2}\int_{y-0}^{6}\int_{z-x^2}^{4} 2xzdzdydx - j\int_{x-0}^{2}\int_{y-0}^{6}\int_{z-x^2}^{4} xdzdydx$$

$$= +\;k\int_{x-0}^{2}\int_{y-0}^{6}\int_{z-x^2}^{4} y^2dzdydx$$

$$= 128i - 24j + 384k$$

GREEN'S THEOREM AND STOKES THEOREM

GREEN'S THEOREM

Let C be *a* smooth, closed curve in the x y-plane that does not intersect itself and that encloses *a* region R.

If $M(x, y)$ and $N(x, y)$ have continuous first derivative in *a* domain D containing C and R, then

$$\oint_C M\,dx + Ndy = \iint_R \left(\frac{\partial N}{\partial x} - \frac{\partial M}{\partial y}\right) dxdy$$

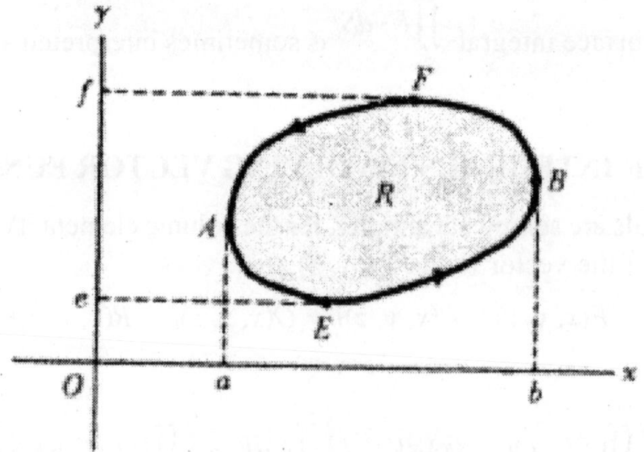

Fig. The Bounded Region

Proof: Let the equations of curves AEB and AFB be $y = y_1(x)$ and $y = y_2(x)$ respectively. If R is the region bounded by C, we have

$$\iint_R \frac{\partial M}{\partial y} dxdy = \int_a^b \left[\int_{y_1(x)}^{y_2(x)} \frac{\partial M}{\partial y} dxdy \right] dx$$

$$= \int_a^b M(x,y) \Big|_{y_1(x)}^{y_2(x)} dx = \int_a^b \left[M(x,y_2) - M(x,y_1) \right] dx$$

$$= -\int_a^b M(x,y) dx - \int_a^b M(x,y_2) - M(x,y_1) dx \; - \oint_C M dx$$

Then

$$1. \quad \oint_C M\,dx = \iint_R \frac{\partial M}{\partial y} dxdy$$

Similarly let the equations of curves EAF and EBF be $x = x_1(x)$ and $x = x_2(x)$ respectively. Then

$$\iint_R \frac{\partial M}{\partial y} dxdy = \int_e^f \left[\int_{x_1(y)}^{x_2(y)} \frac{\partial N}{\partial x} dx \right] dy = \int_e^f \left[N(x_2,y) - M(x_1,y) \right] dy$$

$$= \int_e^f N(x_1,y) dx + \int_e^f N(x_2,y) dx = \oint_C N\,dy$$

Then

2. $\oint_C N\,dy = \iint_R \frac{\partial M}{\partial y} dxdy$

Adding (1) and (2),

$$\oint_C N\,dx + N\,dy = \iint_R \left(\frac{\partial N}{\partial x} - \frac{\partial M}{\partial y} \right) dxdy$$

Example: Evaluate $\oint_C (y - \sin x)\,dx + \cos x dy$ by using the Green theorem, where C is the triangle joining the points from O(0, 0) to A(1, 0), then to B(1, 2) and then back to O(0, 0).

Solution:

$$\oint_C N\,dx + N\,dy = \iint_R \left(\frac{\partial N}{\partial x} - \frac{\partial M}{\partial y} \right) dxdy \iint_R (-\sin x - 1)\,dx\,dy$$

$$= \int_0^1 \left[\int_0^{2x} (-\sin x - 1)\,dy \right] dx = \int_0^1 [-y \sin x - y]_0^{2x}\, dx$$

$$= \int_0^1 \left[\int_0^{2x} (-\sin x - 1)\,dy \right] dx = \int_0^1 (-2x \sin x - 2x)\,dx$$

$$= 2 \cos 1 - 2 \sin 1 - 1$$

THE DIVERGENCE THEOREM

The divergence theorem relates certain surface integrals over surfaces that enclose volumes to triple integrals over the enclosed volume. More precisely, we have the following. LetS be *a* smooth, surface enclosing *a* region V and $A = A_1 i + A_2 j + A3k$ be *a* vector field whose components A_1, A_2 and A_3 have continuous first derivative in *a* domain containing S and V. If n is the unit outer normal to S, then

$$\oiint_S F \cdot ndS = \iiint_V \nabla \cdot F\,dV$$

or

$$\oiint_S (A_1 i + A_2 i + A_3 k) \cdot ndS = \iiint_V \left(\frac{\partial A_1}{\partial x} + \frac{\partial A_2}{\partial y} + \frac{\partial A_3}{\partial z} \right) dV$$

The divergence theorem is also called Gauss's Theorem. The divergence theorem states that the total divergence of A through the volume V is equal to the net flux of A through the surface S.

Proof: Let S be *a* closed surface which is such that any line parallel to the coordinate axes cuts S in at most two points. Assume the equations of the lower and upper portions, S_1 and S_2, to be $z = f_1(x, y)$ and $z = f_2(x, y)$ respectively. Denote the projection of the surface on the *x* y plane by R.

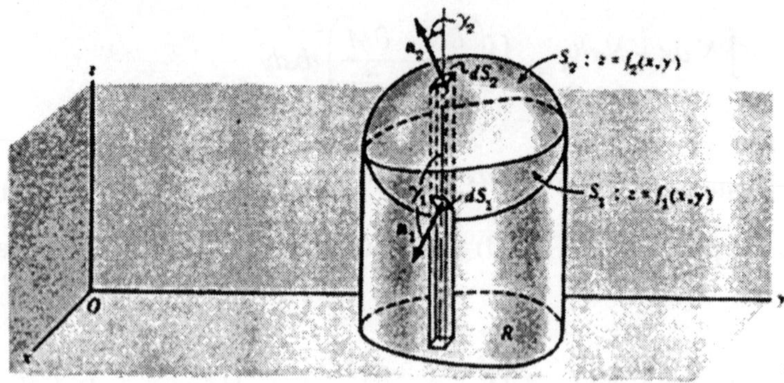

Fig. Integrals Over the Enclosed Volume

Consider

$$\iiint_V \frac{\partial A_3}{\partial z} dV = \iiint_V \frac{\partial A_3}{\partial z} ds\,dy\,dx = \iint_R \left[\int_{f_1(x,y)}^{f_2(x,y)} \frac{\partial A_3}{\partial z} dz \right] dy\,dx$$

$$= \iint_R A_3(x,y,z) \Big|_{f_1}^{f_2} dx = \iint_R [A_3(x,y,f_2) - A_3(x,y,f_1)] dy\,dx$$

For the upper portion S_2, $dydx = \cos\gamma_2 dS_2 = k \cdot n_2 dS_2$ since the normal n_2 to S_2 makes an acute angle γ_2 with k. For the lower portion S_1, $dydx = \cos\gamma_1 dS_1 = k \cdot n_1 dS_1$ since the normal n_1 to S_1 makes an acute angle γ_1 with k. Then

$$\iint_R [A_3(x,y,f_2) dy\,dx = \iint_{S_2} A_3 k \cdot n_2 dS_2$$

$$\iint_R [A_3(x,y,f_1) dy\,dx = -\iint_{S_2} A_3 k \cdot n_1 dS_1$$

and

$$\iint_R A_3(x,y,f_2) dy\,dx - \iint_R A_3(x,y,f_1) dy\,dx = \iint_{S_2} A_3 k \cdot n_2 dS_2 + \iint_{S_2} A_3 k \cdot n_2 dS_2$$

$$= \iint_{S_2} A_3 k \cdot n dS \quad \text{so that}$$

1. $$\iiint_V \frac{\partial A_3}{\partial z} dV = -\iint_R \frac{\partial M}{\partial y} dxdy$$

Similarly, by projection S on the other coordinate planes,

2. $$\iiint_V \frac{\partial A_1}{\partial x} dV = \iint_S A_1 i \cdot n\, dS$$

3. $$\iiint_V \frac{\partial A_2}{\partial y} dV = \iint_S A_2 i \cdot n\, dS$$

Adding (1), (2) and (3)

$$\iiint_V \left(\frac{\partial A_1}{\partial x} + \frac{\partial A_2}{\partial y} + \frac{\partial A_3}{\partial z} \right) dV = \iint_S (A_1 i + A_2 j + A_3 k) \cdot n dS$$

or

$$\iiint_V \nabla \cdot A dV = \iint_S A \cdot n dS$$

Note: Consider *a* point *P* which is enclosed by an infinitesimal volume element DV, we can apply the divergence theorem,

$$\iint_{\Delta S} A \cdot n dS = \iiint_V \text{div } A dV \cong (\text{div } A)\, \Delta V$$

where ΔS is the surface of the volume ΔV. Then taking the limit $\Delta V \to 0$, we have

$$\text{div } A = \lim_{\Delta V \to 0} \frac{1}{\Delta V} \iint_{\Delta S} A \cdot n dS$$

Example: Given that *a* vector function

$$F = (\rho\cos\theta)\hat{\rho} + (\rho\sin\theta)\hat{\theta} - (\rho\sin\theta\cos\phi)\hat{\phi}$$

1. Compute the divergence of F.
2. Check the divergence theorem for *F* by using an inverted hemisphere of radius R and resting on *x* y plane and centered at the origin.

Solution:

1. $\nabla \cdot F = \frac{1}{\rho^2 \partial \rho}\left(\rho^2 F_\rho\right) + \frac{1}{\rho \sin\theta}\frac{\partial}{\partial\theta}(\sin\theta F_\theta) + \frac{1}{\rho\sin\theta}\frac{\partial F_\phi}{\partial\phi}$

$$\therefore\ \nabla \cdot F = \frac{1}{\rho^2}\frac{\partial}{\partial\rho}\left(\rho^3\cos\theta\right) + \frac{1}{\rho\sin\theta}\frac{\partial}{\partial\theta}(\rho\sin^2\theta) - \frac{1}{\rho\sin\theta}\frac{\partial}{\partial\phi}(\rho\sin\theta\cos\phi)$$

$= 3\cos\theta + 2\cos\theta + \sin\phi = 5\cos\theta + \sin\phi$

2. For the curved surface S_1, $dS = R^2\sin\theta d\theta d\phi\tilde{\rho}$ and we have

$$\iint_{S_1} F.dS = \int_{\phi-0}^{2x}\int_{\theta-0}^{\frac{x}{2}}(R\cos\theta)(R^2\sin\theta d\,\theta d\,\phi)$$

$$= \int_{\phi-0}^{2x}\int_{\theta-0}^{\frac{x}{2}} R^3\sin\theta\cos\theta d\,\theta d\,\phi = \pi R^3$$

For the bottom circular surface S_2, $dS = \rho d\rho d\phi\tilde{\theta}$

$$\iint_{S_2} F.dS = \int_{\phi-0}^{2x}\int_{\rho-0}^{R}\left(\rho\sin\frac{\pi}{2}\right)(\rho d\rho d\,\phi)$$

$$= \int_{\phi-0}^{2x}\int_{\rho-0}^{R}\rho^2 d\rho d\phi = \frac{2\pi R^3}{3}$$

Adding, $\oiint F.n\,dS \frac{5\pi R^3}{3}$

On the other hand,

$$\iiint_{\forall} \nabla.F dV = \int_{\phi-0}^{2x}\int_{\theta-0}^{\frac{x}{2}}\int_{\rho-0}^{R}(5\cos\theta + \sin\phi)\rho^2\sin\theta d\rho d\theta d\phi$$

$$= \int_{\phi-0}^{2x}\int_{\theta-0}^{\frac{x}{2}}\int_{\rho-0}^{R}(5\sin\theta\cos\theta + \sin\theta\sin\phi)\rho^2 d\rho d\theta d\theta\phi$$

$$= \frac{5\pi R^3}{3}$$

Hence, the Stokes' theorem is verified.

Example: Use the divergence theorem to prove the Green's identity

$$\iiint_V (T\nabla^2 U - U\nabla^2 T)dV = \iint_S (T\nabla U - U\nabla T).dS$$

where S is the surface of the closed volume V.

Solution:

Let $A = T\nabla U$ in the divergence theorem. Then

$$\iiint_V \nabla.(T\nabla U)dV = \iint_S (T\nabla U).n\, dS = \iint_S \nabla.(T\nabla U).dS$$

But $\nabla.(T\nabla U) = T(\nabla.\nabla U) + (\nabla T).(\nabla U) = T\nabla^2 U + (\nabla T).(\nabla U)$

Thus $\iiint_V \nabla.(T\nabla U)dV = \iiint_V \left[T\nabla^2 U + (\nabla T).(\nabla U)\right]dV$

By the divergence theorem, we have

1. $\iiint_V \left[T\nabla^2 U + (\nabla T).(\nabla U)\right]dV = \iint_S (T\nabla U).dS$

   Interchanging T and U in (1),

2. $\iiint_V \left[U\nabla^2 T + (\nabla U).(\nabla T)\right]dV = \iint_S (U\nabla T).dS$

   Subtracting (2) from (1), we have

$$\iiint_V (T\nabla^2 U - U\nabla^2 T)dV = \iint_S (T\nabla U - U\nabla T).dS$$

STOKES' THEOREM

Stokes' theorem relates certain line integrals around closed curves to surface integrals over surfaces that have the curves as boundaries.

Stokes' Theorem Let C be *a* smooth, closed curve in the x y-plane that does not intersect itself and let S be a piecewise-smooth, orientable surface with C as boundary. Let $A = A_1 i + A_2 j + \mathbf{A}_3 k$ be *a* vector field whose components P, Q and R have continuous first derivative in *a* domain that contain S and C, then

$$\iint_S (\nabla \times A).n dS = \oint_C A.dr$$

or

$$\oint (A_1 dx + A_2 dy + A_3 dz) = \iint_S \left\{ \left(\frac{\partial A_3}{\partial y} - \frac{\partial A_2}{\partial z} \right) i + \left(\frac{\partial A_1}{\partial z} \frac{\partial A_3}{\partial x} \right) j + \left(\frac{\partial A_2}{\partial x} \frac{\partial A_1}{\partial y} \right) k \right\} .n\, dS$$

where n is the unit normal to S chosen in the following way: If when moving along C the surface S ison the left-hand side, then n must be chosen as the unit normal on that side of S. On the other hand,if when moving along C, the surface is on the right, then n must be chosen on the opposite side of S.

Proof: Let S be *a* surface which is such that its projections on the x y, y z and x z planes are regions bounded by simple closed curves. Assume S to have representation $z = f(x, y)$ or $x = g(y, z)$ or $y = h(x, z)$, where f, g, h are differentiable functions.

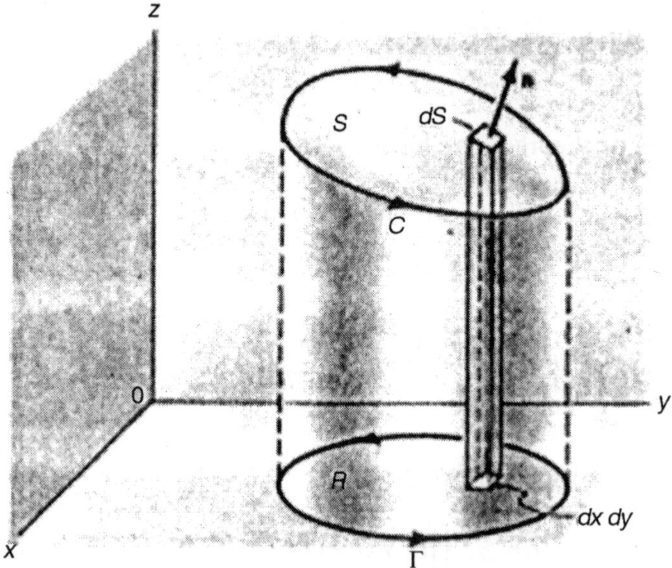

Fig. Projections Planes Bounded by Simple Closed Curves

We must show that

$$\iint_S (\nabla \times A).n\, dS = \iint_S [\nabla \times (A_1 i + A_2 j + A_2 k)].n\, dS$$

$$= \oint_C A.dr$$

where C is the boundary of S. Consider first $\iint_S [\nabla \times (A_1 i)].n\, dS$

Since $\nabla \times (A_1 i) = \frac{\partial A_1}{\partial z} j - \frac{\partial A_1}{\partial y}$,

1. $[\nabla \times (A_1 i)].n\, dS = \left(\frac{\partial A_1}{\partial z} n.j - \frac{\partial A_1}{\partial y} n.k\right) dS$

If $z = f(x, y)$ is taken as the equation of S, then the position vector to any point of S is $r = x\,i + y\,j + zk = x\,i + y\,j + f(x, y)k$ so that $\frac{\partial r}{\partial y} = j + \frac{\partial z}{\partial y} k = j + \frac{\partial f}{\partial y} k$. But $\frac{\partial r}{\partial y}$ is *a* vector tangent to *S* and thus perpendicular to n, so that

$$n.\frac{\partial r}{\partial y} = n.j + \frac{\partial r}{\partial y} n.k = 0 \text{ or } n.j = -\frac{\partial z}{\partial y} n.k$$

Substitute in (1) to obtain

$$\left(\frac{\partial A_1}{\partial z} n.j - \frac{\partial A_1}{\partial y} n.k\right) dS = \left(-\frac{\partial A_1}{\partial z} n.k - \frac{\partial A_1}{\partial y} n.k\right) dS$$

or

2. $[\nabla \times (A_1 i)].n\, dS = -\left(\frac{\partial A_1}{\partial y} + \frac{\partial A_1}{\partial z}\frac{\partial z}{\partial y}\right) n.k\, dS$

Now on *S*, $A_1(x, y, z) = A_1(x, y, f(x, y)) = F(x, y)$; hence $\frac{\partial A_1}{\partial y} + \frac{\partial A_1}{\partial z}\frac{\partial z}{\partial y} = \frac{\partial F}{\partial y}$ and (2) becomes

$$[\nabla \times (A_1 i)].n\, dS = \frac{\partial F}{\partial y} n.k\, dS = -\frac{\partial F}{\partial y} dxdy$$

where *R* is the projection of S on the *x* y plane. By Green's theorem for the plane the last integral equals $\oint_\Gamma F\, dx$ where Γ is the boundary of R. Since at each point (*x*, *y*) of G the value of *F* is the same as the value of A_1 at each point (*x*, *y*, *z*) of *C*, and since *dx* is the same for both curves, we must have

$$\oint_\Gamma F\, dx = \oint_C A_1\, dx$$

or

$$\iint_S [\nabla \times (A_1 i)].n\, dS = \oint_C A_1\, dx$$

Similarly, by projections on the other coordinate planes,

$$\iint_S [\nabla \times (A_2 j)].n\, dS = \oint_C A_2\, dx$$

$$\iint_S [\nabla \times (A_3 k)].n\, dS = \oint_C A_3\, dx$$

Thus by addition

$$\iint_S (\nabla \times A).n\, dS = \oint_C A\, dr$$

Note

If ΔS is *a* surface bounded by *a* closed curve *C*, *P* is any point of ΔS not on *C* and *n* is *a* unit normal to ΔS at P, by Stokes' theorem,

$$\oint_C A.\, dr = \iint_S (\text{curl } A).ndS \cong [(\text{curl } A).n]\Delta S$$

Then taking the limit $\Delta S \to 0$, we have

$$(\text{curl } A) \cdot n = \lim_{\Delta S \to 0} \oint_C A.dr$$

Note:

1. Green's Theorem is *a* special case of Stokes' Theorem. If S and *C* is *a* closed curve in the *x* y-plane, then by Stokes' Theorems

$$\oint_C M\, dx + Ndy = \iint_S \left(\frac{\partial N}{\partial x} - \frac{\partial M}{\partial y}\right) k \cdot ndS$$

$$= \iint_S \left(\frac{\partial N}{\partial x} - \frac{\partial M}{\partial y}\right) dA$$

(we choose S as that part of the *x* y-plane bounded by C, then $n = k$)

THE MEANING OF STOKES' THEOREM

Stokes' theorem states that the total flux of curlA through the surface S is equal to circulation of A round the circuit C i.e.

$$\iint_S (\nabla \times A)\cdot n dS = \oint_C A\cdot dr$$

Example: Verify Stokes' theorem for $A = (2x - y)i - y z^2 j - y^2 zk$, where S is the upper half surface of the sphere $x^2 + y^2 + z^2 = 1$ and C is its boundary.

Solution: The boundary C of S is *a* circle in the xy plane of radius one and centre at the origin.

Let $x = \cos t$, $y = \sin t$ $z = 0$, $0 \le t \le 2\pi$, be parametric equations of C. Then

$$\oint_C A\cdot dr = \oint_C (2x-y)\,dx - yz^2 dy - y^2 z dz$$

$$= \int_0^{2x} (2\cos t - \sin t)(-\sin t)dt = \pi$$

Also,

$$\nabla \times A = \begin{vmatrix} i & j & k \\ \dfrac{\partial}{\partial x} & \dfrac{\partial}{\partial y} & \dfrac{\partial}{\partial z} \\ 2x-y & -yz^2 & -y^2 z \end{vmatrix} = k$$

Then
$$\iint_S (\nabla \times A)\cdot n dS = \iint_S k\cdot n dS = \iint_R d\,xdy$$

where R is the projection of S on the x y plane. This last integral equals

$$\int_{x=-1}^{1}\int_{y=-\sqrt{1-x^2}}^{\sqrt{1-x^2}} dy\,dx = 4\int_0^1\int_0^{\sqrt{1-x^2}} dy\,dx = 4\int_0^1 \sqrt{1-x^2}\,dx = \pi$$

and Stokes' theorem is verified.

Example: Use the Stokes' theorem to show that

$$\iint_S (\nabla \times A)\cdot n dS = \iint_S (A \times \nabla f)\cdot dS + \oint_C fA\cdot dr$$

where S is an open surface and C is its perimeter.

Solution: Applying the Stokes' theorem on the vector function fA, we obtain the identity

$$\oint_C fA\cdot dr = \iint_S \nabla\times(fA)\cdot dS \iint_S f(\nabla\times A)\cdot dS - \iint_S A\times\nabla f\cdot dS$$

Therefore we have established

$$\iint_S f(\nabla\times A)\cdot dS = \oint_C fA\cdot dr + \iint_S A\times\nabla f\cdot dS$$

CURVILINEAR COORDINATES

ORTHOGONAL CURVILINEAR COORDINATES

Let the rectangular coordinates (x, y, z) of any point be expressed as functions of (u_1, u_2, u_3) so that

$$x = x\ (u_1, u_2, u_3),\ y = y\ (u_1, u_2, u3),\ z = z\ (u_1, u_2, u3)$$

Suppose that Equation can be solved for u_1, u_2, u_3 in terms x, y, z, i.e.

$$u_1 = u_1(x, y, z),\ u_2 = u_2(x, y, z),\ u_3 = u_3(x, y, z)$$

Let $r = x\,i + y\,j + z\,k$ be the position vector of *a* point P. Then Equation can be written as r = r(u_1, u_2, u_3). *A* tangent vector to the u_1 curve at *P* (for which u_2 and u_3 are constants) is $\frac{\partial r}{\partial u_1}$. Then *a* unit tangent vector in this direction is

$$e_1 = \frac{\frac{\partial r}{\partial u_1}}{\left|\frac{\partial r}{\partial u_1}\right|} = \frac{1}{h_1}\frac{\partial r}{\partial u_1}$$

so that $\frac{\partial r}{\partial u_1} = h_1 e_1$ where $h_1 = \left|\frac{\partial r}{\partial u_1}\right|$. Similarly, if e_2 and e_3 are unit tangent vectors along u_2 and u_3 curves at *P* respectively, then

$$\frac{\partial r}{\partial u_2} = h_2 e_2 \text{ and } \frac{\partial r}{\partial u_3} = h_3 e_3$$

where $h_2 = \left|\frac{\partial r}{\partial u_2}\right|$ and $h_3 = \left|\frac{\partial r}{\partial u_3}\right|$.

The quantities h_1, h_2, h_3 are called scalar factors. The unit vectors e_1, e_2, e_3 are in the directions of increasing u_1, u_2, u_3 respectively.

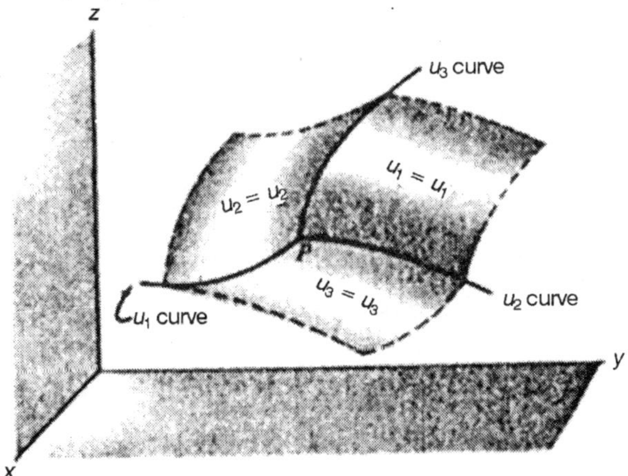

Fig. Unit tangent vector

Definition: The coordinates system (u_1, u_2, u_3) is said to be orthogonal curvilinear coordinates if and only if their unit vectors e_1, e_2, e_3 are orthogonal.

N.B. In this unit, we only consider the orthogonal curvilinear coordinates.

ARC LENGTH AND VOLUME ELEMENTS

In an orthogonal curvilinear coordinates (u_1, u_2, u_3), from $r = r(u_1, u_2, u_3)$, we have

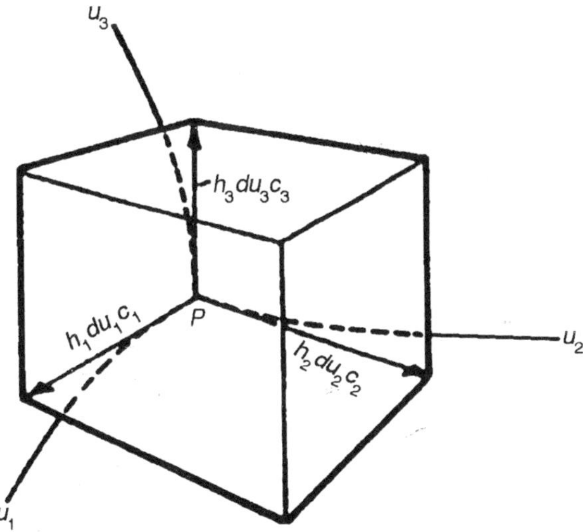

Fig. Orthogonal Curvilinear Coordinates

$$dr = \frac{\partial r}{\partial u_1}du_1 + \frac{\partial r}{\partial u_2}du_2 + \frac{\partial r}{\partial u_3}du_3 = h_1du_1e_1 + h_2du_2e_2 + h_3du_3e_3$$

Then the differential of arc length ds is determined from

$$ds^2 = dr \cdot dr = h_1^2 d_1^2 + h_2^2 du_2^2 + h_3^2 du_3^2$$

Referring to Figure the volume element for an orthogonal curvilinear coordinate system is given by

$$dV = |(h_1 du_1 e_1) \cdot (h_2 du_2 e_2) \times (h_3 du_3 e_3)| = h_1 h_2 h_3 du_1 du_2 du_3$$

since.

OPERATOR IN ORTHOGONAL CURVILINEAR COORDINATES

GRADIENT IN ORTHOGONAL CURVILINEAR COORDINATES

For any scalar function Φ, we can express its gradient in orthogonal curvilinear coordinate system (u_1, u_2, u_3) as

$$\nabla\Phi = f_1 e_1 + f_2 e_2 + f_3 e_3$$

where e_1, e_2, e_3 are unit vectors in the directions of increasing u_1, u_2, u_3 respectively. Since

$$dr = \frac{\partial r}{\partial u_1} du_1 + \frac{\partial r}{\partial u_2} du_2 + \frac{\partial r}{\partial u_3} du_3 = h_1 du_1 e_1 + h_2 du_2 e_2 + h_3 du_3 e_3$$

we have

1. $d\Phi = \nabla\Phi \cdot dr = f_1 h_1 du_1 + f_2 h_2 du_2 + f_3 h_3 du_3$

   But

2. $d\Phi = \dfrac{\partial\Phi}{\partial u_1} du_1 + \dfrac{\partial\Phi}{\partial u_2} du_2 + \dfrac{\partial\Phi}{\partial u_3} du_3$

Equating (1) and (2) $f_1 = \dfrac{1}{h_1}\dfrac{\partial\Phi}{\partial u_1}, f_2 = \dfrac{1}{h_2}\dfrac{\partial\Phi}{\partial u_2}, f_3 = \dfrac{1}{h_3}\dfrac{\partial\Phi}{\partial u_3},$

Then

$$\nabla\Phi = \frac{e_1}{h_1}\frac{\partial\Phi}{\partial u_1} + \frac{e_2}{h_2}\frac{\partial\Phi}{\partial u_2} + \frac{e_3}{h_3}\frac{\partial\Phi}{\partial u_3}$$

This indicates the operator equivalence

$$\nabla \equiv \frac{e_1}{h_1}\frac{\partial}{\partial u_1} + \frac{e_2}{h_2}\frac{\partial}{\partial u_2} + \frac{e_3}{h_3}\frac{\partial}{\partial u_3}$$

DIVERGENCE IN ORTHOGONAL CURVILINEAR COORDINATES

Consider the volume element ΔV having edges $h_1\Delta u_1$, $h_2\Delta u_2$, $h_3\Delta u_3$. Let $A = A_1e_1 + A_2e_2 + A_3e_3$ and let n be the outward drawn unit normal to the surface ΔS of ΔV.

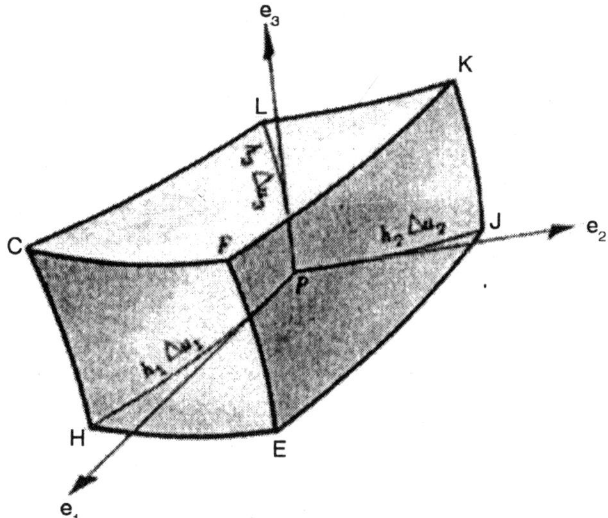

Fig. Divergence in Orthogonal Curvilinear Coordinates

On the face JKLP, $n = -e_1$. Then we have approximately,

$$\iint_{JKLP} A\cdot n\,dS = (A.\ n \text{ at point } P)\ (\text{Area of JKLP})$$

$$= \left[(A_1e_1 + A_2e_2 + A_3e_3)\cdot(-e_1)\right](h_2h_3\Delta u_2\Delta u_3)$$

$$= -\,A_1h_2h_3\Delta u_2\Delta u_3$$

On face EFGH, the surface integral is

$$A_1h_2h_3\Delta u_2\Delta u_3 + \frac{\partial}{\partial u_1}(A_1h_2h_3\Delta u_2\Delta u_3)\Delta u_1$$

apart from infinitesimal of order higher than $\Delta u_1\Delta u_2\Delta u_3$. Then the net contribution to the surface integral from these two faces is

$$\frac{\partial}{\partial u_1}(A_1h_2h_3\Delta u_2\Delta u_3)\Delta u_1 = \frac{\partial}{\partial u_1}(A_1h_2h_3)\Delta u_1\Delta u_2\Delta u_3$$

The contribution from six faces of DV is. Dividing this by the volume $h_1h_2h_3\Delta u_1\Delta u_2\Delta u_3$ and taking the limit as Δu_1, Δu_2, Δu_3 approach *z* ero, we find

$$\text{div } A = \nabla\cdot A = \frac{1}{h_1h_2h_3}\left[\frac{\partial}{\partial u_1}(A_1h_2h_3) + \frac{\partial}{\partial u_2}(A_2h_1h_3) + \frac{\partial}{\partial u_3}(A_3h_1h_2)\right]$$

CURL IN ORTHOGONAL CURVILINEAR COORDINATES

Let us first calculate (curl A) $\cdot$ e_1. To do this consider the surface S_1 normal to e_1 at P, as shown in Figure.

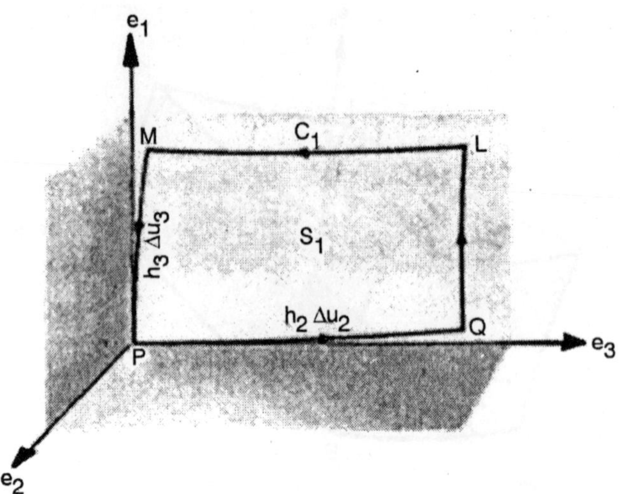

Fig. Orthogonal Curvilinear Coordinates

Denote the boundary of S_1 by C_1. Let $A = A_1e_1 + A_2e_2 + A_3e_3$, we have

$$\oint_{C_1} A \cdot dr = \int_{PQ} A \cdot dr + \int_{QL} A \cdot dr + \int_{LM} A \cdot dr + \int_{MP} A \cdot dr$$

The following approximation holds

1. $\int_{PQ} A \cdot dr = (A \text{ at } P) \times (h_2\ \Delta u_2 e_2) = (A_1e_1 + A_2e_2 + A_3e_3).\ (h_2\Delta u_2 e_2) = A_2h_2\Delta u_2$

   Then

$$\int_{ML} A \cdot dr = A_2h_2\Delta u_2 + \frac{\partial}{\partial u_3}(A_2h_2\Delta u_2)\Delta u_3$$

   or

2. $\int_{ML} A \cdot dr = A_2h_2\Delta u_2 - \frac{\partial}{\partial u_3}(A_2h_2\Delta u_2)\Delta u_3$

   Similarly,

$$\int_{ML} A \cdot dr = (A \text{ at } P) \cdot (h_3\ \Delta u_3 e_3) = (A_3h_3\Delta u_3)$$

   or

3. $\int_{MP} A \cdot dr = -A_3 h_3 \Delta u_3$

and

4. $\int_{QL} A \cdot dr = A_3 h_3 \Delta u_3 + \frac{\partial}{\partial u_2}(A_3 h_3 \Delta u_3)\Delta u_2$

Adding (1), (2), (3) and (4) we have

$$\oint_{C_1} A \cdot dr = \frac{\partial}{\partial u_2}(A_3 h_3 \Delta u_3)\Delta u_2 - \frac{\partial}{\partial u_3}(A_2 h_2 \Delta u_2)\Delta u_3$$

$$= \left[\frac{\partial}{\partial u_2}(A_3 h_3) - \frac{\partial}{\partial u_3}(A_2 h_2)\right]\Delta u_2 \Delta u_3$$

apart from infinitesimal of order higher than $\Delta u_2 \Delta u_3$. Dividing by he area of S_1 equal to $h_2 h_3 \Delta u_2 \Delta u_3$ and taking the limit as Δu_2 and Δu_3 approach zero,

$$(\text{curl } A).\ e_1 = \frac{1}{h_2 h_3}\left[\frac{\partial}{\partial u_2}(A_3 h_3) - \frac{\partial}{\partial u_3}(A_2 h_2)\right]$$

Similarly, by choosing area S_2 and S_3 perpendicular to e_2 and e_3 at P respectively, we find (curl A). e_2 and (curl A). e_2.

This leads to the required result

$$\text{curl } A = \frac{e_1}{h_2 h_3}\left[\frac{\partial}{\partial u_2}(A_3 h_3) - \frac{\partial}{\partial u_3}(A_2 h_2)\right]$$

$$+ \frac{e_2}{h_3 h_1}\left[\frac{\partial}{\partial u_3}(A_1 h_1) - \frac{\partial}{\partial u_1}(A_3 h_3)\right]$$

$$+ \frac{e_1}{h_1 h_2}\left[\frac{\partial}{\partial u_1}(A_2 h_2) - \frac{\partial}{\partial u_2}(A_1 h_1)\right]$$

$$= \frac{1}{h_1 h_2 h_3}\begin{vmatrix} h_1 e_1 & h_2 e_2 & h_3 e_3 \\ \frac{\partial}{\partial u_1} & \frac{\partial}{\partial u_2} & \frac{\partial}{\partial u_3} \\ h_1 A_1 & h_2 A_2 & h_3 A_3 \end{vmatrix}$$

LAPLACIAN IN ORTHOGONAL CURVILINEAR COORDINATES

To evaluate the Laplacian of *a* scalar function Φ in orthogonal curvilinear coordinates, we are making use the *Equation* by taking $A = \nabla\Phi = f_1e_1 + f_2e_2 + f_3e_3$

$$\nabla^2 \Phi = \nabla \cdot A = \frac{1}{h_1h_2h_3}\left[\frac{\partial}{\partial u_1}(f_1h_2h_3) + \frac{\partial}{\partial u_2}(f_2h_2h_3) + \frac{\partial}{\partial u_3}(f_3h_2h_3)\right]$$

$$= \frac{1}{h_1h_2h_3}\left[\frac{\partial}{\partial u_1}\left(\frac{1}{h_1}\frac{\partial\Phi}{\partial u_1}h_2h_3)\right) + \frac{\partial}{\partial u_2}\left(\frac{1}{h_2}\frac{\partial\Phi}{\partial u_2}h_1h_3)\right) + \frac{\partial}{\partial u_3}\left(\frac{1}{h_3}\frac{\partial\Phi}{\partial u_3}h_1h_2)\right)\right]$$

$$= \frac{1}{h_1h_2h_3}\left[\frac{\partial}{\partial u_1}\left(\frac{h_2h_3}{h_1}\frac{\partial\Phi}{\partial u_1}\right) + \frac{\partial}{\partial u_2}\left(\frac{h_3h_1}{h_2}\frac{\partial\Phi}{\partial u_2}\right) + \frac{\partial}{\partial u_3}\left(\frac{h_1h_2}{h_3}\frac{\partial\Phi}{\partial u_3}\right)\right]$$

DIVERGENCE, CURL AND LAPLACIAN IN CURVILINEAR COORDINATES

SPHERICAL COORDINATE

The spherical coordinates (ρ, θ, ϕ) of *a* point have the following relationships with rectangular coordinates:

$$\begin{cases} x = \rho\sin\theta\cos\phi \\ y = \rho\sin\theta\sin\phi \\ z = \rho\cos\theta \end{cases}$$

We have the following relations

$$\begin{cases} dx = \frac{\partial x}{\partial\rho}d\rho + \frac{\partial x}{\partial\theta}d\theta + \frac{\partial x}{\partial\phi}d\phi = (\sin\theta\cos\phi)d\rho + (\rho\cos\theta\cos\phi)d\theta - (\rho\sin\theta\sin\phi)d\phi \\ dy = \frac{\partial y}{\partial\rho}d\rho + \frac{\partial y}{\partial\theta}d\theta + \frac{\partial y}{\partial\phi}d\phi = (\sin\theta\cos\phi)d\rho + (\rho\cos\theta\cos\phi)d\theta + (\rho\sin\theta\sin\phi)d\phi \\ dz = \frac{\partial z}{\partial\rho}d\rho + \frac{\partial z}{\partial\theta}d\theta + \frac{\partial z}{\partial\phi}d\phi = (\cos\theta)d\rho - (\rho\sin\theta)d\theta \end{cases}$$

Hence the unit vector along the r-direction is

$$h_\rho\left|\frac{\partial r}{\partial\rho}\right| = \sqrt{\left(\frac{\partial x}{\partial\rho}\right)^2 + \left(\frac{\partial y}{\partial\rho}\right)^2 + \left(\frac{\partial z}{\partial\rho}\right)^2} = \sqrt{(\sin\theta\cos\phi)^2 + (\sin\theta\sin\phi)^2 + (\cos)^2} = 1$$

$$\tilde{\rho} = \frac{\frac{\partial r}{\partial \rho}}{\left|\frac{\partial r}{\partial \rho}\right|} = \frac{\frac{\partial x}{\partial \rho} i + \frac{\partial y}{\partial \rho} j + \frac{\partial z}{\partial \rho} k}{\sqrt{\left(\frac{\partial x}{\partial \rho}\right)^2 + \left(\frac{\partial y}{\partial \rho}\right)^2 + \left(\frac{\partial z}{\partial \rho}\right)^2}}$$

$$= \frac{(\sin\theta\cos\phi)i + (\sin\theta\sin\phi)j + (\cos\theta)k}{\sqrt{(\sin\theta\cos\phi)^2 + (\sin\theta\sin\phi)^2 + (\cos\theta)^2}}$$

$$= (\sin\theta\cos\phi)i + (\sin\theta\sin\phi)j + (\cos\theta)k$$

Similarly the unit vector along the q -direction is

$$h_\theta = \left|\frac{\partial r}{\partial \theta}\right| = \sqrt{\left(\frac{\partial x}{\partial \theta}\right)^2 + \left(\frac{\partial y}{\partial \theta}\right)^2 + \left(\frac{\partial z}{\partial \theta}\right)^2} = \sqrt{(\cos\theta\cos\phi)^2 + (\cos\theta\sin\phi)^2 + (\rho\sin)^2} = \rho$$

$$\tilde{\theta} = \frac{\frac{\partial r}{\partial \theta}}{\left|\frac{\partial r}{\partial \theta}\right|} = \frac{\frac{\partial x}{\partial \theta} i + \frac{\partial y}{\partial \theta} j + \frac{\partial z}{\partial \theta} k}{\sqrt{\left(\frac{\partial x}{\partial \theta}\right)^2 + \left(\frac{\partial y}{\partial \theta}\right)^2 + \left(\frac{\partial z}{\partial \theta}\right)^2}}$$

$$= \frac{(\rho\cos\theta\cos\phi)i + (\rho\cos\theta\sin\phi)j - (\rho\sin\theta)k}{\sqrt{(\rho\cos\theta\cos\phi)^2 + (\rho\cos\theta\sin\phi)^2 + (\rho\sin\theta)^2}}$$

$$= \frac{(\rho\cos\theta\cos\phi)i + (\rho\cos\theta\sin\phi)j - (\rho\sin\theta)k}{\rho}$$

$$= (\rho\cos\theta\cos\phi)i + (\rho\cos\theta\sin\phi)j - \sin\theta k$$

and the unit vector along the f -direction is

$$h_\phi = \left|\frac{\partial r}{\partial \phi}\right| = \sqrt{\left(\frac{\partial x}{\partial \phi}\right)^2 + \left(\frac{\partial y}{\partial \phi}\right)^2 + \left(\frac{\partial z}{\partial \phi}\right)^2} (\rho\sin\theta\sin\phi)^2 + (\rho\sin\theta\cos\phi)^2 \rho\sin\theta$$

$$\tilde{\phi} = \frac{\frac{\partial r}{\partial \phi}}{\left|\frac{\partial r}{\partial \phi}\right|} = \frac{\frac{\partial x}{\partial \phi} i + \frac{\partial y}{\partial \phi} j + \frac{\partial z}{\partial \phi} k}{\sqrt{\left(\frac{\partial x}{\partial \phi}\right)^2 + \left(\frac{\partial y}{\partial \phi}\right)^2 + \left(\frac{\partial z}{\partial \phi}\right)^2}}$$

$$= \frac{(\rho\sin\theta\sin\phi)i + (\rho\sin\theta\cos\phi)j}{\sqrt{(\rho\sin\theta\sin\phi)^2 + (\rho\sin\theta\cos\phi)^2}}$$

$$= \frac{(\rho\sin\theta\sin\phi)i + (\rho\sin\theta\cos\phi)j}{\rho\sin\theta} = \sin\phi i + \cos\phi j$$

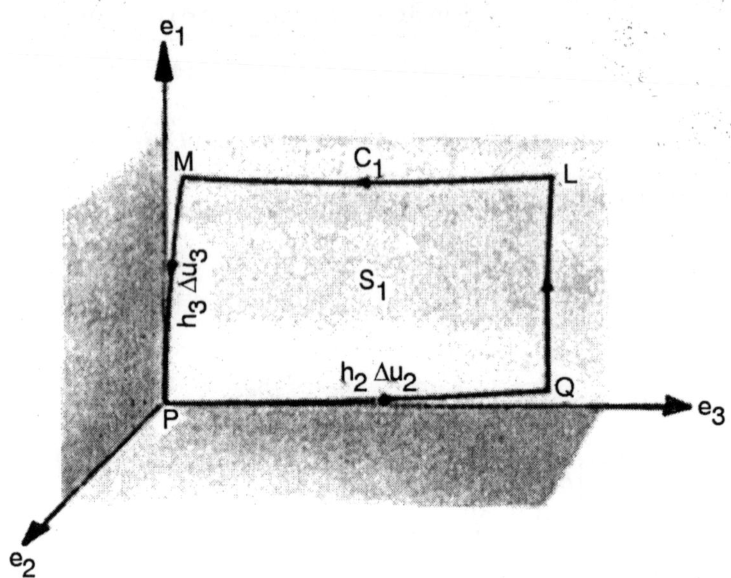

Fig. Spherical coordinate

Therefore

Divergence

$$\nabla \cdot F = \frac{1}{\rho^2}\frac{\partial}{\partial\rho}(\rho^2 F_\rho) + \frac{1}{\rho\sin\theta}\frac{\partial}{\partial\theta}(\sin\theta F_\theta) + \frac{1}{\rho\sin\theta}\frac{\partial F_\phi}{\partial\phi}$$

Curl

$$\nabla \times F = \frac{1}{\rho^2\sin\theta}\left[\frac{\partial}{\partial\rho}(\sin\theta F_\phi) - \frac{\partial F_\phi}{\partial\phi}\right]\tilde{\rho} + \frac{1}{\rho}\left[\frac{1}{\sin\theta}\frac{\partial F_\rho}{\partial\phi} - \frac{\partial}{\partial\rho}(\rho F_\phi)\right]\tilde{\theta} +$$

$$+ \frac{1}{\rho}\left[\frac{\partial}{\partial\rho}(\rho F_\theta)\frac{\partial F_\rho}{\partial\theta}\right]\tilde{\theta}$$

Laplacian

$$\nabla^2 T = \frac{1}{\rho^2}\frac{\partial}{\partial\rho}\left(\rho^2\frac{\partial T}{\partial\rho}\right) + \frac{1}{\rho^2\sin\theta}\frac{\partial}{\partial\theta}\left(\sin\theta\frac{\partial T}{\partial\theta}\right) + \frac{1}{\rho^2\sin^2\theta}\frac{\partial^2 T}{\partial\phi^2}$$

CYLINDRICAL COORDINATE

The spherical coordinates (ρ, θ, *z*) of *a* point have the following relationships with rectangular coordinates:

$$\begin{cases} x = r\cos\theta \\ y = r\sin\theta \\ z = z \end{cases}$$

We have the following relations

$$\begin{cases} dx = \dfrac{\partial x}{\partial r}dr + \dfrac{\partial x}{\partial \theta}d\theta + \dfrac{\partial x}{\partial z}dz = (\cos\theta)dr - (r\sin\theta)d\theta \\ dy = \dfrac{\partial y}{\partial r}dr + \dfrac{\partial y}{\partial \theta}d\theta + \dfrac{\partial y}{\partial z}dz = (\sin\theta)dr - (r\cos\theta)d\theta \\ dz = dz \end{cases}$$

Hence the unit vector along the r-direction is

$$h_r = \left|\frac{\partial x}{\partial r}\right| = \sqrt{\left(\frac{\partial x}{\partial r}\right)^2 + \left(\frac{\partial y}{\partial r}\right)^2 + \left(\frac{\partial x}{\partial r}\right)^2} = \sqrt{\cos^2\theta + \sin^2\theta} = 1$$

$$\hat{r} = \frac{\dfrac{\partial x}{\partial r}}{\left|\dfrac{\partial x}{\partial r}\right|} = \frac{\dfrac{\partial x}{\partial r}i + \dfrac{\partial y}{\partial r}j + \dfrac{\partial z}{\partial r}k}{\sqrt{\left(\dfrac{\partial x}{\partial r}\right)^2 + \left(\dfrac{\partial y}{\partial r}\right)^2 + \left(\dfrac{\partial x}{\partial r}\right)^2}}$$

$$= \frac{\cos\theta\, i + \sin\theta\, j}{\sqrt{\cos^2\theta + \sin^2\theta}}$$

$$= \cos\theta\, i + \sin\theta\, j$$

Similarly the unit vector along the θ-direction is

$$h_\theta = \left|\frac{\partial x}{\partial \theta}\right| = \sqrt{\left(\frac{\partial x}{\partial \theta}\right)^2 + \left(\frac{\partial y}{\partial \theta}\right)^2 + \left(\frac{\partial x}{\partial \theta}\right)^2}$$

$$= \sqrt{(r\sin\theta)^2 + (r\cos\theta)^2} = r$$

$$\hat{\theta} = \frac{\dfrac{\partial r}{\partial \theta}}{\left|\dfrac{\partial r}{\partial \theta}\right|} = \frac{\dfrac{\partial x}{\partial \theta} i + \dfrac{\partial y}{\partial \theta} j + \dfrac{\partial z}{\partial \theta} k}{\sqrt{\left(\dfrac{\partial x}{\partial \theta}\right)^2 + \left(\dfrac{\partial y}{\partial \theta}\right)^2 + \left(\dfrac{\partial x}{\partial \theta}\right)^2}}$$

$$= \frac{(-r\sin\theta)i + (r\cos\theta)j}{\sqrt{(r\sin\theta)^2 i + (r\cos\theta)^2}}$$

$$= \frac{(-r\sin\theta)i + (r\cos\theta)j}{r}$$

$$= -\sin\theta\, i + \cos\theta\, j$$

In the z-direction $h_z = 1$.

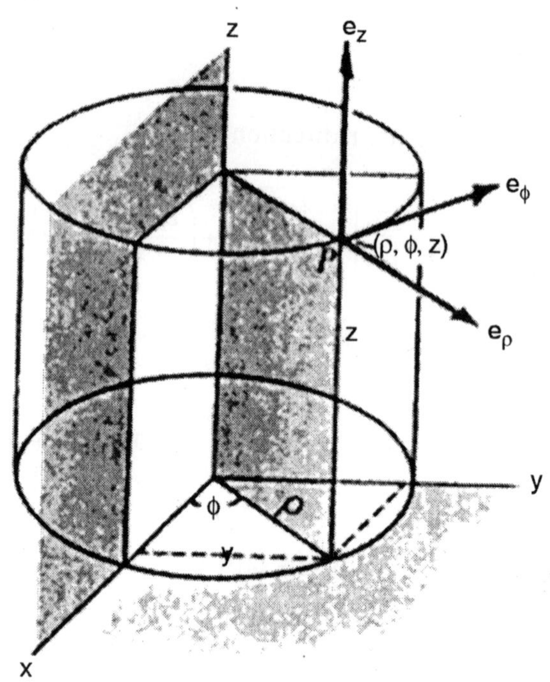

Fig. Cylindrical Coordinate

Therefore

Divergence

$$\nabla \cdot F = \frac{1}{r}\frac{\partial}{\partial r}(rF_\rho) + \frac{1}{r}\frac{\partial F_\phi}{\partial \phi} + \frac{\partial F_z}{\partial z}$$

Curl

$$\nabla \times F = \left(\frac{1}{r}\frac{\partial F_z}{\partial \phi} - \frac{\partial F_\phi}{\partial z}\right)\hat{r} + \left(\frac{\partial F_r}{\partial z} - \frac{\partial F_z}{\partial r}\right)\hat{\phi} + \frac{1}{r}\left[\frac{\partial}{\partial r} - \frac{\partial F_\phi}{\partial z}\right]\hat{z}$$

Laplacian

$$\nabla^2 T = \frac{1}{r}\frac{\partial}{\partial r}\left(r\frac{\partial T}{\partial r}\right) + \frac{1}{r^2}\frac{\partial^2 T}{\partial \phi^2} + \frac{\partial^2 T}{\partial z^2}$$

Example: Find the Laplacian of the function: $T =$ ŕcosqcosf.

Solution: From Equation

$$\nabla^2 T = \frac{1}{\rho^2}\frac{\partial}{\partial \rho}\left(\rho^2\frac{\partial}{\partial \rho}(\rho\cos\theta\cos\phi)\right) + \frac{1}{\rho^2\sin\theta}\frac{\partial}{\partial \theta}\left(\sin\theta\frac{\partial}{\partial \theta}(\rho\cos\theta\cos\phi)\right)$$

$$+ \frac{1}{\rho^2\sin\theta}\frac{\partial^2}{\partial \phi^2}(\rho\cos\theta\cos\phi)$$

$$= \frac{1}{\rho^2}\frac{\partial}{\partial \rho}(\rho^2\cos\theta\cos\phi) + \frac{1}{\rho^2\sin\theta}\frac{\partial}{\partial \theta}(-\rho\sin^2\theta\cos\phi) - \frac{\rho\cos\theta\cos\phi}{\rho^2\sin^2\theta}$$

$$\frac{2\cos\theta\cos\phi}{\rho} - \frac{2\cos\theta\cos\phi}{\rho} - \frac{\cos\theta\cos\phi}{\rho\sin^2\theta}$$

Example:

Define the elliptic cylindrical coordinates

$$x = \text{acoshucosv} \quad y = \text{asinhusinv}$$

1. Find the scaling factors for the elliptical cylindrical coordinates system.
2. Hence write the Laplace's equation

$$\nabla^2\Phi \equiv \frac{\partial^2\Phi}{\partial x^2} + \frac{\partial^2\Phi}{\partial y^2} = 0$$

in elliptical cylindrical coordinates.

Solution:

1. From the elliptical cylindrical coordinates, we have

$$\begin{cases} dx = \dfrac{\partial x}{\partial u}du + \dfrac{\partial x}{\partial y} = (a\sinh u\cos v)\,du - (a\cosh u\sin v)\,dv \\ dy = \dfrac{\partial y}{\partial u}du + \dfrac{\partial y}{\partial v} = (a\cosh u\sin v)\,du + (a\sinh u\cos v)\,dv \end{cases}$$

The scaling factors are

$$h_u = \sqrt{\left(\left(\frac{\partial x}{\partial u}\right)^2 + \left(\frac{\partial y}{\partial u}\right)^2\right)}$$

$$= \sqrt{a^2 \sinh^2 u \cos^2 v + a^2 \cosh^2 u \sin^2 v}$$

$$= \sqrt{a^2 \sinh^2 u(1 - \sin^2 v) + a^2 \cosh^2 u \sin^2 v}$$

$$= \sqrt{a^2 \sinh^2 u - a^2 \sinh^2 u \sin^2 v + a^2 \cosh^2 u \sin^2 v}$$

$$= \sqrt{a^2 \sinh^2 u + a^2 \sin^2 v(\cosh^2 u - \sin^2 u)}$$

$$= a\sqrt{\sinh^2 u + \sin^2 v}$$

$$h_v = \sqrt{\left(\left(\frac{\partial x}{\partial v}\right)^2 + \left(\frac{\partial y}{\partial v}\right)^2\right)}$$

$$= \sqrt{a^2 \cosh^2 u \sin^2 v + a^2 \sinh^2 u \cos^2 v}$$

$$= a\sqrt{\sinh^2 u + \sin^2 v}$$

2. From Equation, we have

$$\nabla^2 \Phi = \frac{1}{h_u h_v}\left[\frac{\partial}{\partial u}\left(\frac{h_v}{h_u}\frac{\partial \Phi}{\partial u}\right) + \frac{\partial}{\partial v}\left(\frac{h_u}{h_v}\frac{\partial \Phi}{\partial v}\right)\right]$$

$$= \frac{1}{a^2(\sinh^2 u + \sin^2 v)}\frac{\partial^2 \Phi}{\partial u^2} + \frac{1}{a^2(\sinh^2 u + \sin^2 v)}\frac{\partial^2 \Phi}{\partial v^2}$$

$$= \frac{1}{a^2(\sinh^2 u + \sin^2 v)}\left(\frac{\partial^2 \Phi}{\partial u^2} + \frac{\partial^2 \Phi}{\partial v^2}\right)$$

Therefore the Laplace's equation can be rewritten as

$$\frac{1}{a^2(\sinh^2 u + \sin^2 v)}\left(\frac{\partial^2 \Phi}{\partial u^2} + \frac{\partial^2 \Phi}{\partial v^2}\right) = 0 \Rightarrow \frac{\partial^2 \Phi}{\partial u^2} + \frac{\partial^2 \Phi}{\partial v^2} = 0$$

Chapter 15

Curvilinear Co-ordinates

ORTHOGONAL CURVILINEAR COORDINATES

Let the rectangular coordinates (x, y, z) of any point be expressed as functions of (u_1, u_2, u_3) so that

$$x = x\,(u_1, u_2, u_3),\ y = y\,(u_1, u_2, u_3),\ z = z\,(u_1, u_2, u_3)$$

Suppose that Equation can be solved for u_1, u_2, u_3 in terms x, y, z, i.e.

$$u_1 = u_1(x, y, z),\ u_2 = u_2(x, y, z),\ u_3 = u_3(x, y, z)$$

Let $r = x\,i + y\,j + z\,k$ be the position vector of *a* point P.

Then Equation can be written as $r = r(u_1, u_2, u_3)$. *A* tangent vector to the u_1 curve at P (for which u_2 and u_3 are constants) is $\frac{\partial r}{\partial u_1}$. Then *a* unit tangent vector in this direction is

$$e_1 = \frac{\frac{\partial r}{\partial u_1}}{\left|\frac{\partial r}{\partial u_1}\right|} = \frac{1}{h_1}\frac{\partial r}{\partial u_1}$$

so that $\frac{\partial r}{\partial u_1} = h_1 e_1$ where $h_1 = \left|\frac{\partial r}{\partial u_1}\right|$. Similarly, if e_2 and e_3 are unit tangent vectors along u_2 and u_3 curves at P respectively, then

$$\frac{\partial r}{\partial u_2} = h_2 e_2 \text{ and } \frac{\partial r}{\partial u_3} = h_3 e_3$$

where $h_2 = \left|\frac{\partial r}{\partial u_2}\right|$ and $h_3 = \left|\frac{\partial r}{\partial u_3}\right|$.

The quantities h_1, h_2, h_3 are called scalar factors. The unit vectors e_1, e_2, e_3 are in the directions of increasing u_1, u_2, u_3 respectively.

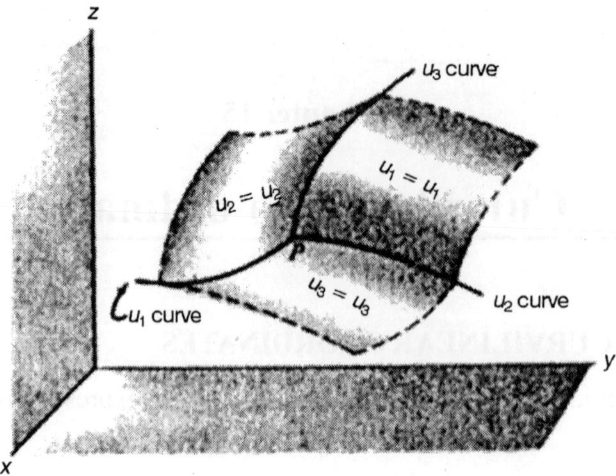

Fig. Unit tangent vector

Definition: The coordinates system (u_1, u_2, u_3) is said to be orthogonal curvilinear coordinates if and only if their unit vectors e_1, e_2, e_3 are orthogonal. N.B. In this unit, we only consider the orthogonal curvilinear coordinates.

ARC LENGTH AND VOLUME ELEMENTS

In an orthogonal curvilinear coordinates (u_1, u_2, u_3), from $r = r(u_1, u_2, u_3)$, we have

$$dr = \frac{\partial r}{\partial u_1} du_1 + \frac{\partial r}{\partial u_2} du_2 + \frac{\partial r}{\partial u_3} du_3 = h_1 du_1 e_1 + h_2 du_2 e_2 + h_3 du_3 e_3$$

Then thè differential of arc length ds is determined from

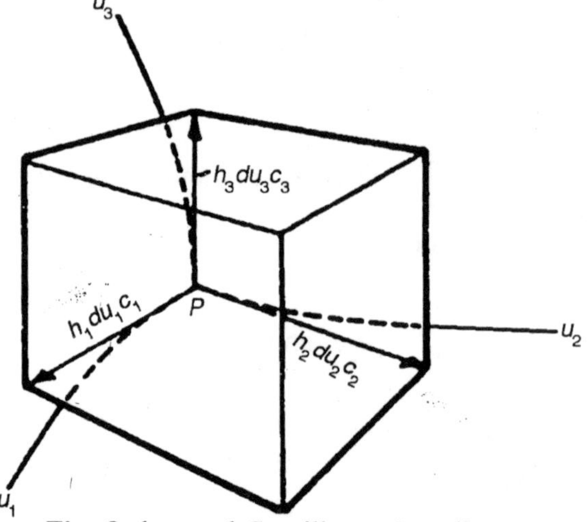

Fig. Orthogonal Curvilinear Coordinates

$$ds^2 = dr \cdot dr = h_1^2 d_1^2 + h_2^2 du_2^2 + h_3^2 du_3^2$$

Referring to Figure the volume element for an orthogonal curvilinear coordinate system is given by

$$dV = |(h_1 du_1 e_1) \cdot (h_2 du_2 e_2) \times (h_3 du_3 e_3)| = h_1 h_2 h_3 du_1 du_2 du_3$$

since.

OPERATOR IN ORTHOGONAL CURVILINEAR COORDINATES

(1) GRADIENT IN ORTHOGONAL CURVILINEAR COORDINATES

For any scalar function Φ, we can express its gradient in orthogonal curvilinear coordinate system (u_1, u_2, u_3) as

$$\nabla\Phi = f_1 e_1 + f_2 e_2 + f_3 e_3$$

where e_1, e_2, e_3 are unit vectors in the directions of increasing u_1, u_2, u_3 respectively. Since

$$dr = \frac{\partial r}{\partial u_1} du_1 + \frac{\partial r}{\partial u_2} du_2 + \frac{\partial r}{\partial u_3} du_3 = h_1 du_1 e_1 + h_2 du_2 e_2 + h_3 du_3 e_3$$

we have

1. $d\Phi = \nabla\Phi \cdot dr = f_1 h_1 du_1 + f_2 h_2 du_2 + f_3 h_3 du_3$

   But

2. $d\Phi = \dfrac{\partial\Phi}{\partial u_1} du_1 + \dfrac{\partial\Phi}{\partial u_2} du_2 + \dfrac{\partial\Phi}{\partial u_3} du_3$

   Equating (1) and (2)

$$f_1 = \frac{1}{h_1}\frac{\partial\Phi}{\partial u_1}, f_2 = \frac{1}{h_2}\frac{\partial\Phi}{\partial u_2}, f_3 = \frac{1}{h_3}\frac{\partial\Phi}{\partial u_3},$$

Then

$$\nabla\Phi = \frac{e_1}{h_1}\frac{\partial\Phi}{\partial u_1} + \frac{e_2}{h_2}\frac{\partial\Phi}{\partial u_2} + \frac{e_3}{h_3}\frac{\partial\Phi}{\partial u_3}$$

This indicates the operator equivalence

$$\nabla \equiv \frac{e_1}{h_1}\frac{\partial}{\partial u_1} + \frac{e_2}{h_2}\frac{\partial}{\partial u_2} + \frac{e_3}{h_3}\frac{\partial}{\partial u_3}$$

DIVERGENCE IN ORTHOGONAL CURVILINEAR COORDINATES

Consider the volume element ΔV having edges $h_1\Delta u_1$, $h_2\Delta u_2$, $h_3\Delta u_3$. Let $A = A_1e_1 + A_2e_2 + A_3e_3$ and let n be the outward drawn unit normal to the surface ΔS of ΔV.

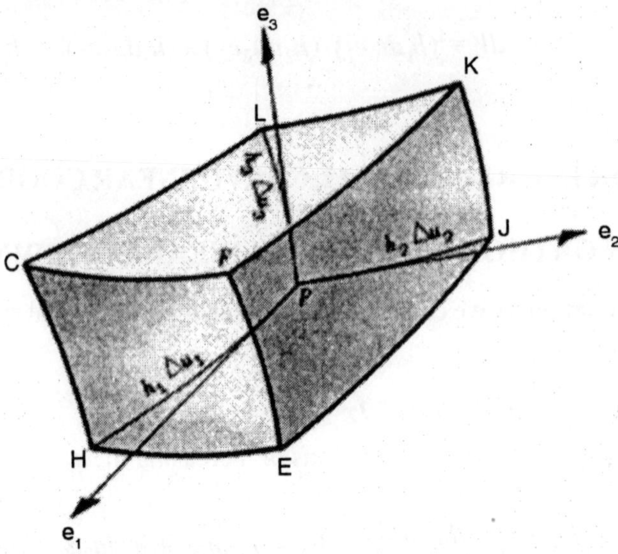

Fig. Divergence in Orthogonal Curvilinear Coordinates

On the face JKLP, $n = -e_1$. Then we have approximately,

$$\iint_{JKLP} A \cdot ndS = (A.\, n \text{ at point } P)\ (\text{Area of JKLP})$$

$$= \left[(A_1e_1 + A_2e_2 + A_3e_3)\cdot(-e_1)\right](h_2h_3\Delta u_2\Delta u_3)$$

$$= -\, A_1h_2h_3\Delta u_2\Delta u_3$$

On face EFGH, the surface integral is

$$A_1h_2h_3\Delta u_2\Delta u_3 + \frac{\partial}{\partial u_1}(A_1h_2h_3\Delta u_2\Delta u_3)\Delta u_1$$

apart from infinitesimal of order higher than $\Delta u_1\Delta u_2\Delta u_3$.

Then the net contribution to the surface integral from these two faces is

$$\frac{\partial}{\partial u_1}(A_1h_2h_3\Delta u_2\Delta u_3)\Delta u_1 = \frac{\partial}{\partial u_1}(A_1h_2h_3)\Delta u_1\Delta u_2\Delta u_3$$

The contribution from six faces of DV is

Dividing this by the volume $h_1h_2h_3\Delta u_1\Delta u_2\Delta u_3$ and taking the limit as Δu_1, Δu_2, Δu_3 approach z ero, we find

$$\text{div } A = \nabla \cdot A = \frac{1}{h_1h_2h_3}\left[\frac{\partial}{\partial u_1}(A_1h_2h_3) + \frac{\partial}{\partial u_2}(A_2h_1h_3) + \frac{\partial}{\partial u_3}(A_3h_1h_2)\right]$$

CURL IN ORTHOGONAL CURVILINEAR COORDINATES

Let us first calculate (curl A) $\cdot$ e_1. To do this consider the surface S_1 normal to e_1 at P, as shown in Figure.

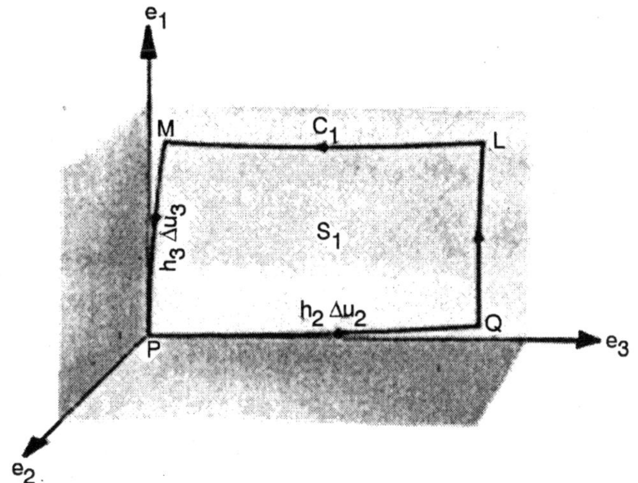

Fig. Orthogonal Curvilinear Coordinates

Denote the boundary of S_1 by C_1. Let $A = A_1e_1 + A_2e_2 + A_3e_3$, we have

$$\oint_{C_1} A \cdot dr = \int_{PQ} A \cdot dr + \int_{QL} A \cdot dr + \int_{LM} A \cdot dr + \int_{MP} A \cdot dr$$

The following approximation holds

1. $\int_{PQ} A \cdot dr = (A \text{ at } P) \times (h_2\, \Delta u_2 e_2) = (A_1e_1 + A_2e_2 + A_3e_3).\ (h_2\Delta u_2 e_2) = A_2h_2\Delta u_2$

   Then

$$\int_{ML} A \cdot dr = A_2h_2\Delta u_2 + \frac{\partial}{\partial u_3}(A_2h_2\Delta u_2)\Delta u_3$$

or

2. $\int_{ML} A \cdot dr = A_2h_2\Delta u_2 - \frac{\partial}{\partial u_3}(A_2h_2\Delta u_2)\Delta u_3$

Similarly,

$$\int_{ML} A\cdot dr = (A \text{ at } P)\cdot(h_3\,\Delta u_3 e_3) = (A_3 h_3 \Delta u_3)$$

or

3. $$\int_{MP} A\cdot dr = -A_3 h_3 \Delta u_3$$

and

4. $$\int_{QL} A\cdot dr = A_3 h_3 \Delta u_3 + \frac{\partial}{\partial u_2}(A_3 h_3 \Delta u_3)\Delta u_2$$

Adding (1), (2), (3) and (4) we have

$$\oint_{C_1} A\cdot dr = \frac{\partial}{\partial u_2}(A_3 h_3 \Delta u_3)\Delta u_2 - \frac{\partial}{\partial u_3}(A_2 h_2 \Delta u_2)\Delta u_3$$

$$= \left[\frac{\partial}{\partial u_2}(A_3 h_3) - \frac{\partial}{\partial u_3}(A_2 h_2)\right]\Delta u_2 \Delta u_3$$

apart from infinitesimal of order higher than $\Delta u_2 \Delta u_3$. Dividing by he area of S_1 equal to $h_2 h_3 \Delta u_2 \Delta u_3$ and taking the limit as Δu_2 and Δu_3 approach zero,

$$(\text{curl } A).\ e_1 = \frac{1}{h_2 h_3}\left[\frac{\partial}{\partial u_2}(A_3 h_3) - \frac{\partial}{\partial u_3}(A_2 h_2)\right]$$

Similarly, by choosing area S_2 and S_3 perpendicular to e_2 and e_3 at P respectively, we find (curl A). e_2 and (curl A). e_2 . This leads to the required result

$$\text{curl } A = \frac{e_1}{h_2 h_3}\left[\frac{\partial}{\partial u_2}(A_3 h_3) - \frac{\partial}{\partial u_3}(A_2 h_2)\right]$$

$$+ \frac{e_2}{h_3 h_1}\left[\frac{\partial}{\partial u_3}(A_1 h_1) - \frac{\partial}{\partial u_1}(A_3 h_3)\right]$$

$$+ \frac{e_1}{h_1 h_2}\left[\frac{\partial}{\partial u_1}(A_2 h_2) - \frac{\partial}{\partial u_2}(A_1 h_1)\right]$$

$$= \frac{1}{h_1 h_2 h_3}\begin{vmatrix} h_1 e_1 & h_2 e_2 & h_3 e_3 \\ \dfrac{\partial}{\partial u_1} & \dfrac{\partial}{\partial u_2} & \dfrac{\partial}{\partial u_3} \\ h_1 A_1 & h_2 A_2 & h_3 A_3 \end{vmatrix}$$

LAPLACIAN IN ORTHOGONAL CURVILINEAR COORDINATES

$$\nabla^2 \Phi = \nabla \cdot A = \frac{1}{h_1 h_2 h_3}\left[\frac{\partial}{\partial u_1}(f_1 h_2 h_3) + \frac{\partial}{\partial u_2}(f_2 h_2 h_3) + \frac{\partial}{\partial u_3}(f_3 h_2 h_3)\right]$$

$$= \frac{1}{h_1 h_2 h_3}\left[\frac{\partial}{\partial u_1}\left(\frac{1}{h_1}\frac{\partial \Phi}{\partial u_1} h_2 h_3)\right) + \frac{\partial}{\partial u_2}\left(\frac{1}{h_2}\frac{\partial \Phi}{\partial u_2} h_1 h_3)\right) + \frac{\partial}{\partial u_3}\left(\frac{1}{h_3}\frac{\partial \Phi}{\partial u_3} h_1 h_2)\right)\right]$$

$$= \frac{1}{h_1 h_2 h_3}\left[\frac{\partial}{\partial u_1}\left(\frac{h_2 h_3}{h_1}\frac{\partial \Phi}{\partial u_1}\right) + \frac{\partial}{\partial u_2}\left(\frac{h_3 h_1}{h_2}\frac{\partial \Phi}{\partial u_2}\right) + \frac{\partial}{\partial u_3}\left(\frac{h_1 h_2}{h_3}\frac{\partial \Phi}{\partial u_3}\right)\right]$$

DIVERGENCE, CURL AND LAPLACIAN IN CURVILINEAR COORDINATES

SPHERICAL COORDINATE

The spherical coordinates (ρ, θ, φ) of *a* point have the following relationships with rectangular coordinates:

$$\begin{cases} x = \rho \sin\theta \cos\phi \\ y = \rho \sin\theta \sin\phi \\ z = \rho \cos\theta \end{cases}$$

We have the following relations

$$\begin{cases} dx = \dfrac{\partial x}{\partial \rho} d\rho + \dfrac{\partial x}{\partial \theta} d\theta + \dfrac{\partial x}{\partial \phi} d\phi = (\sin\theta\cos\phi) d\rho + (\rho\cos\theta\cos\phi) d\theta - (\rho\sin\theta\sin\phi) d\phi \\ dy = \dfrac{\partial y}{\partial \rho} d\rho + \dfrac{\partial y}{\partial \theta} d\theta + \dfrac{\partial y}{\partial \phi} d\phi = (\sin\theta\cos\phi) d\rho + (\rho\cos\theta\cos\phi) d\theta + (\rho\sin\theta\sin\phi) d\phi \\ dz = \dfrac{\partial z}{\partial \rho} d\rho + \dfrac{\partial z}{\partial \theta} d\theta + \dfrac{\partial z}{\partial \phi} d\phi = (\cos\theta) d\rho - (\rho\sin\theta) d\theta \end{cases}$$

Hence the unit vector along the r-direction is

$$h_\rho \left|\frac{\partial r}{\partial \rho}\right| = \sqrt{\left(\frac{\partial x}{\partial \rho}\right)^2 + \left(\frac{\partial y}{\partial \rho}\right)^2 + \left(\frac{\partial z}{\partial \rho}\right)^2} = \sqrt{(\sin\theta\cos\phi)^2 + (\sin\theta\sin\phi)^2 + (\cos)^2} = 1$$

$$\tilde{\rho} = \frac{\dfrac{\partial r}{\partial \rho}}{\left|\dfrac{\partial r}{\partial \rho}\right|} = \frac{\dfrac{\partial x}{\partial \rho} i + \dfrac{\partial y}{\partial \rho} j + \dfrac{\partial z}{\partial \rho} k}{\sqrt{\left(\dfrac{\partial x}{\partial \rho}\right)^2 + \left(\dfrac{\partial y}{\partial \rho}\right)^2 + \left(\dfrac{\partial z}{\partial \rho}\right)^2}}$$

$$= \frac{(\sin\theta\cos\phi)i + (\sin\theta\sin\phi)j + (\cos\theta)k}{\sqrt{(\sin\theta\cos\phi)^2 + (\sin\theta\sin\phi)^2 + (\cos\theta)^2}}$$

$$= (\sin\theta\cos\phi)i + (\sin\theta\sin\phi)j + (\cos\theta)k$$

Similarly the unit vector along the q -direction is

$$h_\theta = \left|\frac{\partial r}{\partial\theta}\right| = \sqrt{\left(\frac{\partial x}{\partial\theta}\right)^2 + \left(\frac{\partial y}{\partial\theta}\right)^2 + \left(\frac{\partial z}{\partial\theta}\right)^2} = \sqrt{(\cos\theta\cos\phi)^2 + (\cos\theta\sin\phi)^2 + (\rho\sin)^2} = \rho$$

$$\tilde{\theta} = \frac{\frac{\partial r}{\partial\theta}}{\left|\frac{\partial r}{\partial\theta}\right|} = \frac{\frac{\partial x}{\partial\theta}i + \frac{\partial y}{\partial\theta}j + \frac{\partial z}{\partial\theta}k}{\sqrt{\left(\frac{\partial x}{\partial\theta}\right)^2 + \left(\frac{\partial y}{\partial\theta}\right)^2 + \left(\frac{\partial z}{\partial\theta}\right)^2}}$$

$$= \frac{(\rho\cos\theta\cos\phi)i + (\rho\cos\theta\sin\phi)j - (\rho\sin\theta)k}{\sqrt{(\rho\cos\theta\cos\phi)^2 + (\rho\cos\theta\sin\phi)^2 + (\rho\sin\theta)^2}}$$

$$= \frac{(\rho\cos\theta\cos\phi)i + (\rho\cos\theta\sin\phi)j - (\rho\sin\theta)k}{\rho}$$

$$= (\rho\cos\theta\cos\phi)i + (\rho\cos\theta\sin\phi)j - \sin\theta k$$

and the unit vector along the f-direction is

$$h_\phi = \left|\frac{\partial r}{\partial\phi}\right| = \sqrt{\left(\frac{\partial x}{\partial\phi}\right)^2 + \left(\frac{\partial y}{\partial\phi}\right)^2 + \left(\frac{\partial z}{\partial\phi}\right)^2} (\rho\sin\theta\sin\phi)^2 + (\rho\sin\theta\cos\phi)^2 \rho\sin\theta$$

$$\tilde{\phi} = \frac{\frac{\partial r}{\partial\phi}}{\left|\frac{\partial r}{\partial\phi}\right|} = \frac{\frac{\partial x}{\partial\phi}i + \frac{\partial y}{\partial\phi}j + \frac{\partial z}{\partial\phi}k}{\sqrt{\left(\frac{\partial x}{\partial\phi}\right)^2 + \left(\frac{\partial y}{\partial\phi}\right)^2 + \left(\frac{\partial z}{\partial\phi}\right)^2}}$$

$$= \frac{(\rho\sin\theta\sin\phi)i + (\rho\sin\theta\cos\phi)j}{\sqrt{(\rho\sin\theta\sin\phi)^2 + (\rho\sin\theta\cos\phi)^2}}$$

$$= \frac{(\rho\sin\theta\sin\phi)i + (\rho\sin\theta\cos\phi)j}{\rho\sin\theta} = \sin\phi i + \cos\phi j$$

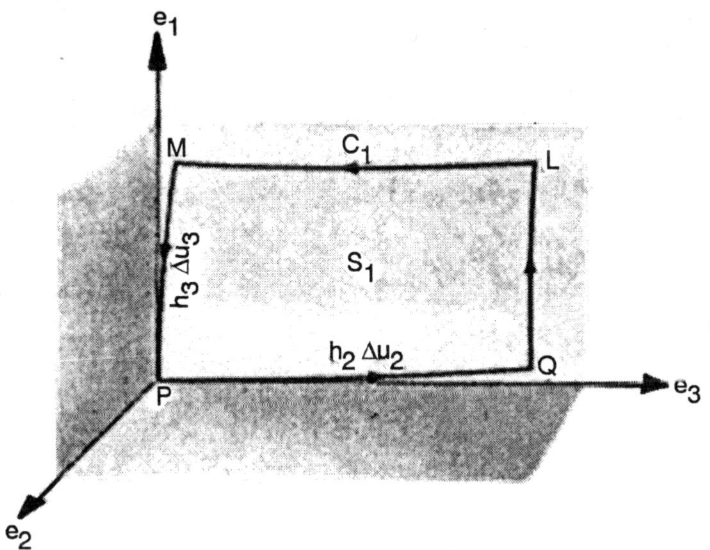

Fig. Spherical coordinate

Therefore

Divergence:

$$\nabla \cdot F = \frac{1}{\rho^2}\frac{\partial}{\partial \rho}(\rho^2 F_\rho) + \frac{1}{\rho \sin\theta}\frac{\partial}{\partial \theta}(\sin\theta F_\theta) + \frac{1}{\rho \sin\theta}\frac{\partial F_\phi}{\partial \phi}$$

Curl:

$$\nabla \times F = \frac{1}{\rho^2 \sin\theta}\left[\frac{\partial}{\partial \rho}(\sin\theta F_\phi) - \frac{\partial F_\phi}{\partial \phi}\right]\tilde{\rho} + \frac{1}{\rho}\left[\frac{1}{\sin\theta}\frac{\partial F_\rho}{\partial \phi} - \frac{\partial}{\partial \rho}(\rho F_\phi)\right]\tilde{\theta} +$$

$$+ \frac{1}{\rho}\left[\frac{\partial}{\partial \rho}(\rho F_\theta)\frac{\partial F_\rho}{\partial \theta}\right]\tilde{\theta}$$

Laplacian:

$$\nabla^2 T = \frac{1}{\rho^2}\frac{\partial}{\partial \rho}\left(\rho^2 \frac{\partial T}{\partial \rho}\right) + \frac{1}{\rho^2 \sin\theta}\frac{\partial}{\partial \theta}\left(\sin\theta \frac{\partial T}{\partial \theta}\right) + \frac{1}{\rho^2 \sin^2\theta}\frac{\partial^2 T}{\partial \phi^2}$$

CYLINDRICAL COORDINATE

The spherical coordinates (ρ, θ, z) of *a* point have the following relationships with rectangular coordinates:

$$\begin{cases} x = r\cos\theta \\ y = r\sin\theta \\ z = z \end{cases}$$

We have the following relations

$$\begin{cases} dx = \dfrac{\partial x}{\partial r}dr + \dfrac{\partial x}{\partial \theta}d\theta + \dfrac{\partial x}{\partial z}dz = (\cos\theta)dr - (r\sin\theta)d\theta \\ dy = \dfrac{\partial y}{\partial r}dr + \dfrac{\partial y}{\partial \theta}d\theta + \dfrac{\partial y}{\partial z}dz = (\sin\theta)dr - (r\cos\theta)d\theta \\ dz = dz \end{cases}$$

Hence the unit vector along the r-direction is

$$h_r = \left|\frac{\partial x}{\partial r}\right| = \sqrt{\left(\frac{\partial x}{\partial r}\right)^2 + \left(\frac{\partial y}{\partial r}\right)^2 + \left(\frac{\partial x}{\partial r}\right)^2} = \sqrt{\cos^2\theta + \sin^2\theta} = 1$$

$$\hat{r} = \frac{\dfrac{\partial x}{\partial r}}{\left|\dfrac{\partial x}{\partial r}\right|} = \frac{\dfrac{\partial x}{\partial r}i + \dfrac{\partial y}{\partial r}j + \dfrac{\partial z}{\partial r}k}{\sqrt{\left(\dfrac{\partial x}{\partial r}\right)^2 + \left(\dfrac{\partial y}{\partial r}\right)^2 + \left(\dfrac{\partial x}{\partial r}\right)^2}}$$

$$= \frac{\cos\theta\, i + \sin\theta\, j}{\sqrt{\cos^2\theta + \sin^2\theta}}$$

$$= \cos\theta\, i + \sin\theta\, j$$

Similarly the unit vector along the θ-direction is

$$h_\theta = \left|\frac{\partial x}{\partial \theta}\right| = \sqrt{\left(\frac{\partial x}{\partial \theta}\right)^2 + \left(\frac{\partial y}{\partial \theta}\right)^2 + \left(\frac{\partial x}{\partial \theta}\right)^2}$$

$$= \sqrt{(r\sin\theta)^2 + (r\cos\theta)^2} = r$$

$$\hat{\theta} = \frac{\dfrac{\partial r}{\partial \theta}}{\left|\dfrac{\partial r}{\partial \theta}\right|} = \frac{\dfrac{\partial x}{\partial \theta}i + \dfrac{\partial y}{\partial \theta}j + \dfrac{\partial z}{\partial \theta}k}{\sqrt{\left(\dfrac{\partial x}{\partial \theta}\right)^2 + \left(\dfrac{\partial y}{\partial \theta}\right)^2 + \left(\dfrac{\partial x}{\partial \theta}\right)^2}}$$

$$= \frac{(-r\sin\theta)i + (r\cos\theta)j}{\sqrt{(r\sin\theta)^2 i + (r\cos\theta)^2}}$$

$$= \frac{(-r\sin\theta)i + (r\cos\theta)j}{r}$$

$$= -\sin\theta\, i + \cos\theta\, j$$

In the z-direction $h_z = 1$.

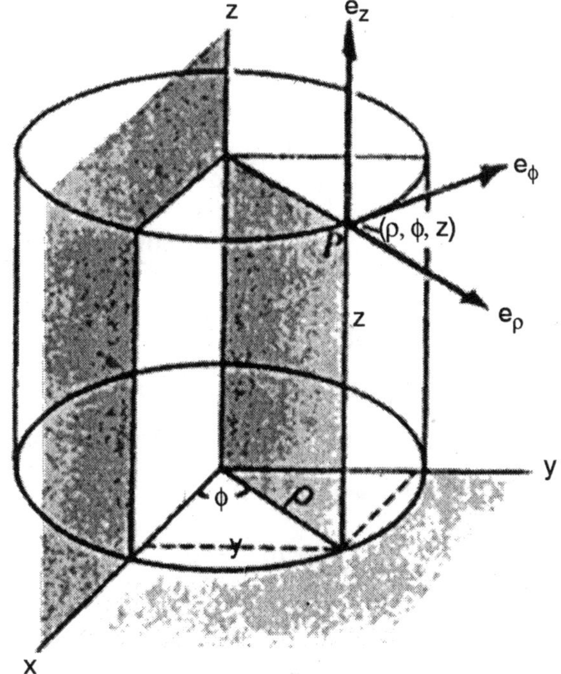

Fig. Cylindrical Coordinate

Therefore

Divergence:

$$\nabla \cdot F = \frac{1}{r}\frac{\partial}{\partial r}(rF_\rho) + \frac{1}{r}\frac{\partial F_\phi}{\partial \phi} + \frac{\partial F_z}{\partial z}$$

Curl:

$$\nabla \times F = \left(\frac{1}{r}\frac{\partial F_z}{\partial \phi} - \frac{\partial F_\phi}{\partial z}\right)\hat{r} + \left(\frac{\partial F_r}{\partial z} - \frac{\partial F_z}{\partial r}\right)\hat{\phi} + \frac{1}{r}\left[\frac{\partial}{\partial r} - \frac{\partial F_\phi}{\partial z}\right]\hat{z}$$

Laplacian:

$$\nabla^2 T = \frac{1}{r}\frac{\partial}{\partial r}\left(r\frac{\partial T}{\partial r}\right) + \frac{1}{r^2}\frac{\partial^2 T}{\partial \phi^2} + \frac{\partial^2 T}{\partial z^2}$$

Example: Find the Laplacian of the function: T = rcosqcosf.

Solution: From Equation

$$\nabla^2 T = \frac{1}{\rho^2}\frac{\partial}{\partial\rho}\left(\rho^2\frac{\partial}{\partial\rho}(\rho\cos\theta\cos\phi)\right) + \frac{1}{\rho^2\sin\theta}\frac{\partial}{\partial\theta}\left(\sin\theta\frac{\partial}{\partial\theta}(\rho\cos\theta\cos\phi)\right)$$

$$+\frac{1}{\rho^2\sin\theta}\frac{\partial^2}{\partial\phi^2}(\rho\cos\theta\cos\phi)$$

$$= \frac{1}{\rho^2}\frac{\partial}{\partial\rho}(\rho^2\cos\theta\cos\phi) + \frac{1}{\rho^2\sin\theta}\frac{\partial}{\partial\theta}(-\rho\sin^2\theta\cos\phi) - \frac{\rho\cos\theta\cos\phi}{\rho^2\sin^2\theta}$$

$$\frac{2\cos\theta\cos\phi}{\rho} - \frac{2\cos\theta\cos\phi}{\rho} - \frac{\cos\theta\cos\phi}{\rho\sin^2\theta}$$

Example: Define the elliptic cylindrical coordinates

x = acoshucosv y = asinhusinv

1. Find the scaling factors for the elliptical cylindrical coordinates system.
2. Hence write the Laplace's equation

$$\nabla^2\Phi \equiv \frac{\partial^2\Phi}{\partial x^2} + \frac{\partial^2\Phi}{\partial y^2} = 0$$

in elliptical cylindrical coordinates.

Solution:

1. From the elliptical cylindrical coordinates, we have

$$\begin{cases} dx = \dfrac{\partial x}{\partial u}du + \dfrac{\partial x}{\partial y} = (a\sinh u\cos v)\,du - (a\cosh u\sin v)\,dv \\ dy = \dfrac{\partial y}{\partial u}du + \dfrac{\partial y}{\partial v} = (a\cosh u\sin v)\,du + (a\sinh u\cos v)\,dv \end{cases}$$

The scaling factors are

$$h_u = \sqrt{\left(\left(\frac{\partial x}{\partial u}\right)^2 + \left(\frac{\partial y}{\partial u}\right)^2\right)}$$

$$= \sqrt{a^2\sinh^2 u\cos^2 v + a^2\cosh^2 u\sin^2 v}$$

$$= \sqrt{a^2 \sinh^2 u(1-\sin^2 v) + a^2 \cosh^2 u \sin^2 v}$$

$$= \sqrt{a^2 \sinh^2 u - a^2 \sinh^2 u \sin^2 v + a^2 \cosh^2 u \sin^2 v}$$

$$= \sqrt{a^2 \sinh^2 u + a^2 \sin^2 v(\cosh^2 u - \sin^2 u)}$$

$$= a\sqrt{\sinh^2 u + \sin^2 v}$$

$$h_v = \sqrt{\left(\left(\frac{\partial x}{\partial v}\right)^2 + \left(\frac{\partial y}{\partial v}\right)^2\right)}$$

$$= \sqrt{a^2 \cosh^2 u \sin^2 v + a^2 \sinh^2 u \cos^2 v}$$

$$= a\sqrt{\sinh^2 u + \sin^2 v}$$

2. From Equation, we have

$$\nabla^2\Phi = \frac{1}{h_u h_v}\left[\frac{\partial}{\partial u}\left(\frac{h_v}{h_u}\frac{\partial\Phi}{\partial u}\right) + \frac{\partial}{\partial v}\left(\frac{h_u}{h_v}\frac{\partial\Phi}{\partial v}\right)\right]$$

$$= \frac{1}{a^2(\sinh^2 u + \sin^2 v)}\frac{\partial^2\Phi}{\partial u^2} + \frac{1}{a^2(\sinh^2 u + \sin^2 v)}\frac{\partial^2\Phi}{\partial v^2}$$

$$= \frac{1}{a^2(\sinh^2 u + \sin^2 v)}\left(\frac{\partial^2\Phi}{\partial u^2} + \frac{\partial^2\Phi}{\partial v^2}\right)$$

Therefore the Laplace's equation can be rewritten as

$$\frac{1}{a^2(\sinh^2 u + \sin^2 v)}\left(\frac{\partial^2\Phi}{\partial u^2} + \frac{\partial^2\Phi}{\partial v^2}\right) = 0 \Rightarrow \frac{\partial^2\Phi}{\partial u^2} + \frac{\partial^2\Phi}{\partial v^2} = 0$$

Chapter 16

Series Solution to Differential Equations

In this chapter we will finally be looking at nonconstant coefficient differential equations. While we won't cover all possibilities in this chapter we will be looking at two of the more common methods for dealing with this kind of differential equation.-

The first method that we'll be taking a look at, series solutions, will actually find a series representation for the solution instead of the solution itself. You first saw something like this when you looked at Taylor series in your Calculus class. As we will see however, these won't work for every differential equation.-

The second method that we'll look at will only work for a special class of differential equations. This special case will cover some of the cases in which series solutions can't be used.

POWER SERIES

Before looking at series solutions to a differential equation we will first need to do a cursory review of power series. A power series is a series in the form,

$$f(x) = \sum_{n=0}^{\infty} a_n (x - x_0)^n$$

where, x_0 and a_n are numbers. We can see from this that a power series is a function of x. The function notation is not always included, but sometimes it is so we put it into the definition above.

Before proceeding with our review we should probably first recall just what series really are. Recall that series are really just summations. One way to write our power series is then,

$$\begin{aligned} f(x) &= \sum_{n=0}^{\infty} a_n (x - x_0)^n \\ &= a_0 + a_1(x - x_0) + a_2(x - x_0)^2 + a_3(x - x_0)^3 + \cdots \end{aligned}$$

Notice as well that if we needed to for some reason we could always write the power series as,

$$\begin{aligned} f(x) &= \sum_{n=0}^{\infty} a_n (x - x_0)^n \\ &= a_0 + a_1(x - x_0) + a_2(x - x_0)^2 + a_3(x - x_0)^3 + \cdots \\ &= a_0 + \sum_{n=0}^{\infty} a_n (x - x_0)^n \end{aligned}$$

All that we're doing here is noticing that if we ignore the first term (corresponding to $n = 0$) the remainder is just a series that starts at $n = 1$. When we do this we say that we've stripped out the $n = 0$, or first, term. We don't need to stop at the first term either. If we strip out the first three terms we'll get,

$$\sum_{n=0}^{\infty} a_n (x - x_0)^n = a_0 + a_1(x - x_0)^2 + \sum_{n=3}^{\infty} a_n (x_n - x_0)^n.$$

There are times when we'll want to do this so make sure that you can do it.

Now, since power series are functions of x and we know that not every series will in fact exist, it then makes sense to ask if a power series will exist for all x. This question is answered by looking at the convergence of the power series. We say that a power series converges for $x = c$ if the series,

$$\sum_{n=0}^{\infty} a_n (c - x_0)^n$$

converges. Recall that this series will converge if the limit of partial sums,

$$\lim_{n \to \infty} \sum_{n=0}^{\infty} a_n (c - x_0)^n$$

exists and is finite. In other words, a power series will converge for $x = c$ if

$$\sum_{n=0}^{\infty} a_n (c - x_0)^n$$

is a finite number.

Note that a power series will always converge if $x = x_0$. In this case the power series will become

$$\sum_{n=0}^{\infty} a_n (x_0 - x_0)^n = a_0.$$

With this we now know that power series are guaranteed to exist for at least one value of x. We have the following fact about the convergence of a power series.

FACT

Given a power series, there will exist a number $0 \leq \rho \leq \infty$so that the power series will converge for $|x-x_0|< \rho$ and diverge for $|x-x_0|< \rho$. This number is called the radius of convergence.

Determining the radius of convergence for most power series is usually quite simple if we use the ratio test.

RATIO TEST

Given a power series compute,

$$L =| x - x_0 | \lim_{n\to\infty}\left|\frac{a_{n+1}}{a_n}\right|$$

then,

$L < 1 \Rightarrow$ the series converges

$L < 1 \Rightarrow$ the series diverges

$L = 1 \Rightarrow$ the series may converge or diverge

Let's take a quick look at how this can be used to determine the radius of convergence.

Example: Determine the radius of convergence for the following power series.

$$\sum_{n=0}^{\infty}\frac{(-3)}{n7^{3+1}}(x-5)^n.$$

Solution: So, in this case we have,

$$a_n \frac{(-3)}{n7^{3+1}} \qquad a_{n+1} = \frac{(-3)^{n+1}}{(n+1)7^{n+2}}.$$

Remember that to compute a_{n+1} all we do is replace all the n's in a_n with $n+1$. Using the ratio test then gives,

$$L = |x-5| \lim_{n\to\infty}\left|\frac{a_{n+1}}{a_n}\right|$$

$$= |x-5| \lim_{n\to\infty}\left|\frac{(-3)^{n+1}}{(n+1)7^{n+2}}\,\frac{n7^{n+n}}{(-3)^n}\right|$$

$$= |x-5| \lim_{n\to\infty}\left|\frac{-3}{(n+1)7}\,\frac{n}{1}\right| = \frac{3}{7}|x-5|.$$

Now we know that the series will converge if,

$$\frac{3}{7}|x-5|<1 \qquad \Rightarrow \qquad |x-5|<\frac{7}{3}$$

and the series will diverge if,

$$\frac{3}{7}|x-5|>1 \qquad \Rightarrow \qquad |x-5|>\frac{7}{3}$$

In other words, the radius of the convergence for this series is,

$$\rho=\frac{7}{3}.$$

As this last example has shown, the radius of convergence is found almost immediately upon using the ratio test. So, why are we worried about the convergence of power series? Well in order for a series solution to a differential equation to exist at a particular x it will need to be convergent at that x. If it's not convergent at a given x then the series solution won't exist at that x. So, the convergence of power series is fairly important.

Next we need to do a quick review of some of the basics of manipulating series. We'll start with addition and subtraction.

There really isn't a whole lot to addition and subtraction. All that we need to worry about is that the two series start at the same place and both have the same exponent of the $x-x_0$. If they do then we can perform addition and/or subtraction as follows,

$$\sum_{n=n_0}^{\infty} a_n(x-x_0)^n \pm \sum_{n=n_0}^{\infty} b_n(x-x_0)^n = \sum_{n=n_0}^{\infty} (a_n \pm b_n)(x-x_0)^n.$$

In other words all we do is add or subtract the coefficients and we get the new series.-

One of the rules that we're going to have when we get around to finding series solutions to differential equations is that the only x that we want in a series is the x that sits in. This means that we will need to be able to deal with series of the form,

$$(x-x_0)^c \sum_{n=0}^{\infty} a_n(x-x_0)^n$$

where c is some constant. These are actually quite easy to deal with.

$$\begin{aligned}(x-x_0)^c \sum_{n=0}^{\infty} a_n(x-x_0)^n &= (x-x_0)^c(a_0+a_1(x-x_0)+a_1(x-x_0)^2+\cdots)\\ &= a_0(x-x_0)^c+a_0(x-x_0)^{1+c}+a_1(x-x_0)^{2+c}+\cdots\\ &\sum_{n=0}^{\infty} a_n(x-x_0)^{n+c}.\end{aligned}$$

So, all we need to do is to multiply the term in front into the series and add exponents. Also note that in order to do this both the coefficient in front of the series and the term inside the series must be in the form x-x_0. If they are not the same we can't do this, we will eventually see how to deal with terms that aren't in this form.

Next we need to talk about differentiation of a power series. By looking it should be fairly easy to see how we will differentiate a power series. Since a series is just a giant summation all we need to do is differentiate the individual terms. The derivative of a power series will be,

$$f'(x) = a_1 + 2a_2(x - x_0) + 3a_3(x - x_0)^2 + \cdots$$

$$= \sum_{n=1}^{\infty} na_n(x - x_0)^{n-1}$$

$$= \sum_{n=1}^{\infty} na_n(x - x_0)^{n-1}.$$

So, all we need to do is just differentiate the term inside the series and we're done. Notice as well that there are in fact two forms of the derivative. Since the n=0 term of the derivative is zero it won't change the value of the series and so we can include it or not as we need to. In our work we will usually want the derivative to start at n=*1*, however there will be the occasion problem were it would be more convenient to start it at n=0.

Following how we found the first derivative it should make sense that the second derivative is,

$$f'(x) = \sum_{n=1}^{\infty} n(n-1)a_n(x - x_0)^{n-2}$$

$$= \sum_{n=1}^{\infty} n(n-1)a_n(x - x_0)^{n-2}$$

$$= \sum_{n=1}^{\infty} n(n-1)a_n(x - x_0)^{n-2}.$$

In this case since the n=0 and n=*1* terms are both zero we can start at any of three possible starting points as determined by the problem that we're working.

Next we need to talk about index shifts. As we will see eventually we are going to want ourpow erseriesw ritten in term sof $(x - x_0)^n (x - x_0)^n$ $(x - x_0)^n$ and they often won't, initially at least, be in that form. To get them into the form we need we will need to perform an index shift. Index shifts themselves really aren't concerned with the exponent on the x term, they instead are concerned with where the series starts as the following example shows.

Example: Write the following as a series that starts at n=0 instead of n=3.

$$\sum_{n=1}^{\infty} n^2 a_{n-1}(x+4)^{n+2}.$$

Solution: An index shift is a fairly simple manipulation to perform. First we will notice that if we define $i=n-3$ then when $n=3$ we will have i=0. So what we'll do is rewrite the series in terms of i instead of n. We can do this by noting that $n=i+3$. So, everywhere we see an n in the actual series term we will replace it with an $i+3$. Doing this gives,

$$\sum_{n=1}^{\infty} n^2 a_{n-1}(x+4)^{n+2} = \sum_{n=1}^{\infty} (i+3)^2 a_{i+3-1}(x+4)^{i+3+2}$$

$$= \sum_{n=1}^{\infty} (i+3)^2 a_{i+2}(x+4)^{i+5}.$$

The upper limit won't change in this process since infinity minus three is still infinity.

The final step is to realize that the letter we use for the index doesn't matter and so we can just switch back to n's.

$$\sum_{n=1}^{\infty} n^2 a_{n+1}(x+4)^{n+2} = \sum_{n=0}^{\infty} (n+3)^2 a_{n+2}(x+4)^{n+5}.$$

Now, we usually don't go through this process to do an index shift. All we do is notice that we dropped the starting point in the series by 3 and everywhere else we saw an n in the series we increased it by 3. In other words, all the n's in the series move in the opposite direction that we moved the starting point.

Example: *3* Write the following as a series that starts at $n=5$ instead of n=3.

$$\sum_{n=1}^{\infty} n^2 a_{n+1}(x+4)^{n+2}.$$

Solution: To start the series to start at $n=5$ all we need to do is notice that this means we will increase the starting point by 2 and so all the other n's will need to decrease by 2. Doing this for the series in the previous example would give,

$$\sum_{n=1}^{\infty} n^2 a_{n+1}(x+4)^{n+2} = \sum_{n=5}^{\infty} (n-2)^2 a_{n-3}(x+4)^{n}.$$

Now, as we noted when we started this discussion about index shift the whole point is to get our series into terms of. We can see in the previous example that we did exactly that with an index shift. The original exponent on the (x+4) was n+2. To get this down to an n we needed to decrease the exponent by 2. This can be done with an index that increases the starting point by 2.

Let's take a look at a couple of more examples of this.

Example: Write each of the following as a single series in terms of $(x - x_0)^n$.

1. $(x+2)^2 \sum_{n=3}^{\infty} na_n (x+2)^{n-4} - \sum_{n=1}^{\infty} na_n (x+2)^{n+1}$.

2. $x \sum_{n=0}^{\infty} (n-5)^2 b_{n+1} (x-3)^{n+3}$.

Solution: 1. $(x+2)^2 \sum_{n=3}^{\infty} na_n (x+2)^{n-4} - \sum_{n=3}^{\infty} na_n (x+2)^{n-1}$.

First, notice that there are two series here and the instructions clearly ask for only a single series. So, we will need to subtract the two series at some point in time. The vast majority of our work will be to get the two series prepared for the subtraction. This means that the two series can't have any coefficients in front of them (other than one of course...), they will need to start at the same value of *n* and they will need the same exponent on the x-x_0. We'll almost always want to take care of any coefficients first. So, we have one in front of the first series so let's multiply that into the first series. Doing this gives,

$$\sum_{n=3}^{\infty} na_n (x+2)^{n-2} - \sum_{n=3}^{\infty} na_n (x+2)^{n+1}.$$

Now, the instructions specify that the new series must be in terms of $(x - x_0)^n$, so that's the next thing that we've got to take care of. We will do this by an index shift on each of the series. The exponent on the first series needs to go up by two so we'll shift the first series down by 2. On the second series will need to shift up by 1 to get the exponent to move down by 1. Performing the index shifts gives us the following,

$$\sum_{n=3}^{\infty} (n+2)a_{n+2} (x+2) - \sum_{n=3}^{\infty} (n-1)a_{n-1} (x+2)^n.$$

Finally, in order to subtract the two series we'll need to get them to start at the same value of *n*. Depending on the series in the problem we can do this in a variety of ways. In this case let's notice that since there is an *n-1* in the second series we can in fact start the second series at *n=1* without changing its value. Also note that in doing so we will get both of the series to start at *n=1* and so we can do the subtraction. Our final answer is then,

$$\sum_{n=3}^{\infty} (n+2)a_{n+2} (x+2)^n - \sum_{n=1}^{\infty} (n-1)a_{n-1} (x+2)^n = \sum_{n=1}^{\infty} [(n+2)a_{n+2} - (n-1)a_{n-1}](x+2)^n$$

2. $x \sum_{n=3}^{\infty} (n-5)^2 b_{n+1} (x-3)^{n+3}$.

In this part the main issue is the fact that we can't just multiply the coefficient into the series this time since the coefficient doesn't have the same form as the term inside the series. Therefore, the first thing that we'll need to do is correct the coefficient so that we can bring it into the series. We do this as follows,

$$x\sum_{n=3}^{\infty}(n-5)^2 b_{n+1}(x-3)^{n+3} = (x-3+3)\sum_{n=0}^{\infty}(n-5)^2 b_{n+1}(x-3)^{n+3}$$
$$= (x-3)\sum_{n=0}^{\infty}(n-5)^2 b_{n+1}(x-3)^{n+3} + 3\sum_{n=0}^{\infty}(n-5)^2 b_{n+1}(x-3)^{n+3}.$$

We can now move the coefficient into the series, but in the process of we managed to pick up a second series. This will happen so get used to it. Moving the coefficients of both series in gives,

$$\sum_{n=0}^{\infty}(n-5)^2 b_{n+1}(x-3)^{n+4} + \sum_{n=0}^{\infty}3(n-5)^2 b_{n+1}(x-3)^{n+3}.$$

We now need to get the exponent in both series to be an n. This will mean shifting the first series up by 4 and the second series up by 3. Doing this gives,

$$\sum_{n=0}^{\infty}(n-9)^2 b_{n-3}(x-3)^{n} + \sum_{n=3}^{\infty}3(n-8)^2 b_{n-2}(x-3)^{n}.$$

In this case we can't just start the first series at $n=3$ because there is not an $n-3$ sitting in that series to make the $n=3$ term zero. So, we won't be able to do this part as we did in the first part of this example. What we'll need to do in this part is strip out the $n=3$ from the second series so they will both start at $n=4$. We will then be able to add the two series together. Stripping out the $n=3$ term from the second series gives,

$$\sum_{v=4}^{\infty}(n-9)^2 b_{n-3}(x-3)^{n} + 3(-5)^2 b_1(x-3)^3 + \sum_{n=4}^{\infty}3(n-8)^2 b_{n-2}(x-3)^{n}.$$

We can now add the two series together.

$$75b_1(x-3)^3 + \sum_{n=4}^{\infty}[(n-9)^2 b_{n-3} + 3(n-8)^2 b_{n-2}](x-3)^{n}.$$

This is what we're looking for. We won't worry about the extra term sitting in front of the series. When we finally get around to finding series solutions to differential equations we will see how to deal with that term there. There is one final fact that we need take care of before moving on. Before giving this fact for power series let's notice that the only way for

$$a+bx+cx^2=0$$

to be zero for all x is to have $a=b=c=0$.

We've got a similar fact for power series.

FACT

If,

$$\sum_{n=4}^{\infty} a_n (x - x_0)^n = 0$$

for all x then,

$$a_n = 0, \quad n = 0,1,2,...$$

This fact will be key to our work with differential equations so don't forget it.-

TAYLOR SERIES

We are not going to be doing a whole lot with Taylor series once we get out of the review, but they are a nice way to get us back into the swing of dealing with power series. By time most students reach this stage in their mathematical career they've not had to deal with power series for at least a semester or two. Remembering how Taylor series work will be a very convenient way to get comfortable with power series before we start looking at differential equations.-

If $f(x)$ is an infinitely differential function then the Taylor Series of $f(x)$ about $x=x_0$ is,-

$$f(x) = \sum_{n=0}^{\infty} \frac{f^{(n)}(x_0)}{n!}(x - x_0)^n.$$

Recall that

$$f^{(0)}(x) = f(x) \quad f^{(n)}(x) = n^{\text{th}} \text{ derivative of } f(x).$$

Let's take a look at an example.

Example: Determine the Taylor series for $f(x) = e^x$about $x= 0$.

Solution: This is probably one of the easiest functions to find the Taylor series for. We just need to recall that,

$$f(^n)\ (x) = e^x\ n = 0,\ 1,\ 2,...$$

and so we get,

$$f(^n)\ (x) = e^x\ n = 0,\ 1,\ 2,...$$

The Taylor series for this example is then,

$$e^x = \sum_{n=0}^{\infty} \frac{x^n}{n!}.$$

Of course, it's often easier to find the Taylor series about $x=0$ but we don't always do that.

Example: Determine the Taylor series for $f(x) = e^x$ about $x = -4$.

Solution: This problem is virtually identical to the previous problem. In this case we just need to notice that,

$$f^{(n)}(-4) = e^{-4} \quad n = 0, 1, 2, \ldots$$

The Taylor series for this example is then,

$$e^x = \sum_{n=0}^{\infty} \frac{e^{-4}}{n!}(x+4)^n.$$

Let's now do a Taylor series that requires a little more work.

Example: Determine the Taylor series for $f(x) = \cos(x)$ about $x = 0$.

Solution: This time there is no formula that will give us the derivative for each n so let's start taking derivatives and plugging in $x = 0$.

$$\begin{array}{ll} f^{(0)}(x) = \cos(x) & f^{(0)}(0) = 1 \\ f^{(1)}(x) = -\sin(x) & f^{(1)}(0) = 0 \\ f^{(2)}(x) = -\cos(x) & f^{(2)}(0) = -1 \\ f^{(3)}(x) = \sin(x) & f^{(3)}(0) = 0 \\ f^{(4)}(x) = \cos(x) & f^{(4)}(0) = 1 \\ \vdots & \vdots \end{array}$$

Once we reach this point it's fairly clear that there is a pattern emerging here. Just what this pattern is has yet to be determined, but it does seem fairly clear that a pattern does exist.

Let's plug what we've got into the formula for the Taylor series and see what we get.

$$\begin{aligned} \cos(x) &= \sum_{n=0}^{\infty} \frac{f^{(n)}(0)}{n!} x^n \\ &= \frac{f^{(n)}(0)}{0!} + \frac{f^{(1)}(0)}{1!} x + \frac{f^{(2)}(0)}{2!} x^2 + \frac{f^{(3)}(0)}{3!} x^3 + \cdots \\ &= \frac{1}{0!} + 0\frac{x^2}{2!} + 0 + \frac{x^4(0)}{4!} + 0 - \frac{x^6}{6!} + 0 + \frac{x^8}{8!} + \cdots \end{aligned}$$

So, every other term is zero.

We would like to write this in terms of a series, however finding a formula that is zero every other term and gives the correct answer for those that aren't zero would be unnecessarily complicated. So, let's rewrite what we've got above and while were at it renumber the terms as follows,

$$\cos(x) = \underbrace{\frac{1}{0!}}_{n=0} - \underbrace{\frac{x^2}{2!}}_{n=1} + \underbrace{\frac{x^4}{4!}}_{n=2} - \underbrace{\frac{x^6}{6!}}_{n=3} + \underbrace{\frac{x^8}{8!}}_{n=4} + \cdots$$

With this "renumbering" we can fairly easily get a formula for the Taylor series of the cosine function about $x = 0$.

$$\cos(x) = \sum_{n=0}^{\infty} \frac{(-1)^n x^{2n}}{(2n)!}.$$

For practice you might want to see if you can verify that the Taylor series for the sine function about $x = 0$ is,

$$\sin(x) = \sum_{n=0}^{\infty} \frac{(-1)^n x^{2n+1}}{(2n+1)!}.$$

We need to look at one more example of a Taylor series. This example is both tricky and very easy.

Example: Determine the Taylor series for $f(x) = 3x^2 - 8x + 2$about $x = 2$.

Solution: There's not much to do here except to take some derivatives and evaluate at the point.

$$f(x) = 3x^2 - 8x + 2 \qquad f(2) = -2$$

$$f'(x) = 6x - 8 \qquad f'(2) = 4$$

$$f''(x) = 6 \qquad f''(2) = 6$$

$$f^{('')}(x) = 0, n \geq 3 \qquad f^{('')}(2) = 0, n \geq 3.$$

So, in this case the derivatives will all be zero after a certain order. That happens occasionally and will make our work easier. Setting up the Taylor series then gives,

$$3x^2 - 8x + 2 = \sum_{n=0}^{\infty} \frac{f^{(n)}(2)}{n!}(x-2)^n$$

$$= \frac{f^{(n)}(2)}{0!} + \frac{f^{(1)}(2)}{1!}(x-2) + \frac{f^{(2)}(2)}{2!}(x-2) + \frac{f^{(3)}(2)}{3!}(x-2)^3 + \cdots$$

$$= -2 + 4(x-2) + \frac{6}{2}(x-2)^2 + 0$$

$$= -2 + 4(x-2) + 3(x-2)^2.$$

In this case the Taylor series terminates and only had three terms. Note that since we are after the Taylor series we do not multiply the 4 through on the second term or square out the third term. All the terms with the exception of the constant should contain an $x-2$.

Note in this last example that if we were to multiply the Taylor series we would get our original polynomial. This should not be too surprising as both are polynomials and they should be equal.

We now need a quick definition that will make more sense to give here rather than in the next section were we actually need it since it deals with Taylor series.

Definition: A function, $f(x)$, is called analytic at $x = a$ if the Taylor series for $f(x)$ about $x=a$ has a positive radius of convergence and converges to $f(x)$.

We need to give one final note before proceeding into the next section. We started this section out by saying that we weren't going to be doing much with Taylor series after this section. While that is correct it is only correct because we are going to be keeping the problems fairly simple. For more complicated problems we would also be using quite a few Taylor series.

SERIES SOLUTIONS TO DIFFERENTIAL EQUATIONS

Before we get into finding series solutions to differential equations we need to determine when we can find series solutions to differential equations. So, let's start with the differential equation,

$$p(x)y''+q(x)y'+r(x)y=0.$$

This time we really do mean nonconstant coefficients. To this point we've only dealt with constant coefficients. However, with series solutions we can now have nonconstant coefficient differential equations. Also, in order to make the problems a little nicer we will be dealing only with polynomial coefficients.-

Now, we say that $x=x_0$ is an ordinary point if provided both

$$\frac{q(x)}{p(x)} \quad \text{and} \quad \frac{r(x)}{p(x)}$$

are analytic at $x=x_0$. That is to say that these two quantities have Taylor series around $x=x_0$. We are going to be only dealing with coefficients that are polynomials so this will be equivalent to saying that

$$p(x_0)\neq 0$$

for most of the problems.

If a point is not an ordinary point we call it a singular point. The basic idea to finding a series solution to a differential equation is to assume that we can write the solution as a power series in the form,

$$\sum_{n=0}^{\infty} a_n (x - x_0)^n$$

and then try to determine what the a_n's need to be. We will only be able to do this if the point $x=x_0$, is an ordinary point. We will usually say that series solution is around $x=x_0$.

Let's start with a very basic example of this. In fact it will be so basic that we will have constant coefficients. This will allow us to check that we get the correct solution.

Example: Determine a series solution for the following differential equation about $x_0 = 0$.

$$y'' + y = 0.$$

Solution: Notice that in this case $p(x)=1$ and so every point is an ordinary point. We will be looking for a solution in the form,

$$y(x) = \sum_{n=0}^{\infty} a_n x^n.$$

We will need to plug this into our differential equation so we'll need to find a couple of derivatives.

$$y'(x) = \sum_{n=0}^{\infty} n a_n x^{n-1} \qquad y''(x) = \sum_{n=0}^{\infty} n(n-1) a_n x^{n-2}.$$

Recall from the power series review section on power series that we can start these at $n=0$ if we need to, however it's almost always best to start them where we have here. If it turns out that it would have been easier to start them at $n=0$ we can easily fix that up when the time comes around.

So, plug these into our differential equation. Doing this gives,

$$\sum_{n=2}^{\infty} n(n-1)\, a_n x^{n-2} + \sum_{n=0}^{\infty} a_n x^n = 0.$$

The next step is to combine everything into a single series. To do this requires that we get both series starting at the same point and that the exponent on the x be the same in both series.

We will always start this by getting the exponent on the x to be the same. It is usually best to get the exponent to be an n. The second series already has the proper exponent and the first series will need to be shifted down by 2 in order to get the exponent up to an n. If you don't recall how to do this take a quick look at the first review section where we did several of these types of problems.

Shifting the first power series gives us,

$$\sum_{n=2}^{\infty}(n+2)(n+1)\,a_n x^n + \sum_{n=0}^{\infty} a_n x^n = 0.$$

Notice that in the process of the shift we also got both series starting at the same place. This won't always happen, but when it does we'll take it. We can now add up the two series. This gives,

$$\sum_{n=2}^{\infty}[(n+2)(n+1)\,a_n x^n + a_n]\,x^n = 0.$$

Now recalling the fact from the power series review section we know that if we have a power series that is zero for all x (as this is) then all the coefficients must have been zero to start with. This gives us the following,

$$(n+2)(n+1)a_{n+2} + a_n = 0, \qquad n = 0,\ 1,\ 2,...$$

This is called the recurrence relation and notice that we included the values of n for which it must be true. We will always want to include the values of n for which the recurrence relation is true since they won't always start at $n = 0$ as it did in this case. Now let's recall what we were after in the first place. We wanted to find a series solution to the differential equation. In order to do this we needed to determine the values of the a_n's. We are almost to the point where we can do that. The recurrence relation has two different a_n's in it so we can't just solve this for a_n and get a formula that will work for all n. We can however, use this to determine what all but two of the a_n's are. To do this we first solve the recurrence relation for the a_n that has the largest subscript. Doing this gives,

$$a_{n+2} = -\frac{a_n}{(n+2)(n+1)} \qquad n = 0,1,2,....$$

Now, at this point we just need to start plugging in some value of n and see what happens,

$$n = 0 \quad a_2 = \frac{-a_0}{(2)(1)} \qquad\qquad n = 1 \quad a_3 = \frac{-a_1}{(3)(2)}$$

$$n = 2 \quad a_4 = -\frac{a_2}{(4)(3)} \qquad\qquad n = 3 \quad a_5 = -\frac{-a_3}{(5)(4)}$$

$$= \frac{a_0}{(4)(3)(2)(1)} \qquad\qquad = \frac{a_1}{(5)(4)(3)(2)}\left(\frac{\pi}{2} - \theta\right)$$

$$n = 4 \quad a_6 = \frac{a_4}{(6)(5)} \qquad\qquad n = 5 \quad a_1 = -\frac{a_5}{(7)(6)}$$

$$= \frac{-a_0}{(6)(5)(4)(3)(2)(1)} \qquad\qquad \frac{-a_1}{(7)(6)(5)(4)(3)(2)}$$

$$\vdots \qquad\qquad \vdots \qquad\qquad \vdots$$

$$n = 2k \quad a_{2k} = \frac{(-1)^k a_0}{(2k)!}, k = 1, 2... \qquad n = 2k+1 \quad a_{2k+1} = \frac{(-1)^k a_1}{(2k+1)!}, k = 1, 2, ...$$

Notice that at each step we always plugged back in the previous answer so that when the subscript was even we could always write the a_n in terms of a_0 and when the coefficient was odd we could always write the a_n in terms of a_1. Also notice that, in this case, we were able to find a general formula for a_n's with even coefficients and a_n's with odd coefficients. This won't always be possible to do.-

There's one more thing to notice here. The formulas that we developed were only for k=1,2,... however, in this case again, the will also work for k=0. Again, this is something that won't always work, but does here.-

Do not get excited about the fact that we don't know what a_0 and a_1 are. As you will see, we actually need these to be in the problem to get the correct solution.

Now that we've got formulas for the a_n's let's get a solution. The first thing that we'll do is write out the solution with a couple of the a_n's plugged in.

$$\begin{aligned} y(x) &= \sum_{n=0}^{\infty} a_n x^n \\ &= a_0 + a_1 x + a_1 x^2 + a_3 x^3 + \cdots + a_{2k} x^{2k} + a_{2k+1} x^{2k+1} + \cdots \\ &= a_0 + a_1 x - \frac{a_0}{2!} x^2 - \frac{a_1}{3!} x^3 + \cdots + \frac{(-1)^k a_0}{(2k)!} x^{2k} + \frac{(-1)^{k+1} a_1}{(2k+1)!} x^{2k+1} + \cdots \end{aligned}$$

The next step is to collect all the terms with the same coefficient in them and then factor out that coefficient.

$$\begin{aligned} y(x) &= a_0 \left\{ 1 - \frac{a_0}{2!} \cdots + \frac{(-1)^k x^{2k}}{(2k)!} + \cdots \right\} + a_1 \left\{ x - \frac{x^3}{3!} + \cdots + \frac{(-1)^{k+1}}{(2k+1)!} x^{2k+1} + \cdots \right\} \\ &= a_0 \sum_{k=0}^{\infty} \frac{(-1)^k x^{2k}}{(2k)!} + a_1 \sum_{k=0}^{\infty} \frac{(-1)^k x^{2k+1}}{(2k+1)!}. \end{aligned}$$

In the last step we also used the fact that we knew what the general formula was to write both portions as a power series. This is also our solution. We are done.-

Before working another problem let's take a look at the solution to the previous example. First, we started out by saying that we wanted a series solution of the form,-

$$y(x) = \sum_{n=0}^{\infty} a_n x^n$$

and we didn't get that. We got a solution that contained two different power series. Also, each of the solutions had an unknown constant in them. This is not a problem. In fact, it's what we want to have happen. From our work with second order constant coefficient differential equations we know that the solution to the differential equation in the last example is,

$$y(x) = c_1 \cos(x) + c_2 \sin(x).$$

Solutions to second order differential equations consist of two separate functions each with an unknown constant in front of them that are found by applying any initial conditions. So, the form of our solution in the last example is exactly what we want to get. Also recall that the following Taylor series,

$$\cos(x) = \sum_{(2n)!}^{\infty} \frac{(-1)^n x^{2n}}{(2n)!} \qquad \sin(x) = \sum_{n=0}^{\infty} \frac{(-1)^n x^{2n+1}}{(2n+1)!}.$$

Recalling these we very quickly see that what we got from the series solution method was exactly the solution we got from first principles, with the exception that the functions were the Taylor series for the actual functions instead of the actual functions themselves.

Now let's work an example with nonconstant coefficients since that is where series solutions are most useful.

Example: Find a series solution around $x_0 = 0$for the following differential equation.-

$$y'' - xy = 0.$$

Solution: As with the first example $p(x)$=1 and so again for this differential equation every point is an ordinary point. Now we'll start this one out just as we did the first example. Let's write down the form of the solution and get its derivatives.

$$y(x) = \sum_{n=0}^{\infty} a_n x^n \quad y'(x) = \sum_{n=1}^{\infty} n a_n x^{n-1} \quad y''(x) = \sum_{n=2}^{\infty} (n-1)\, a_n x^{n-2}.$$

Plugging into the differential equation gives,

$$\sum_{n=2}^{\infty} n(n-1) a_n x^{n-2} - x \sum_{n=1}^{\infty} a_n x^n = 0.$$

Unlike the first example we first need to get all the coefficients moved into the series.

$$\sum_{n=2}^{\infty} n(n-1) a_n x^{n-2} - \sum_{n=1}^{\infty} a_n x^{n-1} = 0.$$

Now we will need to shift the first series down by 2 and the second series up by 1 to get both of the series in terms of x^n.

$$\sum_{n=0}^{\infty} (n+2)(n+1) a_{n+2} x^n - \sum_{n=1}^{\infty} a_{n-1} x^n = 0$$

Next we need to get the two series starting at the same value of n. The only way to do that for this problem is to strip out the n=0 term.

$$(2)(1)a_2x^0 + \sum_{n=1}^{\infty}[(n+2)(n+1)a_{n+2} - a_{n-1}]x^n = 0$$

$$2a_2 + \sum_{n=0}^{\infty}[(n+2)(n+1)a_{n+2} - a_{n-1}]x^n = 0.$$

We now need to set all the coefficients equal to zero. We will need to be careful with this however. The n=0 coefficient is in front of the series and the n=1,2,3… are all in the series. So, setting coefficient equal to zero gives,

$$n = 0: \qquad 2a_2 = 0$$

$$n = 1,2,3,... \quad (n+2)(n+1)a_{n+2} - a_{n+2} = 0$$

Solving the first as well as the recurrence relation gives,

$$n = 0: \qquad a_2 = 0$$

$$n = 1,2,3,... \qquad a_{n+2} = \frac{a_{n-1}}{(n+2)(n+1)}.$$

Now we need to start plugging in values of n.

$$a_3 = \frac{a_0}{(3)(2)} \qquad a_4 = \frac{a_1}{(4)(3)} \qquad a_5 = \frac{a_2}{(5)(4)} = 0$$

$$a_6 = \frac{a_3}{(6)(5)} \qquad a_1 = \frac{a_1}{(7)(6)} \qquad a_8 = \frac{a_5}{(8)(7)} = 0$$

$$= \frac{a_0}{(6)(5)(3)(2)} \qquad = \frac{a_4}{(7)(6)(5)(4)}$$

$$\vdots \qquad \vdots \qquad \vdots$$

$$a_{3k} = \frac{a_0}{(2)(3)(5)(6)...(3k-1)(3k)} \qquad a_{3k+1} = \frac{a_1}{(3)(4)(6)(7)...(3k)(3k+1)} \qquad a_{3k+2} = 0$$

$$k = 1,2,3,.... \qquad k = 1,2,3,.... \qquad k = 1,2,3,....$$

There are a couple of things to note about these coefficients. First, every third coefficient is zero. Next, the formulas here are somewhat unpleasant and not all that easy to see the first time around.

Finally, these formulas will not work for k=0 unlike the first example.

Now, get the solution,

$$y(x) = a_0 + a_1x + a_2x^2 + a_3x^3 + a_4x^4 + \cdots + a_{3k}x^{3k} + a_{3k+1}x^{3k+1}$$

$$= a_0 + a_1x + \frac{a_0}{6}x^3 + \frac{a_1}{12}x^4 \cdots + \frac{a_0x^{3k}}{(2)(3)(5)(6)\cdots(3k-1)(3k)} + \frac{a_0x^{3k}}{(3)(4)(6)(7)\cdots(3k)(3k+1)} + \cdots$$

Again, collect up the terms that contain the same coefficient, factor the coefficient out and write the results as a new series,

$$y(x) = a_0\left\{1 + \sum_{k=1}^{\infty}\frac{x^{3k}}{(2)(3)(5)(6)\cdots(3k-1)(3k)}\right\} + a_1\left\{x + \sum_{\kappa=1}^{\infty}\frac{a_0x^{3k}}{(3)(4)(6)(7)\cdots(3k)(3k+1)}\right\}$$

We couldn't start our series at $k=0$ this time since the general term doesn't hold for $k=0$.

Now, we need to work an example in which we use a point other that $x=0$. In fact, let's just take the previous example and rework it for a different value of x_0. We're also going to need to change up the instructions a little for this example.

Example: Find the first four terms in each portion of the series solution around $x_0 = -2$for the following differential equation.

$$y'' = -xy = 0.$$

Solution: Unfortunately for us there is nothing from the first example that can be reused here. Changing to $x_0 = -2$completely changes the problem. In this case our solution will be,

$$y(x) = \sum_{n=0}^{\infty} a_n(x+2)^n.$$

The derivatives of the solution are,

$$y'(x) = \sum_{n=1}^{\infty} na_n(x+2)^{n-1} \qquad y''(x) = \sum_{n=2}^{\infty} n(n-1)a_n(x+2)^{n-2}.$$

Plug these into the differential equation.

$$y'(x) = \sum_{n=0}^{\infty} na_n(x+2)^{n+2} - x\sum_{n=0}^{\infty} a_n(x+2)^n = 0.$$

We now run into our first real difference between this example and the previous example. In this case we can't just multiply the x into the second series since in order to combine with the series it must be $x+2$. Therefore we will first need to modify the coefficient of the second series before multiplying it into the series.

$$\sum_{n=2}^{\infty} n(n-1)a_n(x+2)^{n-2} - (x+2-2)\sum_{n=0}^{\infty} a_n(x+2)^n = 0$$

$$\sum_{n=2}^{\infty} n(n-1)a_n(x+2)^{n-2} - (x+2)\sum_{n=0}^{\infty} a_n(x+2)^n + 2\sum_{n=0}^{\infty} a_n(x+2)^n = 0$$

$$\sum_{n=2}^{\infty} n(n-1)a_n(x+2)^{n-2} - \sum_{n=0}^{\infty} a_n(x+2)^{n+1} + \sum_{n=0}^{\infty} 2a_n(x+2)^n = 0.$$

We now have three series to work with. This will often occur in these kinds of problems. Now we will need to shift the first series down by 2 and the second series up by 1 the get common exponents in all the series.

$$\sum_{n=2}^{\infty} (n+2)(n+1)a_{n+2}(x+2)^n - \sum_{n=0}^{\infty} a_{n-1}(x+2)^n + \sum_{n=0}^{\infty} 2a_n(x+2)^n = 0.$$

In order to combine the series we will need to strip out the n=0 terms from both the first and third series.

$$2a_2 + 2a_0 + \sum_{n=2}^{\infty} (n+2)(n+1)a_{n+2}(x+2)^n - \sum_{n=1}^{\infty} a_{n-1}(x+2)^n + \sum_{n=1}^{\infty} 2a_n(x+2)^n = 0$$

$$2a_2 + 2a_0 + \sum_{n=2}^{\infty} [(n+2)(n+1)a_{n+2} - a_{n-1} + 2a_n](x+2)^n = 0.$$

Setting coefficients equal to zero gives,

$$n = 0 \qquad 2a_2 + 2a_0 = 0$$

$$n = 1, 2, 3, \ldots \qquad (n+2)(n+1)a_{n+2} - a_{n-1} + 2a_n = 0.$$

We now need to solve both of these. In the first case there are two options, we can solve for a_2 or we can solve for a_0. Out of habit I'll solve for a_0. In the recurrence relation we'll solve for the term with the largest subscript as in previous examples.

$$n = 0 \qquad a_2 = -a_0$$

$$n = 1, 2, 3, \ldots. \qquad a_{n+2} = \frac{a_{n-1} - 2a}{(n+2)(n+1)}.$$

Notice that in this example we won't be having every third term drop out as we did in the previous example.

At this point we'll also acknowledge that the instructions for this problem are different as well. We aren't going to get a general formula for the a_n's this time so we'll have to be satisfied with just getting the first couple of terms for each portion of the solution. This is often the case for series solutions. Getting general formulas for the a_n's is the exception rather than the rule in these kinds of problems. To get the first four terms we'll just start plugging in terms until we've got the required number of terms. Note that we will already be starting with an a_0 and an a_1 from the first two terms of the solution so all we will need are three more terms with an a_0 in them and three more terms with an a_1 in them.

$$n=0 \quad a_2=-a_0.$$

We've got two a_0's and one a_1.

$$n=1 \qquad a_3=\frac{a_0-2a_1}{(3)(2)}=\frac{a_0}{6}-\frac{a_1}{3}.$$

We've got three a_0's and two a_1's.

$$n=2 \qquad a_4=\frac{a_1-2a_2}{(4)(3)}=\frac{a_1-2(-a_0)}{(4)(3)}=\frac{a_0}{6}+\frac{a_1}{12}.$$

We've got four a_0's and three a_1's. We've got all the a_0's that we need, but we still need one more a_1'. So, we'll need to do one more term it looks like.

$$n=3 \qquad a_5=\frac{a_2-2a_3}{(5)(4)}=\frac{a_0}{20}-\frac{1}{10}\left(\frac{a_0}{6}-\frac{a_1}{3}\right)=-\frac{31a_0}{60}+\frac{a_1}{30}.$$

We've got five a_0's and four a_1's. We've got all the terms that we need.

Now, all that we need to do is plug into our solution.-

$$\begin{aligned} y(x) &= \sum_{n=0}^{\infty} a_n(x+2)^n \\ &= a_0+a_1(x+2)a_2(x+2)^2+a_3(x+2)^3+a_4(x+2)^4+a_5(x+2)^5+\cdots \\ &= a_0+a_1(x+2)-a_0(x+2)^2+\left(\frac{a_0}{6}-\frac{a_1}{3}\right)(x+2)^3+\left(\frac{a_0}{6}+\frac{a_1}{12}\right)(x+2)^4 \\ &\quad +\left(\frac{31a_0}{60}+\frac{a_1}{30}\right)(x+2)^5+\cdots \end{aligned}$$

Finally collect all the terms up with the same coefficient and factor out the coefficient to get,

$$y(x)=a_0\left\{1-(x+2)^2+\frac{1}{6}(x+2)^3+\frac{1}{6}(x+2)^4-\frac{31}{60}(x+2)+\cdots\right\}+$$
$$a_1\left\{(x+2)-\frac{1}{3}(x+2)^3+\frac{1}{12}(x+2)^4+\frac{1}{30}(x+2)^5+\cdots\right\}.$$

That's the solution for this problem as far as we're concerned. Notice that this solution looks nothing like the solution to the previous example. It's the same differential equation, but changing x_0 completely changed the solution.

Let's work one final problem.

Example:Find the first four terms in each portion of the series solution around for the following differential equation.

$$(x^2+1)y''-4xy'+6y=0.$$

Solution: We finally have a differential equation that doesn't have a constant coefficient for the second derivative.

$$p(x)=x^2+1 \qquad p(0)=1\neq 0.$$

So is an ordinary point for this differential equation. We first need the solution and its derivatives,

$$y(x)=\sum_{n=0}^{\infty}a_nx^n \qquad y'(x)=\sum_{n=0}^{\infty}na_nx^{n-1} \qquad y''(x)=\sum_{n=0}^{\infty}n(n-1)a_nx^{n-2}.$$

Plug these into the differential equation.

$$(x^2+1)\sum_{n=0}^{\infty}n(n-1)a_nx^{n-2}-4x\sum_{n=0}^{\infty}na_nx^{n-1}+6\sum_{n=0}^{\infty}a_nx^n=0.$$

Now, break up the first term into two so we can multiply the coefficient into the series and multiply the coefficients of the second and third series in as well.

$$\sum_{n=2}^{\infty}n(n-1)a_nx^n+\sum_{n=2}^{\infty}(n-1)a_nx^{n-2}-\sum_{n=1}^{\infty}4na_nx^n+\sum_{n=1}^{\infty}6a_nx^n=0.$$

We will only need to shift the second series down by two to get all the exponents the same in all the series.

$$\sum_{n=2}^{\infty}n(n-1)a_nx^n+\sum_{n=2}^{\infty}(n+2)(n+1)a_{n+2}x^n-\sum_{n=1}^{\infty}4na_nx^n+\sum_{n=1}^{\infty}6a_nx^n=0.$$

At this point we could strip out some terms to get all the series starting at $n=2$, but that's actually more work than is needed. Let's instead note that we could start the third series at $n=0$ if we wanted to because that term is just zero. Likewise the terms in the first series are zero for both $n=1$ and $n=0$ and so we could start that series at $n=0$. If we do this all the series will now start at $n=0$ and we can add them up without stripping terms out of any series.

$$\sum_{n=0}^{\infty}[n(n-1)a_n+(n+2)(n+1)a_{n+2}4na_n+6a_n]x^n=0.$$

$$\sum_{n=0}^{\infty}[(n^2-5n+6)a_n+(n+2)(n+1)a_{n+2}]\,x^n=0$$

$$\sum_{n=0}^{\infty}[(n^2-2)(n-3)a_n+(n+2)(n+1)a_{n+2}]\,x^n=0.$$

Now set coefficients equal to zero.

$$(n-2)(n-3)a_n+(n+2)(n+1)a_{n+1}, n=0,1,2,...$$

Solving this gives,

$$a_{n+2}=-\frac{(n-2)(n-3)a_n}{(n+2)(n+1)}, \quad n=0,1,2,...$$

Now, we plug in values of *n*.

$$n=0: \qquad a_2=-3a_0$$

$$n=1: \qquad a_3=-\frac{1}{3}a_1$$

$$n=2: \qquad a_4=-\frac{0}{12}a_2=0$$

$$n=3: \qquad a_5=-\frac{0}{12}a_3=0.$$

Now, from this point on all the coefficients are zero. In this case both of the series in the solution will terminate. This won't always happen, and often only one of them will terminate. The solution in this case is,

$$y(x)=a_0\{1-3x^2\}+a_1\left\{x-\frac{1}{3}x^3\right\}.$$

EULER EQUATIONS

In this section we want to look for solutions to

$$ax^2y+bxy'+cy=0$$

around $x_0=0$. These type of differential equations are called Euler Equations.

Recall from the previous section that a point is an ordinary point if the quotients,

$$\frac{bx}{ax^2}=\frac{b}{ax} \quad \text{and} \quad \frac{c}{ax^2}$$

have Taylor series around. However, because of the *x* in the denominator neither of these will have a Taylor series around and so is a singular point. So, the method from the previous section won't work since it required an ordinary point.

However, it is possible to get solutions to this differential equation that aren't series solutions. Let's start off by assuming that $x>0$ (the reason for this will be apparent after we work the first example) and that all solutions are of the form,

$$y(x)=x'.$$

Now plug this into the differential equation to get,

$$ax^2(r)(r-1)x^{r-2} + bx(r)x^{r-1} + cx^r = 0$$

$$ar(r-1)x^r + b(r)x^r + cx^r = 0$$

$$(ar(r-1) + b(r) + c)x^r = 0$$

Now, we assumed that $x>0$ and so this will only be zero if,

$$ar(r-1) + b(r) + c = 0.$$

This equation is a quadratic in r and so we will have three cases to look at: Real, Distinct Roots, Double Roots, and Quadratic Roots.

REAL, DISTINCT ROOTS

There really isn't a whole lot to do in this case. We'll get two solutions that will form a fundamental set of solutions (we'll leave it to you to check this) and so our general solution will be,

$$y(x) = c_1 x^{r_1} + c_2 x^{r_2}.$$

Example: Solve the following IVP

$$2x^2 y'' + 3xy' - 15y = 0, \qquad y(1) = 0 \;\; y'(1) = 1.$$

Solution: We first need to find the roots

$$2r(r-1) + 3r - 15 = 0$$

$$2r^2 + r - 15 = (2r-5)(r+3) = 0 \;\;\Rightarrow\;\; r_1 = \frac{5}{2}, r_2 = -3.$$

The general solution is then,

$$y(x) = c_1 x^{\frac{5}{2}} + c_2 x^{-3}.$$

To find the constants we differentiate and plug in the initial conditions as we did back in the second order differential equations chapter.

$$y'(x) = \frac{5}{2} c_1 x^{\frac{5}{2}} - 3c_2 x^{-4}$$

$$\left.\begin{aligned} 0 &= y(0) = c_1 + c_2 \\ 1 &= y'(0) = \frac{5}{2}c_1 - 3c_2 \end{aligned}\right\} \;\;\Rightarrow\;\; c_1 = \frac{2}{11}, c_2 = -\frac{2}{11}.$$

The actual solution is then,

$$y(x) = \frac{2}{11} x^{\frac{5}{2}} - \frac{2}{11} x^{-3}.$$

With the solution to this example we can now see why we required $x>0$. The second term would have division by zero if we allowed $x=0$ and the first term would give us square roots of negative numbers if we allowed $x<0$.

DOUBLE ROOTS

This case will lead to the same problem that we've had every other time we've run into double roots (or double eigenvalues). We only get a single solution and will need a second solution. In this case it can be shown that the second solution will be,

$$y_2(x) = x^r \ln x$$

and so the general solution in this case is,

$$y(x) = c_1 x^r + c_2 x^r + \ln x = x^r (c_1 + c_2 \ln x).$$

We can again see a reason for requiring $x>0$. If we didn't we'd have all sorts of problems with that logarithm.

Example: Find the general solution to the following differential equation.

$$x^2 y'' - 7xy' + 16y = 0.$$

Solution: First the roots

$$r(r-1) - 7r + 16 = 0$$

$$r^2 - 8r + 16 = 0$$

$$(r-4)^2 = 0 \Rightarrow r = 4$$

So the general solution is then,

$$y(x) = c_1 x^4 + c_2 x^4 \ln x.$$

COMPLEX ROOTS

In this case we'll be assuming that our roots are of the form,

$$r_{1,2} = \lambda \pm \mu i \qquad r_{12} = \lambda \pm \mu i r_{1,2} = \lambda \pm \mu i.$$

If we take the first root we'll get the following solution.

$$x^{\lambda+\mu i} \qquad x^{\lambda+\mu i} x^{\lambda+\mu i}.$$

This is a problem since we don't want complex solutions, we only want real solutions. We can eliminate this by recalling that,

$$x^r = e^{\ln x^r} = e^{r \ln x}.$$

Plugging the root into this gives,

$$x^{\lambda+\mu i} = e^{(\lambda+\mu i)\ln x} = e^{\lambda \ln x} e^{\mu i \ln x}$$

$$= e^{\ln x^\lambda} (\cos(\mu \ln x) + i \sin(\mu \ln x))$$

$$= x^\gamma \cos(\mu \ln x) + i x^\lambda \sin(\mu \ln x)$$

Note that we had to use Euler formula as well to get to the final step. Now, as we've done every other time we've seen solution like this we can take the real part and the imaginary part and use those for our two solutions.

So, in the case of complex roots the general solution will be,

$$y(x) = c_1 x^{\lambda} \cos(\mu \ln x) + c_2 x^{\lambda} \sin(\mu \ln x) = x^{\lambda}(c_1 \cos(\mu \ln x) + c_2 x \sin(\mu \ln x)).$$

Once again we can see why we needed to require $x>0$.

Example: Find the solution to the following differential equation.

$$x^2 y'' + 3xy' + 4y = 0.$$

Solution: Get the roots first as always.

$$r(r-1) + 3r + 4 = 0$$

$$r^2 + 2r + 4 = 0 \quad \Rightarrow \quad r_{1,2} = -1 \pm \sqrt{3}i.$$

The general solution is then,

$$y(x) = c_1 x^{-1} \cos\left(\sqrt{3} \ln x\right) + c_2 x^{-1} \sin\left(\sqrt{3} \ln x\right).$$

We should now talk about how to deal with $x<0$ since that is a possibility on occasion. To deal with this we need to use the variable transformation,

$$\eta = -x.$$

In this case since $x<0$ we will get ç>0. Now, define,

$$u(\eta) = y(x) = y(-\eta).$$

Then using the chain rule we can see that,

$$u'(\eta) = -y'(x) \quad \text{and} \quad u''(\eta) = y''(x).$$

With this transformation the differential equation becomes,

$$a(-\eta)^2 u'' + b(-\eta)(-u') + cu = 0$$

$$a\eta^2 u'' + b\eta u' + cu = 0$$

In other words, since ç>0 we can use the work above to get solutions to this differential equation. We'll also go back to x's by using the variable transformation in reverse.

$$\eta = -x.$$

Let's just take the real, distinct case first to see what happens.

$$u(\eta) = c_1 \eta^{r_1} + c_2 \eta^{r_2}$$

$$y(x) = c_1(-x)^{r_1} + c_2(-x)^{r_2}.$$

Now, we could do this for the rest of the cases if we wanted to, but before doing that let's notice that if we recall the definition of absolute value,

$$|x| = \begin{cases} x & \text{if } x \geq 0 \\ -x & \text{if } x < 0 \end{cases}$$

we can combine both of our solutions to this case into one and write the solution as,

$$y(x) = c_1 |x|^{r_1} + c_2 |x|^{r_2}, \quad x \neq 0.$$

Note that we still need to avoid $x=0$ since we could still get division by zero. However this is now a solution for any interval that doesn't contain $x = 0$.

We can do likewise for the other two cases and the following solutions for any interval not containing $x = 0$,.

$$y(x) = c_1 |x|^{r_1} + c_2 |x|^{r} \ln|x|$$

$$y(x) = c_1 |x|^{\lambda} \cos(\mu \ln|x|) + c_2 |x|^{\lambda} \sin(\mu \ln|x|).$$

We can make one more generalization before working one more example. A more general form of an Euler Equation is,

$$a(x - x_0)^2 y'' + b(x - x_0)y' + cy = 0$$

and we can ask for solutions in any interval not containing $x = x_0$. The work for generating the solutions in this case is identical to all the above work and so isn't shown here.

The solutions in this general case for any interval not containing $x=a$ are,

$$y(x) = c_1 |x - a|^{r_1} + c_2 |x - a|^{r_2}$$
$$y(x) = |x - a|^{r} (c_1 + c_2 \ln|x - a|)$$
$$y(x) = |x - a|^{\lambda} (c_1 \cos(\mu \ln|x - a|) + c_2 \sin(\mu \ln|x - a|)).$$

Where the roots are solutions to

$$ar(r - 1) + b(r) + c = 0.$$

Example: Find the solution to the following differential equation on any interval not containing $x = -6$.

$$3(x + 6)^2 y'' + 25(x + 6)y' - 16y = 0.$$

Solution: So we get the roots from the identical quadratic in this case.

$$3r(r - 1) + 25r - 16 = 0$$

$$3r^2 + 22r - 16 = 0$$

$$(3r - 2)(r + 8) = 0 \quad \Rightarrow \quad r_1 = \frac{2}{3}, r_2 = -8.$$

The general solution is then,

$$y(x) = c_1 |x - a|^{\frac{2}{3}} + c_2 |x - a|^{-8}.$$

Chapter 17

Degree Theory

In this chapter we shall introduce an important tool for the study of non-linear equations, the degree of a mapping. We shall mainly follow the analytic development commenced by Heinz in and Nagumo.

Definition of the Degree of a Mapping

Let Ω be a bounded open set in $\mathbb{R}^n$ and let $f:\Omega \to \mathbb{R}^n$ a mapping which satisfies

1. $f \in C^1(W,\mathbb{R}^n) \cap C(W,\mathbb{R}^n)$,
2. $y \in \mathbb{R}^n$ is such that

   $y \notin f(\partial\Omega)$,
3. If $x \in \Omega$ is such that $f(x) = y$ then

   $f'(x) = Df(x)$

   is nonsingular.

Proposition: If f satisfies (1), (2), (3), then the equation

$$f(x) = y$$

has at most a finite number of solutions in Ω.

Definitoin: Let f satisfy (1), (2), (3). Define

$$d(f,\Omega,y) = \sum_{i=1}^{k} \operatorname{sgn}\det f'(x_i)$$

where $x_1,\ldots, x_k$ are solution of (4) in Ω and

$$\operatorname{sgn}\det f'(x_i) = \begin{cases} +1, \text{if } \det f'(x_i) > 0 \\ -1, \text{if } \det f'(x_i) < 0, i = 1,\ldots,k. \end{cases}$$

If equation has no solution in Ω we let d(f, Ω, y) to be defined for mappings $f \in C(\Omega, \mathbb{R}^n)$ which satisfy will coincide with the number just, deifined in case $f \in C(\Omega, \mathbb{R}^n)$ which satisfy will coincide with the number just define4d in case f satisfies. In order to give this definition in the more general case we need a sequence of auxiliary results.

The proof of the first, result, which follows readily be making suitable changes of variables, will be left as an exercise.

Lemma: Let $f:[0,\infty) \to \mathbb{R}$ be continuous and satisfy

$$\phi(0)=0, \phi(t) \equiv 0, t \geq r > 0, \int_{\mathbb{R}^n} f(|x|)dx = 1.$$

Let f satisfy the conditions. Then

$$d(f,\Omega,y) = \int_{\Omega} \phi(|f(x)-y|)\det f'(x)dx,$$

provided r is sufficiently small.

Lemma: Let f satisfy the condition and let $r > 0$ be such that $|f(x) - y| > r$, $x \in \partial\Omega$. *Let* $\phi:[0,\infty) \to \mathbb{R}$ *be continuous and satisfy:*

$$\phi(s) = 0, s = 0, r \leq s, \text{ and}$$

$$\int_0^{\infty} s^{n-1}\phi(s)ds = 0$$

Then

$$\int_{\Omega} \phi(|f(x)-y|)\det f'(x)dx$$

$$= \int_{\Omega'} \phi(|f(x)-y|)\det f'(x)dx$$

where Ω' is any domain with smooth boundary containing $\overline{\Omega}$.

We let

$$\psi(s) = \begin{cases} s^{-n} \int_0^s \rho^{n-1}\phi(\rho)d\rho, 0 < s < \infty \\ 0, s = 0. \end{cases}$$

Then ψ, so defined is a C^1 function, it vanishes in a neighborhood of 0 and in the interval (r,∞), Further ψ satisfies the differential equation

$$s\psi'(s)+n\psi(s)=\phi(s).$$

It follows that the functions

$$g^j(x)=\psi(|x|)x_j, j=1,...,n$$

belong to class C^1 and $g^j(x)=0,|x|\geq r,$

and furthermore that for j = 1,..., n the function $g^j(f(x)=y)$ are C^1 functions which vanish in neighborhood of $\partial\Omega$. If we denote by $a_{ji}(x)$ the cofactgor of the element $\frac{\partial f_i}{\partial x_j}$ in the Jacobian matrix $f'(x)$, it follows that

$$\text{div}\,(aj_1)(x)\, a_{j2}(x)..., a_{jn}(x)) = 0, \text{j} = 1,..., n.$$

We next define for i = 1,..., n

$$v_i(x)=\sum_{j=1}^{n}a_{ji}(x)g^j(f(x)-y)$$

and show that the function $v = (v_1, v_2,..., v_n)$ has the property that

$$divv=\phi(|f(x)-y|)\det f'(x),$$

and hence the result follows from the divergence theorem.

Lemma: Let f satisfy the equation and let $\phi:[0,\infty)\to\mathbb{R}$ be continuous, $\phi(0)=0, f(s)\equiv 0$ for $s\geq r,$ where

$$0<r\leq \min_{x\in\partial\Omega}|f(x)=y|, \int_{\mathbb{R}^n}\phi(|x|)dx=1.$$

Then for all such ϕ, the integrals

$$\int_{\Omega}\phi(|f(x)-y|)\det f'(x)dx$$

have a common value.

Proof: Let $\Phi=\{\phi\in C([0,\infty),\mathbb{R}):f(0)\ =0, f(s)\equiv 0, s\geq r\}$. Put

$$L\phi=\int_0^{\infty}S^{n-1}\phi(s)ds$$

$$M\phi=\int_{\mathbb{R}^n}\phi(|x|)dx$$

$$N\phi=\int_{\Omega}\phi(|f(x)-y|)\det f'(x)dx.$$

Then L, M, N are linear funcationals. It follows from Lemma 4 that $M\phi = 0$ and $N\phi = 0$, whenever $L\phi = 0$. Let $\phi_1, \phi_2, \in \Phi$ with $M\phi_1 = M\phi_2 = 1$, then

$$L((L\phi_2)\phi_1 - (L\phi_1)\phi_2) = 0.$$

It follows that

$$(L\phi_2)(M\phi_1) - (L\phi_1)(M\phi_2) = 0,$$

$$L\phi_2 - L\phi_1 = L(\phi_2 - \phi_1) = 0,$$

and

$$N(\phi_2 - \phi_1) = 0,$$

i.e.

$$N\phi_2 = N\phi_1$$

Lemma: Let f_1 and f_2 satisfy the condition and let $\varepsilon > 0$ be such that

$$| fi(x) - y | > 7\varepsilon, x \in \partial\Omega, \; i = 1, 2,$$

$$| f_1(x) - f_2(x) | < \varepsilon, \; x \in \Omega,$$

then

$$d(f_1, \Omega, y) = d(f_2, \Omega, y).$$

Proof: We may, without loss, assume that $y = 0$, since by Definition

$$d(f, \Omega, y) = d(f - y, \Omega, 0).$$

let $g \in C^1[0, \infty)$ be such that

$$g(s) = 1, 0 \le s \le 2\varepsilon$$

$$0 \le g(r) \le 1, 2\varepsilon \le r < 3\varepsilon$$

$$g(r) = 0, 3\varepsilon \le r < \infty.$$

Consider

$$f_3(x) = [1 - g(| f_1(x) |)] f_1(x) |) f_2(x),$$

then

$$f_3 \in C^1(W, \mathbb{R}^n) \cap C(\Omega \mathbb{R}^n)$$

and

$$| f_i(x) - f_k(x) | < \varepsilon, i, k = 1, 2, 3, x \in \Omega$$

$$| f_i(x) | > 6\varepsilon, x \in \partial\Omega, \; i = 1, 2, 3,$$

Let $f_i \in C[0,\infty)$, i = 1, 2 be continuous and be such that

$$\phi_1(t) = 0, 0 \le t \le 4\varepsilon,\ 5\varepsilon \le t, \le \infty$$

$$\phi_2(t) = 0, \varepsilon \le t < \infty,\ \phi_2(0) = 0$$

$$\int_{\mathbb{R}^n} \phi_i(|x|)dx = 1, \quad i = 1.,2.$$

We note that

$$f_3 \equiv f_1, \text{if } |f_1| > 3\varepsilon$$

$$f_3 \equiv f_2, \text{if } |f_1| > 2\varepsilon$$

Therefore

$$f_1(|f_3(x)|)\det f_3'(x) = f_1(|f_1(x)|)\det f_1'(x)$$

$$f_2(|f_3(x)|)\det f_3'(x) = f_2(|f_2(x)|)\det f_2'(x).$$

Integrating both sides of over Ω and using Lemmas 4 and 5 we obtain the desired conclusion.

Corollary: Let f satisfy conditions, then for $\varepsilon > 0$ sufficiently small if they are sufficientlyi "close" then they are the same degree. In order to extend this definition to a broader class of functions, namely those which we need a version of Sard's Theorem.

Lemma: If Ω is a bounded open set in $\mathbb{R}^n$, f satisfies the condition and

$$E = \{x \in \Omega : \det f'(x) = 0\}.$$

Then $f(E)$ does not contain a sphere fo the from $\{z : |z - y| < r\}$.

Corollary: Let

$$F = \{h \in \mathbb{R}^n : y + h \in f(E)\},$$

where E is given, then F is dense in a neighborhood of $0 \in \mathbb{R}^n$ and

$$f(x) = y + h, x \in \Omega, h \in F$$

implies that $f'(x)$ is nonsingular.

We thus conclude that for all $\varepsilon > 0$, sufficiently small, there exists $h \in F, 0 < |h| < \varepsilon$, such that $d(f, \Omega, y + h) = d(f - h, \Omega, y)$ is defined by Definition. It also follows from Lemma that for such h, $d(f, \Omega, y + h)$ is constant, This justifies the following definition.

Definition: Let f satisfy. We define

$$d(f,\Omega,y)=\lim_{\substack{h\to 0\\ h\in F}} d(f-h,\Omega,y).$$

Where F is given by abd d(f–h, Ω, ·y) is defined by Definition. We next assume that $f\in C(\Omega,\mathbb{R}^n)$ and satisfies. Then for ε > 0 sufficiently small there exists $g\in C^1(\Omega,\mathbb{R}^n)\cap C(\Omega,\mathbb{R}^n)$ such that $y\notin g(\partial\Omega)$ and

$$\| f-g\|=\max_{x\in\cap}| f(x)-g(x)<\varepsilon/4$$

and there exists, by Lemma, $\tilde{g}$ satisfying the such that $\| g-\tilde{g}\|<\varepsilon/4$ and if $\tilde{h}$ satisfies and $\|g-\tilde{h}\|<\varepsilon/4$, $\|\tilde{g}-\tilde{h}\|<\varepsilon/2$, then d(g , Ω, y) = d ($\tilde{h}$, Ω, y) provided ε is small enough. Thus f may be approximated by functions $\tilde{g}$ satisfying the and d ($\tilde{g}$, Ω, y) = constant provided $\|f-\tilde{g}\|$ is small enough. We therefore may define d(f, Ω, y) as follows.

Definition: Let $f\in C(\Omega,\mathbb{R}^n)$ be such $y\notin f(\partial\Omega)$. Let

$$d(f,\Omega,y)=\lim_{g\to f} d(g,\Omega,y)$$

where g satisfies the condition.

The number defined is called the Brouwer degree of f at y relative to Ω. It follows from our considerations above that d(f, Ω, y) is also given by formula for any ϕ which satisfies:

$$f\in C([0,\infty),\mathbb{R}),\phi(0)=0,\phi(s)\equiv 0,$$

$$s\geq r>0,\int_{\mathbb{R}^n} f(|x|)dx=1,$$

where $r<\min_{x\in d\partial\Omega}| f(x)-y+.$

Properties of the Brouwer Degree

We next proceed to establish some properties of the Brouwer degree of a mapping which will be of use in computing the degree and also in extending the definition to mappings defined in infinite dimensional spaces and in establishing globaql solution, results for parameter dependent equations.

Proposition: (Solution property) Let f $f\in C(\Omega,\mathbb{R}^n)$ be such that $y\notin f(\partial\Omega)$ and assume that $d(f,\Omega,y)\neq 0$. Then the equation

$$f(x)=y$$

has as solution in Ω.

The proof is a straightforward consequence of Definition 11 and is left as an exercise.

Proposition: (Continuity property) Let $f \in C(\Omega, \mathbb{R}^n)$ and $y \in \mathbb{R}^n$ b e such that $d(f, \Omega, y)$ is defined. Then there exists $\varepsilon > 0$ such that for all $g \in C(\Omega, \mathbb{R}^n)$ and $\hat{y} \in \mathbb{R}$ with $\|f - g\| + |y - \hat{y}| < \varepsilon$

$$d(f, \Omega, y) = d(g.\Omega, \hat{y})$$

The proof again is left as an exercise.

The proposition has the following important interpretation.

Remark: If we let

$$\{C = \{f \in C(\Omega, \mathbb{R}^n) : y \notin f(\partial\Omega)\}$$

then C is a metric space with metric ρ *defined by* $\rho(f, g) = \| f - g \|$. If we define the mapping $d : C \to \mathbb{N}$ (integers) by $d(f) = d(f, \Omega, y)$, then the theorem asserts that d is a continuous function from C to $\mathbb{N}$ (equipped with the discrete topology). Thus d will be constant on connected components fo *C*.

Using this remark one may establish the following results:

Proposition: (Homotody invariance property) Let $f, g \in C(\Omega, \mathbb{R}^n) f(x)$ and $f(x) \neq y$ for $x \in \partial\Omega$ and $h : [a,b] \times \Omega \to \mathbb{R}^n$ be continuous such that $h(t, x) \neq$ y. $(t,x) \in [a,b] \times \partial\Omega.$ Further let $h(a,x) = f(x), h(b,x) = g(x) x \in \Omega.$ Then

$$d(f, \Omega, y) = d(g, \Omega, y);$$

more generally, d(h,Ω,y); = constant for $a \leq t \leq b.$ the next corollary may be viewed as an extenstion fo Rouche;s theorem concerning the equal number of zeros of certain analytic function. This extenstion will be the content of one of the exercises at the end of this chapter.

Corollary: Let $f \in C(\Omega, \mathbb{R}^n)$ be such that d(f, Ω, y) is defined. Let $g \in C(\Omega, \mathbb{R}^n)$ be such that $| f(x) - g(x) | < | f(x) - y |, x \in \partial\Omega.$ Then $d(f, \Omega, y) = d(g, \Omega, y)$

Proof: For $0 \leq t \leq 1$ and $x \in \partial\Omega$ we have that

$$| y - tg(x) - (1-t) f(x) | = | (y - f(x)) - t(g$$

$$(x) - f(x))]$$

$$> | y - f(x) | - t | g(x) - f(x) |$$

$$> 0 \text{ since } 0 \leq t \leq 1,$$

Hence $h:[0,1]\times\Omega\to R^n$ given by $h(t, x,) = tg(x) + (1-t)(x)$ satisfics the conditions of Proposition 15 and the dfdfdkknclustion follows from the proposition. As an immediate corollary we have following:

Corollary: Assume that f and g are mapping such that $f(x) = g(x),. x\, x\in\partial\Omega$, then $d(f,\Omega,y) = d(g,\Omega,y)$ if the degree is defined, i.e. the degree only depends on the boundary data.

Proposition: (Additivity property) Let Ω be a bounded open set which is the union of m disjoint open sets $\Omega_1...\Omega_m$, and let $f\in C(\Omega,\mathbb{R}^n)$ and $y\in\mathbb{R}^n$ be such that $y\notin f(\partial\Omega_i), i=1,...,m$ Then

$$d(f,\Omega,y) = \sum_{i=1}^{m} d(f,\Omega_i,y)$$

Proposition: (Excision property) Let $f\in C(\Omega,\mathbb{R}^n)$ and let K be a closed subset of Ω such that $y\notin f(\partial\Omega\cup K)$. Then

$$d(f,\Omega,y) = d(f,\Omega\setminus K,y).$$

Proposition: (Cartesion property) Assume that $\Omega=\Omega_1\times\Omega_2$ is a bounded open set in $\mathbb{R}^n$ with Ω_1 open in $\mathbb{R}^p$ and Ω_2 open in $\mathbb{R}^p$, $p+q=n$. For $x\in\mathbb{R}^n$ write $f_1 = (x_1), f_2\,(x_2))$ where $f_1:\Omega_1\to\mathbb{R}^p : f_2:\Omega_2\to\mathbb{R}^q$ are continuous. Suppose $y=(y_1,y_2)\in\mathbb{R}^n$ is such that $y_i\notin f_i(\partial\Omega_i), i=1,2.$ Then

$$\begin{aligned}
d(f,\Omega,y) &= \sum_{x\in f^{-1}(y)} \operatorname{sgn}\det f'(x) \\
&= \sum_{x\in f^{-1}(y)} \operatorname{sgn}\det\begin{pmatrix} f_1'(x_1) & 0 \\ & 0 & f_2'(x_2)\end{pmatrix} \\
&= \sum_{\substack{x\in f^{-1}(y_i)\\ i=1,2}} \operatorname{sgn}\det f_1'(x_1)\operatorname{sgn}\det_2'(x_2) \\
&= \prod_{i=1}^{2}\sum_{x_i\in f_i^{-1}(y_i)} \operatorname{sgn}\det f_i'(x_i) \\
&= d(f_1,\Omega_1,y_1)d(f_2,\Omega_2,y_2).
\end{aligned}$$

To give an example to show how the above properties may be used we prove Borsuk's theorem and the Brouwer fixed point theorem.

The theorems of Borsuk and Brouwer

Theorem: (Borsuk) Let Ω be a symmetric bounded open neighborhood of $0 \in \mathbb{R}^n$ (i.e. if $x \in \Omega$, then $-x \in \Omega$ and let $f \in C(\Omega, \mathbb{R}^n)$ be an odd mapping (i.e. $f(x) = -f(-x)$). Let $0 \notin f(\partial\Omega)$, then $d(f, \Omega, 0)$ is an odd integer.

Proof: Choose $\varepsilon > 0$ such that $B_i(0) = \{x \in \mathbb{R}^n : |x| < \varepsilon\} \subset \Omega$. Let $\alpha : \mathbb{R}^n \to \mathbb{R}$ be a continuous function such that

$$\alpha(x) \equiv 1, |x| \leq \varepsilon, \alpha(x) = 0,$$

$$x \in \partial\Omega, 0 \leq \alpha(x) \leq 1, x \in \mathbb{R}^n$$

and put

$$g(x) = \alpha(x)x + (1 - \alpha(x))f(x)$$

$$h(x) = \frac{1}{2}[g(x) - g(-x)],$$

then $h \in C(W, \mathbb{R}^n)$ is odd and $h(x) + f(x)$, $x \in \partial\Omega, h(x) = x, |x| < \varepsilon$. *Thus by Corollary 16 and the remark following it*

$$d(f, \Omega, 0) = d(h, \Omega, 0)$$

On the other hand, the excision and additivity property imply that

$$d(f, \Omega, 0) = d(h, B_\varepsilon(0), 0) + d(h, W \setminus \overline{B_\varepsilon(0)}, 0),$$

where $\partial B_\varepsilon(0)$ has been excised. It follows from Definition that d(h,B_ε(0),0 = 1, and it therefore suffices to show that (letting $\Theta = \Omega \setminus B_\varepsilon(0)) d(h.\Theta, 0)$ is an even integer.

Since Θ is symmetric, $0 \notin \Theta, 0 \notin h(\partial\Theta)$, and h is odd one may show that there exists $h \in C(\Omega, \mathbb{R}^n)$ which is odd, $h(x) = x \in \partial\Theta$ and is such that $\tilde{h}(x) \neq 0$, for those $x \in \Theta$ with $x_n = 0$. Hence

$$d(h, \Theta, 0) = d(\tilde{h}, \Theta, 0) = d(\tilde{h}, \Theta) \setminus \{x : x_n = 0\} 0)$$

Where we have used the excision property. We let

$$\Theta_1 = \{x \in \Theta : x_n > 0\} = d(\tilde{h}, \Theta_1, 0) \neq d(\tilde{h}, \Theta_2, 0)$$

Since $\Theta_2 = \{x: -x \in \Theta_1\}$ and $\tilde{h}$ is odd one may now employ approximation arguments to conclude that $d(\tilde{h},\Theta_1,0)=d(\tilde{h},\Theta_2,0)$, and hence conclude that the integer given by is even.

Theorem: (Brouwer fixed point theorem) Let $f \in C(\Omega,\mathbb{R}^n), \Omega = \{x \in \mathbb{R}^n : |x| < 1\}$, be such that $f:\Omega \to \Omega$. Then *f* has a fixed in Θ, i.e. there exists $x \in \Omega$ such that $f(x) = x$.

Proof: Assume *f* has no fixed points in $\partial\Omega$. Let $h(t,x) = x - tf(x), \leq t \leq 1$. Then $h(t,x) \neq 0 - 0 \leq t \leq 1, x \in \partial\Omega$ and thus $d(h(t,o),\Omega,0) = d(h(0,0),\Omega,0) = 1$ it follows from the solution property that the equation $x - f(x) = 0$ has solution in Ω.

Theorem remains valid if thc unite balll of $\mathbb{R}^n$ is replaced by any set home omorphic to the unite ball (replaqce *f* by $g^{-1}fg$ where g is the homeomorphisim). That the Theorem also remains valid if the unit ball is replaced in arbitrary compact convex set (or a set homeomorphic to it) may be proved using the extenstion theorem of Dugundji.

Completely Continuous Perturbations of the Identity in a Banach Space

Definition of the degree

Let E be a real Banach space with nomr ||.|| and let $\Omega \subset E$ be a bounded openj set. Let $F:\Omega \to E$ be continuous and let $F(\Omega)$ be contained in a finite dimensional subspace of *E*. the mapping

$$f(x) = x + F(x) = (id + F)(x)$$

is called a fintie dimensional perturbation of the identity in *E*.

Let *y* be a point in *E* and let *E* be a finite dimensional subspace of *E* containing *y* and $F(\Omega)$ and assume

$$y \notin f(\partial\Omega)$$

Select basisi $e_1,\ldots,e_n$ of $\tilde{E}$ and define the linear homeomorphism T: $\tilde{E} \to \mathbb{R}^n$ by

$$T\left(\sum_{i=1}^{n} C_i e_i\right) = (C_1 \ldots C_n) \in \mathbb{R}^n.$$

Consider the mapping

$$TFT^{-1} : T(\Omega \cap E) \to \mathbb{R}^n,$$

then, since $y \notin f(\partial\Omega)$, it follows that

$$T(y) \notin f(T^{-1}(T(\partial\Omega \cap \tilde{E})).$$

Let Ω_0 denote the bounded open set $(\Omega \cap \tilde{E})$ in $\mathbb{R}^n$ and let $\tilde{f} = TFT^{-1}, y_0 = T(y)$. Then $d(\tilde{f}, \Omega_0, y_0)$ is defined.

It is an easy exercise in linear algebra to show that the following lemma holds.

Lemma: The integer $d(\tilde{f}, \Omega_0, y_0)$ calculated above is independent of the choice the finite dimensional space $\tilde{E}$ containing y and F(Ω) and the choiece of basis of $\tilde{E}$.

We hence may define

$$d(f, W, y) = d(\tilde{f}, \Omega_0, y_0),$$

Where $\tilde{f}, \Omega_0, y_0$ are as above

Recall that a mapping $F : \Omega \to E$ is called completely continuous if F is continuous and $F(\Omega)$ is precompact in E(i.e. $F(\Omega)$ is compact. More generally if Δ is any subset of E and $F \in C(D,E)$, then F is called completely continuous if $F(V)$ is precompact for any bounded subset V of D.

We shall now demonstrate that if $f = id + F$, with F completely continuous (f is called a completely continuous pertubation of the identity) and $y \notin f(\partial\Omega)$, then an integer valued function d(*f*, Ω, *y*) (the Schauder degree of *f* at *y* relative to Ω) may be defined having much the same properties as the Brouwer degree, In order to accomplish this we need the following lemma.

Lema: Let $f : \Omega \to E$ be a completely continuous perturbation of the identity and let $y \notin f(\partial\Omega)$. Then there exists an integer d with the following property: If $h : \Omega \to E$ is a finite dimensional continuous perturbation of the identity such that

$$\sup_{x \in \Omega} \| f(x) - h(x) \| < \inf_{x \in \partial\Omega} \| f(x) - y \|,$$

then $y \notin h(\partial\Omega)$ and d(*h*, Ω, *y*).

Proof: That $y \notin h(\partial\Omega)$ follows from (26). Let h_1 and h_2 be any two such that $k(t, x)$ it follows that

$$\begin{aligned} \|f(x) - y\| &= \|f(x) - th_1(x) - (1-t)h_2(x)\| \\ &= \| t(f-(x) - h_1(x)) + (1-t)(f(x) - h_2(x))\| \\ &\le t\|f(x) - h_1(x)\| + (1-t)\|f(x) - h_2(x))\| \\ &< \inf_{x \in \partial\Omega} \|f(x) - y\|, \end{aligned}$$

from which follows that $x \notin \partial\Omega$. Let $\tilde{E}$ be a finite dimensinal subspace containing y and $(h_i - \text{id})(\Omega)$, then $(h_{i,}\ \Omega, y) = d(Th_i t^{-1}, T(\Omega \cap \tilde{E}), T(y))$, where T is given as above, Then $(k(,.) - \text{id})\ (\Omega)$ is contained in $\tilde{E}$ and $Tk(t,.)T^{-1}(x) \neq T(y), x \in T(\partial\Omega \cap \tilde{E})$. Hence we may use the homotopy invariance property of Brouwer degree to conclude that

$$d(Tk(t,.)^{-1}, T(\partial\Omega \cap \tilde{E}), T(y)) = \text{constant},$$

i.e.,

$$d(h_1\Omega, y) = d(h_2\Omega, y).$$

It follows from Lemma that if a finite dimensional perturbation of the identity h exists which satisfies, then we may define d(f, Ω, y) = d, where d is the integer whose existence follows from this lemma, In order to accomplish this we need an approximation results.

Let M be compact subset of E. Then fro every $\varepsilon > 0$ there exists a finite covering of M by spheres of radius ε with centers at $y_1,...y_n \in M$. Define $\mu i : M \to [0,\infty)$ by

$$\mu i(y) = \varepsilon - \| y - y_i \|, \text{ if} \| y - y_i \| \leq \varepsilon\ = 0, \text{ otherwise}$$

and let

$$\lambda_i(y) = \frac{\mu_i(y)}{\sum_{i=1}^{n} \mu_j(y) = 1.}\ 1 \leq i \leq n.$$

Since not all μ_i vanish simultaneously, $\lambda_i(y)$ is non-negative and continuous on M and further $\sum_{i=1}^{n} \lambda_i(y) = 1$. The operator P_ε defined by

$$P_\varepsilon(y) = \sum_{i=1}^{n} \lambda_i(y) y_i$$

is called a Schauder projection operator on M determined by ε, and $y_1,...,y_n$. Such an operator has the following properties:

Lemma: $P\varepsilon : M \to co\{y_1,...y_n\}$ (the convex hull of $y_1,... y_n$) is continuous.

1. $P_\varepsilon(M)$ is contained in a finite dimensional subspace of E.
2. $\|P_\varepsilon y - y\| \leq \varepsilon, y \in M$.

Lemma: $f : \Omega \to E$ be a completely perturbation of the identity. Let $y \notin f(\partial\Omega)$. Let $\varepsilon > 0$ be such that $\varepsilon < ind_{x \in \partial\Omega} \| f(x) - y \|$. Let P_ε be be Schauder projection operator determined by ε and points $\{y_1,...,y_n\} \subset (f - id(\Omega)$. Then

$$\text{d}(\text{id} + P_\varepsilon F, \Omega, y) = d,$$

where d is the integer whose existence is established by Lemma 24.

Proof: The properties of Pε (of Lemma 25) imply that for each $x \in \Omega$

$$\| P_i F(x) - F(x) \| \leq e \inf_{x \varepsilon \partial\Omega} \| f(x) - y \|,$$

and the mapping $\text{id} + P_\varepsilon F$ is a continuous finite dimensional perturbation of the identity.

Definition: The interger d whose existence has been established by Lemma 26 is called the Leray-Schauder degree of *f* relative to Ω and the point *y* and is denoted by d(*f*, Ω, *y*).

Properties of the Degree

Proposition: The Leray-Schauder degree ahs the solution, continuity, homotopy invariance, additivity, and excision properties similar to the Brouwer degree; the Cartesiona product formula also holds.

Proof: (Solution property). We let Ω be a bounde open set in *E* and $f : \Omega \to E$ be a completely continuous perturbation of the identity, *y* a point in *E* with $y \notin f(\partial\Omega)$ and $d(f, \Omega, y) \neq 0$. We claim that the equation *f*(*x*) =*y* has a solution in *E*.

To see this let $\{e_n\}_{n=1}^{\infty}$ be a decreasing sequence of positive numbers with $\lim_{n \to \infty} \varepsilon_n = 0$ and $\varepsilon_1 < \inf_{x \in \partial W} \| f(x) - y \|$. Let $P\varepsilon_i$ be associated Schauder projection operators. Then

$$d(f, \Omega, y) = \text{d}(\text{id} + P\varepsilon_n F, \Omega, y), n = 1, 2, \ldots,$$

where

$$\text{d}(\text{id} + P\varepsilon_n F, \Omega, y) = \text{d}(\text{id} + T_n P_{\varepsilon_n} F T_n^{-1}, T_n(\Omega \cap \tilde{E}_n), T_n(y)),$$

and the spaces $\tilde{E}_n$ and finite dimensional. Hence the solution property of brouwer degree implies the existence of a solution $z_n \in T_n(\Omega \cup \tilde{E}_n)$ of the equation

$$z_n \in T_n P_{\varepsilon_n} F T_n^{-1}(z) = T(y),$$

or equivalently a solution $x_n \in \Omega \tilde{E}_n$ of

$$x + P_{\varepsilon_n} F(x) = y$$

The sequence $\{x_n\}_{n=1}^{\infty} \subset \Omega)$, thus, since *F* is completely continuous there exists a subsequence $\{x_n\}_{n=1}^{\infty}$. such that $F(x_{n_i}) \to u \in F(\Omega)$. We relabel the subsequence and call it again $\{x_n\}_{n=1}^{\infty}$. Then

$$\| x_n - x_m \| = \| P_{\varepsilon_n} F(x_n) - P_{\varepsilon_m} F(x_m) \|$$
$$\leq \| P_{\varepsilon_n} F(x_n) - F(x_n) \| + \| P_{\varepsilon_m} F(x_m) - F(x_m) \|$$
$$+ \| F(x_n) - F(x_m) \| < \varepsilon_n + \varepsilon_m + \| F(x_n) - F(x_m) \|$$

We let $\varepsilon > 0$ be given and choose N such that $n, m \geq N$ imply $\varepsilon_n, \varepsilon_m < \varepsilon/3$ and $\| F(x_n) - F(F_m) \| < \varepsilon/3$. Thus $\| x_n - x_m \| < \varepsilon$. The sequence $\{x_n\}_{n=1}^{\infty}$ therefore is a Cauchy sequence, hence has a limit, say x. It now follows that $x \in \Omega$ and solves the equation $f(x) = y$, hence, since $y \notin f(\partial\Omega)$ we have that $x \in \Omega$ and solves the equation $f(x) = y$, hence, since $y \notin f(\partial\Omega)$ we have that $x \in \Omega$.

Borsuk's Theorem and Fixed Point Theorems

Theorem (Borsuk's theorem): Let Ω be a bounded symmetric open neighborhood of 0 $\in E$ and let $f : \Omega \to E$ be completely continuous odd perturbation of the identity with $0 \notin f(\partial\Omega)$. Then $d(f, \Omega, 0)$ is an odd integer.

Proof: Let $\varepsilon > 0$ be such that $\varepsilon < \inf_{x \in \partial\Omega} \| f(x) \|$ and let P_ε be an associated Schauder projection operator. Let $f_\varepsilon = \mathrm{id} + P_\varepsilon f$ and put $h(x) = 1/2[f_\varepsilon(x) - f_\varepsilon(-x)]$ Then h is a finite dimensional perturbation of the identity which is odd and

$$\| h(x) - f(x) \| \leq \varepsilon.$$

Thus $d(f, \Omega, 0) = d(h, \Omega \cap \tilde{E}), 0)$.

On the order hand $T(\Omega \cap \tilde{E})$ is a symmetric bounded open neighborhood of $0 \in T(\Omega \cap \tilde{E})$ and $The T^{-1}$ is an odd mapping, hence the result follows from Theorem.

We next establish extensions of teh Brouwer fixed point theorem to Banach spaces.

Theorem: (Schauder fixed point theorem) Let K be a compact convex subset of E and let $F : K \to K$ be continuous. Then F has a fixed point in K.

Proof: Since K is compact there exists $r > 0$ such that $K \subset B_r(0) = \{x \in E : \| x \| < r\}$ Using the Extension Theorem we may continuously exten F to $\overline{B_r(0)}$. Call the extension $\tilde{F}$. *Then* $\tilde{F}\overline{(B_r(0))} \subset \mathrm{co}F(K) \subset K$, where co $F(K)$ is the convex hull of $F(K)$, i.e. the smallest convex set containing $F(K)$. Hence $\tilde{F}$ is completely continuous. Consider the homotopy

$$k(t, x) = x - t\tilde{F}(x), 0 \leq t \leq 1.$$

Since $t\tilde{F}(K) \subset tK \subset B_r(0), 0 \leq t \leq 1,\ x \in \overline{B_r(0)}$, it follows that $k(t,x) \neq 0,\ 0 \leq t \leq 1, x \in \partial B_r(0)$. Hence by the homotopy invariance property of the Leray–Schauder degre

$$\text{constant} = d(k(t,.), B_r(0)0)\ = d(id, B_r(0), 0) = 1.$$

The solution property of degree therefore implies that the equation

$$x - \tilde{F}(x) = 0$$

has solution in $x \in B_r(0)$ in K, i.e

$$x - F(x) = 0$$

In many applications the mapping F is know to be completely containuous but it is difficult to find a compact convex set K such that $F : K \to K$, whereas closed convex sets K having this property are more easily found. In such as case the following result may be applied.

Theorem: (Schauder) let K is be a closed, bounded, convex subset of E and let F be completely continuous mapping such that $F : K \to K$, . Then F has a fixed point in K.

PROOF: Let $\tilde{K} = \overline{\text{co}\overline{F(K)}}$. Then since $\overline{F(K)}$ is compact it follows from a theorem of Mazur that $\tilde{K}$ is compact and $\tilde{K} \subset K$. Thus $F : \tilde{K} \to \tilde{K}$ and F has fixed point in $\tilde{K}$ by Theorem. Therefore F has a fixed point in K.

THE METHOD OF LYPUNOV-SCHMIDT

In this topic we shall develop an approach to bifurcation theory whic, is one of the original approaches to thetheory. The results obtained, since the implicit function thorem plays an important role, will be of a local nature.

We first develop the method of Liapunov-Schmmidt and then use it to obtain a local bifurcation result. We then use this result in several examples. Later in the text it will be used to derive a Hopf bifurcation theorem.

Splitting Equations

Let X and Y be real Banach spaces and let F be a mapping

$$F : X \times \mathbb{R} \to Y$$

and let F satisfy the following conditions:

$$F : (0, \lambda) = 0 \forall \lambda \in \mathbb{R},$$

and

$$F \text{ is } C^2 \text{ in a neighborhood of } \{0\} \times \mathbb{R}.$$

We shall be interested in obtaining existence of nontrivial solutions (i.e. $u \neq 0$) of the equation

$$F(u,\lambda) = 0$$

We call λ_0 a bifurcation value or $(0, \lambda_0)$ a bifurcation point for provided every neighborhood of $(0, \lambda_0)$ in $X \times \mathbb{R}$ contains solutions with u ≠ 0.

It then follows from the implicit function theorem that the following holds.

Theorem: If the point $(0, \lambda_0)$ is a bifurcation point for the equation

$$F(u,\lambda) = 0$$

then the Frechet derivative $F_u(0,\lambda_0)$ cannot be a linear homeomorphism of X to Y.

The types of linear operators $F_u(0, \lambda_0)$ we shall consider are so- called Fredholm operators.

Definition: A linear operator $L : X \to Y$ is called a Fredholm operator Provided:

1. The kernel of *L*, ker*L*, *is finite dimensional.*
2. The range of L, imL, is closed in Y.
3. The cokernel of L, cokerL, is finite dimensional.

The following lemma which is a basic result in functional analysis will be important for the development to follow, its proof may be found in any standard text.

Lemma: Let $F_u(0,\lambda_0)$ be a Fredholm operator from *X* to *Y* with kernel *V* and cokernel *Z*. Then there exists a closed subspace *W* of *X* and a closed subspace *T* of *Y* such that

$$X = V \oplus W$$

$$Y = Z \oplus T$$

The operator $F_u(0,\lambda_0)$ restricted to w, $F_u(0,\lambda_0) \mid w$: $W \to T$, is bijective and since *T* is closed it has continuous inverse. Hence $F_u(0,\lambda_0)\,|_W$ is a linear homeomorphism of *W* onto *T*.

We recall that *W* and *Z* are not uniquely given.

Using Lemma 3 we may now decompose every u ∈ X and *F* uniquely as follows:

$$u = u_1 + u_2, \;\; u_1 \in V, u2 \in W,$$

$$F = F_1 + F_2, \;\; F_1 : X \to Z, F_2 : X \to T.$$

Hence equation is equivalent to the system of equations

$$F_1 = (u_1, u_2, \lambda) = 0,$$

$$F_2 = (u_1, u_2, \lambda) = 0.$$

We next let $L = F_u(0, \lambda_0)$ and using a Taylor expansion we may write

$$F(u, \lambda) = F(0, \lambda_0) + F_u(0, \lambda_0)u + N(u, \lambda)$$

or $$Lu + N(u, \lambda) = 0$$

where $$N : X \times \mathbb{R} \to Y.$$

Using the decompostion of X we may write equation as

$$Lu_2 + N(u_1 + u_2, \lambda) = 0.$$

Let $Q : N \to Z$ and $I - Q : Y \to T$ be projections determined by the decomposition, then equation (10) implies that

$$QN(u, \lambda) = 0.$$

Since by Lemma 3, $L|_w : W \to T$ has an inverse $L^{-1} : T \to W$ we obtain from equation the equivalent system

$$u_2 + L^{-1}(1 - Q)N(u_1 + u_2, \lambda) = 0$$

We note, that since Z is finite dimensional, equation is an equation in a finite dim ensional space, hence if u_2 can be determined as a function of u_1 and λ, this equation will be a finite set of equations in finitely many variables ($u_1 \in V$, which is alos assumed finte dimensional!). Concerning equation we have the following result.

Lemma: Assume that $F_u(0, \lambda_0)$ is a Fredholm operator with W nontrivial. Then there exist $\varepsilon > 0.\delta > 0$ and a unique solution $u_2(u_1, \lambda)$ of equation defined for $|\lambda - \lambda_0| + \|u_1\| < \varepsilon$ with $\|u_2(u_1, \lambda)\| < \delta$. This function solves the equation $F_2(u_1, u_2(u_1, \lambda), \lambda) = 0$.

Proof: We employ the implicit function theorem to analyze equation. That this may be done follows from the fact that at $u_1 = 0$ and $\lambda = \lambda_0$ equation has the unique solution $u_2 = 0$ and the Frechet derivative at this point with respect to $u_2 = 0$ and the Frechet derivative at this point with respect to u_2 is simply the identity mapping on W.

Hence, using Lemma, we will have nontrivial solutions of equation once we can solve

$$F_1(u_1, u_2(u_1, \lambda), \lambda) = QF(u_1 + u_2(u_1, \lambda), \lambda) = 0$$

for u_1, whenever $|\lambda - \lambda_0| + \|u_1\| < \varepsilon$ This latter set of equations, usually referred to as the set of bifurcation equations, is, even though a finte set of equations in fintely many unknowns, the more difficulty part in the solution of equation

The next section present situation where these equations may be solved.

Remark: We note that in the above considerations at no point was it required that l be a one dimensinal parameter.

Bifurcation at a Simple Eigenvalue

In this section we shall consider the analysis of the bifurcation equation in the particular case that the kernel V and the cokernel Z of $F_u(0,\lambda_0)$ both have dimension 1. We have the following theorem.

Theorem: In the notation of the previous section assume that the kernel V and the cokernel Z of $F_u(0,\lambda_0)$ both have dimension 1. Let $V = \text{span}\{\phi, \}$and let Q be a projection of Y onto Z. furthermore assume that the second Frechet derivative $F_{u\lambda}$ *satisfies*

$$QF_{u\lambda}(0,\lambda_0)(\phi,1)\neq 0.$$

Then $(0,\lambda_0)$ is a bifurcation point and there exists a unique curve

$$u = u(\alpha), \lambda = \lambda(\alpha),$$

defined for $\alpha \in \mathbb{R}$ in a neighborhood of 0 so that

$$u(0) = 0, u(\alpha) \neq 0, \alpha \neq 0, \lambda(0) = \lambda_0$$

and

$$F(u(\alpha), \lambda(\alpha)) = 0.$$

Proof: Since V is one dimensional $u_1 = \alpha\phi$. Hence for $|\alpha|$ small and l near λ_0 there exists a unique $u_2(\alpha, \lambda)$ such that

$$F_2(\alpha\phi, u_2(\alpha,\lambda), \lambda) = 0$$

We hence need to solve for $\lambda = \lambda\ (\alpha)$ in the equation

$$QF(\alpha\phi + u_2(\alpha,\lambda), \lambda) = 0$$

We let $\eta = \lambda - \lambda_0$ and define

$$g(\alpha,\mu) = QF(\alpha\phi + u_2(\alpha,\lambda), \lambda) = 0$$

Then g maps sa neighborhood of the origin of $\mathbb{R}^2$ into $\mathbb{R}$. Using Taylor's theorem we may write

$$F(u,\lambda)F_u u + \frac{1}{2}\{F_{uu}(u,u) + 2F_{u\lambda}(u,\lambda) \ + F_{\lambda\lambda}(\mu,\mu)\} + R,$$

where R contains higher order remainder terms and all Frechet derivatives above are evaluated at 0, λ_0

Because we have that F_λ and $F_{\lambda\lambda}$ in the above are the zero operators, hence, by applying Q to we obtain

$$QF(u,\lambda)=\frac{1}{2}\{Qf_{uu}(u,u)+2QF_{u_\lambda}(u,\mu)\}+QR,$$

and for $\alpha \neq 0$

$$\frac{g(\alpha,u)}{\alpha}=\frac{1}{2}\left\{QF_{uu}(\phi+\frac{u_2(\alpha,\lambda)}{\alpha},\alpha\phi+u_2(\alpha,\lambda))\right.$$

$$\left.+2QF_{u\lambda}(\phi+\frac{u_2(\alpha,\lambda)}{\alpha},\mu)\right\}+\frac{1}{\alpha}QR.$$

It follows from Lemma that the term $\frac{u_2(\alpha,\lambda)}{\alpha}$ is bounded for α in a heighborhood of 0. The remainder fromula of Taylor's theorem implies a similar statement for the term $\frac{1}{\alpha}QR$. Hence

$$h(\alpha,\mu)=\frac{g(\alpha,\mu)}{\alpha}=O(\alpha), \text{ as } \alpha\to 0.$$

We note that in fact $h(0,0)=0$, and

$$\frac{\partial h(0,0)}{\partial\mu}=QF_{u\lambda}(o,\lambda_0)(f,1)\neq 0.$$

We hence conclude by the implicit function thorem that there exists a unique function $\mu=\mu(\alpha)$ defined in a neighborhood of 0 such that

$$h(\alpha,\mu(\alpha))=0.$$

We net set

$$u(\alpha)=\alpha\phi+u_2(\alpha,\lambda_0+\mu(\alpha)),\lambda=\lambda_0+\mu(\alpha).$$

This proves the theorem.

The following example will serve to illustrate the theorem just established.

Example: The point (0, 0) is a bifurcation point for the ordinary differential equation

$$u''+l(u+u^3)=0$$

subject to the periodic boundary conditions

$$u(0) = u(2\pi), u'(0) = u'(2\pi).$$

To see this, we choose

$$X = C^2[0, 2\pi] \cap \{u : u(0) = u(2\pi), u'(0)$$
$$= u'(2\pi), u''(0) = u''(2\pi)\},$$
$$Y = C[0, 2\pi] \cap \{u : u(0) = u(2\pi)\},$$

both equipped with the usual noms, and

$$F : X \times \mathbb{R} \to Y$$
$$(u, \lambda) \mapsto u'' + l(u + u^3).$$

Then F belongs to class C^2 with Frechet derivative

$$F_u(0, \lambda_0)u = u'' + \lambda_{0u.}$$

This linear operator has nontrivial kernel whenever $\lambda_0 = n^2$, n = 0, 1,.... The kernel being one dimensional if and only if $\lambda_0 = 0$.

We see that h belongs to the range of $F_u(0,0)$ if and only if $\int_0^{2\pi} h(s)ds = 0$, and hence the cokernel will have dimension I also. A Projection $Q : Y \to Z$ then is given by $Qh = \frac{1}{2\pi}\int_0^{2\pi} h(s)ds$. Computing further, we find that $F_{u\lambda}(0,0)(u,\lambda) = \lambda u$, and hence, since we may choose $\phi = 1, F_{u\lambda}(0,0)(1,1) = 1$. Applying Q we get Q1 = 1. We may therefore conclude by Theorem that equation has a solution u satisfying the boundary conditions which is of the form

$$u(\alpha) = \alpha + u_2(\alpha, \lambda(a))..$$

$$u' = Au \mid g(t,u)$$

and show how to construct Lyapunov functionals to test the stability of the trivial solution of this system. The type of Lyapunov functional we shall be looking for are of the form

$$v(x) = x^T Bx,$$

where B is a constant $N \times N$ matrix, i.e. we are looking for v as a quadratic form. If u is a solution of then

$$\frac{dv^*}{dt} = \frac{dv(t, u(t))}{dt}$$
$$= u^T(A^T B + BA)u + g^T(t,u)Bu + u^T Bg(t,u).$$

Hence, given, A if B can be found so that $C = A^T B + BA$ has certain definniteness properties, then the results of the trivial solution. To proceed along these lines we need some linear algebra results.

Proposition: Let A be a constant $N \times N$ matrix having the property that for any eigenvalue λ of A, $= \lambda$ is not an eigenvalue. Then for any $N \times N$ matrix C. there exists a unique $N \times N$ matrix B such that $C = A^T B + BA \cdot$

Proof: On the space of $N \times N$ matrices define the bounded linear operator L by

$$L(B) = A^T B + BA \cdot$$

Then L may be viewed as a bounded linear operator of $\mathbb{R}^{N \times N}$ to itself, hence it will be a bijection provided it does not have 0 as an eigenvalue, Once we show the latter, the result follows. Thus let μ be an eigenvalue of L, i.e., there exists a nonzero matrix B such that

$$L(B) = A^T B +, \mu B$$

Hence

$$A^T B + B(A - \mu I) = 0$$

From this follows

$$B(A - \mu I)^n = (-A^T)^n B,$$

for any integer $n \geq 1$, hence for any polynomial p

$$Bp(A - \mu I)^n = p(-A^T)B.$$

Since, on the other hand if F and G are two matrices with no common eigenvalues, there exists a polynomial p such that $p(F) = I, p(G) = 0$, implies that $A - \mu I$ and $_{-}A^T$ must have a common eigenvalue,. From which follows taht μ is the sum of two eigenvalues of A, which, by hypothesis connot equal 0.

This proposition has the following corollary.

Corollary: Let A be a constant $N \times N$ matrix. Then for any $N \times N$ matrix C, there exists $\mu > 0$ and a unique $N \times N$ matrix B such that $2\mu B + C = A^T B + BA$.

Proof: Let

$$S = \{\lambda \in \mathbb{C} : \lambda = \lambda_1 + \lambda_2\},$$

where λ_1 and λ_2 are eigenvalues of A. Since S is a finite set, there exists $r_0 > 0$. such that $\lambda(\neq 0) \in S$ implies that $|\lambda| > r_0$. Choose $0 < \mu \leq r_0$ and consider the matrix $A_1 = A - \mu I$.

We may now apply Proposition 21 to the matrix A_1 and find for a given matrix C, a unique matrix B such that $C = A_1^T B + BA_1$, i.e. $2\mu BC = A^T B + BA$.

Corollary: Let A be a constant $N \times N$ matrix haivng the property that all eigenvalues l of A have negative real parts. Then for any negative definite $N \times N$ matrix B such that $C = A^T B + BA.$

Proof: Let C be a negative definite matrix and let B be given by Proposition, which may be applied since all eigenvalues of A have negative real part. Let $v(x) = x^T Bx$, and let u be a solution of $u' = Au, u(0) = x_0 \neq 0$ Then

$$\frac{dv^*}{dt} = \frac{dv(u(t))}{dt} = u^T (A^T B + BA)u$$

$$= u^T Cu \leq -\mu \,|\, u \,|^2,$$

since C is negative definite. Since $\lim_{t\to\infty} u(t) = 0$ (all eigenvalues of A have negative real part!), it follows that $\lim_{t\to\infty} v(u(t)) = 0$. We also have

$$v(u(t)) \leq v(x_0) - \int_0^t \mu \,|\, u(s) \,|^2 \, ds,$$

from which follows that $v(x_0) > 0$. Hence B is positive definite.

The next corollary follows from stability theory equations and what has just been discussed.

Corollary: A necessary and sufficient condition that an $N \times N$ matrix A have all of its eigenvalues with negative real part is that there exists a unique positive definite matrix B such that

$$A^T B + BA = -I.$$

We next consider the nonlinear problem with

$$g(t,x) = o(|\, x \,|),$$

uniformly with respect to $t \in [t_0, \infty)$, and show that for certain types of matrices A the trivial solution of the perturbed system has the same stability property as that of the unperturbed problem. The class of matrices we shall consider is the following.

Definition: We call an $N \times N$ matrix A critical if all its eigenvalues have non-positive real part and there exists at least one eigenvalue with zero real part. We call it noncritical otherwise.

Theorem: Assume A is a noncritical $N \times N$ matrix and let g satisfy. Then the stability behavior of the trivial solution of is uniformly asymptotically stable if all eigenvalues of A have negative real part and it is unstable if A has an eigenvalue with positive real part.

Proof: Assume all eigenvalues of A have negative real part. By the above exists a unique positive definite matrix B such that

$$A^T B + BA = -I.$$

Let $v(x) = x^T Bx$. Then v is positive definite and if u is a solution it satisfies

$$\frac{dv^*}{dt} = \frac{dv(u(t))}{dt}$$

$$= -[u(t)|^2 | g^T(t,u)Bu + u^T Bg(t,u).$$

Now

$$| gT(t,u)bu + u^T Bg(t,u) | \le 2 | g(t,u) || B || u |.$$

Choose $r > 0$ such that $| x | \le r$ implies

$$| g(t,u) | \le \frac{1}{4} | B |^{-1} | u |,$$

then

$$\frac{dv^*}{dt} \le -\frac{1}{2} | u(t) |^2,$$

as long as $| u(t) | \le r$.

Let A have an eigenvalue with positive real part. Then there exists $\mu > 0$ such that $2\mu < | \lambda_i + \lambda_j |$, for all eigenvalues λ_i, λ_j of A, and a unique matrix B such that

$$A^T B + BA = 2\mu B = I,$$

as follows from Corollary 22. We note that B connot be positive definite nor positive semidefinite for otherwise we must have, letting $v(x) = x^T Bx$,

$$\frac{dv^*}{dt} = 2\mu v^*(t) - | u(t) |^2,$$

or

$$e^{-2\mu t} v^*(t) - v^*(0) = -\int_0^t e^{-2\mu s} | u(s)^2 \, ds$$

i.e.

$$0 \le v^*(0) - \int_0^t e^{-2\mu s} |u(s)|^2 \, ds$$

for any solution, u, contradicting the fact that solution u exist for which

$$\int_0^t e^{-2\mu s} |u(s)|^2 \, ds$$

becomes unbounded as $t \to \infty$. Hence there exists $x_0 \neq 0$, of arbitrarily small norm, so that $v(x_0) < 0$. Let u be a solution. If the trivial solution were stable, then $|u(t)| \le r$ for some $r > 0$. Again letting $v(x) = x^T Bx$ we obtain

$$\frac{dv^*}{dt} = 2\mu v^*(t) - |u(t)|^2 + g^T(t,u) \quad Bu + u^T Bg(t,u)$$

We can choose r so small that

$$2|g(t,u)||B||u| \le \frac{1}{2}|u|^2, |u| \le r,$$

hence

$$e^{-2\mu t} v^*(t) - v^*(0) \le -\frac{1}{2}\int_0^t e^{-2\mu s} |u(s)|^2 \, ds,$$

i.e.

$$v^*(t) \le e^{-2\mu t} v^*(0) \to -\infty,$$

contradicting that $v^*(t)$ is bounded for bounded u. Hence u connot stay bounded, and we have instability.

It it is the case that the matrix A is a critical matrix, the trivial solution of the linear system may still be stable or it may be unstable.

In either case, one may construct examples, where the trivial solution of the perturbed problem has either the same or opposite stability behavior as the unperturbed system.

Chapter 18

Stability Theory

In the earlier Chapter we studied in detials linear and perturbed linear systems of differential equaions. In the case of constant or periodic coefficients we found criteria which describe the asymptotic behavior of solutioins. In this chapter we shall consider similar problems for general systems of the form

$$u' = f(t,u),$$

where

$$f : \mathbb{R} \times \mathbb{R}^N \to \mathbb{R}^N$$

is a continuous functions.

If $u : [t_0, \infty) \to \mathbb{R}^N$ is a given solution of the equation, then discussing the behavior of another solution v of this equation relative to the solution u, i.e.e discussing thte behavior of the difference $v - u$ is equivlent to studying the behavior of the solution $z = v - u$ of the equation

$$z' = f(t, z + u(t)) = f(t, u(t)),$$

relative to the trivial solution $z \equiv 0$. Thus we may, without loss in geenrality, assume taht has the trivial solution as a reference solution, i.e.

$$f(t,0) \equiv 0,$$

an assumption we shall henceforth make.

Stability Concepts

There are various stability concepts which are important in the asymptotic behavior of systems of differential equations. We shall discuss here some of them and their interrelatioinships.

Definition: We say that the trivial solution is:

1. Stable (s) on $[t_0,\infty)$, if for every $\varepsilon > 0$ there exists $\delta > 0$ such that any solution v of equation with $|v(t_0)| < \delta$ exists on $[t_0,\infty)$ and satisfies $|v(t)| < \varepsilon, t_0 \le t < \infty$;
2. Asymptotically stable (a.s) on $[t_0,\infty)$, if if is stable and $\lim_{t\to\infty} v(t) = 0$, where v is as in (i);
3. Unstable (us), if it is not stable;
4. Uniformly stable (u.s.) on $[t_0,\infty)$, if for every $\varepsilon > 0$ there exists $\delta > 0$ such that any solution v of (1) with $|v(t)_1| < \delta, t_1 \ge t_0$ exists on $[t_0,\infty)$ and satisfies $|v(t)| < \varepsilon, t_1 \le t < \infty$;
5. Uniformly asymptotically stable (u.a.s), if it is uniformly stable and there exists $\delta > 0$ such that for all $\varepsilon > 0$ there exists T = T(ε) such that any solution v with (v) $|v(t)_1| < \delta$, $t_1 \ge t_0$ exists on $[t_0,\infty)$ and satisfies $|v(t)| < \varepsilon, t_1 + T \le t < \infty$;
6. strongly stable (s.s) on $[t_0,\infty)$, if for every $\varepsilon > 0$ there exists $\delta > 0$ such that any solution of with $|v(t)_1| < \delta$ exists on $[t_0,\infty)$ and satisfies $|v(t)| < \varepsilon, t_0 \le t < \infty$;

Stability of Linear Equations

In the case of a linear system $(A \in C(\mathbb{R} \to £(\mathbb{R}^N, \mathbb{R}^N)))$

$$u' = A(t)u,$$

a particular stability property of any solutions is equivlent to that stability property of the trivial solution. Thus one may ascribe that property to the equation and talk about the equation being stable, uniformly stable, etc.

The stability concepts may be expressed in terms of conditions imposed on a fundamental matraix Φ.

Theorem: Let F be a fundamental matrix solution. Then equation is:

1. Stable if and only if there exists $K > 0$ such that
$$|\Phi(t)| \le K, t_0 \le t < \infty;$$
2. Uniformly stable if and only if there exists $K > 0$ such that
$$|\Phi(t)\Phi^{-1}(s)| \le K, t_0 \le s \le t < \infty;$$
3. Strongly stable if and only if there exists $K > 0$, such that
$$|\Phi(t)| \le K\,|\Phi^{-1}(t)| \le K, t_0 \le t < \infty;$$

4. Asymptotically stable if and only if

$$\lim_{t\to\infty} |\Phi(t)| = 0;$$

5. Uniformly asymptotically stable if and only if there exist $K > 0$, a > 0 such that

$$|\Phi(t)\Phi^{-1}(s)| \le Ke^{-\alpha(t-s)}, t_0 \le s \le t\infty.$$

Proof: We shall demonstrate the last part of the theorem and leave the demonstration of the remaining parts as an exercise. We may assume without loss in generality that $\Phi(t_0) - I$, the $N \times N$ identity matrix, since conditions will hold for any fundamental matrix if and only if if they hold for a particular one.

Thus, let us assume that holds. then the second part of the theorem guarantees that the equation is uniformly stable, Moreover, for each $\varepsilon, 0 < \varepsilon < K$, if we put $T = -\frac{1}{\alpha}\log\frac{\varepsilon}{K}$, then for $\xi \in \mathbb{R}^N, |\xi| \le 1$, we have $|\Phi(t)\Phi^{-1}(t_1)\xi| \le Ke^{-\alpha(t-t_1)}$, if $t_1 + T < t$. Thus we have uniform asymptotic stability. Conversely, if the equation is uniformly asymptotically stable, then there exists $\delta > 0$ and fro all e$\varepsilon, < \varepsilon < \delta$, there exists $T = t(\varepsilon) > 0$ such that if $|\xi| < \delta$, then

$$\Phi(t)\Phi^{-1}(t_1)\xi| < \varepsilon, t \ge t_1 + T, t_1 \ge t_0$$

In particluar

$$|\Phi(t+T)\Phi^{-1}(t)\xi) < \varepsilon, |\xi| < \delta,$$

or

$$|\Phi(t+T)\Phi^{-1}(t)\frac{\xi}{\delta}| < \frac{\varepsilon}{\delta}$$

and thus

$$|\Phi(t+T)\Phi^{-1}(t)| \le \frac{\varepsilon}{\delta} < 1, t \ge t_0.$$

Further more, since we have uniform stability,

$$|\Phi(t+h)\Phi^{-1}(t)| \le K, t_0 \le t, 0 \le h \le T.$$

If $t \ge t_1$, we obtain for some integer *n*, that $t_1 + nT \le t < t_1 + (n+1)T$, and

$$|\Phi(t)\Phi^{-1}(t_1)| \le |\Phi(t)\Phi^{-1}(t_1 + nT)|\ |\Phi(t_1 + nT)\Phi^{-1}(t_1)|$$

$$\le K|\Phi(t_1 + nT)\Phi^{-1}(t_1 + (n-1)T)|$$

$$\ldots|\Phi(t_1 + T)\Phi^{-1}(t_1)|$$

$$\le K\left(\frac{\varepsilon}{\delta}\right)^n$$

Letting $-\alpha = -\frac{1}{T}\log\frac{\varepsilon}{\delta}$, we get

$$|\Phi(t)\Phi^{-1}(t_1)| \leq Ke^{-n\alpha T} = Ke^{-n\alpha T}e^{-aT}\frac{\delta}{\varepsilon}$$

$$< K\frac{\delta}{\varepsilon}e^{-\alpha(t-t_1)}, t \geq t_1 \geq t_0.$$

In the case that the matrix A is independent of t one obtains the following corollary.

Corollary: Equation is stable if and only if every eigenvalue of A has nonpositive real part and those with zero real part are semisimple. It is strongly stable if and only if all eigenvalues of A have zero real part and are semisimple. It is asymptotically stable if and only if all eigenvalues have negative real part.

Using the Abel-Liouville formula, we obtain the following result.

Theorem: Equation is unstable whenever

$$\limsup_{t\to\infty} \int_{t_0}^{t} \text{trace}A(s)ds = \infty.$$

If the equation is stable, then it is strongly stable if and only if

$$\liminf_{t\to\infty} \int_{t_0}^{t} \text{trace}A(s)ds > -\infty.$$

Additional stability criteria for linear systems abound. The following concept of the measure of a matrix due to Lozinskii and Dahlquist is particularly useful in numerical computations. We provide a brief discussion.

Definition: For an $N \times N$ matrix A we define

$$\mu(A) = \lim_{h\to 0+} \frac{|I + hA| - |I|}{h},$$

where $|.|$ is a marix norm induced by a norm $|.|$ in $\mathbb{R}^N$ and I is the $N \times N$ identity matrix.

We have the following proposition;

Proposition: For any $N \times N$ matrix $A, m(A)$ exists and satisfies:

1. $\mu(\alpha A) = \alpha\mu(A), \alpha \geq 0;$
2. $|\mu(A)| \leq |A|;$
3. $\mu(A + B) \leq \mu(A) + \mu(B);$
4. $|\mu(A) - \mu(B)| \leq |A - B.$

If u is a solution of equation, then the function $r(t) = |u(t)|$ has a right derivation $r_+^{'}(t)$ at every point t for every norm $|.|$ in $\mathbb{R}^N$ and $r_+^{'}(t)$ satisfies

$$r_+^{'}(t) - \mu(A(t))r(t) \leq 0.$$

Using this inequality we obtain the following proposition.

Proposition: Let $A:[t_0,\infty) \to \mathbb{R}^{N\times N}$ be a continuous matrix and let u be a solution on $[t_0,\infty)$ Then

$$| u(t)e^{-\int_{t_0}^{t} \mu(A(s))ds} \quad ,t_0 \leq t < \infty$$

is a nonincreasing function of t and

$$| u(t)e^{\int_{t_0}^{t} \mu(-A(s))ds} \quad ,t_0 \leq t < \infty$$

is a nondecreasing fucntion of t. Furthermore

$$| u(t_0)e^{-\int_{t_0}^{t} \mu(-A(s))ds} \leq| u(t) | u(t_0) |^{\int_{t_0}^{t} \mu(A(s))ds}$$

This proposition has, for constant matraices A, the immediate eorollary.

Corollary: For any $N \times N$ constant matrix A, the following inequalities hold.

$$e^{-t\mu(-A)} \leq| e^{tA)} |\leq e^{t\mu(A)}$$

The following theorem provides stability criteria for the system in terms of conditions on the measure of the coefficient matrix. Its proof is again left as an exercise.

Theorem: The system is:

1. Unstable, if

$$\liminf_{t\to\infty} \int_{t_0}^{t} \mu(-A(s))ds = -\infty;$$

2. Stable, if

$$\limsup_{t\to\infty} \int_{t_0}^{t} \mu(A(s))ds < \infty;$$

3. Asymptotically stable, if

$$\lim_{t\to\infty} \int_{t_0}^{t} \mu(A(s))ds = -\infty;$$

4. Uniformly asymptotically stable, if

$$\mu(A(t)) \leq -\alpha < 0, t \geq t_0.$$

5. Stability of Nonlinear Equations

In this section we shall consider stability properties of nonlinear equations of the form

$$u' = A(t)u + f(t,u),$$

where

$$f : \mathbb{R} \times \mathbb{R}^N \to \mathbb{R}^N$$

is a continuous function with

$$|f(t,x)| \leq \lambda(t)|x|, x \in \mathbb{R}^N,$$

where λ is some positive continuous function and A is a continuous $N \times N$ matrix defined on $\mathbb{R}$. The results to be discussed are consequences of the variatioin of constants formula and the stability theorem. We shall only present a sample of results.

Throughout this section $\Phi(t)$ will denote a fundamental matrix solutioin of the homogeneous (unperturbed) linear problem. We hence know that if u solution of equation , then u satisfies the integral equation

$$u(t) = \Phi(t)\left(\Phi - 1(t_0)u(t_0) + \int_{t_0}^{t} \Phi^{-1}(s)f(s,u(s))ds\right).$$

The following theorem will have uniform and strong stability as a consequence.

Theorem: Let f satisfythe condition with $\int^{\infty} \gamma(s)ds < \infty$. Further assume that

$$|\Phi(t)\Phi^{-1}(s)| \leq K, t_0 \leq s \leq t, \infty.$$

Proof: If u is a solution of it also satisfies. Hence

$$|u(t)| \leq K|u(t_1)| + \int_{t_1}^{t} \gamma(s)|u(s)ds, t \geq t_1,$$

and

$$|u(t)| \leq K|u(t_1)e^{K\int_{t_1}^{t}\gamma(s)ds} \leq L|u(t_1)|, t \geq t_1,$$

where

$$L = Ke^{\int_{t_1}^{t}\gamma(s)ds}$$

To prove the remaining part of the theorem, we note from that

$$|u(t)| \leq |\Phi(t)|\left(|\Phi^{-1}(t_0)u(t_0)\ | + |\int_{t_0}^{t} \Phi^{-1}(s)f(s,u(s))ds|\right)$$

$$+ |\int_{t_1}^{t} \Phi(t)\Phi^{-1}(s)f(s,u(s))ds|$$

Now

$$\left|\int_{t_1}^{t} \Phi(t)\Phi^{-1}(s)f(s,u(s))ds\right| \le KL\,|\,u(t_0)\,|\;\int_{t_1}^{\infty}\gamma(s)ds.$$

This together with the fact $\Phi(t) \to 0$ as $t \to \infty$ completes the ***Proof:***

Remark: It follows from Theorem 4 that the conditions of the theorem imply that the perturbed system is uniformly (asymptotically) satble whenever the unperturbed system is uniformly (asymptotically) stable. the conditions of Theorem also imply that the zero solution of the perturbed system is strongly stable, whenever the unperturbed system is strongly stable.

A further strability results is the following. Its proof is delegated to the exercises.

Theorem: Assume that there exist K > 0, a > 1 such that

$$|\,\Phi(t)\Phi^{-1}(s)\,| \le Ke^{-\alpha(t-s)}, t_0 \le s \le t < \infty.$$

and *f* satisfies with $\gamma < \dfrac{\alpha}{K}$ a constant. Then any solution *u* of is defined for $t \ge t_0$ and satisfies

$$|\,u(t)\,| \le Ke^{-\beta(t-t_1)}\,|\,u(t_1)\,|, tk.t_1 \ge t_0.$$

where $\beta = \alpha - \gamma K > 0$

Remark: It again follows from Theorem 4 that the conditions of the above theorem imply that the perturbed system is uniformly asymptotically stable

LYAPUNOV STABLITY

In this section we shall introduce some geometric ideas, which were first formulated by Poincare and Lyapunov, about the stability of constant soslutions of systems of the form

$$u' = f(t,u),$$

where

$$f : \mathbb{R} \times \mathbb{R}^N \to \mathbb{R}^N$$

is a continuous function. As observed earlier, we may assume that $f(t, 0) = 0$ and we discuss the stabilityl properties of the trivial solution $u \equiv 0$.

The geometric ideas amount to constructing level surfaces in $\mathbb{R}^N$ which shrink to 0 and which have the property that orbits associated with cross these level surfaces transversally toward the orgin, thus being guided to the origin. To illustrate, let us consider the following example.

$$x' = ax - y + kx(x^2 + y^2)$$

$$y' = x - ay + ky(x^2 + y^2)$$

where α is a constant |a| < 1. Obviously $x = 0 = y$ is a stationary solution. Now consider the family of curves in $\mathbb{R}^2$ given by

$$v(x, y) = x^2 = 2axy + y^2 = \text{constant},$$

a family of ellipses which share the origin as a common center. It we consider an orbit $\{(x(t), y(t)) : t \geq t_0\}$ associated, then as t varies, the orbit will cross members of the above family of ellipses. Computing the scalar product of the tangent vector of an orbit with the gradient to an ellpise we find:

$$\nabla v(x', y') = 2k(x^2 + y^2)(x^2 + y^2 - 2axy),$$

which is clearly negative, whenever k is (and positive if k is). Of course, we may view $v(x(t), y(t))$ as a norm of the point $(x(t), y(t))$ and $\nabla v(x', y') = \frac{d}{dt} v(x(t), y(t))$. Thus, if $k < 0, v(x(t), y(t))$ will be a strictly decreasing function and hence should approach a limit as $t \to \infty$. That this limit must be zero follows by an indrect argument. That is, the orbit tends to the origin and the zero solution attracts all orbits, and hence appears asymptotically stable.

LYAPUNOV FUNCTIONS

Let

$$v : \mathbb{R} \times \mathbb{R}^N \to \mathbb{R}, v(t, 0) = 0, t \in \mathbb{R}$$

be a continuous functional. We shall introduce the following terminology.

Definition: The functional v is called:

1. Positive definite, if there exists a continuous nondecreasing function f:

$$[0, \infty) \to [0, \infty), \text{ with } \phi(0) = 0, \phi(r) \neq 0, \quad r \neq 0 \text{ and}$$

$$\phi(|x|) \leq v(t, x), x \in \mathbb{R}^N, t \geq t_0;$$

2. Redially unbounded, if it is positive definite and

$$\lim_{r \to \infty} f(r) = \infty;$$

3. Decrescent, if it is positive definite and there exists a continuous increasing fucntion $\psi : [0, \infty) \to [0, \infty)$, with $\psi(0) = 0$,

$$\psi(|x|) \geq v(t, x), x \in \mathbb{R}^N, t \geq t_0.$$

We have the following stability criteria.

Theorem: Let there exist a positive definite functional v and $\delta_0 > 0$ such that for every solution u with $|u(t_0)| \le \delta_0$, the function $v^*(t) = v(t, u(t))$ is nonincreasing with respect to t, and hence

$$v^*(t) \le v^*(t_0) = v(t_0, u_0).$$

Therefore

$$\phi(|u(t)|) \le v(t, u(t)) = v^*(t) \le v^*(t_0) = v(t_0, u_0) < \phi(\varepsilon).$$

Since ϕ is nondecreasing, the result follows.

If v is also decrescent we obtain the stronger result.

Theorem: Let there exist a positive definite functional v which is decrescent and $\delta_0 > 0$ such that for every solution u with $|u(t_1)| \le \delta_0$, $t_1 \ge t_0$ the function $v^*(t) = v(t, u(t))$ is nonincreasing with respect to t, then the trivial solution of is unniformly stable.

Proof: We have already shown that the trivial solution is stable. Let 0 < e $0 < \varepsilon \le \delta_0$ be given and let $\delta = \psi^{-1}(\phi(\varepsilon))$. Let u be a solution of (20) with $u(t_1) = u_0$ with $|u_0| < \delta$. Then

$$\phi(|u(t)|) \le v(t, u(t)) = v^*(t) \le v^*(t_1)$$
$$= v(t_1, u_0) \le \phi(|u_0|)$$
$$< \psi(\delta) = \phi(\varepsilon).$$

The result now follows from the monotonicity assumptioin of ϕ.

As an example we consider the two dimensional system

$$x' = a(t)y \,|\, b(t)x(x^2 + y^2)$$
$$y' = -a(t)x \,|\, b(t)y(x^2 + y^2)$$

where the coefficient functions a and b are continuous with $b \le 0$. We choose $v(x, y) = x^2 + y^2$, then, since

$$\frac{dv^*}{dt} = \frac{\partial v}{\partial t} + \nabla v . f(t, u).$$

we obtain

$$\frac{dv^*}{dt} = 2b(t)(x^2 + y^2)^2 \le 0.$$

Since v is positive definite and decrescent, it follows from Theorem 18 that the trivial solution is uniformly stable.

We next prove an instability theorem.

Theorem: Assume there exists a continuous functional

$$v:\mathbb{R}\times\mathbb{R}^N \to \mathbb{R}$$

with the properties:

1. There exists a continuous increasing function $\psi:[0,\infty)\to[0,\infty)$, such that $\psi(0)=0$ and

$$|v(t,x)|\leq y(|x|);$$

2. For all $\delta>0$, and $t_1\geq t_0$, there exists $x_0, |x_0|<\delta$ such that $v(t_1,x_0)<0$;

3. If u is any solution of (23) with $u(t)=x$, then

$$\lim_{h\to 0+}\frac{v(t+h,u(t+h))-v(t,x)}{h}\leq -c(|x|),$$

where c is a continuous increasing function with $c(0)=0$. Then the trivial solution is unstable.

Proof: Assume the trivial solution is stable. then for every $\varepsilon>0$ there exists $\delta>0$ such that any solution u with $|u(t_0)|<\delta$ exists on $[t_0,\infty)$ and satisfies $|u(t)|<\varepsilon, t_0\leq t<\infty$. Choose $x_0, |x_0|<\delta$ such that $v(t_0,x_0)<0$ and let u be a solution with $u(t_0)=x_0$. Then

$$v(t,u(t))|\leq\psi(|u(t)|)\leq\psi(\varepsilon).$$

The third property above implies that $v(t,u(t))$ is nonincreasing and hence for $t\geq t_0$,

$$v(t,u(t))\leq v(t_0,x_0)<0.$$

Thus

$$|v(t_0,x_0)|\leq\psi(|u(t)|)$$

and

$$\psi^{-1}(|v(t_0,x_0)|)\leq|u(t)|.$$

We therefore have

$$v(t,u(t))\leq v(t_0,x_0)-(t-t_0)c(\psi^{-1}(|vt_0,x_0)|),$$

and thus

$$\lim_{t\to\infty} v(t,u(t))=-\infty, \text{ contradicting.}$$

We finally develop some asymptotic stabililty criteria.

Theorem: Let there exist a positive definite functional $v(t, x)$ such that

$$\frac{dv^*}{dt} = \frac{dv(t,u(t))}{dt} \leq -c(v(t,u(t)),$$

for every solution u with $|u(t_0)| \leq \delta_0$, with c a continuous increasing function and $c(0) = 0$. Then the trivial solution is asymptotically stable. If v is also decrescent. Then the asymptotic stability in uniform.

Proof: The hypotheses imply that the trivial solution is stable, as was proved earlier. Hence, if u is a solution with $|u(t)| \leq \delta_0$, then

$$v_0 = \lim_{t \to \infty} v(t,u(t)),$$

exists. An easy indirect argument shows that $v_0 = 0$. Hence, since v is positive definite,

$$\lim_{t \to \infty} \phi(|u(t)|) = 0,$$

implying that

$$\lim_{t \to \infty} |u(t)| = 0$$

since ϕ is increasing. The proof that the trivial solution is uniformly asymptotically stable, whenever v is also decrescent, is left as an exercise.

In the next section we shall employ the stabililty criteria just derived, to, once more study perturbed linear systems, where the associated linear system is a constant coefficient system.

This we do by showing how appropriate Lyapunov functionals may be constructed. The construction is an ecercise in linearalgebra.

Perturbed linear systems

Let A be a constant $N \times N$ matrix and let $g:[t_0,\infty) \times \mathbb{R}^N \to \mathbb{R}^N$ be a continuous function such that

$$g(t,x) = \alpha(|x|),$$

uniformly with respct to $t \in [t_0,\infty)$. We shall consider the equation

$$u' = Au \mid g(t,u)$$

and show how to construct Lyapunov functionals to test the stability of the trivial solution of this system.

The type of Lyapunov functional we shall be looking for are of the form

$$v(x) = x^T Bx,$$

where B is a constant $N \times N$ matrix, i.e. we are looking for v as a quadratic form. If u is a solution of then

$$\frac{dv^*}{dt} = \frac{dv(t,u(t))}{dt}$$

$$= u^T(A^T B + BA)u + g^T(t,u)Bu + u^T Bg(t,u).$$

Hence, given, A if B can be found so that $C = A^T B + BA$ has certain definniteness properties, then the results of the trivial solution. To proceed along these lines we need some linear algebra results.

Proposition: Let A be a constant $N \times N$ matrix having the property that for any eigenvalue λ of A, $= \lambda$ is not an eigenvalue. Then for any $N \times N$ matrix C. there exists a unique $N \times N$ matrix B such that $C = A^T B + BA$.

Proof: On the space of $N \times N$ matrices define the bounded linear operator L by

$$L(B) = A^T B + BA.$$

Then L may be viewed as a bounded linear operator of $\mathbb{R}^{N \times N}$ to itself, hence it will be a bijection provided it does not have 0 as an eigenvalue, Once we show the latter, the result follows. Thus let μ be an eigenvalue of L, i.e., there exists a nonzero matrix B such that

$$L(B) = A^T B +, \mu B$$

Hence

$$A^T B + B(A - \mu I) = 0$$

From this follows

$$B(A - \mu I)^n = (-A^T)^n B,$$

for any integer $n \geq 1$, hence for any polynomial p

$$Bp(A - \mu I)^n = p(-A^T)B.$$

Since, on the other hand if F and G are two matrices with no common eigenvalues, there exists a polynomial p such that $p(F) = I, p(G) = 0$, implies that $A - \mu I$ and $-A^T$ must have a common eigenvalue,. From which follows taht μ is the sum of two eigenvalues of A, which, by hypothesis connot equal 0.

This proposition has the following corollary.

Corollary: Let A be a constant $N \times N$ matrix. Then for any $N \times N$ matrix C, there exists $\mu > 0$ and a unique $N \times N$ matrix B such that $2\mu B + C = A^T B + BA$.

Proof: Let

$$S = \{\lambda \in \mathbb{C} : \lambda = \lambda_1 + \lambda_2\},$$

where λ_1 and λ_2 are eigenvalues of A. Since S is a finite set, there exists $r_0 > 0$. such that $\lambda(\neq 0) \in S$ implies that $|\lambda| > r_0$. Choose $0 < \mu \leq r_0$ and consider the matrix $A_1 = A - \mu I$. We may now apply Proposition 21 to the matrix A_1 and find for a given matrix C, a unique matrix B such that $C = A_1^T B + BA_1$, i.e. $2\mu BC = A^T B + BA$.

Corollary: Let A be a constant $N \times N$ matrix haivng the property that all eigenvalues l of A have negative real parts. Then for any negative definite $N \times N$ matrix B such that $C = A^T B + BA$.

Proof: Let C be a negative definite matrix and let B be given by Proposition, which may be applied since all eigenvalues of A have negative real part. Let $v(x) = x^T Bx$, and let u be a solution of $u' = Au, u(0) = x_0 \neq 0$ Then

$$\frac{dv^*}{dt} = \frac{dv(u(t))}{dt} = u^T (A^T B + BA)u$$

$$= u^T Cu \leq -\mu |u|^2,$$

since C is negative definite. Since $\lim_{t \to \infty} u(t) = 0$ (all eigenvalues of A have negative real part!), it follows that $\lim_{t \to \infty} v(u(t)) = 0$. We also have

$$v(u(t)) \leq v(x_0) - \int_0^t \mu |u(s)|^2 \, ds,$$

from which follows that $v(x_0) > 0$. Hence B is positive definite.

The next corollary follows from stability theory equations and what has just been discussed.

Corollary: A necessary and sufficient condition that an $N \times N$ matrix A have all of its eigenvalues with negative real part is that there exists a unique positive definite matrix B such that

$$A^T B + BA = -I.$$

We next consider the nonlinear problem with

$$g(t, x) = o(|x|),$$

uniformly with respect to $t \in [t_0, \infty)$, and show that for certain types of matrices A the trivial solution of the perturbed system has the same stability property as that of the unperturbed problem. The class of matrices we shall consider is the following.

Definition: We call an $N \times N$ matrix A critical if all its eigenvalues have non-positive real part and there exists at least one eigenvalue with zero real part. We call it noncritical otherwise.

Theorem: Assume A is a noncritical $N \times N$ matrix and let g satisfy. Then the stability behavior of the trivial solution of is uniformly asymptotically stable if all eigenvalues of A have negative real part and it is unstable if A has an eigenvalue with positive real part.

PROOF: (i) Assume all eigenvalues of A have negative real part. By the above exists a unique positive definite matrix B such that

$$A^T B + BA = -I.$$

Let $v(x) = x^T Bx$. Then v is positive definite and if u is a solution it satisfies

$$\frac{dv^*}{dt} = \frac{dv(u(t))}{dt}$$

$$= -[u(t)|^2| g^T(t,u)Bu + u^T Bg(t,u).$$

Now

$$| gT(t,u)bu + u^T Bg(t,u) | \leq 2 | g(t,u) || B || u |.$$

Choose $r > 0$ such that $| x | \leq r$ implies

$$| g(t,u) | \leq \frac{1}{4} | B |^{-1} | u |,$$

then

$$\frac{dv^*}{dt} \leq -\frac{1}{2} | u(t) |^2,$$

as long as $| u(t) | \leq r$.

Let A have an eigenvalue with positive real part. Then there exists $\mu > 0$ such that $2\mu < | \lambda_i + \lambda_j |$, for all eigenvalues λ_i, λ_j of A, and a unique matrix B such that

$$A^T B + BA = 2\mu B = I,$$

as follows from Corollary 22. We note that B connot be positive definite nor positive semidefinite for otherwise we must have, letting $v(x) = x^T Bx$,

$$\frac{dv^*}{dt} = 2\mu v^*(t) - |u(t)|^2,$$

or

$$e^{-2\mu t} v^*(t) - v^*(0) = -\int_0^t e^{-2\mu s} |u(s)^2 ds$$

i.e.

$$0 \leq v^*(0) - \int_0^t e^{-2\mu s} |u(s)|^2 \, ds$$

for any solution, u, contradicting the fact that solution u exist for which

$$\int_0^t e^{-2\mu s} |u(s)|^2 \, ds$$

becomes unbounded as $t \to \infty$. Hence there exists $x_0 \neq 0$, of arbitrarily small norm, so that $v(x_0) < 0$. Let u be a solution . If the trivial solution were stable, then $|u(t)| \leq r$ for some $r > 0$. Again letting $v(x) = x^T Bx$ we obtain

$$\frac{dv^*}{dt} = 2\mu v^*(t) - |u(t)|^2 + g^T(t,u) \quad Bu + u^T Bg(t,u)$$

We can choose r so small that

$$2|g(t,u)||B||u| \leq \frac{1}{2}|u|^2, |u| \leq r,$$

hence

$$e^{-2\mu t} v^*(t) - v^*(0) \leq -\frac{1}{2}\int_0^t e^{-2\mu s} |u(s)|^2 \, ds,$$

i.e.

$$v^*(t) \leq e^{-2\mu t} v^*(0) \to -\infty,$$

contradicting that $v^*(t)$ is bounded for bounded u. Hence u connot stay bounded, and we have instability.

It it is the case that the matrix A is a critical matrix , the trivial solution of the linear system may still be stable or it may be unstable. In either case, one may construct examples, where the trivial solution of the perturbed problem has either the same or opposite stability behavior as the unperturbed system.

Chapter 19

Periodic Solutions

In this chapter we shall develop, using the linear theory developed in the previous chapter together with the implicit function theorem and degree theory, some fo the basic existence results about periodic solutions ofperiodic nonlinear systems ofordinary differential equations. In particular, we shall mainly be concerned with systems of the form

$$u' A(t)u + f(t,u),$$

where

$$A: \mathbb{R} \to \mathbb{R}^N \times \mathbb{R}^N$$

$$f: \mathbb{R} \to \mathbb{R}^N \times \mathbb{R}^N$$

are continuous and $T-$ periodic with respect to t. We call the equation *nonresanant* provided the linear system

$$u' = A(t)u$$

has as its only $T-$ periodic solution the trivial one; we call it *resonant,* otherwise.

Preliminaries

We recall from Chapter 6 that the set of all solutions of the equation

$$u' = A(t)u + g(t),$$

is given by

$$u(t) = \Phi(t)c + \Phi(t)\int_{t_0}^{t} \Phi^{-1}(s)g(s)ds, c \in \mathbb{R}^N,$$

where Φ is a fundamental matrix solution of the linear system. On the other hand it follows from Floquet theor that Φ has the form $\Phi(t) = C(t)e^{tR}$, where C is a continuous nonsingular periodic matric of period T and R is a constant matrix. As we,may choose Φ such that $\Phi(0) = I$ (the $N \times N$ identity matrix), it follows that (with this choice)

$$C(0) = C(T) = I$$

and

$$u(T) = e^{TR}\left(c + \int_0^T \Phi^{-1}(s)g(s)ds, c \in \mathbb{R}^N\right),$$

hence $u(T) = c$ of and only if

$$e = e^{TR}\left(c + \int_0^T \Phi^{-1}(s)g(s)ds, c \in \mathbb{R}^N\right).$$

We note that equation is uniquely solvable.for every g, if and only if

$$I - e^{-TR}$$

is a nonsingular matrix. That is, we have the following result:

Proposition: Equation has a unique T– periodic solution for every T – periodic forcing term g if and only if $e^{TR} - I$ is a nonsingular matrix. If this is the cae, the peiodic soltution u is given by the following formula:

$$u(t) = \Phi(t)\left((I - e^{TR}) - 1e^{TR} \int_0^T \Phi^{-1}(s)g(s)ds + \int_0^t \Phi^{-1}(s)g(s)ds\right)$$

This proposition allows us to formulate a fixed point equation whose solution will determine T-periodic solutions of equation. The following section is devoted to results of this type.

Perturbations of Nonresonant Equations

In.the following let

$$E - \{u \in C([0,T], \mathbb{R}^N) : u(0) = u(T)\}$$

with $\| u \| = \max_{t \in [0,T]} | u(t) |$ and $S : E \to E$ be given by

$$(Su)(t) = \Phi(t)\left((I - e^{TR})^{-1} eTR \int_0^T \Phi^{-1}(s) f(s,u(s))ds + \int_0^t \Phi^{-1}(s) f(s,u(s))ds\right)$$

Proposition: Assume that $I - e^{TR}$ is nonsingular, then the equation has a T–peiodic solution u, whenever the operator S given by equation has a fixed point in the space E.

For f as given above let us define

$$P(r) = \max\{| f(t,u) | : 0 \le t, \le T, | u | \le r\}.$$

We have the following theorem.

Theorem: Assume that A and f are as above and that $I = e^{TR}$ is nonsingular, then the equation has a T- periodic solution u, whenever

$$\liminf_{x \to \infty} \frac{P(r)}{r} = 0,$$

where P is defined by (8).

Proof: Let us define

$$B(r) = \{u \in E : \| u \| \leq r\},$$

then for $u \in B(r)$ we obtain

$$\| Su \| \leq KP(r),$$

where S is the operator defined by equation and K is a constant that depends only on the matrix A. Hence

$$S : B(r) \to B(r),$$

provided that

$$KP(r) \leq r,$$

which holds, by condition, for some r sufficiently large. Since S is completely continuous the result follows from the Schauder fixed point theorem.

As a corollary we immediately obtain:

Corollary: Assume that A and f are as above and that $I = e^{TR}$ is nonsingular, then

$$u' = A(t)u + ef(t,u)$$

has a T-periodic solution u, provided that ε is sufficiently small.

Proof: Using the operator S associated with equation we obtain for $u \in B(r)$

$$\| Su \| \leq | e | KP(r),$$

thus for given $r > 0$, there exists $\varepsilon \neq 0$ such that

$$| \varepsilon | KP(r) \leq r,$$

and S will have a fixed point in $B(r)$.

The above corollary may be considerably extended using the global continuation theorem. Namely we have the following results.

Theorem: Assume that A and f are as above and that $I - e^{TR}$ is nonsingular.

Let

$$\vartheta^+ = \{(u,\varepsilon) \in E \times [0,\infty) : (u,\varepsilon)\,.$$

$$\vartheta^- = \{(u,\varepsilon) \in E \times (-\infty,0] : (u,\varepsilon)\,.$$

Then there exists a continuum $C^+ \subset \vartheta^+ (C- \subset \vartheta^-)$ such that:

1. $C_0^+ \cap E = \{0\} (C_0^- \cap E = \{10\})$,
2. C$^+$ is unbounded in $E \times [0,\infty) (C^-$ is unbounded in $E \times (-\infty, 0])$.

Proof: The proof follows immediately from Theorem by noting that the existence of $T-$ periodic solutions of equation is equivalent to the existence of solutions of the operator equation

$$u - S(\varepsilon, u) = 0$$

where
$$S(\varepsilon,u)(t) = \Phi(t)\Big((I - e^{TR})^{-1} eTR$$

$$\int_0^T \Phi^{-1}(s)\varepsilon f(s,u(s))ds + \int_0^t \Phi^{-1}(s) f\varepsilon(s,u(s))ds\Big)$$

RESONANT EQUATIONS

Preliminaries

We shall now consider the equation subject to constraint

$$u' = f(t,u),$$

$$u(0) = u(T),$$

where
$$f : \mathbb{R} \times \mathbb{R}^N \to \mathbb{R}^N$$

is continous. should f be T-periodic with respect to t, then a T- periodic extension of a solution will be a $T-$ periodic solution $u' = 0$. i.e. we are in the case of resonance.

We now let

$$E = \{u \in C([0,T], \mathbb{R}^N)\}\ .$$

with $\| u \| = \max_{t \in [o,T]} | u(t) |$ and let $S : E \to E$ be given by

$$(su)(t) = u(T) + \int_0^t f(s,u(s))ds,$$

then clearly $S : E \to E$ is a completely continuous mapping because of the continuity assumptiqn in f,

Lemma: An element $u \in E$ is a solution if and only if it is a fixed point of the operator S. This lemam, whose proof is immediate, gives us an operator equation in the space E whose solutions are solution of the problem.

HOMETOPY METHODS

We shall next impose conditions on the finite dimensional vector field

$$x \in \mathbb{R}^N \mapsto g(x)$$

$$g(x) = -\int_0^T f(s,x)ds,$$

which will guarantee the existence of solutions of an associated problem

$$u' = \varepsilon f(t,u),$$

$$u(0) = u(T),$$

where ε is a small parameter, We have the following theorem.

Theorem: Assume that f is continuous and there exists a bounded open set $\Omega \subset \mathbb{R}^N$ such that the mapping g defined by (13) does not vanish of $\partial\Omega$. Further assume that

$$d(g,\Omega,0) \neq 0,$$

where $d(g,\Omega,0)$ is the brouwer degree. Then problem (14) has a solution for all sufficiently small ε.

Proof: We define the bounded open set $G \subset E$ by

$$G = \{u \in E : u : [0,T] \to \Omega\}.$$

For $u \in G$ define

$$u(t,\lambda) = \lambda u(t) + (1-\lambda)u(T), 0 \leq \lambda \leq 1.$$

and let

$$a(t,\lambda) = \lambda t + (1-\lambda)T, 0 \leq \lambda \leq 1.$$

For $0 \leq \lambda \leq 1, 0 \leq \varepsilon \leq 1$, define S: E × [0,1] × [0,1] $\to E$ bY

$$(S(u,\lambda,\varepsilon)(t) = u(T) + \varepsilon \int_0^{a(t,\lambda)} f(s,u(s,\lambda))ds,$$

Then S is a completely continuous mapping and the theorem will be provel once we show that

$$\text{d}(\text{id} - S)(.,1,\varepsilon), G, 0) \neq 0,$$

for ε sufficiently small, for if this is the case, S(.,1,ε) has a fixed point in G which equivalent to the assertion of the theorem. To show that holds we first show that $S(.,\lambda,\varepsilon)$ has no zeros on ∂G for all $\lambda \in [0,1]$ and ε sufficiently samll. This we argue indirectly and hence obtain sequences $\{u_n\} \subset \partial G, \{\lambda_n\} \subset [0,1]$, and $\{\varepsilon_n\}$, $\varepsilon_n \to 0$, such that

$$u_n(T) + \varepsilon_n \int_0^{a(t,\lambda_n)} f(s, u_n(s,\lambda_n))ds \equiv u_n(t), 0 \le t \le T,$$

and hence

$$\int_0^T f(s, u_n(s,\lambda_n))ds = 0, n = 1,2,.....$$

Without loss in generality, we may assume that the sequences mentioned converge to, say, u and λ_0 and the following must hold:

$$u(t) \equiv u(T) = a \in \partial\Omega.$$

Hence also

$$u_n(t,\lambda_n) \to u(t,\lambda_0) \equiv u(T),$$

which further implies that

$$\int_0^T f(x,a)ds = 0,$$

where $a \in \partial\Omega$, in contradiction to the assumptions of the theorem. Thus

$$\mathrm{d}(\mathrm{id} - S(.,0,\varepsilon), G, 0) = \mathrm{d}(\mathrm{id} - S(.,1,\varepsilon), G, 0)$$

by the homotopy invariance property of Leray–Schauder degree, for all ε sufficiently small. On the other hand

$$\begin{aligned} \mathrm{d}(\mathrm{id} - S(.,0,\varepsilon), G, 0) &= \mathrm{d}(\mathrm{id} - S(.,0,\varepsilon), G \cap \mathbb{R}^N, 0) \\ &= \mathrm{d}(\mathrm{id} - S(.,0,\varepsilon), \Omega, 0) \\ &= \mathrm{d}(\mathrm{g}, \Omega, 0) \neq 0, \end{aligned}$$

if $\varepsilon > 0$ and $(=1)^N \mathrm{d}(\mathrm{g},\Omega,0)$, if $\varepsilon < 0$, completing the proof.

Corollalry: Assume the hypotheses of Theorem and assume that all possible solutions u, for $0 < \varepsilon \le 1$, of equations are such that $u \notin \partial G$, where G is given. Then then the equation has a T– periodic solution.

A Lienard type equation

In this section we apply Corollary to prove the existence of periodic solutions of Lienard type oscillators of the form

$$x''+h(x)x'+x=e(t),$$

where

$$e:\mathbb{R}\to\mathbb{R}$$

is a continuous $T-$ periodic forcing term and

$$h:\mathbb{R}\to\mathbb{R}$$

is a continuous mapping. We shall prove the following result.

Theorem: Assume that $T<2\pi$. Then for every continuous T–Periodic forcing term e, equation has a T–Periodic response x. We note taht, since aside from the continuity assumptioin, nothing else is assumed about h, We may, without loss in generality, assume that $\int_0^T e(s)ds=0$, as follows from the substitution

$$y=x-\int_0^T e(s)ds.$$

We hence shall make that assumption. In order to apply our earlier results, we convert into a system as follows:

$$x'=y$$
$$y'=-h(x)y-x-e(t),$$

and put

$$u=\begin{pmatrix}x\\y\end{pmatrix}, f(t,u)=\begin{pmatrix}y\\-h(x)y-x-e(t)\end{pmatrix}.$$

We next shall show that the hypotheses of Theorem 7 and Corollary 8 may be satisfied by choosing

$$\Omega=\left\{u=\begin{pmatrix}x\\y\end{pmatrix}:|x|<R,|y|<R\right\},$$

where R is a sufficiently large constant. We note that holds for such choices of Ω for any $R>0$. Hence, if we are able to provide a priori bounds for solutions of equation for $0<\varepsilon\le 1$ for f given as above, the results will follow. Now

$$u=\begin{pmatrix}x\\y\end{pmatrix},$$

is a solution of (14) whenever x satisfies

$$x''+\varepsilon h(x)x'|\varepsilon^2 e(t).$$

Integrating f rom 0 to T, we find that

$$\int_0^T x(s)ds = 0.$$

Multiply by x and integrating we get

$$-\| x' \|_{L^2}^2 = \varepsilon^2 \langle x, e \rangle L^2,$$

where $\langle x, e \rangle_{L^2} = \int_0^T x(s)e(s)ds.$ Now, since

$$\| x \|_{L^2}^2 \leq \frac{T^2}{4\pi^2} \| x' \|_{L^2}^2,$$

we obtain

$$\left(1 - \frac{T^2}{4\pi^2}\right) \| x' \|_{L^2}^2 \leq -\varepsilon^2 (x, e)_{L^2},$$

from which follows that

$$\| x' \|_{L^2} \leq \left(\frac{2\pi T}{4\pi^2 - T^2}\right) \| e \|_{L^2},$$

from which, in turn, we obtain

$$\| x' \|_{\infty} \leq \sqrt{T} \left(\frac{2\pi T}{4\pi^2 - T^2}\right) \| e \|_{L^2},$$

providing an a priori bound on $| x \|_{\infty}$.

Then

$$\| x'' \|_{\infty} \leq \varepsilon q \| x' \|_{\infty} + \varepsilon^2 (M + p),$$

Hence, by Landau's inequality, we obtain

$$\| x' \|_{\infty}^2 \leq 4M(\varepsilon q \| x' \|_{\infty} + \varepsilon^2 (M + p)),$$

from which follows a bound on $\| x' \|_{\infty}$ which is independent of ε, for $0, \leq \varepsilon \leq 1$. These considerations complete the proof Theorem.

Partial resonance

This section is a continuation of of what has been discussed in Subsection. We shall impose conditions on the finite dimensional vector field

$$x \in \mathbb{R}^p \mapsto g(x)$$

$$g(x) = -\int_0^T f(s,x,0)ds,$$

which will guaranatee the existence of solutions of an associated přoblem

$$u' = \varepsilon f(t,u,v),$$

$$v' = Bv + \varepsilon h(t,u,v)$$

$$u(0) = u(T), v(0) = v(T),$$

where ε is a shall parameter and

$$f : \mathbb{R} \times \mathbb{R}^p \times \mathbb{R}^q \to \mathbb{R}^p$$

$$h : \mathbb{R} \times \mathbb{R}^p \times \mathbb{R}^q \to \mathbb{R}^q$$

are continuous and *T*- periodic with respect to *t* p + q = N. Further *B* is a q × q constant matrix with the property that the system *v*' =*B*v is nonresonant, i.e. only has the trivial solution as a T–periodic solution. We have the following theorem.

Theorem: Assume the above and there exists a bounded open set $\Omega \subset \mathbb{R}^p$ such that the mapping *g* does not vanish on $\partial\Omega$. Further assume that

$$\mathrm{d}(g,\Omega,0) \neq 0,$$

where $\mathrm{d}(g,\Omega,0)$ is the Brouwer degree. Then problem has a solution for all sufficiently small ε.

Proof: To prove the existence of a *T*- periodic solution (*u, v*) of equation is equivalent to establishing the existence of a solution of

$$u(t) = u(T) + e\int_0^t f(s,u(s),v(s),v(s))ds$$

$$v(t) = e^{Bt}v(T) + \varepsilon e^{Bt}\int_0^t e^{-Bs}h(s,u(s),v(s))ds.$$

We consider equation as an equation in the Banach space

$$E = C([0,T],\mathbb{R}^p) \times C([0,T],\mathbb{R}^q).$$

Let A be a bounded neighborhood of $0 \in \mathbb{R}^q$. We define the bounded open set $G \subset E$ by

$$G = \{(u,v) \in E : u : [0,T] \to \Omega, v : [0,T] \to A\}.$$

For $(u,v) \in G$, define as in Subsection 4.2,

$$u(t,\lambda) = \lambda u(t) + (1-\lambda)u(T), 0 \le \lambda \le 1,$$

and let

$$a(t,\lambda) = \lambda t + (1-\lambda)T, 0 \le \lambda \le 1.$$

For $0 \leq \lambda \leq 1$, $0 \leq \varepsilon \leq 1$ defube $S = (S_1, S_2): E \times [0,1] \times [0,1] \to E$ by

$$S_1(u,v,\lambda,e)(t) = u(T) + \varepsilon \int_0^{a(t,\lambda)} f(x, u(s,\lambda), \lambda v(s))ds$$

$$S_2(u,v,\lambda,e)(t) = e^{Bt} v(T) + l\varepsilon e^{Bt} \int_0^t e^{Bt} h(s, u(s), v(s))ds$$

Then S is a completely continuous mapping and the theorem will be proved once we show that

$$\text{d}(\text{id} - S(.,.,1,\varepsilon), G, 0) \neq 0,$$

for ε sufficiently small, for it this is the case, $S(.,1,\varepsilon)$ has no zeros on ∂G all $\lambda \in [0,1]$ and ε sufficiently small. This we argue in a manner similar to the proof of Theorem. Hence

$$\text{d}(\text{id} - S(.,.,0,\varepsilon), G, 0) = \text{d}(\text{id} - S(.,.,1,\varepsilon), G, 0)$$

by the homotopy invarince property of Leray–Schauder degre, for all ε sufficiently small. On the other hand

$$\text{d}(\text{id} - S(.,.,0,\varepsilon), G, 0) = \text{d}((-\varepsilon \int_0^T f(s,.,0,\varepsilon)G, 0) - S_2(.,.0,\varepsilon)), G, (0,0))$$

$$== \text{d}((-\varepsilon \int_0^T f(s,.,0,)ds, \Omega, 0)\text{d}(\text{id} - S_2(.,.0,\varepsilon), \tilde{G}, 0)$$

$$= \text{sgn}\det(I - e^{BT})\text{d}(-\varepsilon \int_0^T f(s,.,0)ds, \Omega, 0 \neq 0,$$

where $\tilde{G} = C([0,T], A)$, completing the proof.

$$u' = Au \mid g(t,u)$$

and show how to construct Lyapunov functionals to test the stability of the trivial solution of this system.

The type of Lyapunov functional we shall be looking for are of the form

$$v(x) = x^T Bx,$$

where B is a constant $N \times N$ matrix, i.e. we are looking for v as a quadratic form. If u is a solution of then

$$\frac{dv^*}{dt} = \frac{dv(t,u(t))}{dt}$$

$$= u^T(A^T B + BA)u + g^T(t,u)Bu + u^T Bg(t,u).$$

Hence, given, A if B can be found so that $C = A^T B + BA$ has certain definniteness properties, then the results of the trivial solution. To proceed along these lines we need some linear algebra results.

Proposition: Let A be a constant $N \times N$ matrix having the property that for any eigenvalue λ of A, $= \lambda$ is not an eigenvalue. Then for any $N \times N$ matrix C. there exists a unique $N \times N$ matrix B such that $C = A^T B + BA \cdot$

Proof: On the space of $N \times N$ matrices define the bounded linear operator L by

$$L(B) = A^T B + BA \cdot$$

Then L may be viewed as a bounded linear operator of $\mathbb{R}^{N \times N}$ to itself, hence it will be a bijection provided it does not have 0 as an eigenvalue, Once we show the latter, the result follows. Thus let μ be an eigenvalue of L, i.e., there exists a nonzero matrix B such that

$$L(B) = A^T B +, \mu B$$

Hence

$$A^T B + B(A - \mu I) = 0$$

From this follows

$$B(A - \mu I)^n = (-A^T)^n B,$$

for any integer $n \geq 1$, hence for any polynomial p

$$Bp(A - \mu I)^n = p(-A^T)B.$$

Since, on the other hand if F and G are two matrices with no common eigenvalues, there exists a polynomial p such that $p(F) = I, p(G) = 0$, implies that $A - \mu I$ and $_{-}A^T$ must have a common eigenvalue,.

From which follows taht μ is the sum of two eigenvalues of A, which, by hypothesis connot equal 0. This proposition has the following corollary.

Corollary: Let A be a constant $N \times N$ matrix. Then for any $N \times N$ matrix C, there exists $\mu > 0$ and a unique $N \times N$ matrix B such that $2\mu B + C = A^T B + BA$.

Proof: Let

$$S = \{\lambda \in \mathbb{C} : \lambda = \lambda_1 + \lambda_2\},$$

where λ_1 and λ_2 are eigenvalues of A. Since S is a finite set, there exists $r_0 > 0$. such that $\lambda(\neq 0) \in S$ implies that $|\lambda| > r_0$. Choose $0 < \mu \leq r_0$ and consider the matrix $A_1 = A - \mu I$. We may now apply Proposition 21 to the matrix A_1 and find for a given matrix C, a unique matrix B such that $C = A_1^T B + BA_1$, i.e. $2\mu BC = A^T B + BA \cdot$

Corollary: Let A be a constant $N \times N$ matrix haivng the property that all eigenvalues l of A have negative real parts. Then for any negative definite $N \times N$ matrix B such that $C = A^T B + BA$.

Proof: Let C be a negative definite matrix and let B be given by Proposition, which may be applied since all eigenvalues of A have negative real part. Let $v(x) = x^T Bx$, and let u be a solution of $u' = Au, u(0) = x_0 \neq 0$ Then

$$\frac{dv^*}{dt} = \frac{dv(u(t))}{dt} = u^T (A^T B + BA)u$$

$$= u^T Cu \leq -\mu |u|^2,$$

since C is negative definite. Since $\lim_{t \to \infty} u(t) = 0$ (all eigenvalues of A have negative real part!), it follows that $\lim_{t \to \infty} v(u(t)) = 0$. We also have

$$v(u(t)) \leq v(x_0) - \int_0^t \mu |u(s)|^2 \, ds,$$

from which follows that $v(x_0) > 0$. Hence B is positive definite.

The next corollary follows from stability theory equations and what has just been discussed.

Corollary: A necessary and sufficient condition that an $N \times N$ matrix A have all of its eigenvalues with negative real part is that there exists a unique positive definite matrix B such that

$$A^T B + BA = -I.$$

We next consider the nonlinear problem with

$$g(t, x) = o(|x|),$$

uniformly with respect to $t \in [t_0, \infty)$, and show that for certain types of matrices A the trivial solution of the perturbed system has the same stability property as that of the unperturbed problem. The class of matrices we shall consider is the following.

Definition: We call an N × N matrix A critical if all its eigenvalues have non-positive real part and there exists at least one eigenvalue with zero real part. We call it noncritical otherwise.

Theorem: Assume A is a noncritical $N \times N$ matrix and let g satisfy. Then the stability behavior of the trivial solution of is uniformly asymptotically stable if all eigenvalues of A have negative real part and it is unstable if A has an eigenvalue with positive real part.

Proof: Assume all eigenvalues of A have negative real part. By the above exists a unique positive definite matrix B such that

$$A^T B + BA = -I.$$

Let $v(x) = x^T Bx$. Then v is positive definite and if u is a solution it satisfies

$$\frac{dv^*}{dt} = \frac{dv(u(t))}{dt}$$

$$= -[u(t)|^2| g^T(t,u)Bu + u^T Bg(t,u).$$

Now

$$| gT(t,u)bu + u^T Bg(t,u) | \leq 2 | g(t,u) \| B \| u |.$$

Choose $r > 0$ such that $| x | \leq r$ implies

$$| g(t,u) | \leq \frac{1}{4} | B |^{-1} | u |,$$

then

$$\frac{dv^*}{dt} \leq -\frac{1}{2} | u(t) |^2,$$

as long as $| u(t) | \leq r$.

Let A have an eigenvalue with positive real part. Then there exists $\mu > 0$ such that $2\mu < | \lambda_i + \lambda_j |$, for all eigenvalues λ_i, λ_j of A, and a unique matrix B such that

$$A^T B + BA = 2\mu B = I,$$

as follows from Corollary 22. We note that B connot be positive definite nor positive semidefinite for otherwise we must have, letting $v(x) = x^T Bx$,

$$\frac{dv^*}{dt} = 2\mu v^*(t) - | u(t) |^2,$$

or

$$e^{-2\mu t} v^*(t) - v^*(0) = -\int_0^t e^{-2\mu s} | u(s)^2 ds$$

i.e.

$$0 \leq v^*(0) - \int_0^t e^{-2\mu s} | u(s) |^2 \, ds$$

for any solution, u, contradicting the fact that solution u exist for which

$$\int_0^t e^{-2\mu s} | u(s) |^2 \, ds,$$

becomes unbounded as $t \to \infty$. Hence there exists $x_0 \neq 0$, of arbitrarily small norm, so that $v(x_0) < 0$. Let u be a solution . If the trivial solution were stable, then $|u(t)| \leq r$ for some $r > 0$. Again letting $v(x) = x^T Bx$ we obtain

$$\frac{dv^*}{dt} = 2\mu v^*(t) - |u(t)|^2 + g^T(t,u) \; Bu + u^T Bg(t,u)$$

We can choose r so small that

$$2|g(t,u)||B||u| \leq \frac{1}{2}|u|^2, |u| \leq r,$$

hence

$$e^{-2\mu t} v^*(t) - v^*(0) \leq -\frac{1}{2}\int_0^t e^{-2\mu s} |u(s)|^2 \, ds,$$

i.e.

$$v^*(t) \leq e^{-2\mu t} v^*(0) \to -\infty,$$

contradicting that $v^*(t)$ is bounded for bounded u. Hence u cannot stay bounded, and we have instability.

It it is the case that the matrix A is a critical matrix, the trivial solution of the linear system may still be stable or it may be unstable.

In either case, one may construct examples, where the trivial solution of the perturbed problem has either the same or opposite stability behavior as the unperturbed system.

Chapter 20

Existence and Uniqueness Theorems

In this chapter, we shall present the basic existence and uniqueness theorems for solution of initial value problems for systems of ordinary differential equaions. To this end let D be an open connected subset of $\mathbb{R} \times \mathbb{R}^N, N \geq 1,$ and let

$$f : D \to \mathbb{R}^N$$

be a continuous mapping.

We consider the differential equaqiton

$$u' = f(t,u),' = \frac{d}{dt}.$$

and seek sufficient condition for the existence of solutions, where $u \in C^1(I, \mathbb{R}^N)$, with I an interval, $I \subset \mathbb{R},$ is called a solution, if $(t,u(t)) \in D, t \in I$ and

$$u'(t) = f(t,u(t)), t \in I.$$

Simple example tell us that a given differential equation may have a multitude of solutions, in general, whereas some constraints on the solutions sought might provide existence and uniquenness of the solution. The most basic such constraints are given by fixing an initial value of a solution. By an *initial value problem* we mean the following:

1. Given a point $(t_0, t_0) \in D$ we seek a solution u such that

$$u(t_0) = u_0.$$

We hanve the following proposition whose proof is straightforward:

Proposition: A funcation $u \in C^1(I, \mathbb{R}^N),$ with $I \subset \mathbb{R},$ and I an interval containg t_0 is a solution of the initial value problem, satisfying the initial condition if and only if $(t,u(t)) \in D,\ t \in I$ and

$$u(t) = u(t_0) + \int_{t_0}^{t} f(s,u(s))ds.$$

We shall now, using Proposition, establish some of the classical and basic existence and existence/uniqueness theorems.

The Picard -Lindelof Theorem

We say that *f* satisfies a *local* Lipschitz *condition on the domain D*, provided for every compact set $K \subset D$, there exists a constant $L = L(K)$, such that for all $(t.\ u_1)$, $(t.\ u_2) \in K$

$$| f(t,u_1) - f(t,u_2) \leq L | u_1 - u_2 |.$$

For such functions, one has the following existence and uniquencdess theorem. This result is usually called the Picard-Lindelof theorem

Theorem: Assume that $f : D \to \mathbb{R}^N$ satisfies a local Lipschitz condition on the domain D, then for every $(t_0, u_0) \in D$ equation has a unique solution satisfying the initial condition on some interval I. We remark that the theorem as stated is a local existence and uniqueness theorem, in the same that the interval *I*, where the solution exists will depend upon the initial condition. Global results will follow from this results, by extending solutions to maximal intervals of existence, as will be seen in a subsequent section.

Proof: Let $(t_0, u_0) \in D$ then, since *D* is open, there exist positive constants *a* and *b* such that

$$Q = \{t,u) : | t - t_0 | \leq a, | u - u_0 | \leq b\} \subset D.$$

Let *L* be the Lepschitz constant for *f* associated with the set Q. Further let

$$m = \max_{(t,u) \in Q} | f(t,u) |$$

$$\alpha = \min\left\{a, \frac{b}{m}\right\}.$$

Let $\tilde{L}$ be any constant $\tilde{L} > L$, and define

$$M = \{u \in C([t_0 - \alpha, t_0 + \alpha], \mathbb{R}^N) : | u(t) \ = u_0 | \leq b, | t - t_0 | \leq \alpha\}.$$

In $C([t_0 - \alpha, t_0 + \alpha], \mathbb{R}^N)$ we define a new norm as follows:

$$\| u \| = \max_{|t - t_0| \leq \alpha} e^{-\tilde{L}|t - t_0|} | u(t) |.$$

Ans we let $\rho(u, v) = \| u - v \|$, then (M, ρ) is a complete metric space. Next define the operator *T* on *M* by:

$$(Tu)(t) = u_0 + \int_{t_0}^{t} f(s, u(s)) ds, | t - t_0 | \leq \alpha.$$

Then

$$(Tu)(t) = u_0 |\leq| + \int_{t_0}^{t} | f(s,u(s)) | \, ds |,$$

and, since $u \in M$,

$$| (Tu)(t) = u_0 | \leq \alpha m \leq b.$$

Hence

$$T : M \to M.$$

Computing further, we obtain, for $u, v \in M$ that

$$| (Tu)(t) - (Tv)(t) | \leq | \int_{t_0}^{t} | f(s,u(s)) \;\; - f(s,v(s)) | \, ds |$$

$$\leq L \int_{t_0}^{t} | u(s) = v(s) | \, ds |,$$

and hence

$$e^{-\bar{L}|t-t_0|}(Tu)(t) - (Tv)(t) | \leq \; e^{-\bar{L}|t-t_0|} L \, | \int_{t_0}^{t} | u(s) - v(s) | \, ds |$$

$$\leq \frac{L}{L} \| u - v \|,$$

and hence

$$\rho(T_u, Tv) \leq \frac{L}{\bar{L}} \rho(u,v),$$

proving that T is a contraction mapping. The result therefore follows from the contraction mapping principle, Theorem I.6.

We remark that, since T is a contracation maping, the contraction mapping theorem gives a constructive means for the solultion of the initial value problem in Theorem 2 and the solution may in fact be obtaqined via an iteration procedure. This procedure is known as *Picard iteratoin.*

In the next section, we shall show, that without the assumption of a local Lipschitz condition, we still get the existence of solution.

The Cauchy-Peano Theorem

The following result, called the Cauchy-Peano theorem provides the local solvability of intial value problems.

Theorem: Assume that $f: D \to \mathbb{R}^N$ is continuous. Then for every $(t_0, u_0) \in D$ the initial value problem (1), (2) has a solution on some interval I, $t_0 \in I$

Proof: Let $(t_0, t_0) \in D$ and let Q, α, m be as in the proof of Theorem. Consider the space $E = C([t_0 - a, t_0 + \alpha], \mathbb{R}^N)$ with norm $\| u \| = \max_{(t - t_0| \leq |\alpha} | u(t) | \; 2$ and note that M is a closed, bounded convex subset of E and further that $T: M \to M$. We hence may apply the Schauder fixed point theorem once we verify that T is completely continuous on M. To see this we note, that, f is continuous, it follows that T is continuous. On the other hand, if $\{u_n\} \subset M$, then

$$| (Tu_n)(t) - (Tu_n)(t) | \leq | \int_t^t | f(8, u_n(s)) | \, ds |$$

Hence $\{Tu_n\} \subset M$, is a uniformly bounded and equicontinuous family in E. It therefore has a uniformly convergent subsequence, showing that $\{Tu_n\}$ is precompact and hence T is completely continuous. This completes the proof.

We note from the above proofs that for a solution u thus obtained, both $(t_0 \pm \alpha, u(t_0 \pm \alpha)) \in D$. We hence may reapply these theorems with initial conditions given at $t_0 \pm \alpha$ and conditions $u(t_0 \pm \alpha)$ and thus continue solutions to larger intervals. We shall prove below that this continuation process leads to maximal intervals of existence and also describes the behavior of solutions as one approaches the endpoints of such maximal intervals.

Caratheodory Equations

In many situaltions the nonlinear terms f is not continuous as assumed above but stisfies the so-called CAratheodory conditions on any parallelepiped $Q \subset D$, where Q is as given in the proof of Theorem, i.e.

1. f is measurable in t for each fixed u and continuous in u and continuous in u for almost, all t,

2. For each Q there exists a function $m \in L^1(t_0 - a, t_0 - a, t_0 + a)$ such that

$$| f(f, u) | \leq m(t), (t, u) \in Q.$$

Under such assumptions we have the following extension of the Cauchy-Peano theorem, Theorem:

Theorem: Let f satisfy the Caratheodory conditions on D. Then for every $(t_0, u_0) \in D$ the initial value problem has a solution on some interval I, $t_0, \in I$, in the sence that there exists an absolutely continuous funciton $u: I \to \mathbb{R}^N$ which satisfies the initial condition and the differential equation a.e. in I.

Proof: Let $(t_0, u_0) \in D$, and choose

$$Q = \{t, u) :| t - t_0 | \le a, | u - u_0 | \le b\} \subset D.$$

Let

$$M = \{u \in C([t_0 - \alpha, t_0 + \alpha], \mathbb{R}^N) :| u(t) \;\; = u_0 | \le b, | t - t_0 | \le \alpha\}.$$

where $\alpha \le a$ is to be determined.

Next define the operator T on M by:

$$(Tu)(t) = u_0 + \int_{t_0}^{t} f(s, u(s)) ds, | t - t_0 | \le \alpha.$$

Then, because of the Caratheodory conditions, T_u is a a continuous function and

$$(Tu)(t) = u_0 | \le | \int_{t_0}^{t} | f(s, u(s)) | ds |$$

Further, since $u \in M$,

$$(Tu)(t) = u_0 | \le | \int_{t_0 - \alpha}^{t_0 + \alpha} m(s) ds \le \| m \|_{L1[t_0 - \alpha, t_0 + \alpha]} \le b,$$

for α small enough, Hence

$$T : M \to M$$

One next shows that T is completely continuous mapping, hence will have a fixed point in M by the Schauder Fixed point Theorem.

That fixed points of T are solutions of the initial value problem, in the sense given in the theorem, is immediate.

Extension Theorems

In the section we establish a basic results about maximal intervals of existence of solutions of initial value problems. We first prove the following lemma.

Lemma: Assume that $f : D \to \mathbb{R}^N$ is continuous and let $\tilde{D}$ be a subdomain of D, with f bounded on $\tilde{D}$.

Further let u be a solution of (1) defined on a bounded interval (a, b) with $(t, u(t)) \in \tilde{D}, t \in (a, b)$. Then the limits

$$\lim_{t \to \alpha +} u(t), \lim_{t \to b -} u(t)$$

exist.

Proof: Let $t_0 \in (a,b)$, then $u(t) = u(t_0) + \int_{t_0}^{t} f(s,u(s))ds$. Hence for $t_1, t_2 \in (a,b)$, we obtain

$$|u(t_1) - u(t_2)| \leq m|t_1 - t_2|,$$

where *m* is bound on *f* on $\tilde{D}$. Hence the above limits exist.

We may there fore, as indicated above, continue the solution beyond the interval (*a*, *b*,) to the lelft of *a* and the right of *b*. We say that a solution *u* has maximal interval of existence (ω_-, ω_+), provided *u* connot be continued as a solution to the right of ω_-. The following theorem holds.

Theorem: Assume that $f : D \to \mathbb{R}^N$ is continuous and let *u* be a solution of the equation defined on some interval I. Then *u* may be extened as a solution of (1) to a maximal interval of existence (ω_-, ω_+) and $(t,u(t)) \to \partial D$ as $(t,u(t)) \to \partial D t \to \omega_{\pm}$.

Proof: We establish the existence of a right maximal interval of existence; a similar argument will yield the existence of a left maximal one and together the two will imply the existence of a maximal interval of existence.

Let *u* be a solution of equation with $u(t_0) = u_0$ defined on an interval $I = [t_0 a_u)$. We say that two solution *v*, *w* satisfy

$$v \preceq w,$$

if and only if:

1. $u \equiv v \equiv w$ on $[t_0, a_u)$
2. *v* is defined on $I_v = [t_0, a_u), a_v \geq a_u$,
3. *w* is defined on $I_w = [t_0, a_w), a_w \geq a_u$,
4. $a_w \geq a_v$,
5. $v \equiv w \text{on} I_v$

We see that $\preceq$ is a partial order on the set of all solutions *S* of which agree with *u* on *I*. One next verifies that the conditions of the Hausdorff maximum principle hold and hence that *S* contains a maximal element, $\tilde{u}$. This maximal element $\tilde{u}$ connot be further extended to the right.

Next let *u* be solution with right maximal interval of existence $[t_0, \omega_+)$. We must show that $(t,u(t)) \to \partial D$ as $t \to \omega_+$, i.e. given any compact set $K \subset D$, there exists t_K, such that $(t,u(t)) \notin K$, for $t > t_K$. If $\omega_+ = \infty$, the conclusion clearly holds. On the Other hand, if $\omega_+ = \infty$, we proceed indirectly. In which case there exists a compact set $K \subset D$, such that for every $n = 1, 2, \ldots$ there exists $tn, 0 < \omega_+ - t_n < \frac{1}{n}$, and $(t_n, u(t_n))\} \in K$. Siknce K is compact, there will be a subsequence, call it again $\{(t_n, u(t_n))\}$ converges to, say, $(\omega_+, u^*) \in K$.

Since $(\omega_+, u^*) \in K$, it is an interior point of D. We may therfore choose a constant a > 0, such that $Q = \{(t,u) \| \omega_+^{-t}, \leq a, | u - u^*\} \leq a\} \subset D$, and thus fo r$n$ large $(t_0, u(t_0)) \in Q$ Let $m = \max_{(t,u)\in Q} | f(t,u)|$, and let n be so largbe that

$$0 < \omega_+ - t_n \leq \frac{a}{2m}, | u(t_n) - u^* | \leq \frac{a}{2}.$$

Then

$$| u(t_n) - u(t) < m(\omega_+ - t_n) \leq \frac{a}{2}, \text{ for } t \leq t < \omega +,$$

as an easy indirect argument shows.

It therefore follows that

$$\lim_{t \to \omega_+} u(t) = u^*,$$

and we may extend u to the right of ω_+ contradicting the maximality of u.

Corollary: Assume that $f : [t_0, t_0 + a] \times \mathbb{R}^N \to \mathbb{R}^N$ is continuous and let u be a solution of the equation defined on some right maximal interval of existence $I \subset [t_0, t_0 + a]$.

Then, either $I = [t_0, t_0 + a]$, or else $I = [t_0, \omega_+), \omega_+ < t_0 + a$, and

$$\lim_{t \to \omega_+} | u(t) | = \infty.$$

We also consider the following corollary, which is of importance for differential equations whose right hand side hanve at most linear growth. I.e., we assume that the following growth condition holds:

$$| f(t,u) \leq \alpha(t) | u | + \beta(t).$$

Corollary: Assume that $f : (a,b) \times \mathbb{R}^N \to \mathbb{R}^N$ is continuous and let f satisfy, where $\alpha, \beta \in L^1(a,b)$ are nonnegative continuous functions. Then the maximal interval of existence (ω_-, ω_+) is $(a, b,)$ for any solution.

$$| u(t) \leq u(t_0) | + | \int_{t_0}^{t} \| \alpha(s) | u(s) | + \beta(x)] ds.$$

Considering the caes $t \geq t_0$, the other case being similar, we let

$$v(t) = \int_{t_0}^{t} \| \alpha(s) | u(s) | + \beta(x)] ds$$

and $c = | u(t_0) |$. Then an easy calculation yields

$$v' - \alpha(t)v \leq \alpha(t)c + \beta(t),$$

and hence

$$v(t) \leq e^{\int_{t_0}^{t} a(r)dr} \int_{t_0}^{t} e^{\int_{t_0}^{t} a(r)dr} [\alpha(t)c + \beta(s)] ds,$$

from which, follows that $\omega + = b$.

Dependence upon Initial Conditions

Let again D be an open connected subset of $\mathbb{R} \times \mathbb{R}^N, N \geq 1$, and let

$$f : D \to \mathbb{R}^N$$

be a continuous mapping.

We consider the initial value problem

$$u' = f(t,u), u(t_0) = u_0,$$

and assume we have conditions which guarantee that the equation has a unique solution

$$u(t) = u(t, t_0, u_0),$$

for every $(t_0, u_0) \in D$. We shall now present condition which guarantee that $u(t, t_0, u_0)$ depends either continuously or smoothly on the initial condition (t_0, u_0)

A somewhat more general situatoin occurs frequently, where the function f also depends upon parameters, $\lambda \in \mathbb{R}^m$, i.e. the equation is replaced by the parameter dependent problem

$$u' = f(t,u,\lambda), u(t_0) = u_0, .$$

and solutions u then are functions of the type

$$u(t) = u(t, t_0, u_0, \lambda),$$

provided that the equation is uniquely solvable, This situation is a special case of the above, as we may augment the original system the equation as

$$u' = f(t,u,\lambda), u(t_0) = u_0,$$

$$\lambda'=0,\lambda(t_0)=\lambda$$

and obtain an initial value problem for a system of equations of higher dimension which does not dpend upon parameters.

Continuous Dependence

We first prove the following proposition.

Theorem: Assume that $f:D\to\mathbb{R}^N$ is a continuous mapping. unique solutiqn $u(t)=u(t,t_0,u_0)$, for every $(t_0,u_0)\in D$. Then the solution depends continuously on (t_0,u_0), in the following sense: If $\{(t_n,u_n)\}\subset D$ converges to $(t_0,u_0)\in D$, then given $\varepsilon>0$, there exists n_ε and an interval I_ε such that for all $n\geq n_\varepsilon$, the solution $u(t)=u(t,(t_n,u_n))$, exists on I_ε and

$$\max_{t\in I_\varepsilon}|u(t)-u_n(t)|\geq\varepsilon.$$

Proof: We relay on the proof of Theorem 3 and find that for given $\varepsilon>0$, there exists n_ε such that $\{(t_n, u_n)\}\subset\tilde{Q}$, where

$$\tilde{Q}=\{(t,u):|t-t_0|\leq\frac{\alpha}{2},|u-u_0|\leq\frac{b}{2}\}\subset Q,$$

and Q is the set given in the proof of Theorems. Using the proof of Theorem we obtains a common compact interval I_ε of existence of the sequence $\{u_n\}$ and $\{(t,u_n(t))\}\subset Q$, for $t\in I_\varepsilon$.The sequence $\{u_n\}$ hence will be uniformly bouned and equicontinuous on I_ε and will therefore have a subsequence converging uniformly on I_ε Employing the integral eqauation we see that the limit must be a solution of abd gebcem by the uniqueness assumption must equal *u*. Since this is true for every subsequence, the whole sequence must converge to *u*, completing the proof.

We have the following corollary, which asserts continuity of solutions with respect to the differential equation. The proof is similar to the above and will hence be omitted.

Corollary: Assume that $f_n:D\to\mathbb{R}^N$, $n=1, 2,\ldots$, are contiunous mappings and that (with $f=f_n$) has a unique solution $u_n(t)=u(t,t_n,u_n)$, for every $(t_n,u_n)\in D$. Then the solution depends continuously on (t_0,u_0), in the following sense: If $\{(t_n, u_n)\}\subset D$ converges to $(t_0,u_0)\in D$, and f_n *converges to f*, uniformly on compact subsets of *D*, then given $\varepsilon>0$, there exists n_ε and an interval I_ε such that for all $n \geq n_\varepsilon$, the solution $u_n(t) = u(t, t_n, u_n)$, exists on I_ε and

$$\max_{t \in I_\varepsilon} | u_n(t) - u_n(t) | \leq \varepsilon.$$

Differentiability with Respect to Initial Conditions

In the following we shall employ the convention $u = (u^1, u^2, ..., u^N)$. We have the following theorem.

Theorem: Assume that $f : D \to \mathbb{R}^N$ is a continuous mapping and that the partial derivatives $\frac{\partial f}{\partial u^i}, 1 \leq i \leq N$ are continuous also. Then the solution $u(t) = u(t, t_0, u_0)$, of class C^1in the variable u_0. Further, if $J(t)$ is the Jacobian matrix

$$J(t) = J(t, t_0, u_0) = \left(\frac{\partial f}{\partial u} \right)_{u = u(t, t_0, u_0)'}$$

then

$$y(t) = \frac{\partial u}{\partial u^i}(t, t_0, u_0)$$

is the solution of the initial value problem

$$y' = J(t)y, y(t_0) = e_i, 1 \leq i \leq N,$$

where $e_i \in \mathbb{R}^n$ is given by $e_i^k = \delta_{ik}$ the Kronecler delta.

Proof: Let e^i be given as above and let $u_h(t) = u(t, t_0, u_0), u_h(t) = u(t, t_0, u_0 + he_i)$, where $| h |$ is sufficiently small so that u_h exists. We note that u and u_h will have a common interval of existence, whenever $| h|$ is sufficiently small. We compute

$$(u_h(t) - u(t))' = f(t, u_h(t) = f(t, u(t)).$$

Letting

$$yh(t) = \frac{uh(t) - u(t)}{h}$$

we get $yh(t_0) = e_i$.. If we let

$$G(t, y_1, y_2) = \int_0^1 \frac{\partial f}{\partial u}(t, sy_1 + (1 - s)y_2)ds,$$

we obtain that yh is the unique solution of

$$y' = G(t, u(t), uh(t))y, y(t_0) = e_i.$$

Since $G(t, u(t), uh(t)) \to J(t)\ h \to 0$, we may apply Corollary to conclude that $yh \to y$ uniformly on a neighborhood of t_0.

Differential Inequalities

We consider in $\mathbb{R}^N$ the following partial orders:

$$x \le y \Leftrightarrow x^i \le y^i, 1 \le i \le N,$$
$$x < y \Leftrightarrow x^i < y^i, 1 \le i \le N,$$

For function $u : I \to \mathbb{R}^N$, where I is an open interval, we consider the Dini derivations

$$D^+ u(t) = \lim\sup_{h \to 0^+} \frac{u(t+h) - u(t)}{h},$$

$$D_+ u(t) = \lim\sup_{h \to 0^+} \frac{u(t+h) - u(t)}{h},$$

$$D\text{-}u(t) = \lim\sup_{h \to 0^+} \frac{u(t+h) - u(t)}{h}$$

$$D_u(t) = \lim\inf_{h \to 0^-} \frac{u(t+h) - u(t)}{h}$$

where lim sup and lim inf are taken componentwise.

Defination: A function $f : \mathbb{R}^N \to \mathbb{R}^N$ is said to be of type *K (after Kamake [23] on a set* $S \subset \mathbb{R}^N$, *lim inf*

Definition: A funcation $f : \mathbb{R}^N \to \mathbb{R}^N$ us sauid to be of type *K*. on a set $S \subset \mathbb{R}^N$, whenever

$$f^i(x) \le f^1(y), \forall x, y \supset, x \le y, x^i = y^i$$

The following theorem on differential inequqalities is of use in obtaining estimates on solutions.

Theorem: Assume that $f : [a, b] \times \mathbb{R}^N \to \mathbb{R}^N$ is a continuous mapping which is of type *K* for each fixed *t*, Let u : [*a*, *b*] $\to \mathbb{R}^N$ is a continuous mapping which is of type *K* for each fixed *t*, Let $u : [a, b] \to \mathbb{R}^N$ be a solution of (1) and let $v : [a, b] \to \mathbb{R}^N$ be continuous and satisfy

$$D^- v(t) > f(t, v(t)), a < t \le b,$$
$$v(a) > u(a),$$

If $z:[a,b] \to \mathbb{R}^N$ is continuous and satisfies

$$D^- z(t) > f(t, z(t)), a < t \le b,$$
$$z(a) > u(a),$$

then $z(t) > u(t), a \le t \le b.$

Proof: We prove the first part of the theorem. The second part follows along the same line of reasoning. by continuity of *u* and *v*, there exists $\delta > 0$, such that

$$v(t) > u(t), a \le t \le b + \delta.$$

If the inequality does not hold throughout [*a*, *b*] there will exist a first point *c* and an index *i* such that

$$v(t) > u(t), a \le t < c, v(c) \ge u(c), v^i(c) = u^i(c)$$

Hence (since *f* is of type K)

$$D^- v^i(c) > f^i(c, v(c)) \ge f^i(c, u(c) = u^i(c).$$

On the other hand,

$$D^- v^i(c) = \limsup_{h \to 0-} \frac{v^i(c+g) - v^i(c)}{h}$$

$$\le \limsup_{h \to 0-} \frac{u^i(c+g) - u^i(c)}{h} = u^{i}{}'(u), \text{ a contradiction.}$$

Definition: A solution u^* is called a right maximal solution on an interval I, if for every $t_0 \in I$ and any solution *u* of (1) such that

$$u(t_0) \le u^*(t_0),$$

it follows that $$u(t_0) \le u^*(t), t_0 \le t \in I.$$

Right minimal solutions are defined similarly.

Theorem: Assume that $f: D \to \mathbb{R}^N$ is a continuous mapping which is of type *K* for each *t*. Then the initial value problem (8) has a unique right maximal (minimal) solution for each $(t_0, u_0) \in D$.

Proof: That maximal and minimal solution are unique follows from the definition Choose $0 < \varepsilon \in \mathbb{R}^N$ and let v_n be any solution of

$$v' = f(f, v, \lambda) + \frac{\varepsilon}{n}, v(t_0) = u_0 + \frac{\varepsilon}{n}$$

Then, given a neighborhood U of () $(t_0,t_0)\in D$, there exists an interval $[t_0,t_1]$ of positive length such that all n sufficiently large. On the other hand it follows from Theorem 13 that

$$v_n(t)>v_m(t), t_0\leq t\leq t_1, n<m.$$

Then, given a neighborhood U of $(t_0,u_0)\in D$, there exists an interval $[t_0,t_1]$ hence will have a subsequence which converges uniformly to a solution u^*. Since that sequence is monotone, the whole sequence will, in fact, converge to u^*. Applying Theorem once more, we obtain that u^* is right maximal of $[t_0,t_1]$, and extending this solution to a right maximal interval of existence as a right maximal solution, completes the proof.

We next prove an existence theorem for initial value problems which allows for estimates of the solution in terms of given solutions of related differential inequalities.

Theorem: Assume that $f:[a,b]\times\mathbb{R}^N\rightarrow\mathbb{R}^N$ is a continuous mapping which is of type K for each fixed t. for each fixed t. Let $v,z:[a,b]\times\rightarrow\mathbb{R}^N$ be continuous and satisfy

$$D^+v(t)\geq f(t,v(t)), a\leq t<b,$$

$$D_+z(t)\leq f(t,z(t)), a\leq t<b,$$

$$z(t)\leq v(t), a\leq t\leq b.$$

Then for every $u_0, z(a)\leq u_0, \leq v(a)$, there exists a solution u (with $t_0=a$) such that

$$z(t)\leq u(t)\leq v(t)\leq b.$$

The functions z and v are called, respectively, sub-and supersolutions.

Proof: Define $\overline{f}(t,x)=f(t,\overline{x})-f(x,\overline{x})$, where for $1\leq i\leq N$,

$$\overline{x}^i=\begin{cases} v^i(t) & ,\text{if} \quad x^i>v^i(t), \\ x^i & ,\text{if} \quad z^i(t)\leq x^i\leq v^i(t), \\ z^i(t) & ,\text{if} \quad x^i<x^i(t). \end{cases}$$

Then $\overline{f}$ is continuous and has at most linear growth, hence the initial value problem, with f replaced by $\overline{f}$ has a solution u that extends to $[a, b]$. We show that $z(t)\leq u(t)\leq v(t), a\leq t\leq b$, and hence may conclude that u solves the original initial value problem.

To see this, we shall prove that fro any $\varepsilon>0$ we have that

$$z^i(t)-e\leq u^i(t)\leq v^i(t)+\varepsilon, \tau\varepsilon[a,b], i=1,...N.$$

This we argue indirectly and suppose there exists $\varepsilon > 0$ a value $c \in (a,b)$ and i such that

$$u^i(c) = v^i(c) + \varepsilon, u^i(t) \geq v^i(t) + \varepsilon, c < t \leq t_1 \leq b.$$

Since $\overline{u}^i(t) = v^i(t)c < t \leq t_1$ and $\overline{u}^k(t) = v^k(t)c < t \leq t_1, k \neq i,$ we get that

$$\overline{f}^i(t,u(t)) = f(t,\overline{u}(t)) = (u^i(t) - v^i(t))$$
$$\leq f^i(t,v(t)) - \varepsilon < D + v^i(t), c < t \leq t_1.$$

On the other hand we have, for $h > 0$, small, that

$$u^i(t+h) - u^i(c) \geq v^i(t+h) + \varepsilon - (v^i(c) + \varepsilon,$$

and hence

$$u^{i\,\prime}(c) \geq D + v^i(c),$$

contradicting what has been concluded above, Since $\varepsilon > 0$ was arbitrary, we conclude that

$$u(t) \leq v(t)$$

That

$$u(t) \geq vz(t)$$

may be proved in a similar way.

Corollary: Assume the hypotheses of Theorem and that f satisfies a local Lipschitz condition. Assume furthermore that

$$z(a) \leq z(b), v(a) \geq v(b).$$

Then the problem

$$u' = f(t,u), u(a) = u(b)$$

has a solution u with

$$z(t) \leq u(t) \leq v(t), a \leq t \leq b.$$

Proof: Since f satisfies a local Lipschitz condition, initial value problems are uniquely solvable. Hence4 for every $u_0, z(a) \leq u_0 \leq v(a)$, there exists a unique solution $u(t,t_0)$ (with $t_0 = a$) such that

$$z(t) \leq u(t) \leq v(t), a \leq t \leq b,$$

as follows from Theorem 16. Define the mapping

$$T : \{x : z(a) \leq x \leq v(t)\} \to \{x : z(b) \leq x \leq v(b)\}$$

by

$$Tx = u(b,x),$$

then, since by hypothesis $\{x : z(a) \le x \le v(a)\} \supset \{x : z(b) \le x \le v(b)\}$ and since $\{x : z(a) \le x \le v(a)\}$ is convex, it follows by Brouwer's fixed point theorem and the fact that T is continuous that T has a fixed poin,t completing the proof.

An important consequence of this corollary is that if in addition f is a function which is periodic in t with period $b-a$, then Corollary asserts the existence of a periodic solution (of period $b-a$). The following results use comparison and differential inequality arguments to provide a priori bounds and extendability results.

Theorem: Assume thaat $F:[a,b]\times\mathbb{R}_+ \to \mathbb{R}_+$ is a continuous mapping and that $f:[a,b]\times\mathbb{R}^N \to \mathbb{R}^N$ is continuous also and

$$|f(t,x)| \le F(t,|x|), a \le t \le b, x \in \mathbb{R}^N$$

Let $u:[a,b]\to\mathbb{R}^N$ be a solution of (1) and let $v:[a,b]\to\mathbb{R}_+$ be the continuous and right maximal solution of

$$v'(t) = F(t,v(t)), a \le t \le b,$$
$$v(a) \ge |u(a)|,$$

then $v(t) \ge |u(t), a \le t \le b.$

Proof: Let $z(t) = |u(t)|$, then z is continuous and $D_z(t) = D^- z(t)$. Further

$$D_z(t) = \liminf_{h\to 0+} \frac{|u(t) - |u(t-h)|}{h}$$
$$\le \lim_{h\to 0+} \frac{|u(t-h) - u(t)|}{h}$$
$$= |f(t,u(t))| \le F(t,z(t)).$$

Hence, by Theorem, we conclude that $u(t) \ge |u(t), a \le t \le b.$

Uniqueness Theorems

In this section we provide supplementary conditions which guarantee the uniqueness of solutions of ivp's

Theorem: Assume that $F:(a,b)\times\mathbb{R}_+ \to \mathbb{R}_+$ is a continuous mapping and that $f:(a,b)\times\mathbb{R}^N \to \mathbb{R}\mathbb{R}^N$ is a continuous also and

$$|f(t,x) - f(t,y)| \le F(t,|x-y|), \quad a \le t \le b, x, y \in \mathbb{R}^N,$$

Let $F(t,0) \equiv 0$ and let, for any $c \in (a,b), w \equiv 0$ be the only solution of $w' F(t,w)$ on (a, b) such that $w(t) = 0(\mu(t)), t \to a$ where μ is a given positive and continuous functions. Then connot have distinct solutions such that $| u(t) - v(t) | = 0(\mu(t)), t \to a.$

Proof: Let u,v be distinct solutions of (1) such that $| u(t) - v(t) | = 0(\mu(t)), t \to a.$ Let $z(t) = | u(t) - v(t) |$. Then z is continuous and

$$D^+(t) \leq | f(t,u(t)) - f(t,v(t)) | \leq F(t,z(t)).$$

The proof is completed by employing arguments like those used in the proof of Theorem.

Remark: Theorem does not require that *f* be defined for $t = a$. The advantage of this may be that $a = -\infty$ or that *f* may be singular there. A similar result, of course, holds for $t \to b-$.

$$u' = Au \mid g(t,u)$$

and show how to construct Lyapunov functionals to test the stability of the trivial solution of this system.

The type of Lyapunov functional we shall be looking for are of the form

$$v(x) = x^T Bx,$$

where B is a constant $N \times N$ matrix, i.e. we are looking for v as a quadratic form. If u is a solution of then

$$\frac{dv^*}{dt} = \frac{dv(t,u(t))}{dt}$$

$$= u^T (A^T B + BA)u + g^T (t,u)Bu + u^T Bg(t,u).$$

Hence, given, A if B can be found so that $C = A^T B + BA$ has certain definniteness properties, then the results of the trivial solution. To proceed along these lines we need some linear algebra results.

Proposition: Let A be a constant $N \times N$ matrix having the property that for any eigenvalue λ of A, $= \lambda$ is not an eigenvalue. Then for any $N \times N$ matrix C. there exists a unique $N \times N$ matrix B such that $C = A^T B + BA$.

Proof: On the space of $N \times N$ matrices define the bounded linear operator L by

$$L(B) = A^T B + BA.$$

Then L may be viewed as a bounded linear operator of $\mathbb{R}^{N\times N}$ to itself, hence it will be a bijection provided it does not have 0 as an eigenvalue, Once we show the latter, the result follows. Thus let μ be an eigenvalue of L, i.e., there exists a nonzero matrix B such that

$$L(B) = A^T B +, \mu B$$

Hence

$$A^T B + B(A - \mu I) = 0$$

From this follows

$$B(A - \mu I)^n = (-A^T)^n B,$$

for any integer $n \geq 1$, hence for any polynomial p

$$Bp(A - \mu I)^n = p(-A^T)B.$$

Since, on the other hand if F and G are two matrices with no common eigenvalues, there exists a polynomial p such that $p(F) = I, p(G) = 0$, implies that $A - \mu I$ and $_{-}A^T$ must have a common eigenvalue,. From which follows taht μ is the sum of two eigenvalues of A, which, by hypothesis connot equal 0.

This proposition has the following corollary.

Corollary: Let A be a constant $N \times N$ matrix. Then for any $N \times N$ matrix C, there exists $\mu > 0$ and a unique $N \times N$ matrix B such that $2\mu B + C = A^T B + BA$.

Proof: Let

$$S = \{\lambda \in \mathbb{C} : \lambda = \lambda_1 + \lambda_2\},$$

where λ_1 and λ_2 are eigenvalues of A. Since S is a finite set, there exists $r_0 > 0$. such that $\lambda (\neq 0) \in S$ implies that $|\lambda| > r_0$. Choose $0 < \mu \leq r_0$ and consider the matrix $A_1 = A - \mu I$. We may now apply Proposition 21 to the matrix A_1 and find for a given matrix C, a unique matrix B such that $C = A_1^T B + BA_1$, i.e. $2\mu BC = A^T B + BA$.

Corollary: Let A be a constant $N \times N$ matrix haivng the property that all eigenvalues l of A have negative real parts. Then for any negative definite $N \times N$ matrix B such that $C = A^T B + BA$.

Proof: Let C be a negative definite matrix and let B be given by Proposition, which may be applied since all eigenvalues of A have negative real part. Let $v(x) = x^T Bx$, and let u be a solution of $u' = Au, u(0) = x_0 \neq 0$ Then

$$\frac{dv^*}{dt} = \frac{dv(u(t))}{dt} = u^T (A^T B + BA)u$$

$$= u^T Cu \leq -\mu \mid u \mid^2,$$

since C is negative definite. Since $\lim_{t \to \infty} u(t) = 0$ (all eigenvalues of A have negative real part!), it follows that $\lim_{t \to \infty} v(u(t)) = 0$. We also have

$$v(u(t)) \leq v(x_0) - \int_0^t \mu \mid u(s) \mid^2 ds,$$

from which follows that $v(x_0) > 0$. Hence B is positive definite.

The next corollary follows from stability theory equations and what has just been discussed.

Corollary: A necessary and sufficient condition that an $N \times N$ matrix A have all of its eigenvalues with negative real part is that there exists a unique positive definite matrix B such that

$$A^T B + BA = -I.$$

We next consider the nonlinear problem with

$$g(t, x) = o(\mid x \mid),$$

uniformly with respect to $t \in [t_0, \infty)$, and show that for certain types of matrices A the trivial solution of the perturbed system has the same stability property as that of the unperturbed problem. The class of matrices we shall consider is the following.

Definition: We call an $N \times N$ matrix A critical if all its eigenvalues have non-positive real part and there exists at least one eigenvalue with zero real part. We call it noncritical otherwise.

Theorem: Assume A is a noncritical $N \times N$ matrix and let g satisfy. Then the stability behavior of the trivial solution of is uniformly asymptotically stable if all eigenvalues of A have negative real part and it is unstable if A has an eigenvalue with positive real part.

Proof: Assume all eigenvalues of A have negative real part. By the above exists a unique positive definite matrix B such that

$$A^T B + BA = -I.$$

Let $v(x) = x^T Bx.$ Then v is positive definite and if u is a solution it satisfies

$$\frac{dv^*}{dt} = \frac{dv(u(t))}{dt}$$

$$= -[u(t)|^2| g^T(t,u)Bu + u^T Bg(t,u).$$

Now

$$| gT(t,u)bu + u^T Bg(t,u) | \leq 2 | g(t,u) || B || u |.$$

Choose $r > 0$ such that $| x | \leq r$ implies

$$| g(t,u) | \leq \frac{1}{4} | B |^{-1} | u |,$$

then $$\frac{dv^*}{dt} \leq -\frac{1}{2} | u(t) |^2,$$

as long as $| u(t) | \leq r.$

Let A have an eigenvalue with positive real part. Then there exists $\mu > 0$ such that $2\mu < | \lambda_i + \lambda_j |,$ for all eigenvalues λ_i, λ_j of A, and a unique matrix B such that

$$A^T B + BA = 2\mu B = I,$$

as follows from Corollary 22. We note that B connot be positive definite nor positive semidefinite for otherwise we must have, letting $v(x) = x^T Bx,$

$$\frac{dv^*}{dt} = 2\mu v^*(t) - | u(t) |^2,$$

or

$$e^{-2\mu t} v^*(t) - v^*(0) = -\int_0^t e^{-2\mu s} | u(s)^2 ds$$

i.e.

$$0 \leq v^*(0) - \int_0^t e^{-2\mu s} | u(s) |^2 \, ds$$

for any solution, u, contradicting the fact that solution u exist for which

$$\int_0^t e^{-2\mu s} | u(s) |^2 \, ds$$

becomes unbounded as $t \to \infty$. Hence there exists $x_0 \neq 0$, of arbitrarily small norm, so that $v(x_0) < 0$. Let u be a solution. If the trivial solution were stable, then $|u(t)| \leq r$ for some $r > 0$. Again letting $v(x) = x^T Bx$ we obtain

$$\frac{dv^*}{dt} = 2\mu v^*(t) - |u(t)|^2 + g^T(t,u) \;\; Bu + u^T Bg(t,u)$$

We can choose r so small that

$$2|g(t,u)|\|B\||u| \leq \frac{1}{2}|u|^2, |u| \leq r,$$

hence

$$e^{-2\mu t}v^*(t) - v^*(0) \leq -\frac{1}{2}\int_0^t e^{-2\mu s}|u(s)|^2\, ds,$$

i.e.

$$v^*(t) \leq e^{-2\mu t}v^*(0) \to -\infty,$$

contradicting that $v^*(t)$ is bounded for bounded u. Hence u connot stay bounded, and we have instability.

It it is the case that the matrix A is a critical matrix, the trivial solution of the linear system may still be stable or it may be unstable. In either case, one may construct examples, where the trivial solution of the perturbed problem has either the same or opposite stability behavior as the unperturbed system.

Chapter 21

Global Solution Theorems

In this chapter we shall consider a globalization of the *implicit function theorem* and provide some global bifurcation results. Our main tools in establishing such global results will be the properties of the Leray-Schauder continuation principle. As will be seen in later section, this result also allows us to derive a globalization of the limplicit function theorem and results about global bifurcation in nonlinear equations.

Let O be a bounded open (in the relative topology) subset of $E \times [a, b]$, where E is a real Banach space, and let

$$F : O \to E$$

be a completely continuous mapping. Let

$$f(u,\lambda) = u - F(u,\lambda)$$

and assume that

$$f(u,\lambda) \neq 0 - (u,\lambda) \in \partial O$$

(here ∂O is the boundary of O in $E \times [a,b]$).

Theorem: (The generalized homotopy principle) Let f be given by equation above and satisfied by condistion. Then for $a \leq l \leq b$,

$$d(f(.,_{\lambda})O_{\lambda},0) = \text{constant},$$

(here $O_{\lambda} = \{u \in E : (u,\lambda) \in O\}$).

Proof: We may assume that O ≠ 0 and that

$$a = \inf\{\lambda : O_{\lambda} \neq 0\}, \text{ b = sup } b = \sup\{\lambda : O_{\lambda} \neq 0\}.$$

We let

$$\hat{O} = O \cup O_a \times (a - \varepsilon, a] \cup O_b \times [b, b + \varepsilon),$$

where $\varepsilon > 0$ is fixed. Then $\hat{O}$ is a bounded open subset of $E \times \mathbb{R}$. Let $\tilde{F}$ be the extension of F to $E \times \mathbb{R}$ whose existence is guaranteed by the Dugundji extension theorem. Let

$$\tilde{f}(u,\lambda) = (u - \tilde{F}(u,\lambda), \lambda - \lambda^*).$$

where $a \leq \lambda^* \leq b$ is fixed. Then $\tilde{f}$ is a completely continuous perturbation of the identity in $E \times \mathbb{R}$

Furthermore for any such λ^*

$$\tilde{f}(u,\lambda) \neq 0, (u,\lambda) \in \partial\hat{O},$$

and hence $d(\hat{f}, \hat{O}, 0)$ is defined and constant (for such λ^*). Let $0 \geq t \geq 1$, and consider the vector field

$$\tilde{f}_t(u,\lambda) = (u - t\tilde{F}(u,\lambda) - 1(1-t)$$

$$\tilde{F}(u,\lambda^*), \lambda - \lambda^*),$$

then $\tilde{f}_t(u,\lambda) = 0$ if and only if $\lambda = \lambda^*$ and $u = \tilde{F}(u,\lambda^*)$. Thus, our hypotheses imply that $\tilde{f}_t(n,\lambda) \neq 0$ for $(u,\lambda) \in \partial\hat{O}$ and $t \in [0,1]$. By the homotopy invariance principle (Proposition III.28) we therefore conclude that

$$d(\tilde{f}_1\hat{O}, 0) = d(\tilde{f}, \hat{O}, 0) = d(\tilde{f}_0\hat{O}, 0).$$

On the other hand,

$$d(\tilde{f}_1\hat{O}, 0) = d(\tilde{f}_0, O_{\lambda^*} \times (a - \varepsilon, b + \varepsilon)0),$$

by the excision property of degree. Using the Cartesian product formula Proposition III. 28), we obtain

$$d(\tilde{f}_0, O_{\lambda^*} \times (a - \varepsilon, b + \varepsilon)0) = d(f(.,\lambda^*), O_{\lambda^*}, 0).$$

This completes the proof.

As an immediate consequence we obtain the continuation principle of LeraySchauder.

Theorem: (Leray–Schauder Continuation Theorem) Let O be a bounded open subset of $E \times [a,b]$ and let $f : O \to E$ be given by equation.

Furthermore assume that

$$d(f(.,a), O_a, 0) \neq 0$$

Let

$$S=\{(u,\lambda)\in O: f(u,\lambda)=0\}.$$

Then there exists a closed connected set C is S such that

$$C_a \cap O_a \neq 0 \neq C_b \cap O_b.$$

Proof: It follows from Theorem 1 that

$$d(f(.,a),O_a,0=d(f.,b),O_b,0).$$

Hence

$$S_u \times \{a\}=A\neq 0\neq B=S_b\times\{b\}.$$

Using the complete continuity of F we may conclude that S is a compact metric subspace of E × [a, b]. We now apply Whyburn's lemma with X = S. If there is no such continuum (as asserted above) there will exist compact sets X_A, X_B in such that

$$A\subset X_A, B\subset X_B, X_A\cap X_B$$

$$=0, X_A\cup X_B=X.$$

We hence may find an open set $U\subset E\times[a,b]$ such that $A\subset U\cap O=V$ and $S\cap\partial V=0=V_b$. Therefore

$$d(f(.,\lambda)V_\lambda,0=\text{constant},\ \ \lambda\geq a.$$

On the other hand, the excision principle imples that

$$d(f(.,a),V_a,0)=d(f(.,a),O_a,0).$$

Since V_b = 0. these equalities yield a contradiction, and there exists a continuum as asserted.

In the following examples we shall develop, as a application of the above results, some basic existence results for the existence of solutions of nonlinear boundary value problems.

Example: Let I = [0, 1] and let $g:[0,1]\times\mathbb{R}\to\mathbb{R}$ be continuous. Consider the nonlinear Dirichlet problem

$$\begin{cases} u''+g(x,u)=0, \text{in } I \\ \qquad u=0, \text{on } \partial I. \end{cases}$$

Let there exist constants $a < 0 < b$ such that

$$g(x,a) > 0 > g(x,b) x \in \Omega.$$

Then the equation has a solution $u \in C^2([0,1],\mathbb{R})$ such that

$$a < u(u) < b, x \in I.$$

$$\begin{cases} u'' + \lambda g(x,u) = 0, \text{in } I \\ \qquad u = 0, \text{ on } \partial I. \end{cases}$$

Let G be defined by

$$G(u)(x) = g(x, u(x)),$$

then the equation is equivalent to the operator equation

$$u = \lambda LG(u), u \in C([0,1],\mathbb{R}) = E,$$

where for each $v \in E$, $w = LG(v)$ is the unique solution of

$$w'' + g(x,u) = 0, in I$$
$$w = 0, \ on \ \partial I.$$

It follows that for each $v \in E, LG(v) \in C^2(I)$ and since $C^2(I)$ is compactly embedded in E that

$$LG(.): E \to E$$

is a completely continuous operator, Let $O = \{(u,l) : u \in E, a < u(x) < b, x \in I,$ $0 \leq \lambda \leq 1\}$. Then O is an open and bounded set is $E \times [0,1]$. If (u, λ) $(u,\lambda) \in \partial O$ is solution of the equation, then there will either exist $x \in I$ such that $u(x) = b$ or there exists $x \in I$ such that $u(x) = a$ and $\lambda > 0$. In either case, yields, via elementary calculus, a contradiction. Hence (4) has no solution in ∂O. Therefore

$$\text{d}(\text{id} - \lambda LG, O_\lambda, 0) = \text{d}(\text{id}, O_0, 0) = 1$$

and Theorem imples the existence of a continuum C of solutioins of equation, hence of the equation, such that $C \cap E \times \{0\} = \{0\}$ and $C \cap E \times \{1\} \neq 0$.

Globalization of the Implicit Funcation The orem

Assume that

$$F: E \times \mathbb{R} \to E$$

is a completely continuous mapping and consider the equation

$$f(u,\lambda) = u - F(u,\lambda) = 0.$$

Let (u_0,λ_0) be a solution of equation such that the condition of the *implicit function theorem* hold at (u_0,λ_0). Then there is a solution curve $\{(u(\lambda),\lambda)\}$ of the equation defined in a neighborhood of λ_0, passing through (u_0,λ_0). Furthermore the conditions of Theorem imply that the solution u_0 is an isolated solution of equation at $\lambda = \lambda_0$, and if O is an isolating neighborhood, we have that

$$d(f(.,\lambda_0),O,0) \neq 0.$$

We shall now show that condition alone suffices to guarantee that equation has global solution branch in the half spaces $E\times[\lambda_0,\infty)$ and $E\times(-\infty,\lambda_0]$.

Theorem: Let O be a bounded open subset of E and assumle that for $\lambda = \lambda_0$ equation has unique solution in O and let hold, Let

$$\vartheta^+ = \{(u,\lambda) \in E\times[\lambda_0,\infty) : (u,\lambda) \text{ solves}$$

and

$$\vartheta^+ = \{(u,\lambda) \in E\times[-\infty,\lambda_0) : (u,\lambda) \text{ solves}$$

Then there exists a continuum $C^+ = \subset \vartheta + (C^- \subset \vartheta^-)$ such sthat:

1. $C^+_{\lambda_0} \cap O = \{u_0\} C^-_{\lambda_0} \cap O\{u_0\})$,
2. $C^+$is either unbounded in $E\times[\lambda_0,\infty)$ (C^-*is* bounded in $E\times[-\infty,\lambda_0)$ or $C^+_{\lambda_0} \cap O(E\setminus\overline{O}) \neq 0 (C^-_{\lambda_0} \cap O(E\setminus\overline{O}) \neq 0)$

Proof: LET C^+ be the maximal connected subset of ϑ^+ su;ch; that 1. above holds, Assume that $C^+ \cap (E\setminus\overline{O}) = 0$ and that C^+ is bounded in $E\times[_0,\infty)$. Then there exists a constant $R > 0$ such that for each $(u, \lambda) \in C^+$ we have that $\| u \| + | y | < R$. Let

$$\vartheta^+_{2R} = \{u,\lambda) \in \vartheta^+ : \| u \| + | \lambda | \leq 2R,$$

then ϑ^+_{2R} is a compact subset of $E\times[\lambda_0,\infty)$, and hence is a compact metric space. There are two possiblities: Either $\vartheta^+_{2R} = C^+$ or else there exists $(u,\lambda) \in \vartheta^+_{2R}$ such that $(u,\lambda) \notin \vartheta^+_{2R}$.

In either case, we may find a bounded open set $U \subset E \times [\lambda_0, \infty)$ such that $U_{\lambda_0} = O$, $\vartheta_{2R}^{+} \cap \partial U = 0C \subset U$. It therefore follows from Theorem 1 that

$$d(f(.,\lambda_0), U_{\lambda_0}, 0) = \text{consant}, \;\; \lambda \geq \lambda_0,$$

where this constant is given by

$$d(f(.,\lambda_0), U_{\lambda_0}, 0) = d(f(.,\lambda_0), O, 0)$$

which is nonzero, because of the equation. On the other hand, there exists $\lambda^* > \lambda_0$ such that U_{λ^*} contains not solutions of the equation and hence $d(f(.,\lambda^*), U_{\lambda^*}, 0) = 0$, contradicting equation.

The existence of C^{-} with the above listed properties is demonstrated in a similar manner.

Remark: The assumption of Theorem that u_0 is the unique solution of equation inside the set O, was made for convenience of proof. If one only assume, one may obtain the conclusion that the set of all such continua is either bounded in the right (left) half space, or else there exists one such continuum which meets the $\lambda = \lambda_0$ hyperplane outside the set $\overline{O}$.

Remark: If the component $C^{+}$of Theorem 4 is bounde dand $\tilde{O}$ is an isolating neighborhood of $C^{+} \cap (E \setminus O) \times k \setminus \{\lambda_0\}$, then it follows from the excision property, Whyburn's lemma, and the generalized homotopy principle that

$$d(f(.,\lambda_0, O, 0) \;\; = -d(f(.,\lambda_0), \tilde{O}, 0).$$

This observation has the following important consequence. If equation has, for $\lambda = \lambda_0$ only isolated solutions and if the integer given by the equation has the same sign with respect to isolating neighborhoods O for all such solutions where equation holds, then all contiua C^{+}must be unbounded.

Example: Let $z \in \mathbb{C}$, be a polynomial of degree n whose leading coefficient is (without loss in generality) assumed to be 1 and let $q(z) = \prod_{i=1}^{n} (z - a_i)$, where $a_1, \ldots, a_n$ are distinct complex numbers. Let

$$f(z, \lambda) = \lambda p(z) + (1 - \lambda) q(z).$$

Then f may be considered as a continuous mapping

$$f : \mathbb{R}^2 \times \mathbb{R} \rightarrow \mathbb{R}^2.$$

Furthermore for $\lambda \in [0,r]$, $r > 0$, there exists a constant R such that any solufion of

$$f(z,\lambda) = 0$$

satisfies $|z| < R$. For all $\lambda \geq 0$, which is unbounded with respect to λ, and must therefore reach of $p(z)$ must be connected to some a_i (apply the above argument backwards from the $\lambda = 1$ – level, if need be).

The Theorem of Krein -Rutman

In this secton we shall employ Theorem to prove an extension of the *Perron-Frobenius* theorem about eigenvalues for postive matrices. The *Krein-Rutman* is a generalization of this classical result to positive compact operators on a not necessarily funite dimensional Banach space.

Let E be a real Banach space and let K be a cone in E, i.e., a closed convex subset of E with the properties:

1. For all $u \in K, t \geq 0, tu \in K$.
2. $K \cap \{-K\} = \{0\}$.

It is an elementary exercise to show that a cone induces a partial orader $\leq$ on E by the convention $u \leq v$ if and only if $v - u \in K$. A linear operator $L : E \to E$ is called *Positive* whenever K is an invariant set for L, i.e. $L : K \to K$. If K is a cone whose interior int$K$ is nonempty, we call L a *strongly positive* operator, whenever $L : K \setminus \{0\} \to \text{int}\, K$.

Theorem: Let E be a real Banach space a cone K and let $L : E \to E$ be a postive compact linear operator, Assume there exists $w \in K$, $w \neq 0$ and a constant $m > 0$ such that

$$w \geq mLw,$$

where $\leq$ is the partial order induced by K. Then there exists $\lambda_0 > 0$ and $u \in K, \|u\| = 1$, such that

$$u = \lambda_0 Lu.$$

Proof: Restrict the operator L to the cone K and denote by $\tilde{L}$ the Dugundji extension of this operator of E. Since L is a compact linear operator, the operator $\tilde{L}$ is a completely; continuous mapping with $\tilde{L}(E) \subset K$. Choose $\varepsilon > 0$ and consider the equation

$$u - \lambda \tilde{L}u + \varepsilon w) = 0$$

For $\lambda = 0$, equatoin (11) has the unique solution $u = 0$ and we may apply Theorem 4 to obtain an unbounded continuum $C_\varepsilon^+ \subset E \times [0, m]$. Since $C_\varepsilon^+ K, \| u_\varepsilon \| = 1$, such that

$$u_\varepsilon - \lambda_\varepsilon L(u_\varepsilon + \varepsilon w).$$

Since L is compact, the set $\{(u_\varepsilon, \lambda\varepsilon)\}$ will contain a convergent subseequence (letting $\varepsilon \to 0$), converging to, say, (λ, λ_0). Since clearly $\|u\| = 1$, it follows that $\lambda_0 > 0$.

If it is the case that L is a strongly positive compact linear operataor, much more can be asserted; this will be done in the theorem of Krein-Rutman which we shall establish as a corollary of Theorem.

Theorem: Let E have a cone K, whose interior, int$K \neq 0$. Let L be a strongly positive compact linear operator. Then there exists a unique $\lambda_0 > 0$.with the following properties.

1. There exists $u \in \operatorname{int} K$, with $u = \lambda_0 L_u$.
2. If $\lambda (\in \mathbb{R}) \neq \lambda_0$ is such that there exist $v \in E$, $v \neq 0$, with $v = \lambda L v$, then $v \notin K \cup \{-K\}$ $\lambda_0 < | \lambda |$.

Proof: Choose $Lw \in \operatorname{int} K$, there exists $\delta w \geq \lambda w$. We therefore many apply Theorem to obtain $\lambda_0 > 0$ and $u \in K$ such that $u = \lambda_0 L_u$. Since L is strongly positive, we must have that $u \in \operatorname{int} K$, If $(v, \lambda) \in (K \setminus \{0\} \times (0, \infty)$ is such that $v = \lambda L v$, then $v \in \operatorname{int} K$. Hence, for all $\delta > 0$, sufficiently small, we have that $u - \delta v \in \operatorname{int} K$. Consequently, there exists a maximal $\delta^* > 0$, such that $u - \delta^* v \in K$, Consequently, there exists a maxiumal $\delta^* > 0$ such that $u - \delta^* v \in K$, i.e.e $u - rv \notin K$, r > r > δ^*. Now

$$L(u) - \delta, v) = \frac{1}{\lambda_0}(u - \frac{\lambda_0}{\lambda}\delta^* v),$$

which imples that $u - \frac{\lambda_0}{\lambda}\delta^* v \in \operatorname{int} K$, unless $u - \delta^* v = 0$. If the latter holds, then $\lambda_0 = \lambda$, if not, then $\lambda_0 = \lambda$, because δ^* is maximal, If $\lambda_0 = \lambda$, a contradiction. Hence it must be the case that $\lambda_0 = \lambda$, We have therefore proved that λ_0 is the only characteristic value of L having an *eigendirection* in the cone K and further that any other eigenvector corresponding to λ_0 must be a constant multiple of u i.e. λ_0 is a characteristic value of L of geometric multiplicity one, i.e. that dimension of the kernel of id- λ_0 L equals one.

Next let $\lambda \neq \lambda_0$ be another characteristic value of L and let $v \neq 0$ be such that $v \notin K \cap \{-K\}$. Again, for $|\delta|$ small, $u - \delta v \in \text{int}\, K$ and there exists $\delta^* > 0$, maximal, such that $u - \delta^* v \in K$, and there exists $\delta^* < 0$ minimal, such that $u - \delta_* v \in K$. Now

$$L(u) - \delta^* v) = \frac{1}{\lambda_0}(u - \frac{\lambda_0}{\lambda}\delta^* v) \in K,$$

and

$$L(u) - \delta_* v) = \frac{1}{\lambda_0}(u - \frac{\lambda_0}{\lambda}\delta_* v) \in K$$

Thus, if $\lambda > 0$, we conclude that $\lambda_0 < \lambda$, whereas, if $\lambda > 0$, we get that $\lambda_0 \delta^* < \lambda \delta_*$, and $\lambda_0 \delta_* > \lambda \delta^*$, i.e. $\lambda_0^2 < \lambda^2$.

As observed above we have that λ_0 is a characteristic value of geometric multiplicity one. Before giving an application of the above result, we shall establish that λ_0, in fact also has algebraic multiplicity one. Recall from the Riesz theory of compact linear operators that the operator $id - \lambda_0 L$ has the following property:

There exists a minimal integer n such that

$$\ker(id - \lambda_0 L)^n = \ker(id - \lambda_0 L)^{n+1}$$

$$= \ker(id - \lambda_0 L)^{n+2} = ...,$$

and the dimension of the generalized eigenspace WithwWith $\ker(id - \lambda_0 L)^n$ is called the algebraic multiplicity of λ_0.

With this terminology, we have the following addition to Theorem.

Theorem: Assume the conditions and let λ_0 is a characteristic value of L of algebraic multiplicity one.

Proof: We assume the contrary. Then, since ker (id$-\lambda_0 L$) has dimension one, it follows that there exists a smallest integer $n > 1$ such that athe generalized eigenspace in given by ker(id$-\lambda_0 L)^n v = 0$ and $(id-\lambda_0 L)^{n-1} v = w \neq 0$. It follows from Theorem and its proof that w = ku, where *u* is given by Theorem and *k* may assumed to be positive. Let $z = (\text{id} - \lambda_0 L)^n$ ^{-2}v, then $z - \lambda_0 Lz = ku$, and hence, by induction, we get that $\lambda_0^m L^m z = z - mku$, for any positive integer *m*. If follows therefore that $z \notin K$, for otherwise $\frac{1}{m}z - ku \in K$, for any integer *m*, implying that $-ku \in K$, a contradiction. Since $u \in \text{int}\, K$, there exist $\alpha > 0$ and $y \in K$ such that $z = \alpha u - y$. Then $y_0^m L^m z = \alpha u - y_0^m L^m y$, or $y_0^m L^m y = y + mku$. Choose $\beta > 0$, such that $y \leq \beta u$, then $y_0^m L^m y \leq \beta u$, and by the above we see that $y + mku \leq \beta u$. Dividing this inequality my *m* and letting $m \to \infty$, we obtain that $ku \in -K$, a contradiction.

Remark: It may be the case that, aside from real characteristic values, *L* also has complex ones, If μ is such a characteristic value, then it may be shown that $|\mu| > \lambda_0$, where λ_0 is as in Theorem 10. We refer the interrested reader to Krasnosel's skii [24] for a verification.

Global Bifurcation

As before, let *E* be a real Banach space and let $f : E \times \mathbb{R} \to E$ have the form

$$f(u, \lambda) = u - F(u, \lambda),$$

where $F : E \times \mathbb{R} \to E$ is completely continuous.We shall now assume that

$$F(0, \lambda) \equiv 0, \lambda \in \mathbb{R},$$

and hence that the equation

$$f(u, \lambda) = 0,$$

has the trivial solution for all values of λ. We shall now condier the questioni of bifurcation from this trivial branch of solutions and demonstrate the existence of global *branches* of nontrivial solutions bifurcating from the trivial branch. Our main tools will again be the properties of the Leray-Schauder degree and Whyburn's lemma.

We shall see that this results is an extension of the local bifurcation theorem, Theorem II. 6.

Theorem: Let there exist $a, b \in \mathbb{R}$ with $a < b$, such that $u = 0$ is an isolated solution of for $\lambda = a$ and $\lambda = b$, are not bifurcation points, furthermore assume that

$$d(f.,a),B_r(0),0) \neq d(f(.,b),B_r(0),0),$$

where $B_r(0) = \{u \in E : \| u \| < r\}$. is an isolating neighborhood of the trivial solution. Let

$$\vartheta = \overline{\{(u,\lambda):(u,\lambda) solves (15) with u \neq 0\}} \cup \{0\}\times[a,b]$$

and let $\varsigma \subset \vartheta$ be the maximal connected subset of ϑ which contains $\{0\}\times[a,b]$. Then either

1. ς is unbounded in $E\times\mathbb{R}$, or else
2. $\varsigma \cap \{0\}\times(\mathbb{R}\setminus[a,b]) \neq 0$.

Proof: Define a class u of subsets of $E\times\mathbb{R}$ *as* follows

$$u = \{\Omega \subset E\times\mathbb{R} : \Omega = \Omega_0 \cup \Omega\infty\},$$

where $\Omega_0 = B_r(0)\times[a,b]$, and Ω_∞ is a bounded open subset of $(E\setminus\{0\})\times\mathbb{R}$. We shall first show that has a nontrivial solution $(u,\lambda)\in\partial\Omega$ for any such $\Omega\in u$. To accomplish this, let us consider the following sets:

$$\begin{cases} K = f-1(0)\cap\bar{\Omega}, \\ A = \{0\}\times[a,b], \\ B = f-1(0)\cap(\partial W\setminus(B_r(0)\times\{a\}\cup B_r(0)\times\{b\})). \end{cases}$$

We observe that K may be regarded as a compact metric space and A and B are compact subsets, of K. We hence may apply Whyburn's lemma to deduce that eitgher there exists a continuum in K connecting A to B or else, there is a separatoin $K_{A'}$ K_B of K, With $A\subset K_A, B\subset K_B$. If the latter holds, we may find open sets U, V in $E\times\mathbb{R}$ such that $K_A\subset U, K_B\subset V$, with $U\cap V = 0$. We let $\Omega^* = \Omega\cap(U\cup V)$ and observe that $\Omega^*\in u$. It follows, by construction, that there are no nontrivial solutions of (15) which belong to $\partial\Omega^*$; this, however, is impossible, since, it would imply, by the generalized homotopy and the excision principle of Leray-Schauder degree, that $d(f(.,a),B_r(0),0) = d(f.,b),B_r(0),0)$, contradicting equation. We hence have that for each $\Omega\in u$ there is a continuum C of solutions of equation which intersects $\partial\Omega$ in a nontrivial solution.

We assume now that neither of the alternatives of the theorem hold, i.e. we assume that ς is bounded and $\varsigma\cap\{0\}\times(\mathbb{R}\setminus[a,b]) = 0$. In this case, we may, using the boundednes of ς, construct a set $\Omega\in u$, containing no nontrivial solutions in its boundary, thus arriving once more at a contradiction.

We shall, throughout this text, apply the above theorem to severaql problems for nonlinear differential equaions. Here we shall, for the sake of illustration provide two simply one dimensional example.

Example: Let $f:\mathbb{R}\times\mathbb{R}\to\mathbb{R}$, be given by

$$f(u,\lambda)=u(u^2+\lambda^2-1)$$

It is easy to see that

$$\vartheta=\{u,\lambda):u^2+\lambda^2=1\}\cup\{0\}\times(-\infty,\infty),$$

and hence that (0,–1) and (0,1) are the only bifurcation points from the trivial solution. Furthemore, the bifurcating continuum is bounded. Also one may quickly check that (16) holds, with *a, b* chosen in a neighborhood of $\lambda=-1$ and also in a neighborhood of $\lambda=1$.

Example: Let $f:\mathbb{R}\times\mathbb{R}\to\mathbb{R}$ be given by

$$f(u,\lambda)=(1-\lambda)u+u\sin\frac{1}{u}.$$

In this case ϑ is given by

$$\vartheta=\{(u,\lambda):\lambda-1=\sin\frac{1}{u}\}\cup\{0\}\times[0,2],$$

which is an unbounded set, and we may check that (16) holds, by choosing $a<0$ and $b<2$.

In many interesting cases the nonlinear mapping *F* is of the special form

$$F(u,\lambda)=\lambda Bu+o(\|u\|), as\ \|u\|\to 0,$$

where *B* is a compact linear operator, In this case bifurcation points from the trivial solution are isolated, in fact one has the following necessary conditions for bifurcation.

Proposition: Assume that *F* has the form the equation, where *B* is the Frechet derivative of *F*. If $(0,\lambda_0)$ is a bifurcation point from the trivial solution for equation, then λ_0 is a characteristic value of *B*.

Using this results, Theorem, and the Leray-Schauder formula for computing the degree of a compact linear perturbation of the identity, we obtain the follwing results.

Theorem: Assume that *F* has the form and let λ_0 be a characteristic value of *B* which is of odd algebraic multiplicity. Then there exists a continuum ς of nontrivial solutions of which bifurcates from the set of trivial solutions at $(0,\lambda_0)$ and ς is either unbounded in $E \times \mathbb{R}$ or else ς also bifurcates from the trivial solution set at $(0,\lambda_1)$ where λ_1 is another characteristic value of *B*.

Proof: Since λ_0 is isolated as a characteristic value, we may find $a < \lambda_0 < b$ such that the interval [*a*,*b*] contains, besides λ_0 no oter characteristic values. It follows that the trivial solution is an isolated solution (in *E*) for $\lambda = a$ and $\lambda = b$. Hence, $\mathrm{d}(f.,a), B_r(0)0)$ and $\mathrm{d}(f.,b), B_r(0)0)$ are defined for *r*, sufficiently small and are, respectively, given by $\mathrm{d}(id - aB, B_r(0), 0)$, amd $\mathrm{d}(id - bB, B_r(0), 0)$. On other hand,

$$\mathrm{d}(\mathrm{id} - aB, B_r(0), 0) = (-1)^{\mathrm{b}}\,\mathrm{d}(\mathrm{id} - bB, B_r(0), 0)$$

where β equals the algebraic multiplicity of λ_0 as a characteristic value of *B*.

The following example serves to demonstrate that, in general, not every characteristic value will yield a bifurcation point.

Example: The system of scalar equations

$$\begin{cases} x = \lambda x + y^3 \\ y = \lambda y - x^3 \end{cases}$$

has only the trivial solution *x* = 0 = *y* for all values of λ. We note, that $\lambda_0 = 1$ is a characteristic value of the Frechet derivative of multiplicity two.

As a further example let us consider a boundary value problem for a second order ordinary differential equation, the pendulum equation.

Example: Consider the boundary value problem

$$\begin{cases} u'' + \lambda \sin u = 0, x \in [0,\pi] \\ u(0) = 0, u(\pi) = 0. \end{cases}$$

As already observed this problem is equivalent to an operator equation

$$i = \lambda F(u)$$

where

$$F : C[0,\pi] \to C[0,\pi]$$

is a completely continuous operator which is continuously Frechet differentiable with

Frechet derivative $F(0)$. On the other hand, to find these eigenvalues is equivalent to finding the values of λ for which

$$\begin{cases} u''+\lambda u = 0, x \in [0,\pi] \\ u(0)=0, u(\pi)=0. \end{cases}$$

has nontrivial solution. These values are given by

$$\lambda = 1, 4 \ldots k^2, \ldots, k \in \mathbb{N}$$

Furthermore we know from elementary differential equations that each such eigenvalue has a one-dimensonal eigenspace and one may convince oneself that the above theorem may be applied at each such eigenvalue and conclude that each value.

$$(0, k^2), k \in \mathbb{N}$$

is a bifurcation point for the equation.

$$u' = Au \mid g(t,u)$$

and show how to construct Lyapunov functionals to test the stability of the trivial solution of this system.

The type of Lyapunov functional we shall be looking for are of the form

$$v(x) = x^T Bx,$$

where B is a constant $N \times N$ matrix, i.e. we are looking for v as a quadratic form. If u is a solution of then

$$\frac{dv^*}{dt} = \frac{dv(t,u(t))}{dt}$$

$$= u^T (A^T B + BA)u + g^T (t,u)Bu + u^T Bg(t,u).$$

Hence, given, A if B can be found so that $C = A^T B + BA$ has certain definniteness properties, then the results of the trivial solution. To proceed along these lines we need some linear algebra results.

Proposition: Let A be a constant $N \times N$ matrix having the property that for any eigenvalue λ of A, $= \lambda$ is not an eigenvalue. Then for any $N \times N$ matrix C. there exists a unique $N \times N$ matrix B such that $C = A^T B + BA \cdot$

Proof: On the space of $N \times N$ matrices define the bounded linear operator L by

$$L(B) = A^T B + BA \cdot$$

Then L may be viewed as a bounded linear operator of $\mathbb{R}^{N\times N}$ to itself, hence it will be a bijection provided it does not have 0 as an eigenvalue, Once we show the latter, the result follows. Thus let μ be an eigenvalue of L, i.e., there exists a nonzero matrix B such that

$$L(B) = A^T B+,\mu B$$

Hence

$$A^T B + B(A - \mu I) = 0$$

From this follows

$$B(A - \mu I)^n = (-A^T)^n B,$$

for any integer $n \geq 1$, hence for any polynomial p

$$Bp(A - \mu I)^n = p(-A^T)B.$$

Since, on the other hand if F and G are two matrices with no common eigenvalues, there exists a polynomial p such that $p(F) = I, p(G) = 0$, implies that $A - \mu I$ and $_{-A^T}$ must have a common eigenvalue,. From which follows taht μ is the sum of two eigenvalues of A, which, by hypothesis connot equal 0.

This proposition has the following corollary.

Corollary: Let A be a constant $N \times N$ matrix. Then for any $N \times N$ matrix C, there exists $\mu > 0$ and a unique $N \times N$ matrix B such that $2\mu B + C = A^T B + BA$.

Proof: Let

$$S = \{\lambda \in \mathbb{C} : \lambda = \lambda_1 + \lambda_2\},$$

where λ_1 and λ_2 are eigenvalues of A. Since S is a finite set, there exists $r_0 > 0$. such that $\lambda(\neq 0) \in S$ implies that $|\lambda| > r_0$. Choose $0 < \mu \leq r_0$ and consider the matrix $A_1 = A - \mu I$. We may now apply Proposition 21 to the matrix A_1 and find for a given matrix C, a unique matrix B such that $C = A_1^T B + BA_1$, i.e. $2\mu BC = A^T B + BA$.

Corollary: Let A be a constant $N \times N$ matrix haivng the property that all eigenvalues l of A have negative real parts. Then for any negative definite $N \times N$ matrix B such that $C = A^T B + BA.$

Proof: Let C be a negative definite matrix and let B be given by Proposition, which may be applied since all eigenvalues of A have negative real part. Let $v(x) = x^T Bx$, and let u be a solution of $u' = Au, u(0) = x_0 \neq 0$ Then

$$\frac{dv^*}{dt} = \frac{dv(u(t))}{dt} = u^T (A^T B + BA)u$$

$$= u^T Cu \leq -\mu |u|^2,$$

since C is negative definite. Since $\lim_{t\to\infty} u(t) = 0$ (all eigenvalues of A have negative real part!), it follows that $\lim_{t\to\infty} v(u(t)) = 0$. We also have

$$v(u(t)) \leq v(x_0) - \int_0^t \mu |u(s)|^2 \, ds,$$

from which follows that $v(x_0) > 0$. Hence B is positive definite.

The next corollary follows from stability theory equations and what has just been discussed.

Corollary: A necessary and sufficient condition that an $N \times N$ matrix A have all of its eigenvalues with negative real part is that there exists a unique positive definite matrix B such that

$$A^T B + BA = -I.$$

We next consider the nonlinear problem with

$$g(t, x) = o(|x|),$$

uniformly with respect to $t \in [t_0, \infty)$, and show that for certain types of matrices A the trivial solution of the perturbed system has the same stability property as that of the unperturbed problem. The class of matrices we shall consider is the following.

Definition: We call an $N \times N$ matrix A critical if all its eigenvalues have non-positive real part and there exists at least one eigenvalue with zero real part. We call it noncritical otherwise.

Theorem: Assume A is a noncritical $N \times N$ matrix and let g satisfy. Then the stability behavior of the trivial solution of is uniformly asymptotically stable if all eigenvalues of A have negative real part and it is unstable if A has an eigenvalue with positive real part.

Proof: (i) Assume all eigenvalues of A have negative real part. By the above exists a unique positive definite matrix B such that

$$A^T B + BA = -I.$$

Let $v(x) = x^T Bx$. Then v is positive definite and if u is a solution it satisfies

$$\frac{dv^*}{dt} = \frac{dv(u(t))}{dt}$$

$$= -[u(t)|^2| g^T(t,u)Bu + u^T Bg(t,u).$$

Now

$$| gT(t,u)bu + u^T Bg(t,u) | \leq 2 | g(t,u) || B || u |.$$

Choose $r > 0$ such that $| x | \leq r$ implies

$$| g(t,u) | \leq \frac{1}{4} | B |^{-1} | u |,$$

then

$$\frac{dv^*}{dt} \leq -\frac{1}{2} | u(t) |^2,$$

as long as $| u(t) | \leq r$.

(ii) Let A have an eigenvalue with positive real part. Then there exists $\mu > 0$ such that $2\mu < | \lambda_i + \lambda_j |$, for all eigenvalues λ_i, λ_j of A, and a unique matrix B such that

$$A^T B + BA = 2\mu B = I,$$

as follows from Corollary 22. We note that B connot be positive definite nor positive semidefinite for otherwise we must have, letting $v(x) = x^T Bx$,

$$\frac{dv^*}{dt} = 2\mu v^*(t) - | u(t) |^2,$$

or

$$e^{-2\mu t} v^*(t) - v^*(0) = -\int_0^t e^{-2\mu s} | u(s)^2 ds$$

i.e.

$$0 \leq v^*(0) - \int_0^t e^{-2\mu s} | u(s) |^2 \, ds$$

for any solution, u, contradicting the fact that solution u exist for which

$$\int_0^t e^{-2\mu s} |u(s)|^2 \, ds$$

becomes unbounded as $t \to \infty$. Hence there exists $x_0 \neq 0$, of arbitrarily small norm, so that $v(x_0) < 0$. Let u be a solution. If the trivial solution were stable, then $|u(t)| \leq r$ for some $r > 0$. Again letting $v(x) = x^T Bx$ we obtain

$$\frac{dv^*}{dt} = 2\mu v^*(t) - |u(t)|^2 + g^T(t,u) \;\; Bu + u^T Bg(t,u)$$

We can choose r so small that

$$2|g(t,u)||B||u| \leq \frac{1}{2}|u|^2, |u| \leq r,$$

hence

$$e^{-2\mu t} v^*(t) - v^*(0) \leq -\frac{1}{2}\int_0^t e^{-2\mu s} |u(s)|^2 \, ds,$$

i.e.

$$v^*(t) \leq e^{-2\mu t} v^*(0) \to -\infty,$$

contradicting that $v^*(t)$ is bounded for bounded u. Hence u connot stay bounded, and we have instability.

It it is the case that the matrix A is a critical matrix, the trivial solution of the linear system may still be stable or it may be unstable. In either case, one may construct examples, where the trivial solution of the perturbed problem has either the same or opposite stability behavior as the unperturbed system.

Chapter 22

The Fundamental Solution of the Heat Equation

In this chapter we return to the subject of the heat equation. The heat equation reads

$$ut = k \nabla^2 u$$

and was first derived by Fourier. The same equation describes the diffusion of a dye or other substance in a still fluid, and at a microscopic level it results from random processes. We shall use this physical insight to make a guess at the *fundamental solution* for the heat equation.

The term fundamental solution is the equivalent of the Green function for a parabolic PDE like the heat equation. Since the equation is homogeneous, the solution operator will not be an integral involving a forcing function. Rather, the solution responds to the initial and boundary conditions. In order to limit the complications, we shall begin with the problem of diffusion in one space dimension, and place the boundaries at infinity, so the conditions will not be looked at explicitly. (Implicitly, there is a condition of boundedness; if we allowed the addition of $u_{unbounded}(t,x) = x$, for example, the problem would be in the second Fredholm alternative.)

Defination: A *well-posed* problem for the one-dimensional heat equation would be of the form:

$$ut = k\ u_{xx}.$$

for $-\infty < x < \infty$ and $0 < t <$, with given initial data:

$$u(0,x) = f(x).$$

The solution depends linearly on the initial data f(x), a homogeneous linear equation, so we can hope that the solution operator is an integral operator of the form

$$u(t,x) = \int K(t : x\xi) f(\xi) d\xi$$

Definition: Suppose that A is a linear operator on a suitable space of functions of x, and that u satisfies

$$u_t = A\ u,\ u(0,x) = f(x).$$

If there is a solution operator of the form, then K(t; x, ξ) is referred to as a *fundamental* solution.

Au usual for Green functions, there is an intuitive physical meaning to the fundamental solution: It is the solution with initial data

$$u(0,x) = (x - \xi).$$

Once we have this solution, any solution with other initial data can be thought of as a continuous superposition of solutions with these simpler, but singular, initial conditions.

There are several ways to derive the fundamental solution of the heat equation in unrestricted space.

For example, we could apply either the Fourier or Laplace transforms, in the spatial variables x, to obtain a differential equation involving only t, and then transform back. Instead, here we use the physics of diffusion as a hint to form an ansatz for the solution and then make a calculation.

We take the opportunity at this stage to scale time to make the constant in the heat equation k =1; the mathematically natural time variable is t' = kt.)

The hint is that the underlying mechanism is some sort of random process which is isotropic, unrestricted in space, and invariant in time, then we might expect that the density u(t,x) could be a Gaussian distribution, and once it has this distribution, it should remain a Gaussian, although it will spread out as time goes by.

A Gaussian density on the x-axis, centered at x = 0, is of the form

$$u(t, x) = \frac{1}{\sqrt{4\pi n}} \exp\left(\frac{-x^2}{4a}\right).$$

The normalization is taken so that the integral of the Gaussian density is 1. If we are right, we should be able to find a function a(t) so that the function solves the one-dimensional heat equation. Let's plug in and see what we get:

$$u_1 (t, x) = \frac{1}{\sqrt{4\pi}} \exp\left(\frac{-x^2}{4a}\right)\left(-\frac{a'}{2a^{3/2}} + \frac{1}{a^{1/2}} \frac{x^2 a'}{4a^2}\right)$$

whereas a slightly longer calculation reveals that

$$u_\pi (t,x) = \frac{1}{\sqrt{4\pi}} \exp\left(\frac{-x^2}{4a}\right)\left(-\frac{1}{2a^{3/2}} + \frac{1}{a^{1/2}} \frac{x^2}{4a^2}\right)$$

These differ only by the factor a'(t). Gratifyingly, the ansatz works with the very simple choice a(t) = t. We now take a closer look at the fundamental solution we have found:

Definition: The function

$$K(t,x\ \xi) = \frac{1}{\sqrt{4\pi}}\exp\left(\frac{(x-\xi)^2}{4t}\right)$$

is known as the one-dimensional *heat kernel.*

We have not yet explained in what sense the heat kernel approaches the delta function as t tends to 0. Notice that the Gaussian distribution of the heat kernel becomes very narrow when t is small, while the height scales so that the integral of the distribution remains one. Hence if we integrate it by any continuous, bounded function f(pix/bfxi.gif) and take the limit, we will in fact get f(x).

In the contrary limit, the distribution becomes very broad, corresponding to our experience that a diffusing substance tends to a widespread, nearly uniform, density.

Example: Suppose k =1 the initial condition is

$$u(0, x) = \begin{cases} x, \ 0 < x < 1 \\ 0, \text{ other rwise.} \end{cases}$$

Solve for u(t,x).

Solution. We perform the integral

$$u(t,x) = \frac{1}{\sqrt{4\pi}}\int_0^1 \exp\left(\frac{(x-\xi)^2}{4t}\right)\xi d\xi$$

with software, and find

$$\frac{1}{2\sqrt{\pi}\sqrt{t}}\left(-e\frac{(-1+x)^2}{4t}\sqrt{t}\right.$$

$$\left(2\sqrt{t}+e\frac{(-1+x)^2}{4t}\sqrt{\pi}\ x\ \text{Erf}\left[\frac{-1+x}{2\sqrt{t}}\right]\right)+$$

$$\left.\frac{x^2}{e\ 4t}\sqrt{t}\left(2\sqrt{t}+e\frac{x^2}{4t}\sqrt{\pi}\ x\ \text{Erf}\left[\frac{x}{2\sqrt{t}}\right]\right)\right)$$

Here Erf denotes the *error function* of statistics,

$$\text{Ef}[z]: = \frac{2}{\sqrt{\pi}}\int_0^1 e^{-t^1}ds.$$

An additional convenient feature of the heat kernel is that the extension to many dimensions is immediate, as can be verified by directly plugging into the heat equation:

Definition. The function

$$K(t,x,\xi) = \frac{1}{(4\pi t)^{\pi/2}} \exp\left(-\frac{|x-\xi|^2}{4t}\right)$$

is the *heat kernel* for R^n.

In addition to helping us solve problems like the solution of the heat equation with the heat kernel reveals many things about what the solutions can be like. For example, if f(x) is any bounded function, even one with awful discontinuities, we can differentiate the expression under the integral sign. In fact, we can differentiate as often as we like. This means that an initially irregular distribution of temperature or a diffusing substance, is instantaneously smoothed out. Perhaps this is not too surprising when we think that the heat kernel itself solves the heat equation, and changes instantaneously from a delta function to a smooth, Gaussian distribution.

Example: Again consider the one-dimensional heat equation with k = 1, but include a source of heat which is constant in time:

$$u_t - u_{xx} = f(x),$$
$$u(0,x) = g(x)$$

for two given functions f(x) and g(x) defined for $-\infty < x < \infty$

Solution. Let's try to reduce this problem to the familiar solved by the fundamental solution we already know, by subtracting something off to make homogeneous.

One way to do this is to write w(t,x) = u(t,x) + F(x), where F″(x) = f(x). That is, F(x) is any second integral of f(x). By substituting, we find that

$$w_t - w_{xx} = u_t - u_{xx} - f(x) = 0,$$
$$w(0,x) = g(x) + F(x).$$

Hence the solution is $u(t,x) = -F(x) \int_{-\infty}^{\infty} K(t,x,\xi)(g(\xi)+F(\xi))d\xi$

DERIVATION OF HEAT EQUATION

Newton articulated some principles of heat flow through solids, but it was Fourier who created the correct systematic theory. Inside a solid there is no convective transfer of heat energy and little radiative transfer, so temperature changes only by conduction, as the energy we now recognize as molecular kinetic energy flows from hotter regions to cooler regions. The first basic principle of heat is:

1. The heat energy contained in a material is proportional to the temperature, the density of the material, and a physical characteristic of the material called the *specific heat capacity*. In mathematical terms,

$$Q = \iiint_{\Omega} \rho \kappa_s u(t,x) d^3x$$

For the other principles of heat transfer, let us do some experiments with the following materials: a hot stove, some iron rods of different, relatively short lengths and various widths, and various ceramic rods of different lengths and widths. Since these will be thought experiments only, it will be safe to use a finger as the probe. Putting your finger right on the stove will convince you that the energy transfer is proportional to the difference in temperature between your finger and the stove.

Using, if necessary, a different, undamaged finger, you will also find that the rate of heat transfer is inversely proportional to the length of an iron rod intervening between your finger to the stove (fixing the cross-sectional area). In other words, the rate of heat flow from one region to another is proportional to the temperature gradient between the two regions.

You will probably also agree that the rate of heat flow will be proportional to the area of the contact; for example, a short pin with one end on a hot stove and the other touching your hand is preferable to putting the palm of your hand on a frying pan. Finally, a ceramic material on the stove being usually more pleasant to the touch than hot iron, we see that the rate of heat transfer depends on the material, as measured with a physical constant known as the *heat conductivity* . The second basic principle is thus:

2. The heat transfer through the boundary of a region is proportional to the heat conductivity, to the gradient of the temperature across the region, and to the area of contact, so if the boundary of the region Ω is wnitten as $\partial\Omega$, with outward normal vector n, then

$$\frac{dQ}{dt} = \iint_{\Omega} \sigma\, n \nabla u(t,x) d^3x.$$

If we differentiate eq. (8.a1) with respect to time and apply Gauss's divergence theorem, we find that dQ/dt can be expressed in two ways as an integral:

$$\iiint_{\Omega} \rho \kappa_s u(t,x) d^3x = \iiint_{\Omega} \nabla.\sigma\, \nabla u(t,x) d^3x.$$

Since the region Ω can be an arbitrary piece of the material under study, the integrands must be equal at almost every point. If the material under study is a slab of a homogeneous substance, then ρ, κ_s, and σ are independent of the position x, and we obtain the heat equation

$$u_t = k \cdot \nabla^2 u$$

where $k = \sigma/\rho\kappa_s$. Ordinary substances have values of k ranging from about 5 to 9000 cm^2/gm.

The one-dimensional heat equation

$$u_t = k\, u_{xx}$$

would apply, for instance, to the case of a long, thin metal rod wrapped with insulation, since the temperature of any cross-section will be constant, due to the rapid equilibration to be expected over short distances.

DERIVATION OF WAVE EQUATION FOR THE VIBRATING STRING

The physics and mathematics of the vibrating string were studied by Jean le Rond d'Alembert, and later by Joseph Louis Lagrange, Leonhard Euler, and Daniel Bernoulli, who gave a satisfactory discussion of the physics of the vibrating string. I have not been able to locate a detailed discussion of Bernoulli's derivation of the wave equation, but it is likely that he based it on an energy principle, somewhat as follows.

First let us recall that in classical mechanics, Newton's equations for the motion of a particle are equivalent to the vanishing of the first variation of the Lagrangian action integral,

$$L := \int_K (KE - PE)dt$$

The kinetic energy is usually of the form

$$KE = \frac{m|\nabla|^2}{2},$$

while the potential energy PE may be a function of position, U(q). (I call the position q to avoid confusion with an x used below). The action integral is calculated along a trajectory C = (q(t)) beginning at position and time coordinates (q_0, t_0) and ending at coordinates (q_1, t_1). The trajectory chosen by physics is one which is stationary with respect to variations of the path. In other words, if we replace C with

$$C(\delta) = (q(t) + \delta\, h(t)),$$

where h(t) is a smooth function with $h(t_0) = h(t_1) = 0$, and we calculate $L(\delta)$, then

$$L'(0) = 0.$$

(You may hear that the physical trajectory *minimizes* the action, but this is only a necessary condition for minimum, and the physical trajectory is not always an actual minimum.) If we expand L in powers of and retain only the first-order term, the stationary condition becomes

$$\int_{t_0}^{t_1}(m\, q(t)\, h(t) - \nabla U(q)(t))h(t))dt = 0$$

and if we integrate by parts, it is:

$$\int_{t_0}^{t_1} h(t)(-m\, q(t) - \nabla U(q)(t)))dt = 0$$

(The boundary terms vanish because of the conditions $h(t_0) = h(t_1) = 0$, which served to fix the beginning and end of the trajectory.) If, now, we let the three components of h range over a complete set such as sin(n t/L), with $L = t_1 - t_0$, we see that the only possibility is for

$$mq(t) = F = -\nabla U\ (q)$$

which is Newton's law.

The big advantage of Lagrangian mechanics is that it allows us relatively easily to find the equations of motion of an extended body, such as a string. Suppose now that we have a taut string along the x-axis between positions a and b, but displaced laterally by an amount u(t,x). If the density is , then the kinetic energy of a small bit of string at position x is

$$dKE = \frac{\rho\,|\,u_t(t,x)\,|^2}{2}dx,$$

so the total kinetic energy is

$$KE = \int_{a}^{b}\frac{\rho\,|\,u_t(t,x)\,|^2}{2}dx.$$

It is plausible that the microscopic force transmitted by the string to a point x is proportional to the amount by which the string is stretched at x, i.e., it depends on the arc length element ds at x. As we know,

$$ds = \sqrt{1+\left(\frac{\partial u}{\partial x}\right)^2}\,dx.$$

Thus we may assume that the differential potential energy depends on u only through u_x. The total potential energy would be of the form

$$PE = \int_{a}^{b}\left(u_x^2 x\right)dx.$$

The Lagrangian action integral will be the integral of KE - PE with respect to time.

Notice that it is u and not x which corresponds to the q used above; the "trajectory" of the string is specified by the function u(t,x), and a variation would entail replacing u(t,x) with

$$u(t,x) + \delta\, h(t,x),$$

where $h(t_0,x) = h(t_1,x) = 0$. If the ends of the string are fixed, we would also have $h(t,a) = h(t,b) = 0$. The action is a double integral,

$$L = \int_{\delta}^{l}\int_{b}^{b} \frac{\rho u_t^2}{2} - F\left(u_x^2 x\right) dx\, dt.$$

and if we use Taylor's theorem to keep only the first-order term in Lagrange's condition reads:

$$0 = \int_{\delta}^{l}\int_{b}^{b} \rho u_t h_t - F\left(u_x^2 x\right) 2u_x h_x dx\, dt.$$

This general formula may be useful in deriving realistic wave equations for non-homogeneous strings, but let us simplify at this stage by assuming that ρ and T: = 2 F′ are constants. If we integrate the first integrand by parts in the t variable and the second by parts in the x variable, we now find that:

$$0 = \int_{\delta}^{l}\int_{b}^{b} h(-\rho u_{tt} + T u_{xx} dx)\, dt.$$

If h is arbitrary enough to run through a complete set, we must conclude that

$$u_{tt} = c^2\, u_{xx}, \qquad \text{(WE)}$$

with $c^2 = T/\rho$ a positive constant with dimensions of velocity^2.

Judging from guitar strings, for which a 1 meter taut string gives a musical note in the mid range of the musical scale, typical values of c for thin metal strings are on the order of 1000 m./sec. The wave equation (WE) also describes one-dimensional acoustic waves (c ~ 344 m/sec. in air at room temperature or 330 m./sec. at 0 C) and light waves (c ~ 300,000,000 m./sec.), although the physical derivation in these cases is very different.

FILTERING AND SIGNAL PROCESSING

An intuitive way to understand projections of functions, which has many applications in signal processing, is in terms of filtering. The word "filtering" refers to an attempt to extract the important part of some data while eliminating random contributions called "noise" or other unwanted features which obscure the ones that matter. Electrical engineering offers many examples of filtering, which you are probably aware of as a consumer if you have a stereo system and a collection of recordings, but the notion is much more widely applicable than that.

To emphasize this, we call upon a less familiar example, drawn from astronomy. In recent years astonomomers have obtained reliable evidence that several nearby stars are orbited by planet-sized companions. The evidence was extracted from observations of the positions of the stars taken over periods of several years. How was this accomplished? As we all know, stars twinkle, which means that the image of the star moves in an erratic way at high frequencies, due to motions in our atmosphere.

Other contributions to the motion of the image include the motion of the earth, which has daily, monthly, and yearly components, and human activities such as calibration, maintenance and cleaning of the facilities. Indeed, the first claimed proofs that a nearby star, Barnard's star, had an orbiting companion, were retracted when it became apparent that the periodic motion which was extracted from observations correlated with regularly scheduled human activities at the Swarthmore College Sproul Observatory.

Many of these contributions are much larger than ones corresponding to planetary motion, and it is the task of the astronomer to filter out the unwanted contributions and see whether the remaining parts of the signal reveal a regular wobble, as would be caused by a planet.

Fortunately, this task is rather straightforward and can be reliably be solved with the tools of linear mathematics. All we need to do is use the same formula which allows us to project a vector onto another vector, in order to separate various contributions to a function.

To illustrate the procedure, we simplify the selection of features which are regarded as good or bad. Suppose that we have a signal f(t) which contains some pure frequencies we are interested in, say corresponding to pure vibratory functions

$\sin(2\pi t), \sin(4\pi t), \ldots, \sin(20\pi t), \cos(2\pi t), \cos(4\pi t), \ldots, \cos(20\pi t)$, with frequencies 1,2,..., 10, plus some noise $f_{noise}(t)$. (The frequency of a sinusoidal function $\sin(\omega t)$ differs from the angular frequency by the factor 2π.) We write the full function in the form:

$$\begin{aligned} f(t) &= a_0 + a_1 \cos(2\pi t) + a_2 \cos(4\pi t) + \ldots a_{10} \cos(20\pi t) \\ &\quad + b_1 \sin(2\pi t) + b_2 \sin(4\pi t) + \ldots b_{10} \sin(20\pi t) \\ &\quad + f_{noise}(t) \\ &=: f_{filtered}(t) + f_{noise}(t) \end{aligned}$$

Often the constant term a_0 is left out of the expansion, since a signal usually oscillates around average zero.

A model problem in signal processing is to take a measured signal f(t) and extract from it the low frequency part while projecting away f_{noise}. The goal, then, is to determine the coefficients

$$a_1, \ldots, a_{10} \text{ and } b_1, \ldots, b_{10}.$$

As explained in Chapter, the projection formula tells us the best choice for these coefficients, in the sense that the r.m.s. error

$$\|f_{noise}(t)\| = \|f - (a_1 \cos(2\pi t) + ... + b_{10} \sin(20\pi t))\|$$

is as small as possible, just as the projection of a vector into a plane is the planar vector whose distance from the original vector is as small as possible.

The projection formula for the orthogonal set on the basic period $0 < t < 1$ tells us that

$$a_k = 2\{\cos(2\pi kt), f\}$$

and similarly for the coefficients b_k.

Notice that if we put these terms together, we see that the filtering is accomplished by an integral operator of the separable type,

$$f = \int_0^1 K(x,t)f(t)dt,$$

where

$$K(x,t) = 2\sum_{n=1}^{10} \cos(2\pi nx)\cos(2\pi nt) + 2\sum_{m=1}^{10} \sin(2\pi nx)\sin(2\pi nt)$$

We discuss the theory of such integral operators in a later chapter.

Example: With the aid of mathematical software, we can see the effect of filtering as applied to some interesting functions on the interval $0 < x < 1$. Here we focus on the function sin(1/(x-.05)), which oscillates strongly near x=0 (graph).

In the first calculation we filter out the frequencies greater than 3 - this would be called a low-pass filter - and compare with the unfiltered function on a common graph:

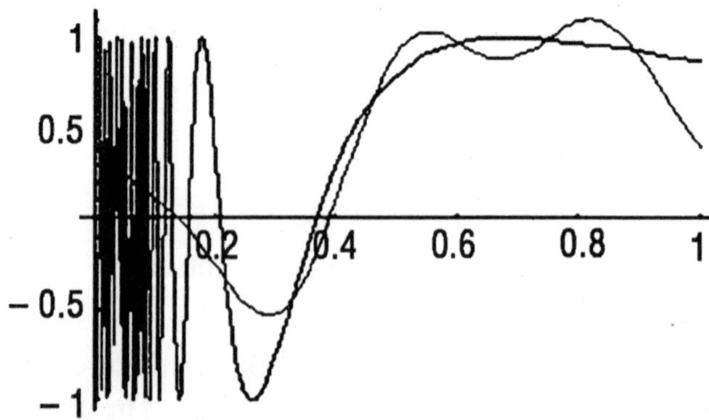

The match is rather good where the original function is varying slowly, but its high-frequency wiggles are filtered out.

For comparison, let's look at the high-frequency part of the same function. A high-pass filter would throw away the result of the low-pass filter and keep the rest:

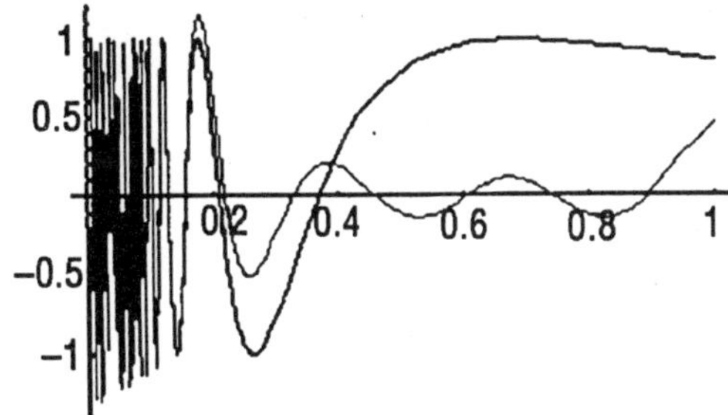

Here, the gently varying part of the function has been filtered out, and the wiggles remain.

Sometimes a signal is not only filtered but also processed, for instance to amplify some part of the signal. As a simple model, suppose that we wish to enhance the low frequency components. A good way to do this would be to select out various components by projection, but then to add them back together with weighting factors.

Example: If we modify the integral kernel K(x,t) as given above to

$$K(x,t) = 6\sum_{n=1}^{10} \frac{\cos(2\pi nx)\cos(2\pi nt)}{n} + 6\sum_{m=1}^{10} \frac{\sin(2\pi nx)\sin(2\pi nt)}{m}.$$

the result would be a signal which has its higher frequencies diminished in comparison with the lower ones (and has lost the frequencies above 10 altogether). Here is a plot of the same function as used along with the function obtained by signal processing with the kernel K:

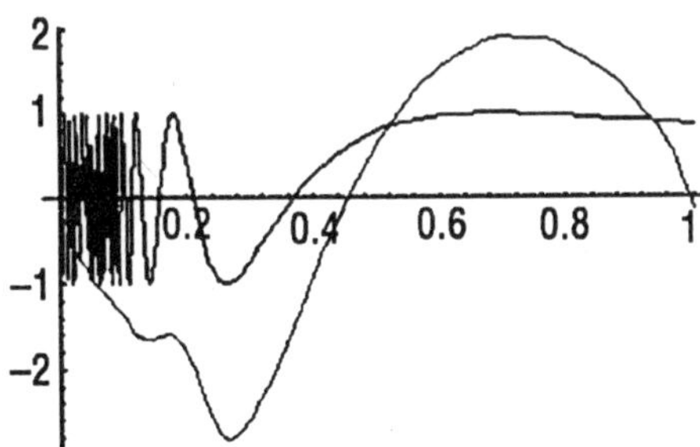

Of course it matches less well than the low-pass filter of ***Example:*** We have emphasized the low frequencies.

Chapter 23

Invariant Sets

In this chapter, we shall present some of the basic results about invariant sets for solutions of initial value problems for systems of autonomous (i.e. time independent) ordinary differential equations. We let D be an open connected subset of $\mathbb{R}^N, N \geq 1$, and let

$$f : D \to \mathbb{R}^N$$

be a locally Lipschitz continuous mapping.

We consider the initial value problem

$$u' = f(u)$$

$$u(0) = u_0 \in D$$

and seek sufficient conditions on subsets $M \subset D$ for solutions we have the property that $\{u(t)\} \subset M, t \in I,$ whenever $u_0 \in M$, where I is the maximal interval of existence of the solution u. we note that the initial value problem (1) is uniquely solvable since f satisfies a local Lipschitz condition (viz. Chapter V). Under these assumptions, we have for each $u_0 \in D$ a maximal interval of existence I_{u0} and a function

$$u : I_{u0} \to D$$

$$t \mapsto u(t, u_0).$$

i.e., we obtain a mapping

$$u : I_{u0} \times D \to D$$

$$(t, u_0) \mapsto u(t, u_0).$$

Thus if we let

$$U = \cup_{v \in D} I_v \times \{v\} \subset \mathbb{R} \times \mathbb{R}^N,$$

we obtain a mapping

$$u : U \to D,$$

which has the following properties:

Lemma:

1. u is continuous,

2. $u(0,u_0) = u_0, \forall u_0 \in D$

3. For each $u_0 \in D, s \in I_{u_0}$ and $t \in I_{u(s,u_0)}$ we have $s+t \in I_{u_0}$ and $u(s+t, u_0) = u(s+u_0))$.

A mapping having the above properties is callled a *flow* on *D* and we shall hencefoorth, in this chapter, use this term freely and call *u* the *flow determind by f.*

Orbits and flows

Let *u* be the flow determined by *f* via the initial value problem. We shall use the following (standard) convention. If $S \subset I_{u_0} = (I_{u_0}, t^-_{u_0}, t^+_{u_0})$,

$$u(S,u_0) = \{v : v = u(t,u_0), t \in S \subset I_{u_0}\}$$

We shall call

$$\gamma(v) = u(I_v, v)$$

the orbit of *v*,

$$\gamma^+(v) = u([0,t_v^+), v)$$

the positive semiorbit of *v* and

$$\gamma^-(v) = u((t_v^+), v)$$

the negative semiorbit of *v*.

Furthermore, we call $v \in D$ a stationary or critical point of the flow, when ever $f(v) = 0$. It is, of course immediate, that if $v \in D$ is a stationary point, then $I_v = \mathbb{R}$ and

$$\gamma(v) = \gamma^+(v) = \gamma^-(v) = \{v\}.$$

We call $v \in D$ a periodic point of peiod *T* of the flow, whenever there exists $T > 0$, such that $u(0,v) = u(T,v)$. It *v* is a periodic point which is not a critical point, one calls *T* > 0 its *minimal* period, provided $u(0;v) \neq u(t,v), 0 < t < T$.

We have the following proposition.

Propostion: Let u be the flow determined by f and let $v \in D$. Then either:

1. v is a stationary point, or
2. v is a periodic point having a minimal positive period, or
3. the flow $u(.,v)$ is injective.

If $\gamma^+(v)$ is relatively compact, then $t^+(v) = +\infty$, if $\gamma^-(v)$ is relatively compact, then $t^-(v) = -\infty$, whereas if $\gamma(v)$ is relatively compact, then $I_v = \mathbb{R}$

Invariant Sets

A subset $M \subset D$ is called positively invariant with respect to the flow u determined by f, whenever

$$\gamma^+(v) \subset M, \forall v \in M,$$

i.e.,

$$\gamma^+(M) \subset M.$$

We similarly call $M \subset D$ negatively invariant provided

$$\gamma^-(M) \subset M$$

and invariant provided

$$\gamma(M) \subset M.$$

We note that a set M is invariant if and only if it is both positvely and negatively invariant.

We have the following proposition:

Proposition: Let u be the flow determined by f and let $V \subset D$. Then there exists a smallest positively invariant subset $M, V \subset M \subset D$ and there exists a largest invariant set $\tilde{M}, \tilde{M} \subset V$. Also there exists a largest negatively invariant subset $M, V \supset M$ and there exists a smallest invariant set $\tilde{M}, \tilde{M} \supset V$. As a consequence V contains a largest invariant subset and is contained in a smallest invariant set,

As a corollary we obtain:

Corollary:

1. If a set, M is positively invariant with respect to the flow u, then so are $\bar{M}$ and int$M$.

2. A closed set M is positively invariant with respect to the flow u if and only if for every $v \in \partial M$ there exists $\varepsilon > 0$ such that $u([0,\varepsilon),v) \subset M$.

3. A set M is positively invariant if and only if comp$M$ (the complement of M) is negatively invariant.

4. If a set M is invariant, then so is ∂M is invariant, then so are $\bar{M}\mathbb{R} \setminus M$, and int$M$.

We now provide a geometric condition of ∂M which will guarantee the invariance of a set M and, in particular will aid us in finding invariant sets.

We have the following theorem, which provides a relationship between the vector field f and the set M (usually called the subtangent condition) in order that invariance holds.

Theorem: Let $M \subset D$ be a closed set, Then M is positively invariant with respect to the flow u determined by f if and only if for every $v \in M$

$$\liminf_{t \to 0+} \frac{dist(v + tf(v), M)}{t} = 0.$$

Proof: Let $v \in M$, then a taylor expansion yields,

$$u(t,v) = v + tf(v) + o(t), t > 0.$$

Hence, if M is positively invariant,

$$\text{dist}(v + tf(v), M) \leq | u(t,v) - v - tf(v) | = o(t),$$

proving the necessity.

Next, let $v \in M$, and assume holds. Choose a compact neighborhood B of v such that $B \subset D$ and choose $\varepsilon > 0$ so that $u([0,\varepsilon], v) \subset B$. Let $w(t) = \text{dist}(u(t,v), M)$, then for each $t \in [0,\varepsilon]$ there exists $v_t \in M$ such that $w(t) = (u(t,v), M)$, and $\lim_{t \to 0}{+}v_t = v$. It follows that for some constant L,

$$\begin{aligned} w(t+s) &= | u(t+s,v) - v_{t+s} | \\ &\leq w(t) + | u(t+s,v) - u(t,v) - sf(u(t,v)) | \\ &\quad + s | f(u(t,v)) - f(v_t) | + | v_t + sf(v_t) - v_{t+s} | \\ &\leq w(t) + sLw(t) + \text{dist}(vt + tf(v_t), M), \end{aligned}$$

because f satisfies a local Lipschitz condition. Hence

$$D_+ w(t) \le Lw(t), 0 \le t < \varepsilon, w(0) = 0,$$

or

$$D_+(e^{-Lt} w(t)) \le 0, 0 \le t < \varepsilon, w(0) = 0,$$

which implies $w(t) \le 0, 0 \le t < \varepsilon, i.e. w(t) \equiv 0$.

We remark here that condition only needs to be checked for point $v \in \partial M$ since it obviously holds for interior points.

We next consider some special cases where the set M is given as a smooth manifold. We have the following theorem.

Theorem: Let $\phi \in C^1(D, \mathbb{R})$ be a function which is such that every value $v \in \phi^{-1}(0)$ is a regular value, i.e. $\nabla\phi(v) \neq 0$. Let $M = \phi^{-1}(-\infty, 0]$, then M is positively invariant with respect to the flow determined by f if and only if

$$\nabla\phi(v).f(v) \le 0, \forall v \in \partial M = \phi^{-1}(0).$$

We leave the proof as an exercise. We remark that the set M given in the previous theorem will be negatively invariant provided the reverse inequality holds and hence invariant if and only if

$$\nabla\phi(v).f(v) = 0, \forall v \in \partial M = \phi^{-1}(0).$$

in which case $\phi(u(t,v)) \equiv 0$, $\forall v \in \partial M$, i.e. f is a first integral for the differential equatioin.

Limit Sets

In this section we shall consider semiorbits and study thier limit sets.

Let $\gamma^+(v), v \in D$ be the positive semiorbit associated with $v \in D$, We define the set $\Gamma^+(v)$ as follows:

$$\Gamma^+(v) = \{w : \exists\{t_n\}, t_n \nearrow t_v^+, u(t_u, v) \to w\}.$$

The set $\Gamma^+(v)$ is the set of limit points of $\gamma^+(v)$ and is called the positive limit set of v. (Note that if $t_v^+ < \infty$, then $\Gamma^-(v) \subset \partial D$.)

For all the results discussed below for positive limit sets there is an analagous result for negative fimit sets; we shall not state these results.

We have the followintg proposition:

Proposition:

1. $\overline{\gamma^+(v)} = \gamma^+(v) \cup \Gamma^+(v)$.

2. $\Gamma^+(v) = \cap_{w \in \gamma^+(v)} \overline{\gamma + (w)}$.

3. If $\gamma^+(v)$ is bounded, then $\Gamma^+(v) \neq 0$ and compact.

4. If $\Gamma^+(v) \neq 0$ and bounded, then

$$\lim_{t \to t_v^+} \text{dist}(u(t,v), \Gamma^+(v)) = 0$$

5. $\Gamma^+(v) \cap D$ is an invariant set.

Proof: We leave most of the proof to the exercises and only discuss the verification of the last part of the proposition. Thus let us assue that $\Gamma^+(v) \cap D \neq 0$. It then follows that $t_v^+ = \infty$. Let $w \in \Gamma^+(v) \cap D$. Then there exists a seqence $\{t_n\}_{n=1}^{\infty}, t_n \nearrow \infty$, such that $u(t_n, v) \to w$. For each $n \geq 1$, the function

$$u_n(t) = u(t + t_n, v)$$

is the unique solution of $\quad u' = f(u), u(0) = u(t_n, v)$,

and hence the maximal interval of existence of u_n will contain the interval $[-t_n, \infty)$. Since $u(t_n, v) \to w$, there will exist a subsequence of $\{u_n(t)\}$, which we relabel as $\{u_n(t)\}$ converging to the solution, call it *y*, of

$$u' = f(u), u(0) = w.$$

We note that given any compact interval [*a*, *b*] the sequence $\{u_n(t)\}$ will be defined on [*a*, *b*] for *n* sufficiently large and hence *y* wil be defined on [*a*, *b*], Since this interval is arbitrary it follows that *y* is defined on $(-\infty, \infty)$. Futhermore for any t_0

$$y(t_0) = \lim_{n \to \infty} u_n(t_0) = \lim_{n \to \infty} u(t_0 + t_n, v),$$

and hence $y(t_0) \in \Gamma^+(v)$. Hence solutions through points of $\Gamma^+(v)$ are defined for all time and their orbit remains in $\Gamma^+(v)$, showing that $\Gamma^+(v) \cap D$ is an invariant set.

Theorem: If $\gamma^+(v)$ is contained in a compact subset $K \subset D$, then $\Gamma^+(v) \neq 0$ is a compact connected set, i.e. a continuum.

Proof: We already know that $\Gamma^+(v)$ is a compact set. Thus we need to show it is connected. Suppose it is not. Then there exist nonempty disjoint compact sets M and N such that

$$\Gamma^+(v) = M \cup N.$$

Let $\delta = \inf\{|v - w| : v \in M, w \in N\} > 0$. Since $M \subset \Gamma^+(v)$ and $N \subset \Gamma^+(v)$, there exist values of t arbitrarily large such that di $\text{dist}(u(t,v), M) < \frac{\delta}{2}$ and hence there exists a sequence $\{t_n \to \infty\}$ such that $\text{dist}(u(t_n,v), M) = \frac{\delta}{2}$. The sequence $\{u(t_n,v)\}$ must have a convergent subsequence and hence has a limit point which is in neither M nor N, a contracdiction.

LaSalle's theorem

In this section we shall return again to invariant sets and consider Lyapunov type functions and their use in determining invariant sets.

Thus let $\phi : D \to \mathbb{R}$ be a C^1 function. We shall use the following notation

$$\phi'(v) := \nabla\phi(v).f(v).$$

Lemma: Assume that $\phi'(v) \leq 0, \forall v \in D$. Then for all $v \in D, \phi$ is constant on the set $\Gamma^+(v) \cap D$.

Theorem: Let there exist a compact set $K \subset D$ such that $\phi'(v) \leq 0, \forall v \in K$.

Let

$$\tilde{K} = \{v \in K : \phi'(v) = 0\}$$

and let M be the largest invariant set contained in $\tilde{K}$. Then for all $v \in D$ such that $\gamma^+(v) \subset K$

$$\lim_{t \to \infty} \text{dist}(u(t,v), M) = 0$$

Proof: Let $v \in D$ such that $\gamma^+(v) \subset K$, then, using the previous lemma, we have that ϕ is constant on $\Gamma^+(v)$, which is an invariant set and hence contained in M.

Theorem: (La Salle's Theorem): Assume that $D = \mathbb{R}^N$ and let $\phi'(x) \leq, \forall x \in \mathbb{R}^N$. Furthermore suppose that f is bounded below and that $\phi(x) \to \infty$ as $|x| \to \infty$. Let $E = \{v : \phi'(v) = 0\}$, then

$$\lim_{t \to \infty} \text{dist}(u(t, v), M) = 0,$$

for all $v \in \mathbb{R}^N$, where M is the largest invariant set contained in E.

As an example ot illustrate Theorem 11 consider the oscillator

$$mx'' + hx' + kx = 0,$$

where m, h, k are positive constants. This equation may be written as

$$x' = y$$

$$y' = -\frac{k}{m}x - \frac{h}{m}y.$$

We choose

$$\phi(x, y) = \frac{1}{2}(my^2 + kx^2)$$

and obtain that

$$\phi'(x, y) = -hy^2.$$

Hence $E = \{(x, y)\} | : y = 0\}$. The largest invariant set contained in E in the origin, hence all solution orbits tend to the origin.

Two Dimensional Systems

In this section we analyze limit sets for two dimensional systems in somewhat more details and prove a classical theorem (Theorem of Poincare-Bendixson) about periodic orbits of such systems. Thus we shall assume throughout this section that $N = 2$.

Let $v \in D$ be a regular (i.e. not critical) point of *f*. We call a compact straight line segment $l \subset D$ throught v a transversal throught v, provided it contains only regular point and if for all $\omega \in l, f(\omega)$ is not parallel to the direction of *I*.

We shall need the following observatioin.

Lemma: Let $v \in D$ be a regular point of *f*. Then there exists a transversal *l* containing

v in its relative interior. Also every orbit associated with f which crosses l, crosses l in the same direction.

Proof: Let v be a regular point of f. Choose a neighborhood V of v consisting of regular point only. Let $\eta \in \mathbb{R}^2$ be any directioin not parallel to $f(v)$, i.e. $\eta \in f(v) \neq 0$, (here $\times$ is the cross product in $\mathbb{R}^3$). We may restricit V further such that $\eta \times f(w) \neq 0, \forall \omega \in V$, and is bounded away from 0 on V. We then may take l to be the intersection of the straight line through v with direction η and and $\overline{V}$. The proof is completed by observing that

$$\eta \times f(\omega) = (0, 0, |\eta|| f(\omega)| \sin \theta).$$

where θ is the angle between η and $f(\omega)$.

Lemma: Let v be an interior point of some transversal l. Then for every $\varepsilon > 0$ there exists a circular disc D_e with center at v such that for every $\omega \in D\varepsilon., u(t, \omega)$ will cross l in time $t, |t| < \varepsilon$.

Proof: Let $v \in \text{int } l$ and let

$$l = \{z : z = v + s\eta, s_0 \leq s \leq s_1\}$$

Let B be a disc centered at v containing only regular points of f. Let $L(t, w) = au_1(t, w) + bu_2 \ (t, w) + c, u = (u_1, u_2)$ where $u(t, w)$ is the solution with initial condition w and $au_1 + bu_2 + c = 0$ is the equation of the straight line containing l. Then $L(0, v) = 0$, and $\dfrac{\partial L}{\partial t}(0, v) = (a, b).f(v) \neq 0$. We hence may apply the implicit function theorem to complete the proof.

Lemma: Let l be a transversal and let $\Gamma = \{w = u(t, v) : a \leq t \leq b\}$ be a closed arc of an orbit u associated with f which has the property that $u(t_1, v) \neq u(t_2, v), a \leq t_1 < t_2 \leq b$. Then if Γ intersects l it does so at a finite number of points whose order on Γ is the same as the order on l. If the orbit is periodic it intersects l at most once.

The proof relies on the Jordan curve theorem and is left as an exercise. The next lemma follows immediately from the previous two.

Lemma: Let $\gamma^+(v)$ be a semiorbit which does not intersect itself and let $\omega \in \Gamma^+(v)$ be a regular point of f. Then any transveral containing ω in its interior contains no other point of $\Gamma^+(v)$ in its interior.

Lemma: Let $\gamma^+(v)$ be a semiorbit which does not intersect it self and which is contained in a compact set $K \subset D$ and let all points in $\Gamma^+(v)$ be regular points of *f*. Then $\Gamma^+(v)$ contains a periodic orbit.

Proof: Let $\omega \in \Gamma^+(v)$. it follows from Proposition that $\Gamma^+(v)$ is an invariant set and hence that $\gamma^+(\omega) \subset \Gamma^+(v)$, and thus also $\Gamma^+(\omega) \subset \Gamma^+(v)$. Let z $z \in \Gamma^+(\omega)$, and let l be a transversal containing z in its relative interior. It follows that the semiorbit $\gamma^+(\omega)$ must intersect l and by the above for an infinite number of values of t.On the other hand, the previous lemma implies all these point of intersection must be same.

It therefore follows from Lemma 16 that every point in $\Gamma^+(v)$ in a point on some periodic orbit of a minimal positive period. On the other hand, it also follows from earlier results that $\Gamma^+(v)$ is a compact connected set. Hence, if for some $\omega \in \Gamma^+(v), \lambda(\omega) \neq \Gamma^+(v)$, then $\Gamma^+(v) \setminus \gamma(\omega)$ must be a relatively open set with $A = \Gamma^+(v) \setminus \gamma(\omega) \cap \lambda(\omega) \neq 0$. One now easily obtains a contradiction by examining transversals through points of A. One hence concludes that in fact under the hypotheses of Lemma, the limit set $\Gamma^+(v)$ is a peiodic orbit. We summarize the above in the following theorem.

Theorem: (Poincare-Bendixson) Let $\gamma^+(v)$ be a semiorbit which does not intersect it self and which is contained in a compact set $K \subset D$ and let all points in $\Gamma^+(v)$ be refular points of *f* Then $\Gamma^+(v)$ is the orbit of a periodic solution u_T with smallest positive period T.

Theorem: Let G be a periodic orbit of (1) which together with its interior is contained in a compact set $K \subset D$. Then there exists at least one singlular point of *f* in the interior of D.

Proof: Let

$$\Omega = \operatorname{int} \Gamma$$

Then *f* is continuous on $\overline{\Omega}$, and does not vanish on $\Gamma = \partial\Omega$. Let us assume that *f* has no stationary points in Ω. Then for each $\omega \in \Omega, \Gamma^+(\omega)$ is a periodic orbit. We partially order the collection $\{T_\alpha\}_{\alpha \in I}$, where I is an index set, of all periodic orbits which are contained in Ω, by saying that

$$\Gamma_\alpha \leq \Gamma_\beta \Leftrightarrow \operatorname{int} \Gamma_\alpha \subset \operatorname{int} \Gamma_\beta.$$

One now employs the Hausdorff minimum principle together with Theorem 17 to arrive at a contradiction.

HOPF BIFURCATION

This topic is devoted to a version of the classical **Hopf biffurcation** theorem which establishes the existence of nontrivial periodic orbits of autonomous systems of differentaïl equations which depend upon a parameter and for which the stability properties of the trivial solution changes as the parameter is varied.

A Hopf Bifurcation Theorem

Let

$$f : \mathbb{R}^n \times \mathbb{R} \to \mathbb{R}^n,$$

be a C^2 mapping which is such that

$$f(0,\alpha) = 0, \text{ all } \alpha \in \mathbb{R}.$$

We consider the system of ordinary differential equations depending on a parameter α

$$\frac{du}{dt} + f(u,\alpha) = 0,$$

and prove a theorem about the existence of nontrivial periodic solutions of this system. Results of the type proved here are referred to as *Hopf bifurcatioin* theorems.

We establish the following ***Theorem:***

1. Theorem *Assume* that f satisfies the following conditioins:

   (a) For some given value of $\alpha = \alpha_0, i = \sqrt{-1}$ and $-\ i$ are not eigenvalues of $n = 0, 2, 3, \ldots$, are not eigenvalues of $f_u(0_{a_0})$;

   (b) In a neighborhood of α_0 there is a curve of eigenvalues and eigenvectors

$$f_u(0,\alpha)\alpha(\alpha) = \beta(\alpha)a(\alpha)$$

$$a(\alpha_0) \neq 0, \beta(\alpha_0) = i, \mathrm{Re}\frac{d\beta}{d\alpha}|_{\alpha_0} \neq 0.$$

Then there exist positive number ε and η and a C^1 function

$$(u,\rho,\alpha) : (-\eta,\eta) \to C^1_{2\pi} \times \mathbb{R} \times \mathbb{R},$$

where $C^1_{2\pi}$ is the space of 2π periodic $C^1\mathbb{R}^n$ – valued functions, such that $(u(s),\rho(s),\alpha(s))$ solves the equation

$$\frac{du}{dr}+\rho f(u,\alpha)=0,$$

nontrivially, i.e. $u(s)\neq 0,\ s\neq 0$ and

$$\rho(0)=1,\alpha(0)=\alpha_0,u(0)=0.$$

Furthemore, if (u_1,α_1) is a nontrivial solution period $2\pi\rho_1$, with $|\rho_1-1|<\varepsilon,|\alpha_1-\alpha_0|<\varepsilon,|u_1(t)|<\varepsilon,t\in[0,2\pi\rho_1]$, then there exists $s\in(-\eta,\eta)$ such that $\rho_1=\rho(s),\alpha_1=\alpha(s)$ and $u_1(\rho_1 t)=u(s)(\tau+\theta)\ \ \tau=p_1 t\in[0,2\pi],q\in[0,2\pi)$.

Proof: We note that $u(t)$ will be solution of period $2\pi\rho$, whenever $u(\tau),\ \tau=\pi t$ is a solution of period 2π. We thus let X = $C^1_{2\pi}$ and Y = $C_{2\pi}$ be Banach spaces of C^1, respectively, continuous 2π- periodic functions endowed with the usual norms and define an operator

$$F:X\times\mathbb{R}\times\mathbb{R}\to Y$$

$$(u,\rho,\alpha)\mapsto u'+pf(u,\alpha),'=\frac{d}{dr}.$$

Then F belongs to class C^2 and we seek nontrivial solution of the equation

$$F(u,\rho,\alpha)=0$$

with values of ρ closed to 1, α closed α_0, and u $u\neq 0$.

We note that

$$f(u,\rho,\alpha)=0 \text{ for all } \rho\in\mathbb{R},\alpha\in\mathbb{R}.$$

Thus the claim is that the value $(1,\alpha_0)$ of the two dimensional parameter (ρ,α) is a bifurcation value.Theorem tells us that

$$u'+fu(0,a_0)u$$

cannot be a linear homeomorphism of X onto Y. This is guaranteed by the assumptions since the functions

$$\phi_0=Re(e^{i\tau}a(a_0)),\phi_1=\mathrm{Im}(e^{i\tau}a(\alpha_0)) \text{ are 2}$$

are $2\pi-$ periodic solution of

$$u'+f_u(0,a_0)u=0.$$

and they span the the kernel of $Fu(0,1,a_0)$,

$$\ker F_u(0,1a_0)0\{\phi_0 m\phi_1-\phi'_0\}$$

It follows from the theory of linear differential equations that the image, imFu(0,1,α0), is closed in Y and that

$$imF_u(0,1,a_0)=\{f\in Y:\langle\phi,\psi_i\rangle=0,i=0,1\}$$

where $\langle .,. \rangle$ denotes the L^2 inner product and $\{\psi_0,\psi_1\}$ forms basis for $\ker\{-u'+f_u^T(0,\alpha_0)u\}$, the diffeential equation adjoint operator of $u'+f_u(0,\alpha_0)u$. In fact $\psi_1=\psi_0'$ and $\langle\psi_i,\psi_j\rangle=\delta_{ij,}$ the Kronecker delta. thus $F_u(0,1,a_0)$ is a linear Fredhohm mapping from *X* to *Y* having a two dimensional kernel as well as a two dimensional cokernel. We now write, as in the beginning of Cahpter II,

$$X=V\oplus W$$
$$y=z\oplus T.$$

We let U be a neighborhood of $(0,1\alpha_0,0)$ in $V\times\mathbb{R}\times\mathbb{R}\times\mathbb{R}$ and define G on U as follows

$$G(v,\rho,\alpha,s=\begin{cases}\dfrac{1}{s}F(s(\phi_0+v),\rho,\alpha) & s\neq 0\\ F_u(0,\rho,\alpha)(\phi_0+v) & s=0\end{cases}$$

We now want to slove the equation

$$G(v,\rho,\alpha,s=0,$$

for v,ρ,α in terms of s in a neighborhood of $0\in\mathbb{R}$. We note that G is C^{-1} and

$$G(v,\rho,\alpha,0)=(\phi_0+v)'+\rho fu(0,\alpha)(\phi_0+v).$$

Hence

$$G(v,\rho,\alpha,0)=\phi_0'+\rho f_u(0,\alpha)(\phi_0).$$

Thus, in order to be able to apply the implicit funciton theorem, we need to differentaite the map

$$(v, \rho, \alpha,) \mapsto G(v, \rho, \alpha, s)$$

evaluate the result at $(0, 1a_0 0)$ and show that this derivative is a linear home-omorphism of $V \times \mathbb{R} \times \mathbb{R}$ on to Y.

Computing the Taylor expansion, we obtain

$$G(v, \rho, \alpha, s) = G(0, 1a_0, s) + G_p(0, 1a_0, s)(\rho - 1)$$

$$G_\alpha(0, 1a_0, s(\alpha - a_0) + G_v(0, 1\alpha_0, s)v + \ldots,$$

and evaluating at s = 0 we get

$$G_{v,\rho,\alpha}(0, 1\alpha_0, 0)(v, \rho - 1, \alpha - \alpha_0) \quad = f_v(0, \alpha_0)\phi_0(\rho - 1)$$

$$+ f_{v\alpha}(0, \alpha_0)\phi_0(\alpha - \alpha_0)$$

$$+ v' + f_v(0, a_0)v).$$

Note that the mapping

$$v \mapsto v' + fv(0, \alpha_0)v$$

is a linear homeorphim of V onto T. thus we need to show that map

$$(\rho, \alpha) \mapsto f_v(0, \alpha_0)\phi_0(\rho - 1) + f_{va}(0, \alpha_0)\phi_0(\alpha - \alpha_0)$$

only belongs to T if $\rho = 1$ and $\alpha - \alpha_0$ and for all $\psi \in Z$ there exists a unique (ρ, α) such that

$$f_v(0, \alpha_0)\phi_0(\rho - 1) + f_{va}(0, \alpha_0)\phi_0(\alpha - \alpha_0) = \psi.$$

By the characterization of T, we have that

$$f_v(0, \alpha_0)\phi_0(\rho - 1) + f_{va}(0, \alpha_0)\phi_0(\alpha - \alpha_0) \in T.$$

if and only if

$$< f_v(0, \alpha_0)\phi_0, \psi_i > (\rho - 1) \; + f_{va} < f_{v\alpha}(0, > (\alpha - \alpha_0) = 0$$

i 1,2

Since

$$(\rho, \alpha) \mapsto < f_{v\alpha}(0, \alpha_0)\phi_0 = f_0' = \phi_1$$

which has only the trivial solution if and only if

$$< f_{v\alpha}(0, \alpha_0)\phi_0, \psi_0 > \neq 0. >$$

Computing this lett expression, one obtains

$$(\rho,\alpha)\mapsto< f_{v\alpha}(0,\alpha_0)\phi_0,\psi_0 >\neq \operatorname{Re} B'(0),'$$

which byi hypothesis is not zero.The unifurcation assrtion we leave as an exercise.

For much further discus $x''+x-\alpha(1-x')x'=0.$ sioin of Hopf bifurcation we refer to [at]. The following examply of the classical *Van* der Pol oscillator fromn nonlinear electrical circuit theory will serve to illustrate the applicabililty of the theorem.

Example: Consider the nonlinear oscilll oscillator

$$x''+x-\alpha(1-x')x'=0.$$

This equation has for for certain small values of a nontrivial periodic solutions with periods close to 2π.

We first transform into a system by setting

$$u=\begin{pmatrix}u_1\\u_2\end{pmatrix}=\begin{pmatrix}x\\x'\end{pmatrix}$$

and obtain

$$u'=\begin{pmatrix}0 & -1\\1 & -\alpha\end{pmatrix}+\begin{pmatrix}0\\u\frac{2}{1}u_2\end{pmatrix}=\begin{pmatrix}0\\0\end{pmatrix}$$

We hence obtain that

$$f_u(0,\alpha)=\begin{pmatrix}0 & -1\\1 & -\alpha\end{pmatrix}$$

whose eigenvalues satisfy

$$f_u(0,\alpha)=\begin{pmatrix}0 & -1\\1 & -\alpha\end{pmatrix}$$

Letting $\alpha_0=0$, we get $\beta(0)=\pm i,$ and computing $\frac{d\beta}{d\alpha}=\beta'$ we obtain $2\beta\beta'+\beta'\alpha+\beta=0, or\, B^t\,\frac{-\beta}{\alpha+2\beta}=-\frac{1}{2}$, for $\alpha = 1$.

STURM -LIOUVILLE BOUNDARY VALUE PROBLEMS

In this chapter we shall study a very classical problem in the theory of ordinary differential equations, namely linear second order differential equations which are parameter dependent and are subject to boundary condtions. While the existence of *eigenvalves* (parameter values for which nontrivial solutions exist and *eigenfunctions*(corresponding nontrivial solutions) follows easily from the abstract riesz spectral theory for compact liinear operators, it is instructive to deduce the smae conclusions using some of the results we have developed up to now for ordinary differential equations. While thte theory presented below is for some rather specific cases, much more general problems and various other eases may be considered and similar theorems may be established. We refer to the books where the subject is studied in some more detial.

Linear boundary vlaue Problems

Let $I = [a,b]$ be a compact interval and let $p,q,r \in C(I,\mathbb{R})$, with *p, r* positive on *I.* Consider the linear differential equation

$$(p(t)x')' + (\lambda r(t) + q(t))x = 0, t \in I,$$

where λ is a complex parameter.

We seek parameter values (*eigenvalues*) for which(1) has nontrivial solutions (*eigensolutions or eigenfunctions*) when it is subject to the set of boundary conditions

$$x(a)\cos\alpha - p(a)x'(a)\sin\alpha = 0$$

$$x(b)\cos\beta - p(b)x'(b)\sin\beta = 0,$$

where α and β are given constants and (without loss in generality, $0 \le \alpha < \pi, 0 < \beta \le \pi$). Such a boundary value problem is called a *Sturm -Liouville* boundary value problem.

We note that is equaivalent to the requirement

$$c_1 x(a) + c_2 x'(a) = 0, |c_1| + |c_2| \neq 0$$

$$c_3 x(b) + c_4 x'(b) = 0, |c_3| + |c_4| \neq 0.$$

We have the following lemma.

Lemma: Every eigenvalue of equationi (1) subject to the boundary conditions (2) is real.

PROOF, Let the differential operator *L* be defined by

$$L(x) = (px') + qx.$$

Then, if λ is an eigenvalue

$$L(x) = -\lambda rx,$$

for some nontrivial x which satisfies the boudary conditions. Hence also

$$L(\overline{x}) = -\overline{\lambda r x} = -\overline{\lambda} r \overline{x}.$$

Therefore

$$\overline{x}L(x) - xL(\overline{x}) = -(\lambda - \overline{\lambda}) r x \overline{x}.$$

Hence

$$\int_a^b \overline{x}L(x) - xL(\overline{x}) dt = -(\lambda - \overline{\lambda}) \int_a^b r x \overline{x} dt.$$

Integrating the latter expression and using the fact that both x and $\overline{x}$ satisfy the boundary conditions we obtain the value 0 for this expression and hence $\lambda - \overline{\lambda}$.

We next let $u(t,\lambda) = u(t)$ be the solution of (1) which satisfies the (initial) conditions

$$u(a) = \sin\alpha, p(a)u'(a) = \cos\alpha,$$

then $u \neq 0$ and satisfies the first set of boundary condtions. We introduce the following transformation (*Priifer transformation*)

$$\rho = \sqrt{u^2 + p^2 (u')^2}, f = \arctan \frac{u}{pu'}.$$

Then ρ and ϕ are solutions of the differential equations

$$\rho' = \left[(\lambda r + q) - \frac{1}{p}\right] \text{ r sin f cosf}$$

and

$$\phi' = \frac{1}{p}\cos^2\phi + (\lambda r + q)\sin^2\phi.$$

Further ϕ(a) = α. (Note that the second equation depends upon ϕ only, hence, once ϕ is know, ρ may be determined by integrating a linear equation and hence u is determined.

We have the following lemma describing the dependence of ϕ upon λ.

Lemma: Let ϕ be the solution of such that ϕ(α) = α. Then ϕ satisfies the following conditions:

1. ϕ(b, λ) is a coninuous strictly increasing funciton of λ;
2. $\lim_{\lambda \to \infty} \phi(b,\lambda) = \infty$;

3. $\lim_{\lambda\to\infty}\phi(b,\lambda)=0.$

Proof: To prove the other parts of the lemma, we find it convenient to make the change of independent variable

$$s=\int_a^t \frac{dr}{p(r)},$$

which transforms equation (1) to

$$x''+p(\lambda r+q)x=0,'=\frac{d}{ds}.$$

Using the above lemma we obtain the existence of eigenvalues, namely we have the following theorem.

Theorem: The boundary value problem has an unbounded infinite sequence of eigenvalues

$$\lambda_0<\lambda_1<\lambda_1<\ldots$$

with

$$\lim_{n\to\infty}\lambda_n=\infty.$$

The eigenspace associated with each eigenvalue is one dimensional and the eigenfunctions associated with λ_k have preciselyk simple zeros in (a, b).

Proof: The equatioin

$$\beta+k\pi=\phi(b,\lambda)$$

has a unique solution λ_k, for $K=0,1,\ldots$. This set $\{\lambda_k\}_{k=0}^{\infty}$ has the desired properties.

We also have the following lemma.

Lemma: Let $u_i, i=j,k, j\neq k$ be eigenfunctions of the boundary value problem corresponding to the eigenvalues λ_j and λ_k, Then u_j are orthogonal with respect to the weight function *r,* i.e.

$$(u_j,u_k)=\int_a^b ru_ju_k=0.$$

In what is to follow we denote by $\{u_i\}_{i=0}^{\infty}$ the set of eigenfunctions whose existence is guaranteed by Theorem with u_i an eigenfunction corresponding to $\lambda_i, i = 0,1,\ldots$ which has been normalized so that

$$\int_a^b ru_i^2 - 1.$$

We also consider the nonhomogeneous boundary value problem

$$(p(t)x')' + l(\lambda r(t) + q(t))x = rh, t \in I,$$

where $h \in L^2(a,b)$ is a given function, the equation being subject to the boundary condtions and solutions being interpreted in the Caratheodory sence.

We have the following resullts.

Lemma: For $\lambda = \lambda_k$ equation has a solution subject to the boundary conditions if and only if

$$\int_a^b ru_k h = 0.$$

If this is the case, and ω is a particulalr solution, then any other solution has the form $\omega + cu_k$, where c is an arbitrary constant.

Proof: Let v be a solution of $(p(t)x')' + (\lambda_k r(t))x = 0$, which is linearly independent of u_k, then

$$(u_k v' - u_k' v) = \frac{c}{p},$$

where c is a nonzero constant, One verifies that

$$w(t) = \frac{1}{c}\left(v(t)\int_a^t ru_k h + u_k \int_t^b rvh\right)$$

is a solution (for $(\lambda = \lambda_k)$, whenver $\int_t^b ru_k h = 0$ holds.

Completeness of Eigenfunctions

We note that it follows from Lemma that equation has a solution for every $\lambda_k, k = 0,1,2,\ldots$ if and only if $\int_a^b ru_k h = 0$ for $k = 0, 1, 2, \ldots$. Hence, since $\{u_i\}_{i=0}^{\infty}$ forms an orthonormal system for the Hilbert space $L_r^2(a,b)(i.e. L^2(a,b)$ with weight function r defining the inner product), $\{u_i\}_{i=0}^{\infty}$ will be a complete orthonormal system, once we can show that $\int_a^b u_k h = 0$ for k = 0,1,2,..., implies that $h=0$. The aim of this section is to prove completeness.

The following lemma will be needed in this discussion.

Lemma: If $\lambda \neq \lambda_k$, k = 0, 1,... has a solution subject to the boundary conditions for every $h \in L^2(a,b)$.

Proof: For $\lambda \neq \lambda_k$, $k = 0$, 1,...we let u be a nontrivial solution which satisfie the first boundary condition and let v be a nontrivial solution which satisfies the second boundary condition. Then

$$uv' - u'v = \frac{c}{p}$$

with c a nonzero constant, Define the Green's function

$$G(t,s) = \frac{1}{c}\begin{cases} v(t)u(s), a \leq s \leq t \\ v(s)u(t), t \leq s \leq b. \end{cases}$$

Then

$$w(t) = \int_a^b G(t,s)r(s)h(s)ds$$

is the unique solution.

We have the following corollary.

Corollary: Lemma defines a continuous mapping

$$G : L^2(a,b) \rightarrow C^1[a,b],$$

by

$$h \mapsto G(h) = w.$$

Further

$$\langle Gh, w \rangle = \langle h, Gw \rangle.$$

Proof: We merely need to examine the definition of $G(t, s)$ as given by equaiton. Let us now let

$$S = \{w \in L^2(a,b) : \langle u_i, h\rangle = 0, i = 0,1,2,\ldots\}.$$

Using the definition of G we obtain the lemma.

Lemma: $G : S \to S.$

We note that S is a linear manifold in $L^2(a,b)$ which is weakly closed, i.e. if $\{x_n\} \subset S$ is a sequence such that

$$\langle x_n, h\rangle \to \langle x, h\rangle, \forall h \in L^2(a,b),$$

then $x \in S$.

Lemma: If $S \neq \{0\}$, then there exists $x \in S$ such that

$$\langle G(x), x\rangle \neq 0.$$

Proof: If $\langle G(x), x\rangle = 0.$ for all $x \in S$, since S is a linear manifold, we have for all $x, y \in S$ and $\alpha \in \mathbb{R}$

$$0 = \langle G(x + \alpha y), x + \alpha y\rangle$$

$$2\alpha\langle G(y), x\rangle,$$

in particluar, choosing $x = G(y)$ we obtain a contradiction, since for $y \neq 0 G(y \neq 0)$.

Lemma: If $S \neq \{0\}$, then there exists $x \in S \setminus \{0\}$ and $\mu \neq 0$ such that

$$G(x) = \mu x.$$

Proof: Since there exists $x \in S$ such that $\langle G(x), x\rangle \neq 0$ we set

$$\mu = \begin{cases} \inf\{\langle G(x), x\rangle : x \in S, \| x \| = 1, \text{if} \langle G(x), x\rangle \leq 0, \forall x \in s\} \\ \sup\{\langle G(x), x\rangle : x \in S, \| x \| = 1, \text{if} \langle G(x), x\rangle > 0, \text{for some} \, x \in S\}. \end{cases}$$

We easily see that there exists $x_0 \in S, \| x_0 \| = 1$ such that $\langle G(x_0), x_0\rangle = \mu \neq 0$. If S is one dimensional, then $G(x_0) = \mu x_0$. If S. Letting $z = \dfrac{x_0 + \varepsilon y}{\sqrt{1+\varepsilon^2}}$ we find that $\langle G(z), z\rangle$ has an extremum at $\varepsilon = 0$ and thus obtain that $\langle G(x_0), y \rangle = 0$, for any $y \in S$ with $\langle y, x_0 \rangle = 0$. Hence since $\langle G(x_0) - \mu x_0, x_0 \rangle = 0$ it follows that $\langle G(x_0), G(x_0) - \mu x_0 \rangle = 0$ and thus $\langle G(x_0), -\mu x_0, G(x_0) - \mu x_0 \rangle = 0$, proving that μ is an eigenvalue.

Combining the above results we obtains we obtain the following completeness theorem.

Theorem: The set of eigenfunctions $\{u_i\}_{i=0}^{\infty}$ forms a complete orthonormal system for the the Hilbert space $L_r^2(a,b)$.

Proof: Following the above reasoning, we merely need to show that $S = \{0\}$. If this is not the case, we obtain, by Lemma, a nonzero element $h \in G(h)$ satisfies the boundary conditions; hence *h* satisfies the boundary conditions and solves the equation

$$(p(t)h')' + (\lambda r(t) + q(t))h = \frac{r}{\mu} h, t \in I,$$

i.e. $\lambda - \frac{1}{\mu} = \lambda_j$ for some *j*. Hence $h = cu_j$ for some nonzero constant *c*, contradicting $h \in S$.

Chapter 24

Laplace Transforms

There are many kinds of transforms out there in the world. Laplace transforms and Fourier transforms are probably the main two kinds of transforms that are usedo solve differential equations. As we will see in later sections we can use Laplace transforms to reduce a differential equation to an algebra problem. The algebra can be messy on occasion, but it will be simpler than actually solving the differential equation directly in many cases. Laplace transforms can also be used to solve IVP's that we can't use any previous method on. For "simple" differential equations such as those in the first few sections of the last chapter Laplace transforms will be more complicated than we need.

In fact, for most homogeneous differential equations such as those in the last chapter Laplace transforms is significantly longer and not so useful. Also, many of the "simple" nonhomogeneous differential equations that we saw in the Undetermined Coefficients and Variation of Parameters are still simpler (or at the least no more difficult than Laplace transforms) to do as we did them there. However, at this point, the amount of work required for Laplace transforms is starting to equal the amount of work we did in those sections.

Laplace transforms comes into its own when the forcing function in the differential equation starts getting more complicated. In the previous chapter we looked only at nonhomogeneous differential equations in which $g(t)$ was a fairly simple continuous function. In this chapter we will start looking at $g(t)$'s that are not continuous. It is these problems were the reasons for using Laplace transforms starts to become clear. We will also see that, for some of the more complicated nonhomogeneous differential equations from the last chapter, Laplace transforms are actually easier on those problems as well.

Laplace transforms (or just transforms) can seem scary when we first start looking at them. However, as we will see, they aren't as bad as they may appear at first. Before we start with the definition of the Laplace transform we need to get another definition out of the way.

A function is called piecewise continuous on an interval if the interval can be broken into a finite number of subintervals on which the function is continuous on each open subinterval (*i.e.* the subinterval without it's endpoints) and has a finite limit at the endpoints of each subinterval. In other words, a piecewise continuous function is a function that has a finite number of breaks in it and doesn't blow up to infinity anywhere.

Definition: Suppose that *f(t)* is a piecewise continuous function. The Laplace transform of *f(t)* is denoted $\mathfrak{L}\{f(t)\}$ and defined as

$$\mathfrak{L}\{f(t)\} = \int_0^\infty e^{-st} f(t)dt.$$

There is an alternate notation for Laplace transforms. For the sake of convenience we will often denote Laplace transforms as,

$$\mathfrak{L}\{f(t)\} = F(s).$$

With this alternate notation, note that the transform is really a function of a new variable, *s*, and that all the *t*'s will drop out in the integration process. Now, the integral in the definition of the transform is called an improper integral and it would probably be best to recall how these kinds of integrals work before we actually jump into computing some transforms.

Example: If c ≠ 0, evaluate the following integral.

$$\int_0^\infty e^{ct} dt.$$

Solution: Remember that you need to convert improper integrals to limits as follows,

$$\int_0^\infty e^{ct} dt \lim_{n\to\infty} \int_0^\infty e^{ct} dt.$$

Now, do the integral, then evaluate the limit.

$$\begin{aligned}\int_0^\infty e^{ct} dt &= \lim_{n\to\infty} \int_0^\infty e^{ct} dt \\ &= \lim_{n\to\infty} \left(\frac{1}{c} e^{ct}\right)\Bigg|_0^n \\ &= \lim_{n\to\infty} \left(\frac{1}{c} e^{cn} - \frac{1}{c}\right).\end{aligned}$$

Now, at this point, we've got to be careful. The value of *c* will affect our answer. We've already assumed that *c* was non-zero, now we need to worry about the sign of *c*. If *c* is positive the exponential will go to infinity. On the other hand, if *c* is negative the exponential will go to zero. So, the integral is only convergent (*i.e.* the limit exists and is finite) provided $c<0$. In this case we get,

$$\int_0^\infty e^{ct} dt = -\frac{1}{c} \text{ provided } c < 0.$$

Now that we remember how to do these, let's compute some Laplace transforms. We'll start off with probably the simplest Laplace transform to compute.

Example: Compute $\mathfrak{L}\{1\}$.

Solution: There's not really a whole lot do here other than plug the function $f(t) = 1$

$$\mathfrak{L}\{1\} = \int_0^{\infty} e^{-s} dt.$$

Now, at this point notice that this is nothing more than the integral in the previous example with $c = -s$. Therefore, all we need to do the appropriate substitution. Doing this gives,

$$\mathfrak{L}\{1\} = \int_0^{\infty} e^{-s} dt = -\frac{1}{-s} \quad \text{provided } -s < 0$$

Or, with some simplification we have,

$$\mathfrak{L}\{1\} = \frac{1}{s} \qquad \text{provided } -s < 0.$$

Notice that we had to put a restriction on s in order to actually compute the transform. All Laplace transforms will have restrictions on s.

At this stage of the game, this restriction is something that we tend to ignore, but we really shouldn't ever forget that it's there.

Let's do another example.

Example: Compute $\mathfrak{L}\{e^{at}\}$

Solution: Plug the function into the definition of the transform and do a little simplification.

$$\mathfrak{L}\{e^{at}\} = \int_0^{\infty} e^{-st} e^{at} dt = \int_0^{\infty} e^{(a-s)t} dt.$$

Once again, notice that we can provide $c = a - s$. So let's do this.

$$\begin{aligned}\mathfrak{L}\{e^{at}\} &= \int_0^{\infty} e^{(a-s)t} dt \\ &= \frac{1}{a-s} \qquad \text{provided } a - s < 0 \\ &= \frac{1}{s-a} \qquad \text{provided } s > a.\end{aligned}$$

Let's do one more example that doesn't come down to an application.

Example : Compute $\mathfrak{L}\{\sin(at)\}$.

Solution: Note that we're going to leave it to you to check most of the integration here. Plug the function into the definition. This time let's also use the alternate notation.

$$\mathfrak{L}\{\sin(at)\} = F(s)$$

$$= \int_0^{\infty} e^{-st}\sin(at)dt$$

$$= \lim_{n\to\infty}\int_0^{\infty} e^{-st}\sin(at)\,dt.$$

Now, if we integrate by parts we will arrive at,

$$F(s) = \lim_{n\to\infty}\left(-\left(\frac{1}{a}e^{-st}\cos(at)\right)\Bigg|_0^n - \frac{s}{a}\int_0^n e^{-st}\cos(at)\,dt\right).$$

Now, evaluate the first term to simplify it a little and integrate by parts again on the integral. Doing this arrives at,

$$F(s) = \lim_{n\to\infty}\left(\frac{1}{a}\left(1-e^{-in}\cos(at)\right) - \frac{s}{a}\left(\frac{1}{a}e^{-st}\sin(at)\right)\Bigg|_0^n + \frac{s}{a}\int_0^n e^{-st}\cos(at)\,dt\right)\Bigg).$$

Now, evaluate the second term, take the limit and simplify.

$$F(s) = \lim_{n\to\infty}\left(\frac{1}{a}(1-e^{-n}\cos(an)) - \frac{s}{a}\left(\frac{1}{a}e^{-sn}\sin(an) + \frac{s}{a}\int_0^n e^{-st}\sin(at)\,dt.\right)\right)$$

$$= \frac{1}{a} - \frac{s}{a}\left(\frac{s}{a}\int_0^{\infty} e^{-st}\sin(at)\,dt\right) = \frac{1}{a} - \frac{s^2}{a^2}\int_0^{\infty} e^{-st}\sin(at)\,dt.$$

Now, notice that in the limits we had to assume that $s>0$ in order to do the following two limits.

$$\lim_{n\to\infty} e^{-sn}\cos(an) = 0$$

$$\lim_{n\to\infty} e^{-sn}\sin(an) = 0.$$

Without this assumption, we get a divergent integral again. Also, note that when we got back to the integral we just converted the upper limit back to infinity.

The reason for this is that, if you think about it, this integral is nothing more than the integral that we started with. Therefore, we now get,

$$F(s) = \frac{1}{a} - \frac{s^2}{a^2}F(s).$$

Now, simply solve for *F(s)* to get,

$$\mathfrak{L}\{\sin(at)\} = F(s) = \frac{a}{s^2 + a} \qquad \text{provided } s > 0.$$

As this example shows, computing Laplace transforms is often messy.

Before moving on to the next section, we need to do a little side note. On occasion you will see the following as the definition of the Laplace transform.

$$\mathfrak{L}\{f(t)\} = \int_0^{\infty} e^{-st} f(t)dt.$$

Note the change in the lower limit from zero to negative infinity. In these cases there is almost always the assumption that the function *f(t)* is in fact defined as follows,

$$f(t) = \begin{cases} 0 & \text{if } t < 0 \\ f(t) & \text{if } t < 0 \end{cases}.$$

In other words, it is assumed that the function is zero if $t<0$. In this case the first half of the integral will drop out since the function is zero. A Heaviside function is usually used to make the function zero for $t<0$. We will be looking at these in a later section

LAPLACE TRANSFORMS

As we saw in the last section computing Laplace transforms directly can be fairly complicated. Usually we just use a table of transforms when actually computing Laplace transforms. The table that is provided here is not an inclusive table, but does include most of the commonly used Laplace transforms and most of the commonly needed formulas pertaining to Laplace transforms.

Before doing a couple of examples to illustrate the use of the table let's get a quick fact out of the way.

FACT

Given *f(t)* and *g(t)* then,

$$\mathfrak{L}\{af(t) + bg(t)\} = aF(s) + bG(s).$$

for any constants *a* and *b*.

In other words, we don't worry about constants and we don't worry about sums or differences of functions in taking Laplace transforms. All that we need to do is take the transform of the individual functions, then put any constants back in and add or subtract the results back up.

So, let's do a couple of quick examples.

Example: Find the Laplace transforms of the given functions.

1. $f(t) = 6e^{-5t} + e^{3t} + 5t^3 - 9$
2. $g(t) = 4\cos(4t) - 9\sin(4t) + 2\cos(10t)$
3. $h(t) = 3\sin(2t) + 3\sin(2t)$
4. $g(t) = e^{st} + \cos(6t) - e^{3t}\cos(6t).$

Solution: We'll do these examples in a little more detail than is typically used since this is the first time we're using the tables.

1. $f(t) = 6e^{-5t} + e^{3t} + 5t^3 - 9$

$$F(s) = 6\frac{1}{s-(-5)} + \frac{1}{s-3} + 5\frac{3!}{s^{3+1}} - 9\frac{1}{s}$$

$$= \frac{6}{s+5} + \frac{1}{s-3} + \frac{30}{s^4} - \frac{9}{s}$$

2. $g(t) = 4\cos(4t) - 9\sin(4t) + 2\cos(10t)$

$$G(s) = 4\frac{s}{s^2+(4)^2} - 9\frac{4}{s^2+(4)^2} + 2\frac{s}{s^2+(10)^2}$$

$$= \frac{4s}{s^2+16} - \frac{36}{s^2+16} + \frac{2s}{s^2+100}$$

3. $h(t) = 3\sinh(2t) + 3\sin(2t)$

$$H(s) = 3\frac{2}{s^2-(2)^2} + 3\frac{2}{s^2+(2)^2}$$

$$= \frac{6}{s^2-4} + \frac{6}{s^2+4}$$

4. $g(t) = e^{3t} + \cos(6t) - e^{3t}\cos(6t)$

$$G(s) = \frac{1}{s-3} + \frac{s}{s^2+(6)^2} - \frac{s-3}{(s-3)^2+(6)^2}$$

$$= \frac{1}{s-3} + \frac{s}{s^2+36} - \frac{s-3}{(s-3)^2+36}.$$

Make sure that you pay attention to the difference between a "normal" trig function and hyperbolic functions. The only difference between them is the "+ a^2" for the "normal" trig functions becomes a "- a^2" in the hyperbolic function! It's very easy to get in a hurry and not pay attention and grab the wrong formula.

Let's do one final set of examples.

Example: Find the transform of each of the following functions.

1. $f(t) = t\cosh(3t)$
2. $h(t) = t^2 \sin(2t)$
3. $g(t) = t^{\frac{3}{2}}$
4. $f(t) = (10t)^{\frac{3}{2}}$
5. $f(t) = tg'(t)$

Solution:

1. $f(t) = t\cosh(3t)$

This function is not in the table of Laplace transforms. If we take *n=1.*

$$F(s) = \mathfrak{L}\{tg(t)\} = -G'(s), \quad \text{where } g(t) = \cosh(3t).$$

So, we then have,

$$G(s) = \frac{s}{s^2 - 9} \qquad G'(s) = -\frac{s^2 + 9}{(s^2 - 9)^2}$$

we then have,

$$F(s) = \frac{s^2 = 9}{\left(s^2 - 9\right)^2}$$

2. $h(t) = t^2 \sin(2t)$

We could use it with n = 1 n = 1n = 1.

$$H(s) = \mathfrak{L}\{tf(t)\} = -F'(s), \quad \text{where } f(t) = t\sin(2t).$$

Or we could use it with $n = 2$.

$$H(s) = \mathfrak{L}\{t^2 f(t)\} = F''(s), \quad \text{where } f(t) = t\sin(2t).$$

Since it's less work to do one derivative, let's do it the first way. So we have,

$$F(s) = \frac{4s}{\left(s^2 + 4\right)^2} \qquad F'(s) = -\frac{12s^2 - 16}{\left(s^2 + 4\right)^2}.$$

The transform is then,

$$H(s) = \frac{12s^2 - 16}{\left(s^2 + 4\right)}$$

3. $g(t) = t^{\frac{3}{2}}$

This part can be with $n = 2$. We need to notice that

$$\int_0^t \sqrt{v}dv = \frac{2}{3}t^{\frac{2}{3}} \qquad \Rightarrow \qquad t^{\frac{2}{3}} = \frac{2}{3}\int_0^t \sqrt{v}dv.$$

Now,

$$f(t) = \sqrt{t} \qquad F(s) = \frac{\sqrt{\pi}}{2s^{\frac{3}{2}}}.$$

we get the following.

$$G(s) = \frac{3}{2}\left(\frac{\sqrt{\pi}}{2s^{\frac{3}{2}}}\right)\left(\frac{1}{s}\right) = \frac{3\sqrt{\pi}}{4s^{\frac{3}{2}}}$$

4. $f(t) = (10t)^{\frac{3}{2}}$

To see this note that if

$$g(t) = t^{\frac{3}{2}}$$

then

$$f(t) = g(10t).$$

Therefore, the transform is.

$$F(s) = \frac{1}{10}G\left(\frac{s}{10}\right)$$

$$= \frac{1}{10}\left(\frac{3\sqrt{\pi}}{4\left(\frac{s}{10}\right)^{\frac{5}{2}}}\right) = 10^{\frac{3}{2}}\frac{3\sqrt{\pi}}{4s^{\frac{5}{2}}}$$

5. $f(t) = tg'(t)$

$$\mathfrak{L}\{tg'(t)\} = -\frac{d}{ds}\mathfrak{L}\{g'\} = -\frac{d}{ds}\{sG(s) - g(0)\}$$

$$= \left(G(s) + sG'(s) - 0\right)$$

$$= G(s) - sG'(s).$$

Remember that *g(0)* is just a constant so when we differentiate it we will get zero!

As this set of examples has show us we can't forget to use some of the general formulas to derive new Laplace transforms for functions that aren't explicitly listed in the table!

INVERSE LAPLACE TRANSFORMS

Finding the Laplace transform of a function is not terribly difficult if we've got a table of transforms in front of us to use as we saw in the last section. What we would like to do now is go the other way.

We are going to be given a transform, *F(s)*, and ask what function (or functions) did we have originally. As you will see this can be a more complicated and lengthy process than taking transforms. In these cases we say that we are finding the Inverse Laplace Transform of *F(s)* and use the following notation.

$$f(t)=\mathfrak{L}^{-1}\{F(s)\}.$$

As with Laplace transforms, we've got the following fact to help us take the inverse transform.

FACT

Given the two Laplace transforms *F(s)* and *G(s)* then

$$\mathfrak{L}^{-1}\{aF(s)+bG(s)\}=a\mathfrak{L}^{-1}\{F(s)\}+b\mathfrak{L}^{-1}\{G(s)\}$$

for any constants a and b.

So, we take the inverse transform of the individual transforms, put any constants back in and then add or subtract the results back up.

Let's take a look at a couple of fairly simple inverse transforms.

Example: Find the inverse transform of each of the following.

1. $F(s)=\dfrac{6}{s}-\dfrac{1}{s-8}+\dfrac{4}{s-3}$

2. $H(s)=\dfrac{19}{s+2}-\dfrac{1}{3s-5}+\dfrac{7}{s^5}$

3. $F(s)=\dfrac{6s}{s^2+25}+\dfrac{3}{s^2+25}.$

4. $G(s)=\dfrac{8}{3s^2+12}+\dfrac{3}{s^2+49}$

Solution: I've always felt that the key to doing inverse transforms is to look at the denominator and try to identify what you've got based on that. If there is only one entry in the table that has that particular denominator, the next step is to make sure the numerator is correctly set up for the inverse transform process. If it isn't, correct it (this is always easy to do) and then take the inverse transform.

If there is more than one entry in the table has a particular denominator, then the numerators of each will be different, so go up to the numerator and see which one you've got. If you need to correct the numerator to get it into the correct form and then take the inverse transform.

So, with this advice in mind let's see if we can take some inverse transforms.

1. $F(s)=\frac{6}{s}-\frac{1}{s-8}+\frac{4}{s+3}$.

From the denominator of the first term it looks like the first term is just a constant. The correct numerator for this term is a "1" so we'll just factor the 6 out before taking the inverse transform. The second term appears to be an exponential with $a = 8$ and the numerator is exactly what it needs to be. The third term also appears to be an exponential, only this time $a = 3$ and we'll need to factor the 4 out before taking the inverse transforms.

So, with a little more detail than we'll usually put into these,

$$F(s) = 6\frac{1}{s}-\frac{1}{s-8}+4\frac{1}{s-3}$$

$$f(t) = 6(1)-e^{8t}+4\left(e^{3t}\right)$$

$$= 6-e^{8t}+4e^{3t}$$

2. $H(s)=\frac{19}{s+2}-\frac{1}{3s-5}+\frac{7}{s^5}$.

The first term in this case looks like an exponential with $a = -2$and we'll need to factor out the 19. Be careful with negative signs in these problems, it's very easy to lose track of them.

The second term almost looks like an exponential, except that it's got a $3s$ instead of just an s in the denominator. It is an exponential, but in this case we'll need to factor a 3 out of the denominator before taking the inverse transform.

The denominator of the third term appears with $n = 4$. The numerator however, is not correct for this. There is currently a 7 in the numerator and we need a $4! = 24$ in the numerator. This is very easy to fix.

Whenever a numerator is off by a multiplicative constant, as in this case, all we need to do is put the constant that we need in the numerator. We will just need to remember to take it back out by dividing by the same constant.

So, let's first rewrite the transform.

$$H(s) = \frac{19}{s-(-2)} - \frac{1}{3\left(s-\frac{5}{3}\right)} + \frac{7}{s^{4+1}}$$

$$= 19\frac{1}{s-(-2)} - \frac{1}{3}\frac{1}{3s-\frac{5}{3}} + \frac{7}{4!s^{4+1}}.$$

So, what did we do here? We factored the 19 out of the first term. We factored the 3 out of the denominator of the second term since it can't be there for the inverse transform and in the third term we factored everything out of the numerator except the 4! since that is the portion that we need in the numerator for the inverse transform process.

Let's now take the inverse transform.

$$h(t) = 19e^{-2t} - \frac{1}{3}e^{\frac{5t}{3}} + \frac{7}{24}t^4$$

3. $F(s) = \dfrac{6s}{s^2+25} + \dfrac{2}{s^2+25}.$

The numerators will tell us which we've actually got. The first one has an s in the numerator and so this means that the first term must be #8 and we'll need to factor the 6 out of the numerator in this case. The second term has only a constant in the numerator and so this term must be #7, however, in order for this to be exactly #8 we'll need to multiply/divide a 5 in the numerator to get it correct for the table.

The transform becomes,

$$F(s) = 6\frac{s}{s^2+(5)^2} + \frac{3\frac{5}{5}}{s^2+(5)^2}$$

$$= 6\frac{s}{s^2+(5)^2} + \frac{3}{5}\frac{5}{s^2+(5)^2}.$$

Taking the inverse transform gives,

$$f(t)\, 6\cos(5t) + \frac{3}{5}\sin(5t)$$

4. $G(s) = \dfrac{8}{3s^2+12} + \dfrac{2}{s^2-49}.$

In this case the first term will be a sine once we factor a 3 out of the denominator, while the second term appears to be a hyperbolic sine. Again, be careful with the difference between these two. Both of the terms will also need to have their numerators fixed up. Here is the transform once we're done rewriting it.

$$G(s) = \frac{1}{3}\frac{8}{s^2+4} + \frac{3}{s^2-49} = \frac{1}{3}\frac{(4)(2)}{s^2+(2)^2} + \frac{3\frac{7}{7}}{s^2-(7)^2}.$$

Notice that in the first term we took advantage of the fact that we could get the 2 in the numerator that we needed by factoring the 8.The inverse transform is then,

$$g(t) = \frac{4}{3}\sin(2t) + \frac{3}{7}\sin h(7t).$$

So, probably the best way to identify the transform is by looking at the denominator. If there is more than one possibility use the numerator to identify the correct one. Fix up the numerator if needed to get it into the form needed for the inverse transform process. Finally, take the inverse transform.

Let's do some slightly harder problems. These are a little more involved than the first set.

Example: Find the inverse transform of each of the following.

1. $F(s) = \dfrac{6s-5}{s^2+7}$
2. $F(s) = \dfrac{1-3s}{s^2+8s+21}$
3. $G(s) = \dfrac{3s-2}{2s^2-6s-2}$
4. $H(s) = \dfrac{s+7}{s^2-3s-10}$

Solution:

1. $F(s) = \dfrac{6s-5}{s^2+7}.$

From the denominator of this one it appears that it is either a sine or a cosine. However, the numerator doesn't match up to either of these in the table. A cosine wants just an s in the numerator with at most a multiplicative constant, while a sine wants only a constant and no s in the numerator. We've got both in the numerator. This is easy to fix however. We will just split up the transform into two terms and then do inverse transforms.

$$F(s) = \frac{6s-5}{s^2+7} - \frac{5\frac{\sqrt{7}}{\sqrt{7}}}{s^2+7}$$

$$f(t) = f(t) = 6\cos\left(\sqrt{7}t\right) - \frac{5}{\sqrt{7}}\sin\left(\sqrt{7}t\right).$$

Do not get too used to always getting the perfect squares in sines and cosines that we saw in the first set of examples. More often than not (at least in my class) they won't be perfect squares!

2. $F(s)=\dfrac{1-3s}{s^2+8s+21}$

In this case there are no denominators in our table that look like this. We can however make the denominator look like one of the denominators in the table by completing the square on the denominator. So, let's do that first.

$$\begin{aligned} s^2+8s+21 &= s^2+8s+16-16+21 \\ &= s^2+8s+16+5 \\ &= (s+4)^2+5 \end{aligned}$$

Recall that in completing the square you take half the coefficient of the s, square this, and then add and subtract the result to the polynomial. After doing this the first three terms should factor as a perfect square.

So, the transform can be written as the following.

$$F(s)=\frac{1-3s}{(s+4)^2+5}.$$

However, note that in order for it to be a #19 we want just a constant in the numerator and in order to be a #20 we need an s_a in the numerator. We've got neither of these so we'll have to correct the numerator to get it into proper form.

In correcting the numerator always get the s_a first. This is the important part. We will also need to be careful of the 3 that sits in front of the s. One way to take care of this is to break the term into two pieces, factor the 3 out of the second and then fix up the numerator of this term. This will work, however it will put three terms into our answer and there are really only two terms.

So, we will leave the transform as a single term and correct it as follows,

$$\begin{aligned} F(s) &= \frac{1-3(s+4-4)}{(s+4)^2+5} \\ &= \frac{1-3(s+4)+12}{(s+4)^2+5} \\ &= \frac{-3(s+4)+13}{(s+4)^2+5}. \end{aligned}$$

We needed an $s + 4$ in the numerator, so we put that in. We just needed to make sure and take the 4 back out by subtracting it back out. Also, because of the 3 multiplying the s we needed to do all this inside a set of parenthesis. Then we partially multiplied the 3 through the second term and combined the constants. With the transform in this form, we can break it up into two transforms each of which are in the tables and so we can do inverse transforms on them,

$$F(s) = -3\frac{s+4}{(s+4)^2+5} + \frac{13\frac{\sqrt{5}}{\sqrt{5}}}{(s+4)^2+5}$$

$$f(t) = -3e^{-4t}\cos\left(\sqrt{5}t\right) + \frac{13}{\sqrt{5}}e^{-4t}\sin\left(\sqrt{5}t\right)$$

3. $G(s) = \dfrac{3s-2}{2s^2-6x-2}$.

This one is similar to the last one. We just need to be careful with the completing the square however. The first thing that we should do is factor a 2 out of the denominator, then complete the square. Remember that when completing the square a coefficient of 1 on the s^2 term is needed! So, here's the work for this transform.

$$G(s) = \frac{3s-2}{2\left(s^2-3s-1\right)}$$

$$= \frac{1}{2}\frac{3s-2}{s^2-3s+\frac{9}{4}-\frac{9}{4}-1}$$

$$= \frac{1}{2}\frac{3s-2}{\left(s^2-\frac{3}{2}\right)-\frac{13}{4}}.$$

Here's the work for that and the inverse transform.

$$G(s) = \frac{1}{2}\frac{\left(s-\frac{3}{2}+\frac{3}{2}\right)-2}{\left(s-\frac{3}{2}\right)^2-\frac{13}{4}}$$

$$= \frac{1}{2}\frac{3\left(s-\frac{3}{2}\right)+\frac{5}{2}}{\left(s-\frac{3}{2}\right)^2-\frac{13}{4}}$$

$$= \frac{1}{2}\left(\frac{3\left(s-\frac{3}{2}\right)}{\left(s-\frac{3}{2}\right)^2-\frac{13}{4}} + \frac{\frac{5}{2}\frac{\sqrt{13}}{\sqrt{13}}}{\left(s-\frac{3}{2}\right)^2-\frac{13}{4}}\right)$$

$$g(t) = \frac{1}{2}\left(3e^{\frac{3t}{2}}\cosh\left(\frac{\sqrt{13}}{2}t\right) + \frac{5}{\sqrt{13}}e^{\frac{3t}{2}}\sin h\left(\frac{\sqrt{13}}{2}t\right)\right).$$

In correcting the numerator of the second term, notice that I only put in the square root since we already had the "over 2" part of the fraction that we needed in the numerator.

4. $H(s)=\dfrac{s+7}{s^2-3s-10}$.

This one appears to be similar to the previous two, but it actually isn't. The denominators in the previous two couldn't be easily factored. In this case the denominator does factor and so we need to deal with it differently. Here is the transform with the factored denominator.

$$H(s)=\frac{s+7}{(s+2)(s-5)}.$$

The denominator of this transform seems to suggest that we've got a couple of exponentials, however in order to be exponentials there can only be a single term in the denominator and no s's in the numerator.

To fix this we will need to do partial fractions on this transform. In this case the partial fraction decomposition will be

$$H(s)=\frac{A}{s+2}+\frac{B}{a-5}.$$

Don't remember how to do partial fractions? In this example we'll show you one way of getting the values of the constants and after this example we'll review how to get the correct form of the partial fraction decomposition.

Okay, so let's get the constants. There is a method for finding the constants that will always work, however it can lead to more work than is sometimes required.

Eventually, we will need that method, however in this case there is an easier way to find the constants.

Regardless of the method used, the first step is to actually add the two terms back up. This gives the following.

$$\frac{s+7}{(s+2)(s-5)}=\frac{A(s-5)+B(s+2)}{(s+2)(s-5)}.$$

Now, this needs to be true for any s that we should choose to put in. So, since the denominators are the same we just need to get the numerators equal. Therefore, set the numerators equal.

$$s+7=A(s-5)+B(s+2).$$

Again, this must be true for ANY value of s that we want to put in. So, let's take advantage of that. If it must be true for any value of s then it must be true for $s = -2$, to pick a value at random. In this case we get,

$$s = A(-7) + B(0) \qquad \Rightarrow \quad A = -\frac{5}{7}.$$

We found A by appropriately picking s. We can B in the same way if we chose $s = 5$.

$$12 = A(0) + B(7) \qquad \Rightarrow \quad B = \frac{12}{7}.$$

This will not always work, but when it does it will usually simplify the work considerably.

So, with these constants the transform becomes,

$$H(s) = \frac{-\frac{5}{7}}{s+2} + \frac{\frac{12}{7}}{s-5}.$$

We can now easily do the inverse transform to get,

$$h(t) = -\frac{5}{7}e^{-2t} + \frac{12}{7}e^{5t}.$$

The last part of this example needed partial fractions to get the inverse transform. When we finally get back to differential equations and we start using Laplace transforms to solve them, you will quickly come to understand that partial fractions are a fact of life in these problems. Almost every problem will require partial fractions to one degree or another.

Note that we could have done the last part of this example as we had done the previous two parts. If we had we would have gotten hyperbolic functions. However, recalling the definition of the hyperbolic functions we could have written the result in the form we got from the way we worked our problem.

However, most students have a better feel for exponentials than they do for hyperbolic functions and so it's usually best to just use partial fractions and get the answer in terms of exponentials. It may be a little more work, but it will give a nicer (and easier to work with) form of the answer.

Be warned that in my class I've got a rule that if the denominator can be factored with integer coefficients then it must be.

So, let's remind you how to get the correct partial fraction decomposition. The first step is to factor the denominator as much as possible. Then for each term in the denominator we will use the following table to get a term or terms for our partial fraction decomposition.

Factor in denominator	*Term in partial fraction decomposition*
$ax+b$	$\frac{A}{ax+b}$
$(ax+b)^k$	$\frac{A_1}{ax+b}+\frac{A_2}{(ax+b)^2}+\cdots+\frac{A_k}{(ax+b)^k}$
ax^2+bx+c	$\frac{Ax+B}{ax^2+bx+c}$
$(ax^2+bx+c)^k$	$\frac{A_1x+B_1}{ax^2+bx+c}+\frac{A_2x+B_2}{(ax^2+bx+c)}+\cdots+\frac{A_kx+B_k}{(ax^2+bx+c)^k}.$

Notice that the first and third cases are really special cases of the second and fourth cases respectively.

So, let's do a couple more examples to remind you how to do partial fractions.

Example: Find the inverse transform of each of the following.

1. $G(s)=\frac{86s-78}{(s+3)(s-4)(5s-1)}$

2. $F(s)=\frac{2-5s}{(s-6)(s^2+11)}$

3. $G(s)=\frac{25}{s^3(s^2+4s+5)}$

Solution:

1. $G(s)=\frac{86s-78}{(s+3)(s-4)(5s-1)}$

Here's the partial fraction decomposition for this part.

$$G(s)=\frac{A}{(s+3)}+\frac{A}{(s-4)}+\frac{A}{(5s-1)}.$$

Now, this time we won't go into quite the detail as we did in the last example.

We are after the numerator of the partial fraction decomposition and this is usually easy enough to do in our heads.

Therefore, we will go straight to setting numerators equal.

$$86s-78=A(s-4)(5s-1)+B(s+3)(5s-1)+C(s+3)(s-4).$$

As with the last example, we can easily get the constants by correctly picking values of s.

$$s=-3 \qquad -336=A(-7)(-16) \quad \Rightarrow \quad A=-3$$

$$s=\frac{1}{5} \qquad -\frac{304}{5}=C\left(\frac{16}{5}\right)\left(-\frac{19}{5}\right) \quad \Rightarrow \quad C=5.$$

$$s=4 \qquad 266=B(7)(19) \quad \Rightarrow \quad B=2$$

So, the partial fraction decomposition for this transform is,

$$G(s)=-\frac{3}{s+3}+\frac{2}{s-4}+\frac{5}{5s-1}.$$

Now, in order to actually take the inverse transform we will need to factor a 5 out of the denominator of the last term. The corrected transform as well as its inverse transform is.

$$G(s)=-\frac{3}{s+3}+\frac{2}{s-4}+\frac{5}{s-\frac{1}{5}}$$

$$g(t)=-3e^{-3t}+2e^{4t}+e^{\frac{t}{5}}$$

2. $F(s)=\dfrac{2-5s}{(s-6)(s^2+11)}.$

So, for the first time we've got a quadratic in the denominator. Here's the decomposition for this part.

$$F(s)=\frac{A}{s-6}+\frac{Bs+C}{s^2+11}.$$

Setting numerators equal gives,

$$2-5s=A(s^2+11)+(Bs+C)(s-6).$$

Okay, in this case we could use $s=6$to quickly find A, but that's all it would give. In this case we will need to go the "long" way around to getting the constants.

Note that this way will always work, but is sometimes more work than is required.

The "long" way is to completely multiply out the right side and collect like terms.

$$\begin{aligned} 2-5s &= A(s^2+11)+(Bs+C)(s-6) \\ &= As^2+11A+Bs^2+6B+Cs-6C \\ &= (A+B)s^2+(-6B+C)s+11A-6C. \end{aligned}$$

In order for these two to be equal the coefficients of the s^2, s and the constants must all be equal. So, setting coefficients equal gives the following system of equations that can be solved.

$$\left.\begin{aligned} s^2 &: \quad A+B=0 \\ s^1 &: -6B+C=-5 \\ s^0 &: 11A-6C=2 \end{aligned}\right\} \Rightarrow \quad A=-\frac{28}{47}, \quad B=\frac{28}{47}, \quad C=-\frac{67}{47}.$$

Notice that I used $s0$ to denote the constants. This is habit on my part and isn't really required, it's just what I'm used to doing.

Also, the coefficients are fairly messy fractions in this case. Get used to that. They will often be like this when we get back into solving differential equations.

There is a way to make our life a little easier as well with this. Since all of the fractions have a denominator of 47 we'll factor that out as we plug them back into the decomposition. This will make dealing with them much easier.

The partial fraction decomposition is then,

$$F(s) = \frac{1}{47}\left(\frac{28}{s-6} + \frac{28s-67}{s^2+11}\right)$$

$$= \frac{1}{47}\left(\frac{28}{s-6} + \frac{28s}{s^2+11} - \frac{67\frac{\sqrt{11}}{\sqrt{11}}}{s^2+11}\right).$$

The inverse transform is then.

$$f(t) = \frac{1}{47}\left(-28e^{6t} + 28\cos\left(\sqrt{11}t\right) - \frac{67}{\sqrt{11}}\sin\left(\sqrt{11}t\right)\right).$$

3. $G(s) = \dfrac{25}{s^3(s^2+4s+5)}$

With this last part do not get excited about the s^3. We can think of this term as

$$s^3 = (s-0)^3$$

and it becomes a linear term to a power. So, the partial fraction decomposition is

$$G(s) = \frac{A}{s} + \frac{B}{s^2} + \frac{C}{s^3} + \frac{Ds+E}{s^2+4s+5}.$$

Setting numerators equal and multiplying out gives.

$$25 = As^2(s^2+4s+5)+Bs(s^2+4s+5)+C(s^2+4s+5)+(Ds+E)s^3$$
$$=(A+D)s^4+(4A+B+E)s^3+(5A+4B+C)s^2(5B+4C)s+5C.$$

Setting coefficients equal gives the following system.

$$\left.\begin{array}{lr} s^4: & A+D=0 \\ s^3: & 4A+B+E=0 \\ s^2: & 5A+4B+C=0 \\ s^1: & 5B+4C=0 \\ s^0: & 5C=25 \end{array}\right\} \Rightarrow A=\frac{11}{5},\ B=-4,\ C=5, D=-\frac{11}{5}, E=-\frac{24}{5}.$$

This system looks messy, but it's easier to solve than it might look. First we get C for free from the last equation. We can then use the fourth equation to find B. The third equation will then give A, *etc.*

When plugging into the decomposition we'll get everything with a denominator of 5, then factor that out as we did in the previous part in order to make things easier to deal with.

$$G(s)=\frac{1}{5}\left(\frac{11}{s}-\frac{20}{s^2}+\frac{25}{s^3}-\frac{11s+24}{s^2+4s+5}\right).$$

Note that we also factored a minus sign out of the last two terms. To complete this part we'll need to complete the square on the later term and fix up a couple of numerators. Here's that work.

$$G(s)=\frac{1}{5}\left(\frac{11}{s}-\frac{20}{s^2}+\frac{25}{s^3}-\frac{11s+24}{s^2+4s+5}\right)$$
$$=\frac{1}{5}\left(\frac{11}{s}-\frac{20}{s^2}+\frac{25}{s^3}-\frac{11(s+2-2)+24}{(s+2)^2+1}\right)$$
$$=\frac{1}{5}\left(\frac{11}{s}-\frac{20}{s^2}+\frac{25\frac{2!}{2!}}{s^3}-\frac{11(s+2)}{(s+2)^2+1}-\frac{2}{(s+2)^2+1}\right).$$

The inverse transform is then.

$$g(t)=\frac{1}{5}\left(11-20t+\frac{25}{2}t^2-11e^{-2t}\cos(t)-2e^{-2t}\sin(t)\right).$$

So, one final time. Partial fractions are a fact of life when using Laplace transforms to solve differential equations. Make sure that you can deal with them.

APPLICATION FOR LAPLACE TRANSFORM

In this Section we discuss some theorem on Laplace transform. Also, we discuss the application of Laplace transform.

THEOREMS OF THE LAPLACE TRANSFORM

The Laplace transform of the function f(t) is denoted by

$$L(f(t)) = \int_0^\omega e^{-st} f(t)\, dt$$

(a) *Superposition theorem:* For arbitrary constants a_1 and a_2 it follows that

$$\{a_1 f_1(t) + a_2 f_2(t)\} = a_1 F_1(s) + a_2 F_2(s)\,.$$

The Laplace transformation is a linear integral transformation.

(b) *Similarity theorem:* For an arbitrary constant a > 0

$$\{f(at)\} = \frac{1}{a} F\left(\frac{s}{a}\right)$$

is valid. This follows from Equation by the substitution of $\tau = at$.

(c) *Real Shifting theorem:* For an arbitrary constant a > 0

$$\{f(t-a)\} = e^{-as} F(s)$$

is valid. This follows directly from Equation by the substitution of $\tau = t - a$.

(d) *Complex Shifting theorem:* For an arbitrary constant a > 0

$$\{e^{-at} f(t)\} = F(s+a)$$

is valid.

(e) *Derivative theorem:* For a causal function of time, $f(t)$, for which the derivative for $t > 0$ exists.

$$\left\{\frac{df(t)}{dt}\right\} = s\ F(s) - f(0+),$$

and in the case of multiple differentiation

$$\left\{\frac{d^n f(t)}{dt^n}\right\} = s^n\ F(s) - \sum_{i=1}^{n} s^{n-i} \left.\frac{d^{(i-1)} f(t)}{dt^{(i-1)}}\right|_{t=0+}$$

(f) *Complex differentiation theorem:* This theorem shows that a differentiation of the mapped function F(s) corresponds to a multiplication with the time t in the time domain:

$$\{t^k = f(t)\} = (-1)^k \frac{d^k F(s)}{ds^k}\,.$$

(g) *Integral theorem:* The integral of a function is mapped by

$$\left\{\int_o^t = f(\tau)\mathrm{d}\tau\right\} = \frac{1}{s}F(s)\,.$$

(h) *Convolution in the time domain:* The convolution of two functions of time $f_1(t)$ and $f_2(t)$, presented by the symbolic notation $f_1(t) * f_2(t)$, is defined as

$$f_1(t) * f_2(t) = \int_o^t f_1(\tau)\, f_2(t-\tau)\,\mathrm{d}\tau\,.$$

In the previous section it is shown that the convolution of the two original functions corresponds to the multiplication of the related mapped functions, that is

$$\{f_1(t) * f_2(t)\} = F_1(s)\, F_2(s)$$

(i) *Convolution in the frequency domain:* Whereas in (h) the convolution of two functions of time was given, a similar result for the convolution of two functions in the frequency domain exists and is given by

$$\{f_1(t)\, f_2(t)\} = \frac{1}{2\pi \mathrm{j}} \int_{c-\mathrm{j}oo}^{c+\mathrm{j}oo} F_1(p)\, F_2(s-p)\,\mathrm{d}p$$

Here $F_1(s) \bullet\!\!-\!\!\circ f_1(t)$ and $F_2(s) \bullet\!\!-\!\!\circ f_2(t)$ is valid. Furthermore, p is the complex variable of integration. According to this theorem the Laplace transform of the product of two functions of time is equal to the convolution of $F_1(s)$ and $F_2(s)$ in the mapped domain.

(j) *Initial and final value theorems:* The *theorem of the initial condition* allows the direct calculation of the function value $f(0+)$ of a causal function of time $f(t)$ from the Laplace transform $F(s)$. If the Laplace transform of $f(t)$ and $f(t)$ exist, then

$$f(0+) = \lim_{t\to 0+} f(t) = \lim_{s\to\infty} s\, F(s)$$

is valid if the $\lim_{t\to 0} f(t)$ exists.

Using the *theorem of the final value* the value of $f(t)$ for $t\to\infty$ can be determined from $F(s)$, if the Laplace transform of $f(t)$ and $f(t)$ exist and the limit $\lim_{t\to\infty} f(t)$ also exists. Then it follows from previous section that

$$f(\infty) = \lim_{t\to\infty} f(t) = \lim_{t\to 0} s\, F(s)$$

One has to observe that

$$\lim_{t\to\infty} f(t) \text{ or } \lim_{t\to 0} f(t)$$

can be calculated only from the corresponding Laplace transform $\{f(t)\}$ by application of the theorems of the initial or final value, if the existence of the related limit in the time domain is a priori assured. The following two examples should explain this:

Example: $f(t) = e^{at}\,(\alpha > 0) \bullet\!\!-\!\!\circ\; F(s) = \dfrac{1}{s-\alpha}$

The limit $\lim\limits_{t\to\infty} e^{\alpha t}$ does not exist so that the final value theorem may not be applied.

Example: $f(t) = \cos w_0 t \bullet\!\!-\!\!\circ\; F(s) = \dfrac{s}{s^2 + w_0^2}$

The limit $\lim\limits_{t\to\infty} \cos w_0 t$ does not exist and therefore the final value theorem may not be applied. It can be concluded from the last two examples that the following general statement is valid: If the Laplace transform $F(s)$ has, apart from a single pole at the origin $s = 0$, poles on the imaginary axis or in the right-half plane, then the initial or final value theorems cannot be applied.

THE INVERSE LAPLACE TRANSFORM

As already mentioned in previous section in many cases a direct evaluation of the complex inverse integral is not necessary. A complicated function, $F(s)$, must be decomposed into a sum of simple functions of s that is

$$F(s) = F_1(s) + F_2(s) + ... + F_n(s),$$

which have a known inverse Laplace transform:

$$L^{-1}\{F(s)\} = L^{-1}\{F_1(s)\} + L^{-1}\{F_2(s)\} + ... + L^{-1}\{F_3(s)\}$$
$$= f_1(t) + f_2(t)_n(t) = f(t).$$

For many problems in control the function $F(s)$ is a ratio of polynomials in s, known as *rational fraction,* that is

$$\mathrm{F}(s) = \frac{n_0 + n_1 s + ...n_m s^m}{d_0 + d_1 s + ... + s^n} = \frac{N(s)}{D(s)},$$

where $N(s)$ and $D(s)$ are the numerator and the denominator, respectively.

If $m > n$, then $N(s)$ is divided by $D(s)$, where a polynomial in s and *a* ratio of polynomials are obtained. The numerator of the fraction $N_1(s)$ has a lower order than n. E.g, if $m = n + 2$, then

$$\frac{N(s)}{D(s)} = k_2 s^2 + k_1 s + k_0 + \frac{N_1(s)}{D(s)},$$

where by degree $N_1(s) < n$ and k_0, k_1 and k_2 are constants.

A rational fraction $F(s)$ given in Equation can be decomposed into more simple functions by application of *partial fraction decomposition*, as shown in Equation. In order to perform this decomposition the denominator polynomial $D(s)$ must be factorised into the form

$$F(s) = \frac{N(s)}{(s-s-1)\,(s-s_2)...(s-s_n)}.$$

For a denominator polynomial of n -th order one obtains roots or zeros $s = s_1, s_2,..., s_n$. The zeros of $D(s)$ are also known as the *poles* of $F(s)$, since they define where $F(^s)$ is infinite. The partial fraction decomposition for different types of poles is shown in the following.

***Case 1**:* $F(s)$ has only *single poles.*

Here $F(s)$ can be expanded into the form

$$F(s) = \sum_{k=1}^{n} \frac{c_k}{s - s_k},$$

where the *residuals* c_k are real or complex constants. Using the table of correspondences one immediately can obtain the corresponding function of time

$$F(t) = \sum_{k=1}^{n} c_k e^{s_k t} \quad \text{for} \qquad t > 0.$$

The values c_k can be determined either by comparing the coefficients or by using the theorem of residuals from the theory of functions according to

$$c_k = \frac{N(s_k)}{D'(s_k)} = (s - s_k)\frac{N(s)}{D(s)}\bigg|_{s=s_k}$$

for $k = 1, 2,..., n$ with $D'(s_k) = \mathrm{d}\,D/\mathrm{d}s\,|_{s=s_k}$

***Case 2**: F(s) has multiple poles.*

For multiple poles of $F(s)$ each with multiplicity $r_k (k = 1, 2,..., l)$ the corresponding partial fraction decomposition is

$$F(s) = \sum_{k=1}^{l} \sum_{v=1}^{rk} \frac{c_{kv}}{(s - s_k)^v} \quad \text{with } n = \sum_{k=1}^{l} r_k$$

The back transformation of Equation into the time domain is

$$f(t) = \sum_{k=1}^{l} e^{s_k t} \sum_{v=1}^{rk} \frac{c_{kv} t^{v-1}}{(v-1)!} \text{ for } t > 0$$

The real or complex coefficients c_kv for v = 1, 2,...,r_k determined by the theorem of residuals are

$$c_{kv} = \frac{1}{(r_k - v)!}\left\{\frac{d(r_k - v)}{ds(r_k - v)!}[F(s)\,(s - s_k)^{r_k}]\right\}_{s=s_k}$$

This general relation also contains the case of single poles of *F*(*s*). The poles may be real or complex.

Case 3: *F*(*s*) has also *conjugate complex poles.*

As both, the numerator *N*(*s*) and the denominator *D*(*s*) of the function *F*(*s*) are rational algebraic functions, complex factors always arise as conjugate complex pairs. If *F*(*s*) has a conjugate complex pair of poles s_1, 2 $= \sigma_1 \pm jw_1$, then for the function F_1,2(*s*) in the partial fraction decomposition of

$$F(s) = \frac{N(s)}{D(s)} = F_{1,2}(s) + F_3(s) + ... + F_n(s)$$

Equation can be applied to give

$$F_{1,2}(s) = \frac{c_1}{s - (\sigma + jw_1)} + \frac{c_2}{s - (\sigma_1 - jw_1)},$$

where the residuals

$$c_{1,2} = \delta_1 \pm j\varepsilon_1$$

are also a conjugate complex pair. Therefore, both fractions of $F_{1,2}(s)$ can be combined, and one obtains

$$F_{1,2}(s) = \frac{\beta_0 + \beta_1 s}{\alpha_0 + \alpha_1 s + s^2}$$

with the real coefficients

$$\left.\begin{aligned} \alpha_0 &= \alpha_1^2 + w_1^2 && ;\ \alpha_1 = -2\sigma_1 \\ \beta_0 &= -2(\sigma_2\beta_1 + w_1\varepsilon_1) && ;\ \beta_1 = 2\delta_1 \end{aligned}\right\}.$$

The determination of the coefficients β_0 and β_1 is performed again using the theorem of residuals by

$$(\beta_0 + \beta_1 s)\,|_{s=s_1} = (s - s_1)\,(s - s_2)\left.\frac{N(s)}{D(s)}\right|_{s=s_1}$$

As s_1 is complex, both sides of this equation are complex. Comparing the real and imaginary parts of both sides one gets two equations for the calculation of β_0 and β_1. This procedure is demonstrated now using the following example.

Example: Find the inverse Laplace transform $f(t)$ of

$$F(s) = \frac{1}{(s^2 + 2s + 2)(s + 2)}$$

The partial fraction decomposition of $F(s)$ is

$$F(s) = F_{1,2}(s) + F_3(s) = \frac{\beta_0 + \beta_1 s}{s^2 + 2s + 2} + \frac{c_3}{s + 2},$$

where the function $F_{1,2}(s)$ contains the conjugate pair of poles

$$s_{1,2} = -1 \pm j.$$

In addition the third pole of $F(s)$ is

$$s_3 = -2$$

For the coefficients β_0 and β_1 it follows from Equation

$$(\beta_0 + \beta_1 s)|_{s=s_1} = \left.\frac{1}{s+2}\right|_{s=s_1}$$

$$(\beta_0 + \beta_1) + j\beta_1 = \frac{1}{-1 + j + 2} = \frac{1}{2} - j\frac{1}{2}.$$

Comparing the real and imaginary parts on both sides one obtains

$$\beta_0 - \beta_1 = \frac{1}{2} \text{ and } \beta_1 = -\frac{1}{2}$$

and from this finally $\beta_0 = 0$.

Using Equation the residual is

$$c_3 = (s+2)\left.\frac{1}{(s^2 + 2s + 2)}\right|_{s=s_3} = \frac{1}{2}.$$

The partial fraction decomposition of $F(s)$ is thus

$$F(s) = -\frac{1}{2}\left[\frac{s}{s^2 + 2s + 2}\right] + \frac{1}{2}\frac{1}{s+2},$$

which can be rearranged in the form

$$F(s) = -\frac{1}{2}\left[\frac{s+1}{(s+1)^2 + 1} - \frac{1}{(s+1)^2 + 1} - \frac{1}{s+2}\right]$$

which can be directly applied to find the inverse transformation.

$$f(t) = -\frac{1}{2}[e^{-t}\cos t - e^{-t}\sin t - e^{-2t}] \text{ for } t > 0,$$

which can be rearranged as

$$f(t) = \frac{1}{2}e^{-t}[e^{-t} + \sin t - \cos t] \text{ for } t > 0.$$

The graphical representation of $f(t)$ is shown in Figure. Figure shows the corresponding poles, marked by a x, for this $F(s)$ in the complex s plane.

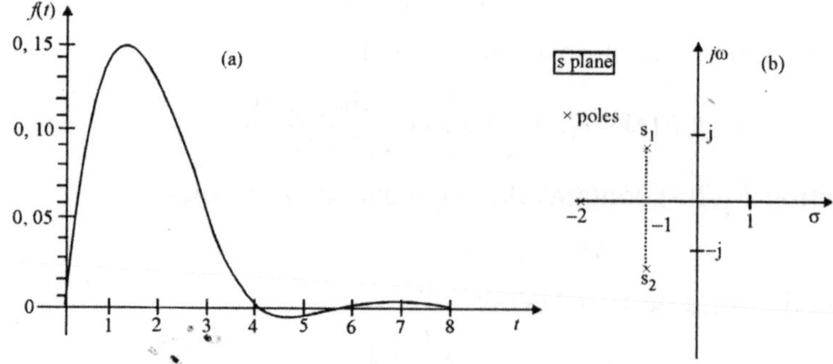

Fig. (a) Graph of the Original Function $f(t)$ (Function in the time Domain) and (b) Position of the Poles of $F(s)$ in the s Plane

It can be seen from this example that the position of the poles s_1, s_2 and s_3 affects the shape of the graph of $f(t)$. In this case all poles of $F(s)$ have negative real parts, therefore the graph of $f(t)$ shows a damped behaviour, i.e. it decreases to zero for $t \to \infty$. If the real part of one pole be positive, then the graph of $f(t)$ would be infinitely large for $t \to \infty$.

Since in control problems the original function $f(t)$ always represents the time behaviour of a system variable, the behaviour of this system variable $f(t)$ can be judged to a large extent by investigation of the positions of the poles of the corresponding mapped function $F(s)$.

SOLVING LINEAR DIFFERENTIAL EQUATIONS USING THE LAPLACE TRANSFORM

The Laplace transform, the basics of which have been introduced in the sections above, is an elegant way for fast and schematic solving of linear differential equations with constant coefficients. In the following the importance of this approach is demonstrated. Instead of solving the differential equation with the initial conditions directly in the original domain, the detour via a mapping into the frequency domain is taken, where only an algebraic equation has to be solved. Thus solving differential equations is performed according to Figure in the following three steps: Transformation of the differential equation into the mapped space, Solving the algebraic equation in the mapped space, Back transformation of the solution into the original space.

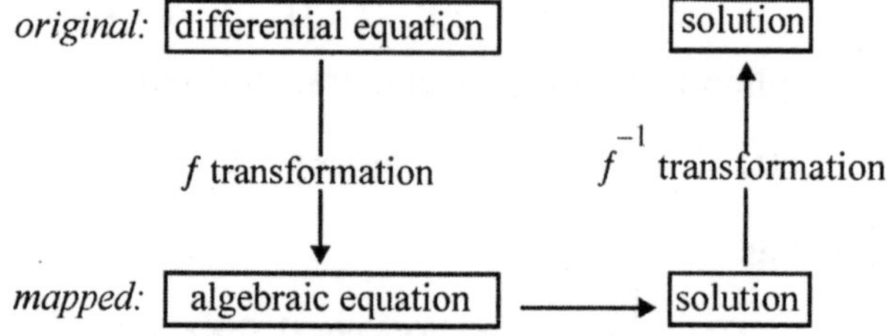

Fig. Schema for Solving Differential Equations using the Laplace Transformation

The procedure will be demonstrated by the following two examples.

Example: Consider the differential equation

$$\ddot{f}(t)+3\dot{f}(t)+2f(t)=e^{-t}$$

with the initial conditions $f(0+)=\dot{f}(0+)=0$

Solution: Proceeding using the steps given above one has

Step 1: $s^2F(s)+3sF(s)+2F(s)=\dfrac{1}{s+1}$

Step 2: $F(s)=\dfrac{1}{s+1}\,\dfrac{1}{s^2+3s+2}$

Step 3: The complex function $F(s)$ must be decomposed into partial fractions in order to use the tables of correspondences. This gives

$$F(s)=\frac{1}{s+1}-\frac{1}{s+1}+\frac{1}{(s+1)^2}.$$

By using the inverse Laplace transformation that the solution of the given differential equation is

$$f(t)=e^{-2t}-e^{-t}+te^{-t}.$$

Example: Given the differential equation

$$\ddot{x}+a_1\dot{x}+a_0x=0,$$

where a_0 and a_1 are constants and the initial conditions $\dot{x}(0+)$ and $x(0+)$ are known. Then

Solution:

Step 1: $s^2X(s)-s\;x(0+)-\dot{x}(0+)+a_2[s\cdot X(s)-x(0+)]+a_0\;X(s)=0$

Step 2: $X(s)=\dfrac{s+a_1}{s^2+a_1s+a_0}x(0+)+\dfrac{1}{s^2+a_1s+a_0}\dot{x}(0+),$

$$X(s)=L_0(s)\;x(0+)+L(s)\;\dot{x}(0+)$$

with the abbreviation

$$L_0(s)=\frac{N_0(s)}{D(s)}=\frac{s+a_1}{s^2+a_1s+a_0}\quad\text{and}\quad L(s)=\frac{N(s)}{D(s)}=\frac{1}{s^2+a_1s+a_0}.$$

Step 3: ***Case (a):*** two *single real zeros of the denominator:*

This means

$$D(s)=s^2+a_1s+a_0=(s-\alpha_1)(s-\alpha_2).$$

For both rational expressions $L_0(s)$ and $L(s)$ it follows by partial fraction decomposition that

$$L_0(s) = \frac{A_1}{s-\alpha_1} + \frac{A_2}{s-\alpha_2} \text{ and } L(s) = \frac{B_1}{s-\alpha_1} + \frac{B_2}{s-\alpha_2}.$$

The coefficients A_i and B_i can now be determined by comparing coefficients or by applying Equation.

$$A_i = \frac{N_0(\alpha_i)}{D'(\alpha_i)};\ B_i = \frac{N(\alpha_i)}{D'(\alpha_i)} \text{ for } i = 1, 2.$$

Thus for Equation follows

$$X(s) = \left[\frac{A_1}{s-\alpha_1} + \frac{A_2}{s-\alpha_2}\right] x(0+) + \left[\frac{B_1}{s-\alpha_1} + \frac{B_2}{s-\alpha_2}\right] \dot{x}(0+),$$

and by applying the formula the solution of the differential equation is

$$x(t) = [A_1 e^{\alpha_1 t} + A_2 e^{\alpha_2 t}]\, x(0+) + [B_1 e^{\alpha_1 t} + B_2 e^{\alpha_2 t}]\, \dot{x}(0+)$$

$$= [A_1 x(0+) + B_i \dot{x}(0+)] e^{\alpha_1 t} + [A_2\, x(0+) + B_2 \dot{x}(0+)] e^{\alpha_2 t}.$$

Case (b): One *double real zero of the denominator:* Here is $D(s) = (s - \alpha\,)^2$

For the two rational expressions $L0(s)$ and $L(s)$ of Equation the partial fraction decomposition is

$$L_0(s) = \frac{A_1}{s-\alpha} + \frac{A_2}{(s-\alpha)^2} \text{ and } L(s) = \frac{B_1}{s-\alpha} + \frac{B_2}{(s-\alpha)^2}.$$

The coefficients ^{Ai} and ^{Bi} are determined by comparing both sides or by evaluation of Equation:

$$A_1 = \left\{\frac{d}{ds}\left[\frac{N_0(s)}{D(s)}(s-\alpha)^2\right]\right\}_{s=\alpha} = 1,\quad A_2 = \left[\frac{N_0(s)}{D(s)}(s-\alpha)^2\right]_{s=\alpha} = \alpha + \alpha_1$$

and $$B_1 = \left\{\frac{d}{ds}\left[\frac{N(s)}{D(s)}(s-\alpha)^2\right]\right\}_{s=\alpha} = 0,\quad B_2 = \left[\frac{N(s)}{D(s)}(s-\alpha)^2\right]_{s=\alpha} = 1$$

From these results one obtains the solution

$$X(s) = \frac{x(0+)}{s-\alpha} + \frac{(\alpha + a_1)\, x(0+) + \dot{x}(0+)}{(s-\alpha)^2}$$

in the mapped space. By applying the inverse Laplace transformation the required solution of the differential equation is

$$x(t) = x(0+)e^{\alpha t} + [(\alpha + \alpha_1)\, x(0+) + \dot{x}(0+)]\, te^{\alpha t}.$$

Case (c): Two *conjugate complex zeroes of the denominator:* Here

$$D(s) = (s - \alpha_1)(s - \alpha_2) \text{ with } \alpha_{1,2} = \sigma_1 \pm jw_1 .$$

Introducing the values of α_1 and α_2 and after multiplication of this expression one obtains

$$D(s) = (s - \sigma_1)^2 + w_1^2 .$$

Comparison with the denominator of the original relation, Equation, gives according to Equation

$$a_0 = \sigma_1^2 + w_1^2 \text{ and } a_1 = -2\sigma_1 .$$

With these coefficients Equation is in the form

$$X(s) = \frac{s - 2\sigma_1}{(s - \sigma_1)^2 + w_1^2} x(0+) + \frac{1}{(s - \sigma_1)^2 + w_1^2} \dot{x}(0+)$$

$$= \left[\frac{s - \sigma_1}{(s - \sigma_1)^2 + w_1^2} - \frac{\sigma_1}{w_1} \frac{w_1}{(s - \sigma_1)^2 + w_1^2} \right] x(0+) + \frac{1}{w_1} \frac{w_1}{(s - \sigma_1)^2 + w_1^2} \dot{x}(0+) ,$$

and from this one gets $X(s)$ the corresponding time function

$$x(t) = e^{\sigma_1 t} \left[\cos w_1 t - \frac{\sigma_1}{w_1} \sin w_1 t \right] x(0+) + \frac{1}{w_1} e^{\sigma_1 t} \dot{x}(0+) \sin w_1 t \quad \text{or rearranged}$$

$$x(t) = e^{\sigma_1 t} \left\{ x(0+) \cos w_1 t + \left[\frac{1}{w_1} \dot{x}(0+) - \frac{\sigma_1}{w_1} x(0+) \right] \sin w_1 t \right\} .$$

Also from Equation of this example the importance of the position of the zeros of $D(s)$, the poles of $X(s)$, on the solution is clear. For all three cases the solution of the differential equation according to Equations are mainly influenced by the position of the poles of $X(s)$. These poles of $X(s)$ are - as one can see from the two examples - only depend on the left side of the corresponding differential equation, i.e. the homogeneous part of it. As is generally known the solution of the homogeneous differential equation describes the *modes* of the system, that is the behaviour, which depends only on the initial conditions. Therefore, consider for the general case only the homogeneous part of an nth-order ordinary homogeneous linear differential equation with constant coefficients that is

$$\sum_{i=0}^{n} a_i \frac{\mathrm{d}^i x_a(t)}{\mathrm{d}t^i} = 0$$

with all n initial conditions $(\mathrm{d}^i x_a(t)/\mathrm{d}t^i)|_{t=0+}$ for $i = 0, 1, ..., n - 1$.

One obtains by Laplace transformation

$$X_a(s)\sum_{i=0}^{n} a_i s^i - \left[\sum_{i=1}^{n} a_i \sum_{v=1}^{i} s^{i-v} \frac{d^{v-1}}{dt^{v-1}} x_a(t)\,|_{t=0+}\right] = 0$$

and a form according to Equation

$$X_a(s) = \frac{\sum_{i=1}^{n} a_i \sum_{v=1}^{i} s^{i-v} \frac{d^{v-1}}{dt^{v-1}} x_a(t)\,|_{t=0+}}{\sum_{i=0}^{n} a_i s^i} \mathrel{\overset{\triangle}{=}} \frac{N(s)}{D(s)},$$

where $N(s)$ and $D(s)$ are polynomials in s and the initial conditions are only in the numerator polynomial $N(s)$. The poles $s_k (k = 1, 2,..., n)$ of $X_a(s)$ can be determined directly from the solution of the equation

$$\sum_{i=1}^{n} a_i s^i = 0$$

After factorisation of this equation one obtains

$$a_n(s - s_1)\,(s - s_2)...\,(s - s_n) = 0.$$

The poles s_k of $X_a(s)$ make it possible to perform a partial fraction decomposition of $X_a(s)$, e.g. for the case of single poles according to Equation. For this case one obtains following Equation the solution of the homogeneous differential equation, in the form

$$x_a(t) = \sum_{k=1}^{n} c_k e^{s_k t} \text{ for } t > 0.$$

From this one can realise that the position of the poles s_k of $X_a(s)$ in the plane completely characterises the modes or inherent behaviour of the system described by Equation.

Thus one obtains for Re $s_k < 0$ (left-half s plane) a decreasing and for Re $s_k > 0$ (right-half s plane) an increasing behaviour of $x_a(t)$, while for pairs of poles with Re $s_k = 0$ permanent oscillations occur. Therefore, Equation is called the *characteristic equation* and the poles s_k of $X_a(s)$ are often called *eigenvalues* of the equation.

Therefore investigation of the characteristic equation provides the most important information about the oscillating behaviour of a system. Laplace transform of the impulse $\delta(t)$ and step $\sigma(t)$ The impulse function $\delta(t)$ is *not a function* in the sense of classical analysis, but a distribution (pseudo-function). Therefore without entering the theory of distributions the integral

$$\{\delta(t)\} = \int_0^{\infty} \delta(t) e^{-st} dt$$

is not defined. The singularity exactly matches with the lower integration limit. The impulse function can be approximately described by the limit

$$\delta(t) = \lim_{e \to 0} r_e(t)$$

with the rectangular impulse function

$$r_e = \begin{cases} 1/\varepsilon & \text{for } 0 \le \text{t} \le \varepsilon \\ 0 & \text{otherwise.} \end{cases}$$

Strictly speaking this representation of $\delta(t)$ is not a distribution, as $r_e(t)$ for $0 \le t \le \infty$ is not arbitrarily often differentiable. Because of the simple description compared with other functions (e.g. Gaussian functions) this approach is preferred here. From Equation it follows that

$$\{\delta(t)\} = \left[\int_0^\infty \lim_{e \to 0} r_e(t) \right] e^{-st} \, \mathrm{d}t.$$

As Equation can also be represented in the form

$$r_e(t) = \frac{1}{\varepsilon}[\sigma(t) - \sigma(t - \varepsilon)],$$

where $\sigma(t)$ is the unit step. Since the integration is independent of ε, the limit and integration can be permuted so that

$$\{\delta(t)\} = \lim_{e \to 0} \left\{ \frac{1}{\varepsilon} \int_0^\infty [\sigma(t) - \sigma(t - \varepsilon)] \, e^{-st} \mathrm{d}t \right\}$$

$$\{\delta(t)\} = \lim_{e \to 0} \left\{ \frac{1}{\varepsilon} \frac{1}{s} (1 - e^{-\varepsilon s}) \right\}.$$

By applying l' Hospital's rule one obtains

$$\{\delta(t)\} = \lim_{e \to 0} \frac{se^{-\varepsilon s}}{s} = 1 .$$

As the impulse $\delta(t)$ has an area of unity it is also called unit impulse.

The solution can be determined in the following three steps:

Step 1: The Laplace transform of the given differential equation is:

$$sY(s) - y(0-) = 1 \text{ with } y(0-) = 0.$$

Step 2: The solution of the algebraic equation is:

$$Y(s) = \frac{1}{s}$$

Step 3: From the back transformation the solution follows as

$$y(t) = \sigma(t)$$

where $\sigma(t)$ is the unit step function.

TRANSFER FUNCTIONS

Transfer functions simplify working with linear differential equations, which describe the transition behavior of linear dynamical systems. As transfer functions contain at least a rational function of a complex variable, the physical interpretation of poles and zeros is discussed. Finally, the most popular operations of combining systems described by transfer functions are presented. The properties of transfer functions are shown by examples and interactive questions allow the readers to test their knowledge of transfer functions.

Definition: Linear, continuous-time time-invariant systems with lumped parameters - as initially a dead time is not being taken into account - will be described by the ordinary differential equation

$$\sum_{i=0}^{n} a_i \frac{d^i x_{3.}(t)}{dt^i} = \sum_{j=0}^{m} bj \frac{djx_e(t)}{dt^j}.$$

If all *initial conditions* are set to *zero* and the Laplace transformation is applied to both sides of this equation, one obtains

$$X_{3.}(s)\sum_{i=0}^{n} a_i s^i = X_e(s)\sum_{j=0}^{m} b_j s^j,$$

or reordering

$$\frac{X_{3.}(s)}{X_e(s)} = \frac{b_0 + b_1 s + \ldots + b_m s^m}{a_0 + a_1 a + \ldots + a_n s^n} = G(s) = \frac{N(s)}{D(s)},$$

where $N(s)$ and $D(s)$ describe the numerator and denominator polynomials, respectively. The quotient of the Laplace-transformed output and input of such a type of system is a rational fraction. The coefficients of this fraction depend only on the structure and parameters of the system. Such a type of function $G(s)$, which describes completely the transfer behaviour of a system, is called the *transfer function* of the system. With such a transfer function the output

$$X_{3.}(s) = G(s)X_e(s)$$

INTERPRETATION OF THE TRANSFER FUNCTION

Comparing Equation with the convolution theorem, it then follows for the representation of Equation in the time domain that

$$x_{3.}(t) = \int_0^t g(t-r)x_e(\tau)\, d\tau,$$

where obviously the inverse Laplace transform of $G(s)$ is the function $g(t)$. This function is generally known as the *weighting function* of the system. In other words, the transfer function is the Laplace-transformed weighting function according to

$$G(s) = \{g(t)\}.$$

If the unit impulse $\delta(t)$ is taken as the input signal $x_{3.}(t)$ for a system described by the transfer function $G(s)$, one obtains according to Equation

$$X_{3.}(s) = G(s)\ \{\delta(t)\}$$

and after using the Laplace transform of the unit impulse $\delta(t)$ from Equation

$$X_{3.}(s) = G(s)$$

or from Equation

$$x_{3.}(t) = g(t).$$

This shows that the response to an unit impulse $\delta(t)$ is the weighting function. Therefore the weighting function is also called the *impulse response*.

Another interpretation is when the system is excited by the input signal

$$x_e(t) = \bar{x}_e e^{\sigma t} \sin(wt).$$

For the steady-state case one obtains for the output signal

$$x_{3S}(t) = |G(\sigma + jw)|\, \bar{x}_e e^{\sigma t} \sin(wt + \phi(\sigma + jw)).$$

This shows that the modulus $|G(\sigma + jw)|$ of the transfer function describes the gain, and that $\phi(\sigma + jw) = \arg|G(\sigma + jw)$ describes the phase shift of a sinusoidal function with the frequency w and with increasing or decreasing amplitude according to $e^{\sigma t}$ can be immediately calculated for a known input signal $x_e(t)$, and therefore $X_e(s)$

REALISABILITY AND PROPERNESS OF TRANSFER FUNCTIONS

It must be mentioned that a transfer function with $m > n$ is physically not realisable. The transfer function of an ideal differentiator is described by $G(s) = s$ according to Equation. Any transfer function with $m > n$ can be decomposed into

$$G(s) = \frac{N(s)}{D(s)} = \frac{N_1(s)}{D(s)} + k_0 + k_1 s + \ldots + k_{m-m} s^{n-n},$$

where degree $N_1(s) = n - 1$ and terms in s with positive powers also occur. Such derivative elements would deliver for input signals of arbitrary high frequency corresponding output signals of arbitrary high amplitude, which are physically not realisable.

The *condition of realisability* of the transfer function according to Equation is

degree $N(s) \leq$ degree $D(s)$ or $m \leq n$.

This condition is also called as the *condition of properness*. If a transfer function does not follow this condition, it is called an *improper* transfer function. It has the property that $G(s) \to \infty$ as $s \to \infty$. A realisable transfer function is called as *proper* and it always follows $G(s) \to k_0$ as $s \to \infty$. A transfer function with $k_0 = 0$ is called as *strictly proper*.

Transfer functions with dead time. If a time delay or *dead time* T_t is introduced in the input signal $x_e(t)$, one obtains instead of Equation the differential equation

$$\sum_{i=0}^{n} a_i \frac{d^i x_{3.}(t)}{dt^i} = \sum_{j=0}^{m} bj \frac{djx_e(t - T_t)}{dt^j}.$$

In this case taking the Laplace transformation gives the *transcendental* transfer function

$$G(s) = \frac{N(s)}{D(s)} e^{-sT_t}.$$

Poles and zeros of the transfer function

In some cases (e.g. stability analysis) it is expedient to represent the rational transfer function $G(s)$ according to Equation in the factorised form

$$G(s) = \frac{N(s)}{D(s)} = k_0 \frac{(s - s_{z_1})(s - s_{z_2})...(s - s_{z_m})}{(s - s_{p_1})(s - s_{p_2})...(s - s_{p_n})}.$$

For physical reasons only real coefficients a_i, b_j occur. Therefore the poles sp_i and the *zeros* s_{z_j} of $G(s)$, respectively, can be *real or complex conjugate* pairs. The terms zeros and poles are chosen, because the transfer function is zero at s_{z_j} and infinite at s_{p_i}. Zeros and poles can be graphically represented in the complex plane as shown in Figure. A linear time-invariant system *without* dead time is described completely by the distribution of its poles and zeros and the gain factor k_0.

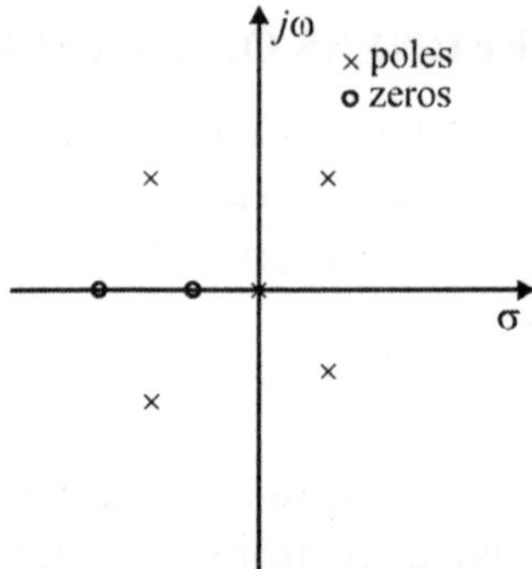

Fig. Example of the Pole and zero Distribution of a Rational Transfer Function in the Complex Plane

Moreover, the poles and zeros of a transfer function have a further significance. Observing a system without input ($x_e(t) \equiv 0$) according to Equation and determining the time response $x_{3.}(t)$ for the given *n*initial conditions, one has to solve the associated homogeneous differential equation

$$\sum_{i=0}^{n} a_i \frac{d^i x_{3.}(t)}{dt^i} = 0,$$

which corresponds exactly to Equation. For the approach $x_{3.}(t) = e^{st}$ of Equation one obtains for the solution in s the *characteristic equation*

$$P(s) = \sum_{i=0}^{n} a_i s^i = 0$$

which was already mentioned in Equation. This relation can be directly determined by setting the denominator of $G(s)$ to zero ($D(s) = 0$), as long as $D(s)$ and $N(s)$ have no common factor. The zeros s_k of the characteristic equation are the poles sp_i of the transfer function. As already shown in previous section the modes (i.e. $x_e(t) \equiv 0$) are described by the characteristic equation, so that the poles s_{p_i} of a transfer function contain all of this information.

The zeros of a transfer function are those values $s = s_{z_j}$ for which $|G(s_{z_j})| = 0$. This means that the output signal $X_{3.}(s)$ does not contain any components which depend on s_{z_j}. In order to explain this in more detail a stable system with a transfer function according to Equation is excited by the input signal

$$x_e(t) = e^{s_{z_j}t}$$

First for simplification the zero $s_{z_j} = \sigma_{z_j}$ is assumed to be real. For this case the input signal is $x_e(t) = e^{\sigma_{z_j}t}$. Because $|G(s_{z_j})| = 0$ one obtains from Equation the steady-state output signal as

$$I_{35}(t) = 0$$

In the case of complex conjugate pairs of zeros $s_{z_j}, s_{z_j+1} = s^*_{z_j}$ both zeros have to be taken into consideration in the input signal

$$\begin{aligned} x_e(t) &= 2e^{sz_it} + e^{s^*_{zj}t} \\ &= 2e^{\sigma z_it} \cos\omega t \\ &= 2e^{\sigma z_it} \sin\left(\omega t + \frac{\pi}{2}\right). \end{aligned}$$

Equation leads also to the result $x_{as}(t) = 0$. This shows that a zero sz_i of a system blocks the transmission of the input signal $e^{\sigma z_i t}$.

Example: The mass-spring-damper mechanical system in Figure with the mechanical constants $c_1 = 1$, $c_2 = 2$, $d = 1.5$, $m_1 = 1$ and $m_2 = 4$ is excited by the force x_e. The transfer function between the force and the position can be shown to be

$$G(s) = \frac{s^2 + 1}{s^1 + 0.5s^3 + 1.75s^2 + 0.5s + 0.5}$$

which has the zeros $sz_{1,2} = \pm j$. If this system is excited by the sinusoidal input signal

$$X_e(s) = \frac{1}{(s - sz_1)(s - sz_2)} = \frac{1}{s^2 + 1}$$

$$^{-1}\{X_e(s)\} = \sin t,$$

which is derived from this pair of zeros, the output signal $x_a(t)$ decays to zero as shown in Figure even though the input signal is a sinusoidal signal and the mass m_1 shows an undamped oscillation. The system does not pass this oscillation to the mass when the frequency matches the zeros.

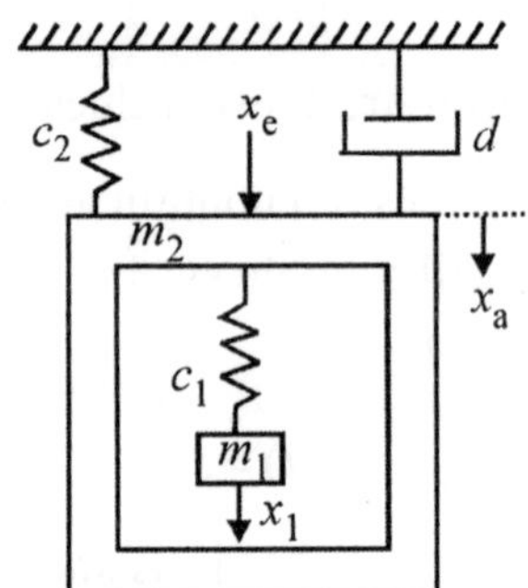

Fig.. Mass-Spring-Damper Mechanical System used for the Interpretation of zeros

Using transfer functions for calculations.

For combinations of transfer functions simple rules for determining the resulting transfer function can be derived. The combinations are of the type that transfer function blocks are connected. In making any transfer function block connection it is assumed that the connection does not load the block to which the connection is being made.

1. *Series connection:* From the diagram in Figure it follows that

$$Y(s) = G_2(s)X_{e2}(s)$$

$$X_{e2}(s) = X_{a1}(s) = G_1(s)\, U(s)$$

$$Y(s) = G_2(s)G_1(s)\, U(s).$$

The total transfer function of this series connection is

$$G(s) = \frac{Y(s)}{U(s)} = G_1(s)G_2(s)$$

$U = X_{e_1}$ → $G_1(s)$ → $X_{a_1} = X_{e_2}$ → $G_2(s)$ → $X_{a_2} = Y$

Fig. Series Connection of two Transfer Functions

2. *Parallel connection:* For the output of both functions it follows from Figure that

$$X_{a1}(s) = G_1(s)\ U(s)$$

$$X_{a2}(s) = G_2(s)\ U(s).$$

The output of the total system is

$$Y(s) = X_a(s) = X_{a_2}(s) + X_{a_2}(s) = [G_1(s) + G_2(s)]U(s),$$

and from this the transfer function of a parallel connection is

$$G(s) = \frac{Y(s)}{U(s)} = G_1(s) + G_2(s)$$

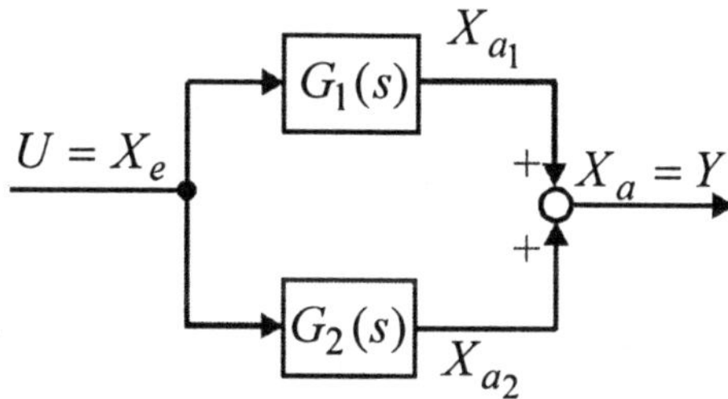

Fig. Parallel Connection of two Transfer Functions

3. *Feedback loop:* From Figure the output is

$$Y(s) = X_a(s) = [U(s) \underset{(+)}{\overset{-}{}} X_{a_2}(s) + G_1(s)$$

Since

$$X_{a_2} = G_2(s)Y(s)$$

one obtains

$$Y(s) = [U(s) \underset{(+)}{\overset{-}{}} G_2(s)Y(s)]G_1(s),$$

and from this

$$Y(s) = \frac{G_1(s)}{1 \underset{(-)}{+} G_1(s)G_2(s)} U(s)$$

The total transfer function of the feedback loop is

$$G(s) = \frac{Y(s)}{U(s)} = \frac{G_1(s)}{1 \underset{(-)}{+} G_1(s)G_2(s)}$$

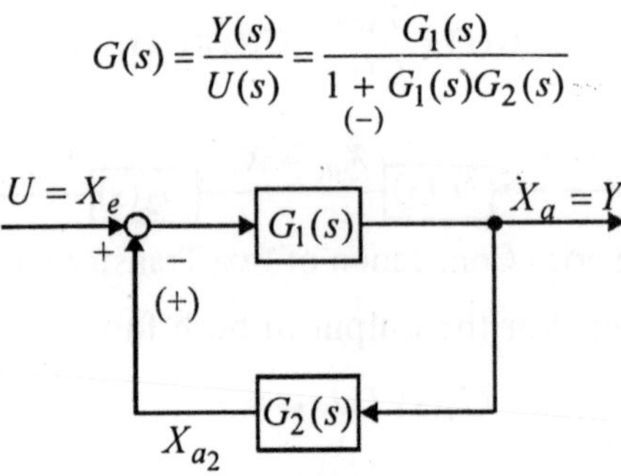

Fig. Feedback using two Elements

As the output of $G_1(s)$ is fed back via $G_2(s)$ to the input, this is called *feedback.* One has to distinguish between a positive feedback for positive adding of $X_{a_2}(s)$ and a negative feedback for negative adding of $X_{a_2}(s)$.

Example: For the special case of $G_1(s)$ being a pure amplifier with a high gain $K \to \infty$, one obtains for negative feedback

$$G(s) = \frac{K}{1 + KG_2(s)} = \frac{1}{\frac{1}{K} + G_2(s)} \approx \frac{1}{G_2(s)}.$$

The entire technique of operational amplifiers is based on this principle. In a feedback loop for which $G_1(s)$ is an amplifier with $K \to \infty$ then an element $G_2(s)$ can be used to realise any transfer function within certain limits.

Chapter 25

Fourier Series

As mentioned in the previous section, perhaps the most important set of orthonormal functions is the set of sines and cosines. These are what is known as a complete orthonormal set for the square-integrable functions on the interval [0,L]. They are also complete and orthonormal on any interval [a,b] with $b - a = L$.

Definition: Square-integrable functions on [a,b] are functions f(x) for which

$$\int_b^a |f(x)|^2 \, dx \text{ exists and is finite.}$$

The set of square-integrable functions is usually denoted L^2, and we shall see that it is an inner-product space. Later, we will also speak of square-integrable functions on regions in two or three dimensions, in which case we have multiple integrals over those regions.

Roughly speaking, a function on a finite interval is square integrable unless it is infinite somewhere. It can be very discontinuous, and in fact can even be slightly infinite - like the function ln(x), or even $|x|^{-1/3}$.

Most familiar functions are square-integrable.

Definition: An orthonormal set $\{e_n(x)\}$ is *complete* (on some fixed set of values of x) if for any square integrable function f(x) and any $\varepsilon > 0$, there is a finite linear combination

$$\sum_{n=1}^{N} \alpha_n e_n(x) \text{ such that } \left\| f(x) \sum_{n=1}^{N} \alpha_n e_n(x) \right\| < \varepsilon.$$

In other words, any reasonable function can be approximated as well as you wish (in the mean-square sense) by finite sums of the set. Indeed, we will say that it is the limit of an infinite series:

$$f(x) = \sum_{n=1}^{\infty} \alpha_n e_n(x)$$

The sense in which this infinite sum converges is, of course, in mean square. An equivalent way to describe completeness is: An orthonormal set $\{e_n(x)\}$ is complete if the statement that

$$\{f,e\} = 0 \text{ for all } e_n$$

Implies f(x) = 0 a.e. This is often a practical way to show that a set is incomplete, because you just have to exhibit a nonzero function which is orthogonal to the entire set.

Unfortunately, it is not as easy to prove completeness as it is to disprove it. It is a theorem that each of the sets of trigonometric functions is complete and orthonormal, but the techniques for proving such a theorem are beyond the scope of this course. To understand the issue better, consider a finite linear combination of an orthonormal basis,

$$f(x) = \sum_{n=1}^{N} \alpha_n e_n(x)$$

What is the norm of f? Because orthogonality makes the cross terms vanish, a little calculation shows us that

$$||f(x)||2 = \sum_{n=1}^{N} |\alpha_n|^2 .$$

This is an extension of the Pythagorean theorem, since it says that the square of the hypotenuse is the sum of the squares of the lengths of the sides, if the sides are at right angles; only in function space, lots of things can all be at right angles. Suppose we had left out some of the components. Then we would have

$$||f(x)||2 \geq \sum_{\text{some of the n}} |\alpha_n|^2 .$$

This inequality is still true if N = infinity, and is known as *Bessel's inequality*. The problem of completeness is that it is not easy to tell if we have included all the basis elements necessary to make both sides equal. We can leave many of them out and still have an infinite number left; in the set of Fourier functions, for example, we could leave out all the sine functions and still have all the cosine functions left.

The set is particularly useful for *periodic* functions, that is, functions such that f(x+L) = f(x) for some fixed length L, called the *period*, and all x. Since each of the functions in the set is periodic with period L, any linear combination of them is also periodic with the same period. The completeness just alluded to means that every periodic function can be resolved into the trigonometric functions of the same period. If the independent variable is time (which you might prefer to denote t rather than x), a periodic function of the form sin(2n t/L) or cos(2n t/L) may be detected by your ear and perceived as a pure musical tone with frequency n/L. *Any periodic sound wave can be resolved into pure musical tones.*

If you are given a sound wave f(t), which is periodic with period L you can extract its components of frequency k/L with formulae, setting m or n = k.

There are two such components, one with the sine function and the other with the cosine. This degree of freedom corresponds to the phase of the sound wave, because of the trig identity:

$$a \cos(\theta) + b \sin(\theta) = \sqrt{a^2 + b^2} \cos(\theta - \varphi),$$

where $$\tan(\varphi) = \frac{b}{a}$$

The intensity (power) carried by the component of a sound wave at pure frequency k/L is proportional to $|a_k|^2 + |b_k|^2$.

In the next chapter this resolution into pure frequencies is carried out for the square wave and the results are plotted, among other things. You may wish to glance at those plots now to get an intuitive feel for how a Fourier series can approximate a function.

THE POWERFUL THEOREM BEHIND THE FOURIER

I will now carefully formulate a theorem which justifies the use of Fourier series for square-integrable functions and tells us several useful facts.

Historically, parts of this theorem were contributed by Fourier, Parseval, Plancherel, Riesz, Fischer, and Carleson, and in most sources it is presented as several theorems. For proofs and further details, refer to Rudin's *Real Analysis* or Rogosinski's *Fourier Series.*

If f(x) is square-integrable on an interval [a,b], then

(a) All of the coefficients a_m and b_n are definite numbers uniquely.

(b) All of the coefficients a_m and b_n depend linearly on f(x).

(c) The series

$$f(x) \cong a_0 \sum_{m=1}^{M} a_m \cos\left(\frac{2\pi mx}{b-a}\right) + \sum_{n=1}^{N} b_n \sin\left(\frac{2\pi nx}{b-a}\right)$$

converges to f(x) in the mean-square sense. In other words,

$$\left\| a_0 \sum_{m=1}^{M} a_m \cos\left(\frac{2\pi mx}{b-a}\right) + \sum_{n=1}^{N} b_n \sin\left(\frac{2\pi nx}{b-a}\right) - f(x) \right\| \to 0.$$

We shall express this by writing

$$f(x) = a_0 + \sum_{m=1}^{\infty} a_m \cos\left(\frac{2\pi mx}{b-a}\right) + \sum_{n=1}^{\infty} b_n \sin\left(\frac{2\pi nx}{b-a}\right).$$

(Remember, however, that this series converges in a mean sense, and not necessarily at any given point.) It is also true that it converges a.e.

$$\|f(x)\|^2 = (b-a)\,|\,a_0\,|^2 + \frac{b-a}{2}\left(\sum_{m=1}^{\infty} |\,a_m\,|^2 + \sum_{n=1}^{\infty} |\,b_n\,|^2\right);$$

and the right side is guaranteed to converge.

If g is a second square-integrable function, with Fourier coefficients

$\tilde{a}_m$ and $\tilde{b}_n$ then

$$\{f, g\} = (b-a)\,\bar{a}_0\tilde{a}_0 + \frac{b-a}{2}\sum_{m=1}^{\infty} \bar{a}_m\tilde{a}_m + \frac{b-a}{2}\sum_{n=1}^{\infty} \bar{b}_n\tilde{b}_n$$

This is known as the *Parseval formula.*

Conversely, given two square-summable sequences a_m and a_n, i.e., real or complex numbers such that

$$\sum_{m=0}^{\infty} |\,a_m\,|^2 + \sum_{n=1}^{\infty} |\,b_n\,|^2$$

is finite, they determine a square integrable function f(x) uniquely a.e. such that and statements b)-d) hold.

There are very similar theorems for the Fourier sine series and the Fourier cosine series series, which are based, respectively, on the orthogonal sets. Let's stand back and think about what this big theorem tells us. Square-integrable functions are very general, so this is telling us that any reasonable function can be approximated arbitrarily well, in the r.m.s. sense, by a "trigonometric series."

We have a formula to generate the coefficients, and in fact a full correspondence between the square-integrable functions and the square-summable sequences. The square-integrable functions L^2 form an inner product vector space. The set of double sequences $\{a_m, b_n\}$ is also a vector space with an inner product. You can think of each object in this space as a vector with an infinite number of components, some of which are denoted a_m and others b_n.

Here are some examples showing why mean-square approximation is not always good enough. At a later stage we shall discuss when Fourier series converge at individual points. We shall see that if f(x) is continuous near x, the Fourier series converges at x.

Examples: If you look at the various Fourier series that are plotted in the next chapter, you will see that the crazy phenomenon of Example 2 doesn't happen. In fact, the convergence is very good except at the ends of the intervals or at places where the function is discontinuous. If you look at the periodic extension of a function, you see that the end of an interval is a place where there is likely to be a discontinuity, and it is when this happens that the series did not converge at the end of the interval. (A good example is f(x) = x on the basic interval [0,L].) What we observe is described by a general theorem, which we now formulate.

A function is said to be *piecewise continuous* (some say *sectionally continuous*) if it is continuous except at a discrete set of jump points, where it at least has an identifiable value on the left and a different one on the right. Here is a formal way to state this:

Definition: A function f(x) is *piecewise continuous* on a finite interval $a <= x <= b$ if it is continuous except at a finite number of points $a = x_0 < x_1 <... < x_n = b$, and all the one-sided limits

$$f\left(x_k^-\right) := \lim_{x \uparrow x_k} f\left(x_k\right),$$

and

$$f\left(x_k^+\right) := \lim_{x \downarrow x_k} f\left(x_k\right),$$

exist (except that we only assume the limit from above at a and the limit from below at b).

Here the up-arrow indicates that the limit is taken for values of x tending to x_k from below, and the down-arrow indicates that the limit is taken for values of x tending to x_k from above.

What we see from the examples is that where a function has a discontinuity, the Fourier series, when truncated to a large but finite number of terms, takes on a value between the right and left limits. The theorem says that the Fourier series finds the average of the two possibilities.

Theorem: Suppose that f(x) and f'(x) are piecewise continuous on a finite interval [a,b]. Then the Fourier series converges at every value of x between a and b as follows:

$$\lim_{M,N \to \infty}\left(a_0 + \sum_{m=1}^{M} a_m \cos\left(\frac{2\pi m x}{b-a}\right) + \sum_{n=1}^{N} b_n \cos\right) = \frac{1}{2}\left(f(x^+) + f(x^-)\right).$$

At the end points, we have:

$$\lim_{M,N\to\infty}\left(a_0+\sum_{m=1}^{M}a_m\cos\left(\frac{2\pi ma}{b-a}\right)+\sum_{n=1}^{N}b_n\cos\left(\frac{2\pi na}{b-a}\right)\right)$$

$$=\lim_{M,N\to\infty}\left(a_0+\sum_{m=1}^{M}a_m\cos\left(\frac{2\pi mb}{b-a}\right)+\sum_{n=1}^{N}b_n\cos\left(\frac{2\pi nb}{b-a}\right)\right)$$

$$=\frac{1}{2}\left(f(a^+)+f(b^-)\right)$$

The limit at the end points is reasonable, because when the function is extended periodically, they are effectively the same point. And a particular consequence is that: *At places where such a function is continuous, the Fourier series does indeed converge to the function.*

In addition, if the function is continuous on the interval [a,b], and f(a) = f(b), then we can state a bit more, namely that the Fourier series converges *uniformly* to the function. This means that the error can be estimated independently of x:

Definition: A sequence of functions $\{f_k(x)\}$ converges *uniformly* on the set to a function g provided that

$$|f_k(x) - g(x)| < c_k,$$

where c_k is a sequence of constants (independent of x in Ω) tending to 0.

The following theorem gives a general condition guaranteeing uniform convergence of Fourier series.

Theorem: Suppose that f'(x) is piecewise continuous, f(x) itself is continuous on a finite interval [a,b], and f(a) = f(b). Then

$$\left|a_0+\sum_{m=1}^{M}a_m\cos\left(\frac{2\pi mb}{b-a}\right)+\sum_{n=1}^{N}b_n\sin\left(\frac{2\pi nb}{b-a}\right)-f(x)\right|\to 0.$$

The condition that f(a) = f (b) is again reasonable if you think of f as a periodic function extending beyond the interval [a,b] - the extended function would be discontinuous at the end points if f(a) did not match f(b).

CALCULATING FOURIER SERIES

In this section we calculate several Fourier series. As we know, to find a Fourier series simply means calculating various integrals, which can often be done with software or with integral tables. Since performing integrals is not much more interesting in the modern age than long division, our goals in this section will be to get a visual and analytic impression of what to expect from Fourier series, and to understand the rôle of symmetry in the calculations.

Let's begin by evaluating the Fourier series for the functions:

$f(x) = 1$ for $0 \le x < L/2$, but 0 for $L/2 \le x \le L$

and

$$g(x) = \le x,\ 0\ x < L.$$

The functions have not been defined at the points of discontinuity, but as we know, the Fourier series will converge there to the average of the limit from the left and the limit from the right. The end point L is essentially a jump point, because the periodic extension of the functions make the values $x = L$ and $x = 0$ equivalent.

Here is a graph of the function f, called a "square pulse" or "square wave" (when extended periodically):

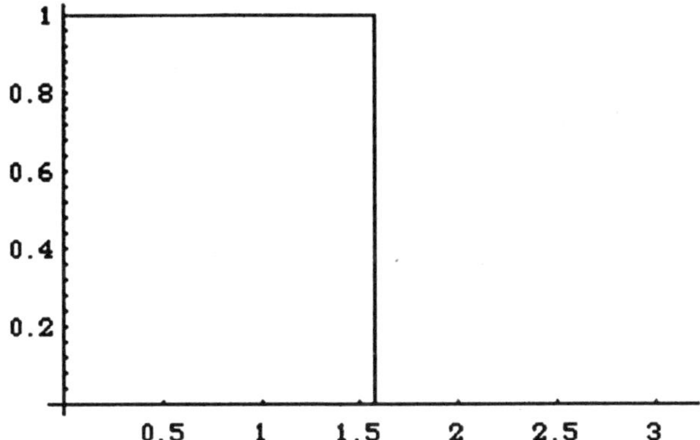

The length L has been chosen as .

We want to represent these functions in the form

$$a_0 + \sum_{m=1}^{\infty} a_m \cos\left(\frac{2\pi mb}{b-a}\right) + \sum_{n=1}^{\infty} b_n \sin\left(\frac{2\pi nb}{b-a}\right),$$

beginning with f(x).

Example: Use software to calculate the full Fourier series for the function f(x) as defined above, and investigate its convergence.

Solution:

$$a_0 = \frac{1}{L}\int_0^x f(x)\,dx,\ m = 1,2,\ldots$$

(i.e., the average of f), and for the other coefficients:

$$a_m = \frac{2}{L}\int_0^x \cos\left(\frac{2\pi mx}{L}\right) f(x)\, dx,\ m = 1,2,...$$

and

$$b_n = \frac{2}{L}\int_0^x \sin\left(\frac{2\pi nx}{L}\right) f(x)\, dx,\ n = 1,2,...$$

Here are the results of the calculation:

$$a_0 = 1/2\ ;$$
$$a_m = 0 \text{ for } m = 1, 2,...;$$
$$b_n = (1 - (-1)^n)/(n)$$

A point to ponder: why are all the $a_m = 0$? Surely this is no coincidence! The sort of expression we got for the b_n simplifies in a way which arises often in Fourier series for elementary functions: If $n = 2k$ is even, $b_{2k} = 0$. Thus only odd terms survive, in which case

$$b_{2k+1} = 2/((2k + 1)).$$

We can plot the Fourier series and watch it converge to the original function as more and more terms are included:

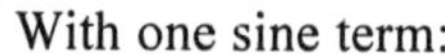
With one sine term:

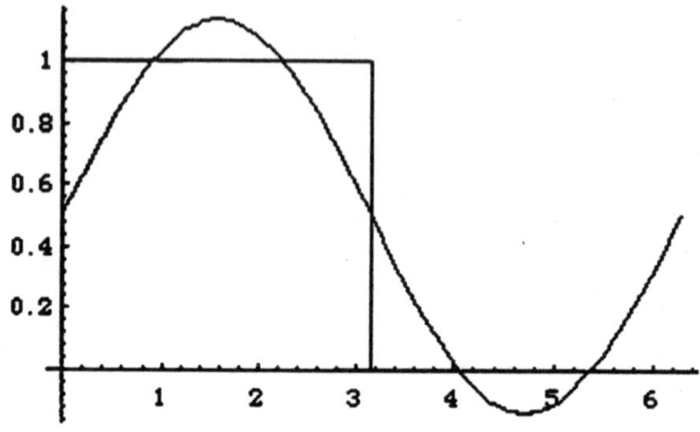

With three sine terms:

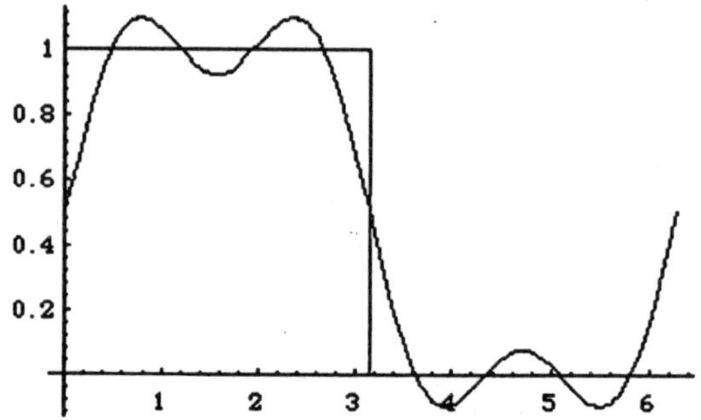

With seven sine terms:

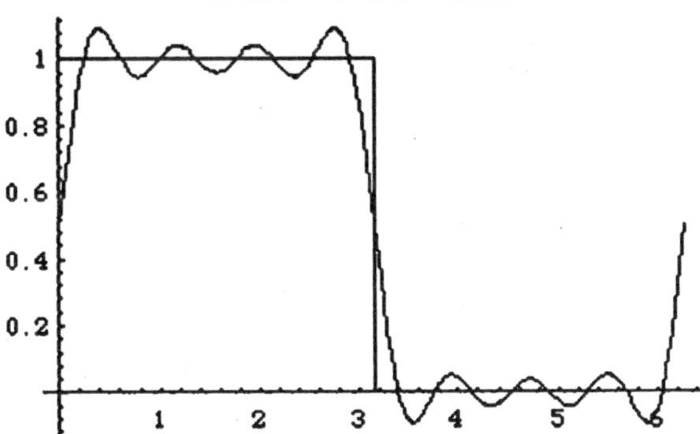

And so on. We call attention to the systematic overshoot that occurs at the edges of the jump. Curiously, the size of this last overshoot does not tend to 0 as we include more terms. The bumps next to the jump get thinner rather than shorter. This is known as the Gibbs phenomenon.

Example: We can now investigate some questions about the excitation of mechanical resonances. Suppose that experiments by K. Battle at the Wiener Staatsoper in Austria show that a crystal goblet will shatter if the intensity of the tone at frequency 1760 Hz (high A) exceeds .01. We use physical units in which the proportionality between the power and the square of the amplitude is 1, i.e., we define the intensity simply as

$$I = a_k^2 + b_k^2.$$

At Georgia Tech, not a particularly musical spot, we can generate a square pulse with amplitude A for 1/704 sec., then amplitude 0 for the same length of time, then A, etc., periodically with period 1/352.

Question: What amplitude A will cause the glass to shatter in our laboratory?

Solution. The issue is essentially to estimate the magnitude of the Fourier coefficient for our function corresponding to the frequency 1760 Hertz.

The mapping from functions to their Fourier coefficients is linear. Hence the coefficients can be obtained from the ones we just calculated.

Step 1. Scale the independent variable by replacing x with t and L with 1/352.

Step 2. Multiply all the coefficients by A.

The harmonic with frequency 1760 is the one with n=5, so the Fourier coefficients corresponding to this frequency are

$$a_5 = 0, \text{ and}$$

$$b_5 = 2A/(5\ )$$

The intensity is $4\ A^2/(25\ ^2)$. The amplitude A above which the glass will shatter is /4, numerically about 0.78539816.

Example: For comparison, let us find another Fourier series, namely the one for the periodic extension of g(x) = x, 0 x 1, sometimes designated x mod 1. Watch it converge.

Solution. For x between 1 and 2, the function is (x-r1L), for x between 2 and 3 it is (x-2), etc. Its graph has a sawtooth shape:

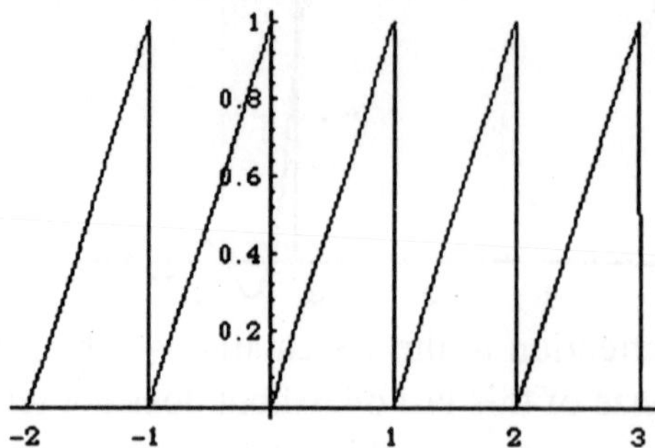

Because the interval has length 1, the Fourier basis functions are 1, cos(2n x), and sin(2n x).

The Fourier coefficients are calculated as:

$$a_0 = 1/2 \;;$$

$$a_m = 0 \text{ for } m = 1, 2,...;$$

$$b_n = (-1)/(n\ )$$

The coefficients a_m are all zero, again. Why?

Let us get an impression of how the series converges, by plotting the contributions of the first two sines, and then of six sines:

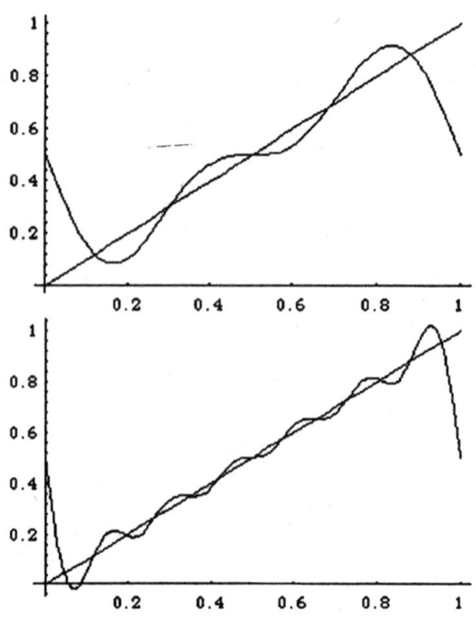

We also recall that on a longer interval, the Fourier series produces a periodic function:

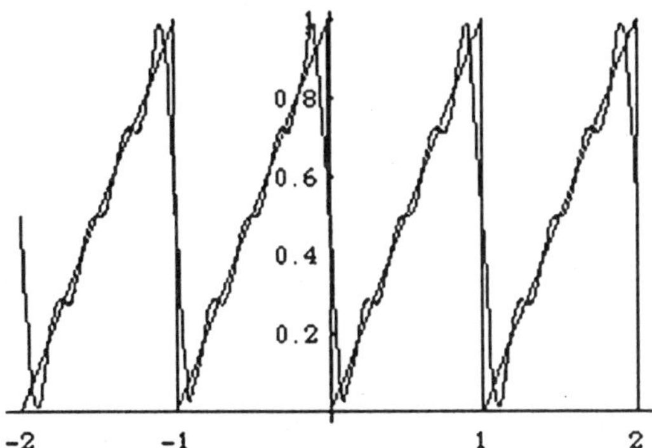

The Fourier series is converging nicely to the function except at the end-points of the interval, which are places where the full periodic saw-tooth function has jumps, and we saw something similar with the square pulse.

In both cases we see numerical evidence for the theorem that the Fourier series converges to f(x) where f(x) is continuous, and where it has a jump, the Fourier series converges to the average of the upper and the lower value at the jump.

Let's try out a different sort of example, where we need to integrate numerically.

Example: Find the Fourier series for the function sin(/x).

Solution. The integrals called for are not in the integral tables. Indeed, *most* integrals are not in the integral tables, even when they are nowhere near as wild as this function. In the information age this is no barrier.

We simply call on the software to do the integrals numerically. Since the function sin(/x) is odd in x, the coefficients associated with even functions, a_m, will all be zero. With some complaints about the convergence, the software calculates the first six sine coefficients as:

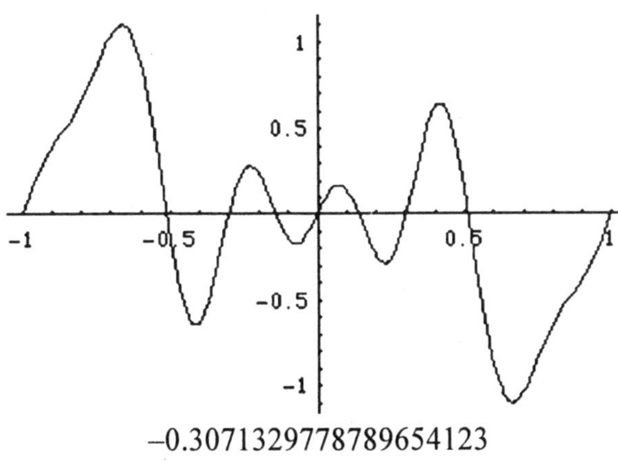

−0.307132977878965412З

0.506250967969952191 2

– 0.271449220623610834

– 0.2561291634161970945

0.1947511859947028479

0.2192911372159977088

Let us plot the Fourier series with these six terms and compare with the function itself:

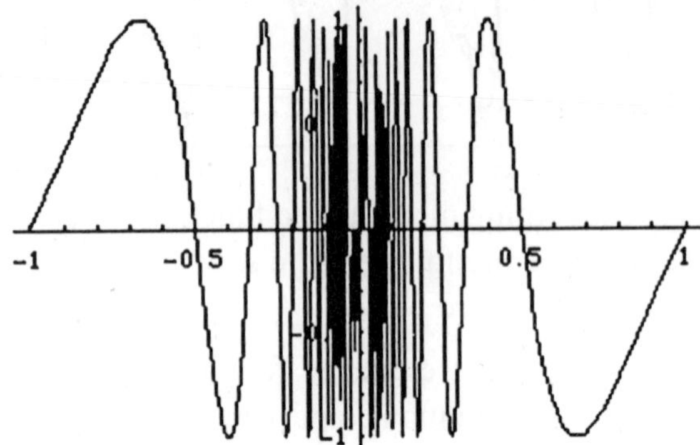

Notice that the convergence is pretty good away from the nasty singularity at x=0. In effect, truncating the Fourier series has filtered out the high-frequency oscillations.

Example: Find the Fourier series for a function which is square-integrable but has an infinite jump.

Solution. The example we look at is $x^{-1/3}$. Like the previous example, the function is odd, and there will only be sine contributions. The first several coefficients are numerically:

1.710689953277251171

0.3736036030974259086

0.7368949814369842807

0.2752161613201822172

0.5010701734296802681

0.2252029789384463762

Here is a comparison of the function and its Fourier approximation:

Think about the series calculation at x = 0 and at x = +/ –1. Did what you expected happen?

As we have seen in computations, symmetries can be used to simplify the computation of Fourier series. Recall that a function is even if f(– x) = f(x) for all x, and odd if f(– x) = – f(x) for all x. If f is a periodic even function, then all of the sine coefficients $b_n = 0$: b_n is the inner product of f with a sine function, but any even function and any odd function are orthogonal on the interval [– A, A].

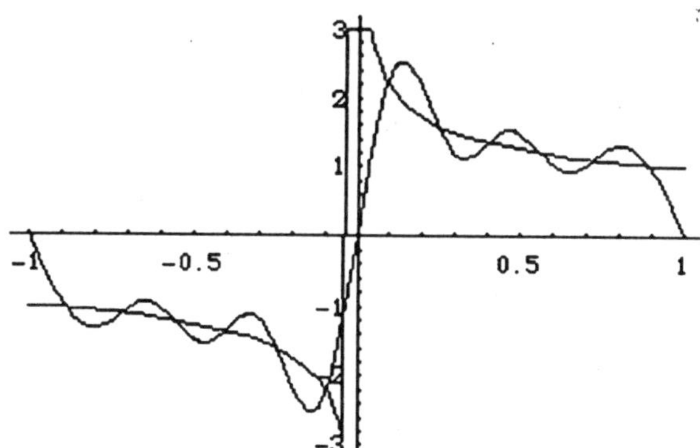

(If f(x) is periodic with period L, then we can analyze it on any interval of length L, so even if it is originally defined on [0,L], we can just as well look at it on [– L/2, L/2]. Look at a graph of a periodic function to understand why.) For similar reasons, if f is periodic and odd, all the coefficients $a_m = 0$.

Any function can be written as the sum of an even function and an odd function, and the Fourier series picks out the two parts. If, for instance, we want to find the Fourier series of a function such as $\pi x - x^2, -\pi \le x \le \pi$, we can save some work by thinking about the symmetries.

Example: Use symmetries to efficiently find the Fourier series for the function $\pi x - x^2$ on the interval $-\pi < x < \pi$.

Solution. This function is neither even nor odd when we change x into – x, but its "odd part" is πx, and its "even part" is $-x^2$. The sine coefficients in the Fourier series will come entirely from the odd part and the even coefficients entirely from the even part; $-x^2$ is orthogonal to all the Fourier cosine functions, and πx is orthogonal to all the Fourier sine functions. Moreover, the symmetry can be used to change the integration range so that it begins at 0: The coefficients b_n are calculated as follows:

$$b_n = \frac{2}{2\pi}\int_{-1}^{1} \pi\, x \sin\left(\frac{2\,\pi n x}{2n}\right) dx = \int_{-1}^{1} x \sin(nx)\, dx.$$

The product of two odd functions is even, so this integral is:

$$b_n = 2\int_0^x x \sin(nx)\, dx. = \frac{\pi(-1)^{n+1}}{n}.$$

The final result was calculated by software. In a calculus class you would have probably calculated it by integration by parts. The calculation of the coefficients an can also be simplified. First, notice that only $-x^2$ contributes:

$$a_n = \frac{2}{2\pi}\int_{-1}^{1}(-x^2)\sin\left(\frac{2\pi n x}{2\pi}\right) dx = -\frac{1}{\pi}\int_{-1}^{1} x^2 \sin(nx)\; dx.$$

The value of this integral is

$$\frac{(-1)^{m+1}4}{m^2}$$

When you combine these, you have the Fourier coefficients for $\pi x - x^2$ on this interval.

Whoever was the marketing expert consulted when they invented "complex numbers" should be fired! Complex numbers almost always make things much simpler, and this is true in particular for Fourier series. The key to the simplification is Euler's formula:

$$e^{ix} := \exp(i x) := \cos(x) + i \sin(x),$$

It is useful to know several special cases, especially,

$$\exp(i n \pi) = (-1)^n.$$

and

$$\exp(i (2k+1) \pi/2) = (-1)^k i,$$

Euler's formula allows us to discuss complex numbers not only in the Cartesian system, $z = x + iy$, but also in the polar system: $z = r \exp(i \theta)$, where θ, the argument of z, is the polar angle in the complex plans, and $r = |z|$ (the modulus of z) is the distance from the origin. One of the truly great things about this formula is that is allows us to replace calculations with trigonometric functions with much easier calculations with exponential functions.

As an example, let us think of how to calculate the Fourier series for the function $f(x) = x$, $0 \le x < L$ (repeated periodically, if you wish, outside this interval - notice that this is not a centered interval, so this calculation is different from the one done above.). a_0 is the average of the function, so, obviously, $a_0 = L/2$. For the other a_m, we have to calculate integrals of the form

$$\frac{2}{L}\int_0^x x \cos\left(\frac{2\pi x}{L}\right) dx \text{ and } \frac{2}{L}\int_0^x x \sin\left(\frac{2\pi x}{L}\right) dx.$$

If you don't have software or integral tables handy, you can do these integrals with integration by parts. Or you can reason as follows:

$$a_m \frac{2}{L}\int_0^x x \cos\left(\frac{2\pi m x}{L}\right) dx = \frac{2}{L}\mathrm{Re}\left(\int_0^x x \exp\left(\frac{2\pi i m x}{L}\right) dx\right)$$

Example: Evaluate this integral by a method different from integration by parts.

Solution. We can use the following useful trick:

$$\int_0^x x \exp(Ax)\, dx = \frac{d}{dA}\int_0^x \exp(Ax) dx = \frac{d}{dA}\left(\frac{1}{A}(\exp(AL) - 1)\right).$$

After the calculation, we set A = 2 π i m/L. Calling the integral in INT, we find

$$\text{INT} = \frac{-1}{A^2}(\exp(AL)-1)+\frac{L}{A}(\exp(AL).$$

which starts looking nice after we substitute A = 2 im/L, which reveals that these exponential terms are both equal to 1:

$$\text{INT} = \frac{L^2}{2\pi m^2 i} = -\frac{L^2}{2\pi m^2} i \text{ in standard form.}$$

This is purely imaginary, so all $a_m = 0$.

Example: Evaluate the coefficients b_n with a similar method.

Solution. STOP! Don't go back to the beginning of the calculation. We don't have to start all over, since

$$\frac{2}{L}\int_0^x x\sin\left(\frac{2\pi x}{L}\right)\,dx = \frac{2}{L}\,\text{Im}\left(\int_0^x \exp\left(\frac{2\pi ix}{L}\right)\,dx\right),$$

which is the integral we already did. From equation we see that this imaginary part is just

$$b_n = -\frac{L^2}{2\pi m^2}.$$

Actually, we can be much more systematic if we simply replace the complete set of functions:

$$\left\{\frac{1}{\sqrt{L}}\exp\left(\frac{2\pi inx}{L}\right)\right\}_{n>-\infty}^{\infty}$$

These are simply related to the sines and cosines by Euler's formula and its easy consequence:

$$\cos(\alpha) = \frac{\exp(i\alpha)+\exp(-i\alpha)}{2} \text{ and } \sin(\alpha) = \frac{\text{esp}(i\alpha)-\text{esp}(-i\alpha)}{2i}$$

Definition: The Fourier exponential series is an expansion (for an arbitrary square-integrable function):

$$f(x) = \sum_{k>-\infty}^{\infty} c_k \exp\left(\frac{2\pi ikx}{L}\right)$$

Since the exponential functions are an orthonormal set, a familiar kind of calculation shows us that the formula for c_k is:

$$c_k \frac{1}{L}\left\langle \exp\left(\frac{2\pi ikx}{L}\right), f\right\rangle = \frac{1}{L}\int_k^b \exp\left(\frac{2\pi ikx}{L}\right) f(x)\, dx.$$

Notice the tricky minus sign - this is a place where the complex conjugate in the inner product is important.

Let's observe how much more convenient this formula is than the one without complex numbers. There is only one sum, not two sums and a constant term. There is only one formula for c_k, not separate ones for a_0, a_m, and b_n. The constant term c_0 fits in with all the others, for instance, and is not put aside as a special case. The Parseval formula is now much simpler:

$$\{f, g\} = L \sum_{k>-\infty}^{\infty} c_k \tilde{c}_k.$$

Here, the coefficients for g have the tildes. In particular,

$$\|f\|^2 = L \sum_{k>-\infty}^{\infty} |c_k|^2 .$$

Notice that the Parseval formula is similar to the Pythagorean theorem, since, other than the normalization factor L, it states that a certain length squared is equal to the sum of the squares of its components in an orthogonal basis. We see here, however, that the space in which f lives is infinite-dimensional.

The important thing to know is that the Fourier exponential series is completely equivalent to the usual "full" Fourier series. We will later look at the Fourier sine and Fourier cosine series, which are truly different series, but the exponential series is not. If we substitute with Euler's formula, the full series becomes the exponential series or *vice versa*. We can recombine c_k with c_{-k} as follows:

$$c_k \exp\left(\frac{2\pi ikx}{L}\right) + c_{-k} \exp\left(\frac{-2\pi ikx}{L}\right) = (c_k + c_{-k})\cos\left(\frac{2\pi kx}{L}\right) + (c_k c_{-k}) i \sin\left(\frac{2\pi kx}{L}\right)$$

From this we get the connecting formulae that:

$$a_0 = c_0,\ a_m = c_m + c_{-m},\ \text{and}\ b_n = i\ (c_n - c_{-n}\}).$$

Another useful fact to know about Fourier series is that you can normally safely differentiate them and integrate them, as long as the functions you are representing are periodic and differentiable. Actually, even if the function is not smooth, these manipulations are still often possible. This topic is explored in the next chapter.

DIFFERENTIATING FOURIER SERIES

One of the most important uses of Fourier series is as a tool in solving differential equations. Because of that we will frequently want to differentiate a Fourier series. Since the Fourier series, like any infinite series, is a limit, questions can arise about whether it is permissible to differentiate in x before summing in n. In your advanced calculus class you should have seen examples where interchanging the order of two limits leads to different answers. This can happen quite easily.

Example: Let f[n,x]:= 0 if –1/n < = x < = 1/n, and 1 otherwise – i.e. the indicator function of the complement of the interval [–1/n, 1/n]. Then f[n,0] = 0 for all n, so if we let x tend to zero before n tends to infinity,

$$\lim_{n\to\infty}\lim_{x\to 0} f[n,x] = \lim_{n\to 0} 0 = 0$$

whereas if n tends to infinity first,

$$\lim_{x\to 0}\lim_{n\to\infty} f[n,x] = \lim_{x\to 0} 1 = 1,$$

One of the great things about Fourier series is that, despite a very reasonable worry, it is usually completely reliable to interchange limits. Indeed, one nice way to calculate Fourier series is to differentiate or integrate other Fourier series. With the formulae for the Fourier coefficients is a routine matter to calculate the Fourier series for the function $x - x^3$ defined for $-1 < x < 1$. We find that there are no cosine contributions, and

$$x - x^3 = \sum_{n=1}^{\infty} \frac{12(-1)^{k+1}}{(k\pi)^3} \sin(k\pi x).$$

To get a feel for this series, let us plot the sum of the first three terms and compare with $x-x^3$:

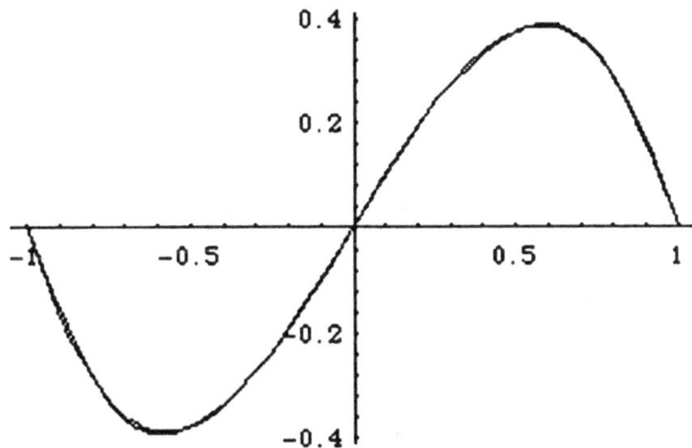

Wonderful! The function and the truncated series match rather closely.

What happens if we differentiate?

Example: Differentiate the terms in the Fourier series for this function, and compare with $1 - 3x^2$.

Solution: If we differentiate the series for $x - x^3$, we get the series

$$\sum_{n=1}^{\infty} \frac{12(-1)^{k+1}}{(k\pi)^2} \cos(k\pi x)$$

Supposing that it is legitimate to differentiate the infinite series term by term. First note that the function $1-3x^2$ is even, so there will only be cosine contributions, as we have found. Next, let us plot the sum of the first four terms in the differentiated series and compare with the exact function:

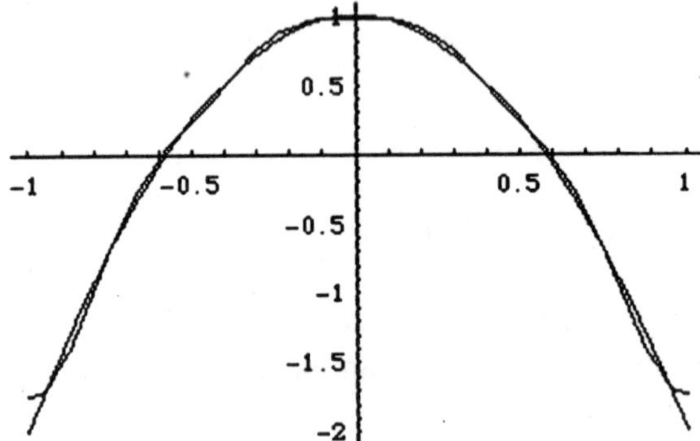

A superb match! Just for fun, let's see the comparison outside the interval where we cut off the polynomials:

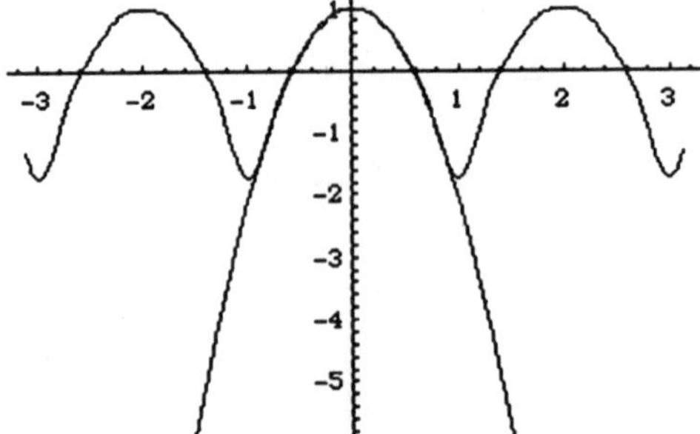

Remember - the Fourier series always corresponds to the periodic extension of the function on the basic interval. The formula $1 - 3x^2$ is not valid for the series outside $[-1,1]$.

Finally, let us calculate the Fourier series using the integral formulae. The result,

$$a_0 = 0,$$

$$a_k = 12\,(-1)^{k+1}/(k\,\pi)^2,\ k = 1, 2, \ldots,$$

As expected. Suppose now that instead of differentiating, we integrate a Fourier series term by term. If a_0 (which = c)$_0$ is different from 0, we get another Fourier series! If a_0 is not 0, then we would only get another Fourier series after replacing the function x with a Fourier series, but we won't consider that case now.

Does the integrated series converge to the integral of the original function? According to our experiment with differentiation, it seems so. Let us try again, by integrating the series for $x - x^3$term by term. We find:

$$\sum_{n=1}^{\infty} \frac{12(-1)^k}{(k\pi)^4} \cos(k\pi x)$$

What does this show about the constant of integration? When we integrated the series we got the integral with average average a_0, chosen here as 0. Remember that a_0 is always the average of the function. The difference in height between the two graphs is just the average of $x^2/2 - x^4/4$, which doesn't happen to be 0.

Chapter 26

Vibrating String

In this chapter we separate variables in order to solve the wave equation. There are three basic models in the subject of partial differential equations, which describe very different physics and have very different mathematical properties. They are simple in form, yet realistic for many problems in scince and engineering. Moreover, they teach us what to expect when we encounter more complicated equations used to model complicated laboratory situations. They are:

The wave equation,

$$u_{tt} = c^2 u_{xx}; \qquad \text{(WE)}$$

the heat equation, also known as the diffusion equation,

$$u_t = k\, u_{xx}; \qquad \text{(HE)}$$

and the potential equation, also known as Poisson's equation,

$$u_{xx} + u_{yy} = \alpha\, \rho\, (x,y), \quad \text{(PE)}$$

where ρ is a given function corresponding in electrostatics to a charge distribution, and the other quantities c, k , and are physical constants. Often there is no charge in the region where you want to measure a potential , and (PE) reduces to Laplace's equation

$$u_{xx} + u_{yy} = 0. \qquad \text{(LE)}$$

One thing you are likely to notice if you take several advanced courses in science or engineering is that these same equations arise in many different contexts. The function u may have a different interpretation in each incarnation, but the mathematics remains the same. For example, the function u in (PE) or (LE) might be an electric potential, a magnetic potential, the height of a taut membrane, or the potential of the velocity field in a fluid.

The reason these equations arise again and again is that they are the most symmetric second order partial differential equations that can be written down. Fundamental physical principles usually do not involve derivatives of higher than second order. If a physical model is isotropic, has a superposition principle, and is based on fundamental physical principles, it is very likely to lead to an equation of one of these forms.

The mathematics and physics of the three types of equations differ in important ways, and lead to a classification of the equations as *hyperbolic* (WE), *parabolic* (HE), and *elliptic* (PE/LE). Other second-order equations of a given type will share the qualitative features of our three models.

We shall begin our study of the method of separation of variables with the one-dimensional wave equation, which controls the vibrations of a taut string and some other phenomena. The wave equation (WE) written can be written with the aid of a wave operator

$$f(t, x)): = \left(\frac{1}{c^2}\frac{\partial^2}{\partial t^2} - \frac{\partial^2}{\partial x^2}\right) f(t,x)),$$

So that (WE) is the equation for the kernel of this operator. This operator is also known as the d'Alembertian. The wave equation is a linear, homogeneous equation, so the superposition principle holds for its solutions. If we can find some solutions $u_1, \ldots, u_k$, then any linear combination of these solutions is again a ***solution.*** We also classify it as "second-order," since derivatives up to second order occur, but no higher-order derivatives. This will be the case for the great majority of PDE's in this class.

We shall solve the wave equation in two entirely different ways. The first method is known as *separation of variables*. Let's set c = 1 for convenience and test out the wave operator by calculating its action on some representative functions:

$$t \sin^2 (x) = 2t \cos^2 (x) + 2t \sin^2 (x);$$

$$\sin (x - t) = 0$$

As we shall see, there are lots of solutions, and several strategies for finding them. With the method of separation of variables, we try to find solutions of a special form, u = T(t) X(x), called product solutions . This is a special assumption of form - what is referred to as an *ansatz*, to give some definiteness to our search for solutions. Some other physical circumstances, called the boundary and initial conditions, will be quite important, and will constrain us in choosing T and X; a partial differential equation coupled with some boundary conditions will be called a *boundary-value problem*. For the vibrating string, the most common boundary conditions are the zero Dirichlet conditions:

$$u(t,0) = 0, \; u(t,L) = 0 \qquad \text{(DBC)}$$

This means that the string is fastened at two ends, x = 0 and x = L. Actually, if we assigned some other fixed values to u besides 0 at the end points, we could just redefine the dependent variable:

If $\qquad u(t,0) = a, \; u(t,L) = b,$

then define $\qquad v: = u(t,x) - a - (b - a)\, x/L.$

The function v will still solve the wave equation, $v_{tt} - v_{xx}$, but it will have the simpler boundary conditions (DBC). When we act on a product T(t)X(x) with the wave operator and divide the result by T(t)X(x), we get

$$\frac{T''(t)}{T(t)} - \frac{X''(x)}{X(x)}.$$

If u solves the wave equation, this expression must be zero. (We could quibble about what happens when T or X = 0, but that would either mean that u is the uninteresting "trivial" solution or else the problem occurs only at isolated values of x and t.) Thus

$$T''/T = X''/X.$$

Since the left side of this is independent of x and the right side is independent of t, both sides must in fact be a constant. We don't know its value yet, so let us call it – μ. It will turn out to be negative.

Hence

$$-X''(x) = X(x)$$

This sort of equation is similar to a matrix-vector equation in linear algebra and is called an *eigenvalue equation.* The function X and the number μ are both unknown, and are called and eigenfunction and eigenvalue. We do not allow the possibility X(x) = 0 for all x. While it is always a solution for all u, it is not useful. By definition, an eigenfunction must be a nonzero solution of an equation. Along with equation we have the boundary conditions

$$X(0) = \mu X(L) = 0$$

and this will restrict the μ's for which has nonzero solutions. Equation is an elementary ordinary differential equation with the familiar solution (also obtainable with software),

$$X(x) := D_1 \cos(\mu^{1/2} x) + D_2 \sin(\mu^{1/2} x).$$

Because of the boundary condition at x = 0, we see that $D_1 = 0$. We are left with only one constant, which we can scale to 1, remembering that we may want to multiply our solution by an arbitrary constant later. Because of the BC at x = 0, we now know that up to a constant multiple, $X(x) = \sin(\mu^{1/2} x)$. The other boundary condition in tells us what the allowable choices for μ are, namely $\mu^{1/2} = n\,\pi/L$, n = 1,2,.... Taking L= 1 for convenience, the eigenfunctions and eigenvalues are:

$$X_n(x) := \sin(n\,\pi\,x)$$

$$\mu_n := (n\,\pi)^2$$

Now let's look at the other part of the PDE which we "separated off":

$$T''(t) = -\,\mu\,T(t)$$

As we saw, there are many possible values of μ which are consistent with the boundary condition for the function X, and the same possibilities occur in equation (multiply μ by c^2 here if c is not set to 1). Therefore

$$T_n(t) = A_n \cos(n \pi t) + B_n \sin(n \pi t).$$

We cannot go further with the T function without being given the *initial condition*, that is, information about what happens at $t = 0$.

We already have something of importance in science and engineering, however: the *normal modes* . Many physical systems, like the vibrating string, preferentially undergo simple motions with pure frequencies. Such a possible motion is exactly what we find by positing a product solution, imposing any boundary conditions, and solving the resulting eigenvalue problem.

Definition: A normal mode of a boundary value problem is a solution of the form

$$T(t)\, X(x),$$

where X(x) incorporates all boundary conditions.

Here the spatial variable x might be more than one-dimensional, as for the problem of the vibrating drum, which will be encountered later. I have generalized the notion of a normal mode so that any time dependence is allowed, although most commonly it will be sinusoidal, $T(t) = \sin(\omega t + \phi)$. When it is sinusoidal, the angular frequency omega is usually a simple function of the eigenvalue. The general solution we come up with is a linear combination of the particular solutions we get by separating the equation. It is not yet obvious that this is a completely general solution, but it is.

$$u(t,x) = \sum_{n=1}^{\infty} (A_n \cos(n\pi t) + B_n \sin(n \pi t)) \sin(n \pi x),$$

The coefficients A and B have to be determined from the initial conditions, that is, by the condition of the string at $t = 0$.

Suppose that

$$u(0,x) = f(x)$$

and

$$u_t(0,x) = g(x)$$

are given, and that $0 < x < 1$. Plugging in to formula shows that:

$$f(x) = \sum_{n=1}^{\infty} A_n \sin(n\pi t),$$

$$g(x) = \sum_{n=1}^{\infty} n \pi B_n \sin(n\pi t),$$

In other words, A[n] are the Fourier sine coefficients for f(x), and the more complicated expression

$$(n\ \pi)\ B_n$$

Gives the Fourier sine coefficients of g(x). Notice that the stuff in the sine is n π x/L and not

2 n π x/L (with L=1). In other words, there are more sines in this series than in the usual Fourier series, but there are no cosines. It is precisely as if we imagine the string to be defined on the interval [--1,1], twice as long as the actual interval, but on the nonphysical part from x = –1 to x = 0 we pretend that the shape of the string is defined to give us an odd function: f(– x) = – f(x). This would give us only sine functions, as we have seen before. Earlier we called those coefficients B_n, but here they happen to be called A_n. The thing to remember is that since they go along with sine functions (of x) the formula for them contains sines. We would thus write

$$An\ [f]: = \int_{-1}^{1} \sin(n\pi x) f(x)\, dx,$$

but we can simplify this a bit by using the symmetry:

$$An: = 2 \int_{0}^{1} \sin(n\pi x) f(x)\, dx,$$

What about the other coefficients? Well, this time it is the whole expression (n π) B_n which is a Fourier sine coefficient for g(x), so we need to divide through by the extra factor:

$$B_n\ [g]: = \frac{2}{n\pi} \int_{0}^{1} \sin(n\pi x) g(x)\, dx.$$

Hermple is an ere is an example to illustrate the situation.

Example: Solve the problem of the vibrating string using Fourier sine series, and initial conditions

$$u(x,0) = 0,\ u_t(x,0) = 1.$$

Solution. We can easily calculate the formulae for the coefficients A and B, either by hand or with software:

$$A_m = 0$$

and

$$Bn = 2\frac{1-(-1)^n}{(n\pi)^n}.$$

If n is even, $B_n = 0$, else $B_n = 4/(n\ \pi)^2$

The general solution is

$$u\ (t,x) = \frac{1}{4\pi^2} \sum_{\substack{n=1 \\ n\,odd}}^{\infty} \frac{\sin(n\pi t)\sin(n\pi x)}{n^2}$$

Next we show some plots of the solution at various times. For computational purposes we keep terms only up to n = 7.

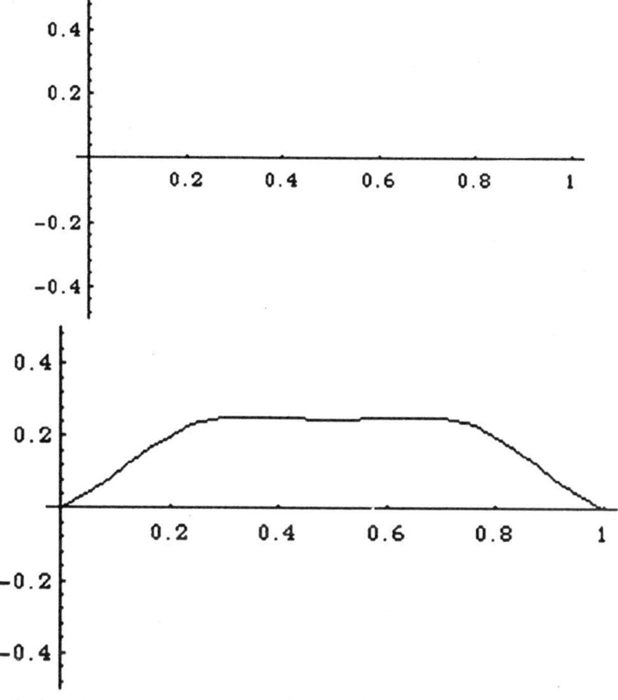

By the way, infinity is larger than 7, but 7 is close enough to give us a reasonable picture of the solution:

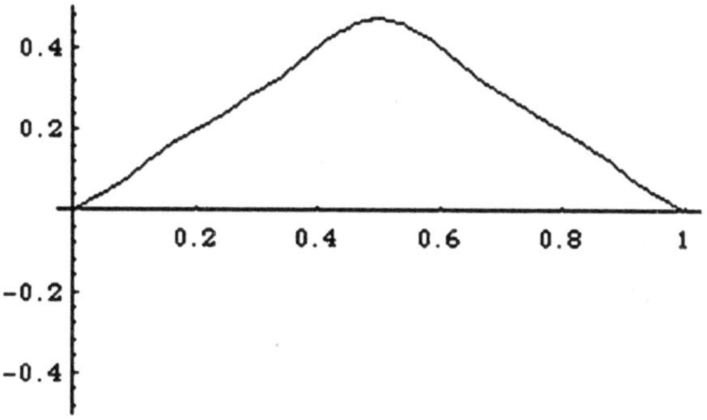

The boundary conditions we have been using are not the only useful ones. Suppose that our string is attached to a pair of tracks in the vertical direction perpendicular to the x-axis using a frictionless roller. Then u(t,0) and u(t,L) are not specified, but instead we have:

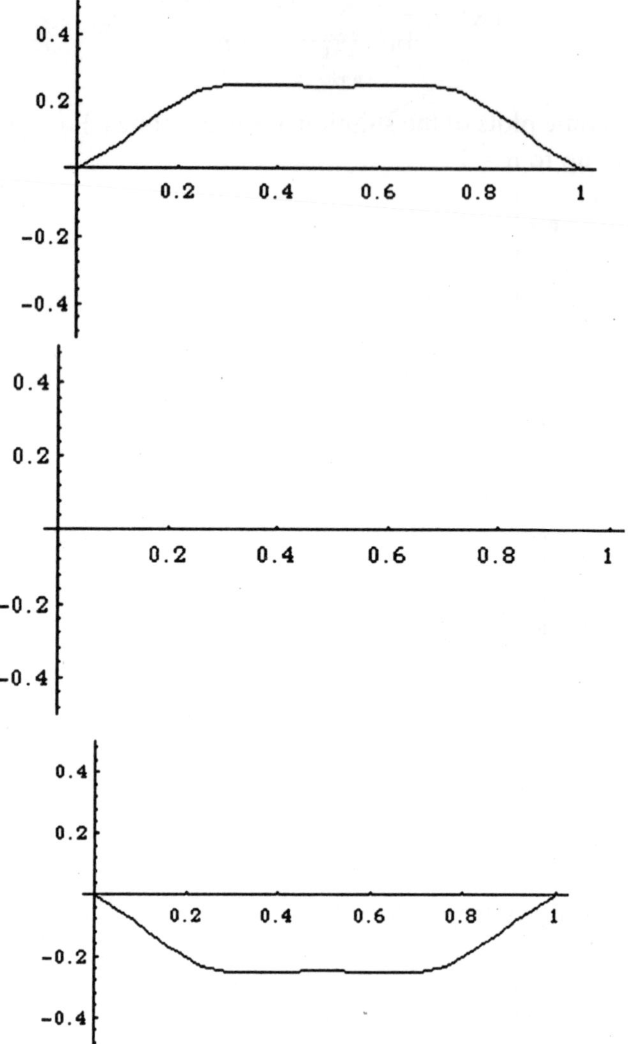

These are called Neumann boundary conditions. Obviously, as with DBC, it would be a minor change to impose them at x=a and x=b where a does not have to be 0. At this stage, we can exactly repeat the analysis of separation of variables, until the point where we first used the boundary conditions.

With NBC, we instead evaluate the x-derivative at 0:

$$\frac{\partial u}{\partial x}(t,0) = 0,$$

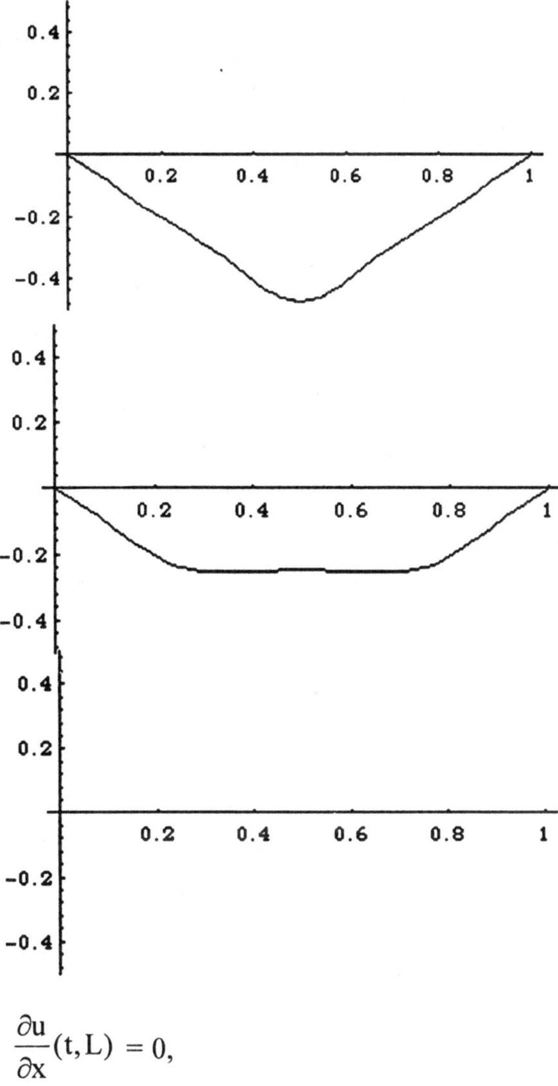

$$\frac{\partial u}{\partial x}(t, L) = 0,$$

Example: Set up the general solution of the vibrating string with Neumann BC., with the simplifications c = L = 1.

Solution. The partial differential equation separates exactly as before, so we still know that when we assume u = X(x) T(t), we get separated equations. The values of μ will be different, however, because they were determined by the BC. Since as before,

$$X(x, \mu) := D_1 \cos(\mu^{1/2} x) + D_2 \sin(\mu^{1/2} x),$$

we see that

$$x'(x, \mu) := -\sqrt{\mu} D_1 \sin(\sqrt{\mu x} x) + \sqrt{\mu} D_2 \cos(\sqrt{\mu} x),$$

and for this to be 0 when x=0, there are two possibilities, namely $D_1 = 0$ or $\mu = 0$. (We ignored the latter possibility before since sin(0) = 0 is not a useful solution, but cos(0) = 1 can be one.)

Possibility 1. $\mu = 0$, which means that the output just given is not really the solution, which has become $X''(x) = 0$. The solutions of this are of the form A + B x, but B has to be 0 so that $X'(0) = 0$. Thus with this possibility, X(x) = 1 (times any constant). Equation now also becomes T''(t) = 0, which leads to the product solution of (WE):

$$(A_0 + B_0\, t)\, 1 = A_0 + B_0\, t.$$

Since the boundary conditions do 0not apply to the t variable, we keep the general linear combination. A_0 and B_0 will be determined by the initial conditions.

Possibility 2. $\mu > 0$. The boundary condition at x = 0 now forces $X(x) = \cos(\mu^{1/2}\, x)$ (again, up to a constant multiple), and the other boundary condition forces one of the values $\mu_n = (n\,\pi)^2$. The product solutions become

$$(A_n \cos(n\,\pi\, t) + B_n \sin(n\,\pi\, t)) \cos(n\,\pi\, t)$$

and the general solution will be

$$u(t, x) = A_0 + B_0 t + \sum_{n=1}^{\infty} (A_n \cos(n\pi t) + B_n \sin(n\pi t)) \cos(n\pi x).$$

The coefficients are determined by the initial conditions because:

$$u(0, x) = A_0 + \sum_{n=1}^{\infty} A_n \cos(n\pi x),$$

and
$$u_t(0,x) = B_0 + \sum_{n=1}^{\infty} n\,\pi\, B_n \cos(n\pi x),$$

The numbers A_n are the Fourier cosine coefficients of u[(0,x) = f(x), while B_0 and n p B_n are the Fourier cosine coefficients of $u_t(0,x) = g(x)$. (Notice the extra factors of n p.) A realistic string or optical fiber may not be uniform, so some of the simplifying assumptions in our derivation fo the wave equation are not valid. For example, the constant c^2 = (spring constant) / (mass density) for the spring may be different at different positions, leading to an equation of the form

$$u_{tt} = (c(x))^2\, u_{xx}$$

or, as a thorough examination of the derivation of the wave equation reveals,we could more generally have:

$$\frac{\partial^2 u}{\partial t^2} = p(x) \frac{\partial}{\partial x}\left(s(x) \frac{\partial u}{\partial x} \right).$$

For some potentially complicated positive functions p(x) and s(x). How well does the method of separation of variables do for problems like this? Rather well, actually, although we may have to encounter some new functions.

TRAVELING WAVES AND THE METHOD OF D'ALEMBERT

This is an interlude from our study of wave equations by the method of separation of variables. For the standard wave equation

$$\frac{\partial^2 u(x)}{\partial t^2} = c^2 \frac{\partial^2 u(t,x)}{\partial x^2},$$

where c is a constant, there is a completely different-looking method of solution, due to the French mathematical physicist Jean le Rond d'Alembert. Indeed, the idea behind the method of this chapter can be the subject of an entire course on *the method of characteristics*.

Suppose ϕ is a function of one variable z, but we set z = x + ct. Then ϕ (x+ct) represents a fixed wave-form which moves to the left at speed x as time goes on.

Provided that can be differentiated twice, it is automatically a solution of the wave equation, which we can rewrite as in chapter VI in terms of a wave operator as u = 0. (Recall that by definition,

$$f(t, x)) := \left(\frac{1}{c^2}\frac{\partial^2}{\partial x^2} - \frac{\partial^2}{\partial x^2}\right) f(t,x)),$$

for any t,x and suitable f.) It is not hard to verify this fact with the chain rule, and you may recall that we saw a particular example.

Since the wave equation is linear, given any two functions of a single variable, which can be differentiated twice, the expression ϕ (x + ct) + ψ (x–ct) is also a *solution.* Suppose now that we have an infinitely long string, and initial conditions of the usual form:

$$u(0, x) = f(x)$$

$$\partial u\,(0, x)/\,\partial t = g(x).$$

We do not need boundary conditions for this problem, although sometimes it is physically important to have "boundary conditions at infinity" that as x –> ± ∞, u –> 0, so that the wave energy is finite.

Can we always find a solution of the form

$$u(t,x) = \phi\,(x + ct) + \psi\,(x - ct),$$

which solves this initial-value problem?

Yes indeed! For simplicity, let us begin with the case where g(x): = 0. Substituting t = 0 shows that

$$\phi(x) + \psi(x) = f(x)$$

while, since $u_t(0,x) = c\phi\,'(x + ct) - c\psi'(x - ct)$,

$$c\phi'(x) - c\,\psi'(x) = 0.$$

This last equation can be integrated to give $\psi(x) = \phi(x) + C$, but we take this constant to be zero (if f(x) – > 0 at infinity, this is forced on us by the BC at infinity). Thus:

$$\phi\,(x) \;= \psi\,(x) = f(x)\,/2.$$

and

$$u(t,x) = f(x + ct)\,/2 + f(x - ct)\,/2.$$

Example: Let the initial displacement f(x) be Exp($-x^2$), and let the initial velocity g(x) = 0. Then the solution in the future is

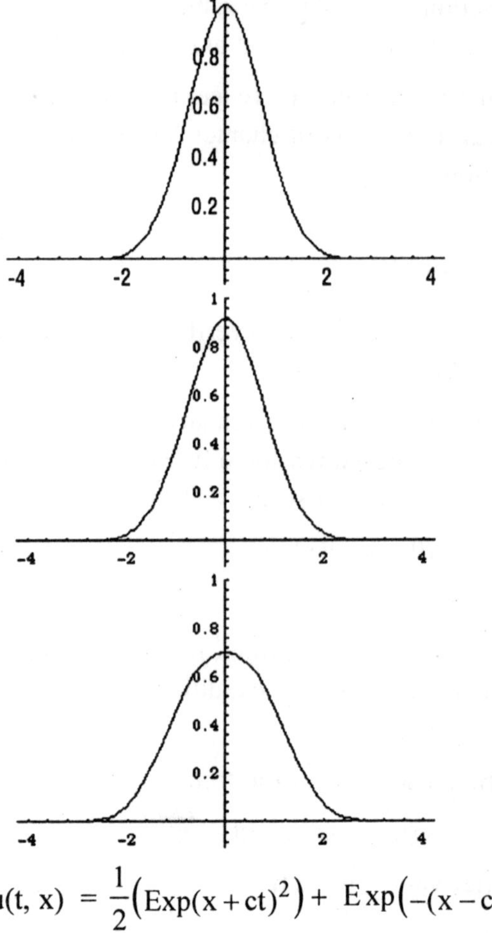

$$u(t,\,x) \;= \frac{1}{2}\left(\text{Exp}(x+ct)^2\right) + \;\text{E}\,\text{xp}\left(-(x-ct)^2\right)$$

We should see the initial bump break into the superposition of two bumps of half the height, one moving to the right and one to the left. Here are some snapshots at different times of the displacement as a function of x:

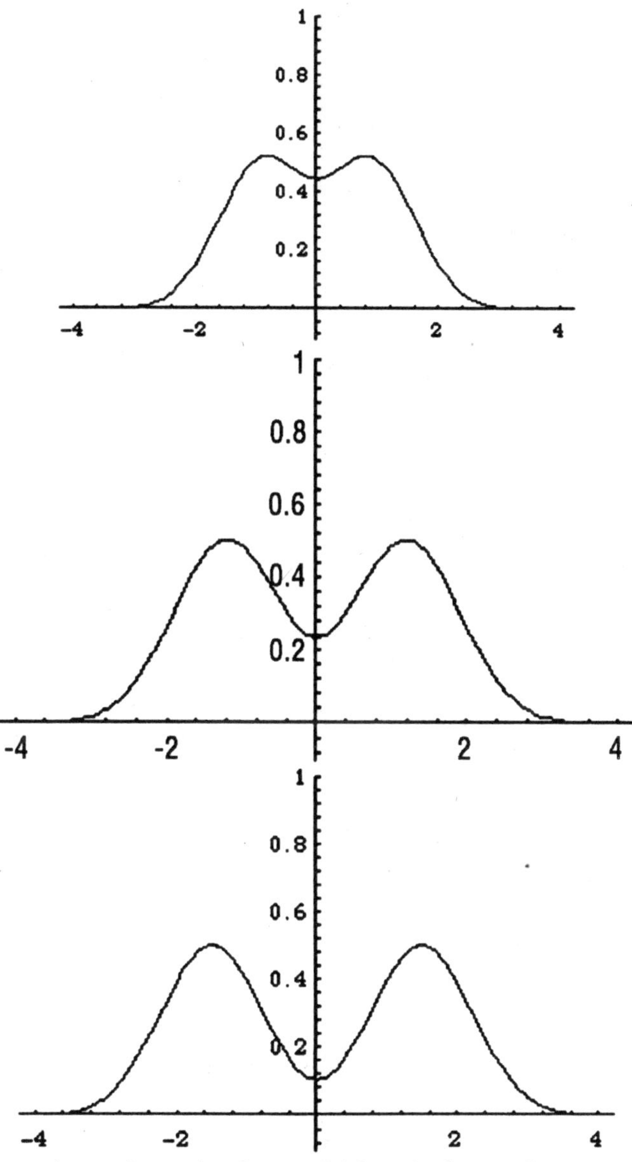

One thing we can learn from the form of this solution and can notice in these graphs is that the integral of u(t,x) with respect to x is independent of time.

Next, let's turn the tables and suppose that f(x) = 0 but g(x) differs from 0. Putting in t = 0, we see that

$$\phi(x) = \psi - (x),$$

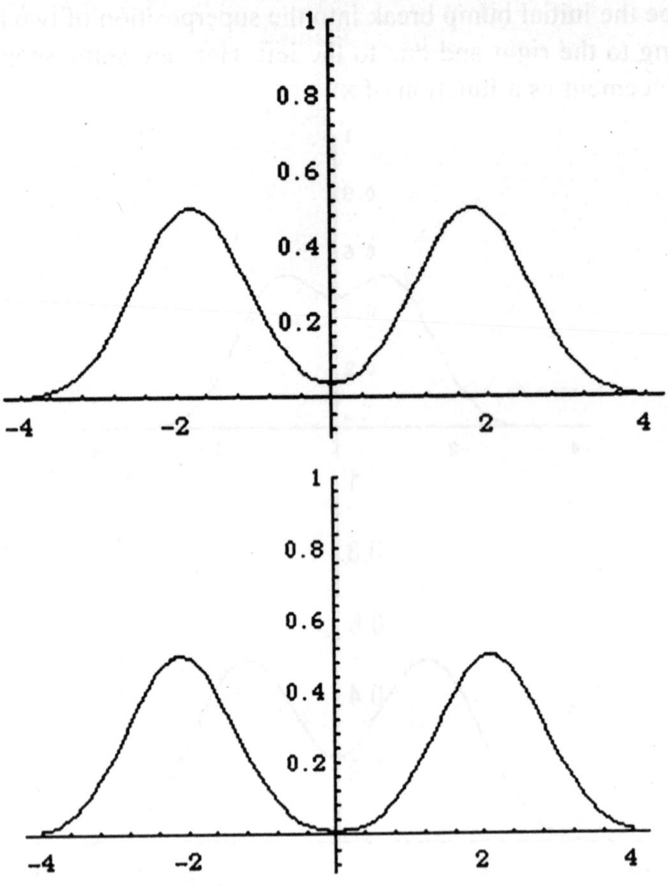

while

$$c\,\phi'(x) - c\,\psi'(x) = 2c\phi'(x) = g(x).$$ Clearly, then,

$$\Phi(x) = \frac{1}{2c}\int^{\infty} g(x)\,dx$$

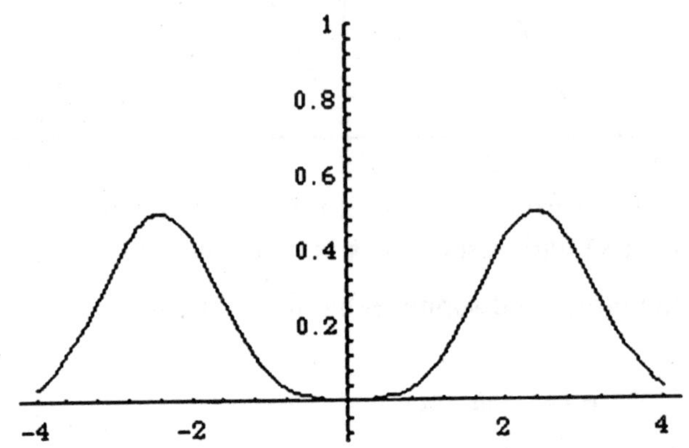

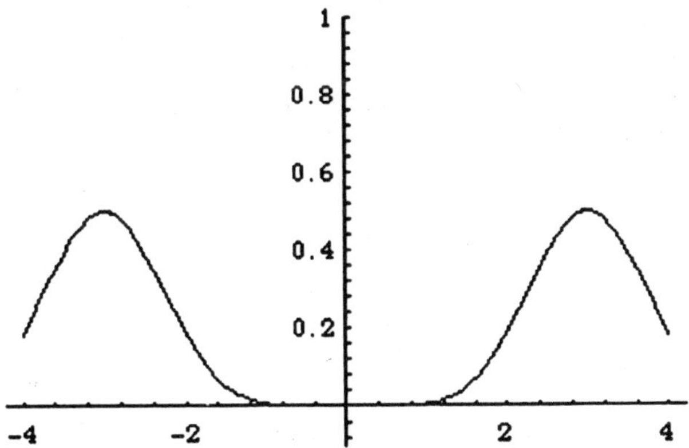

The constant of integration is not specified, but since the full solution is of the form

$$f(x + ct) - f(x - ct),$$

It will drop out of the solution.

Then a calculation shows that

$$u(t, x) = \frac{\arctan(t - x) + \arctan(t + x)}{2}.$$

Here are some snapshots of the solution:

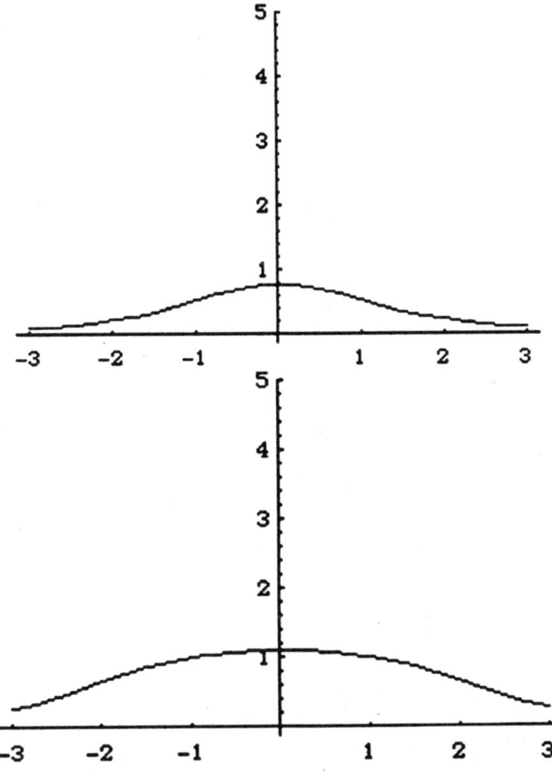

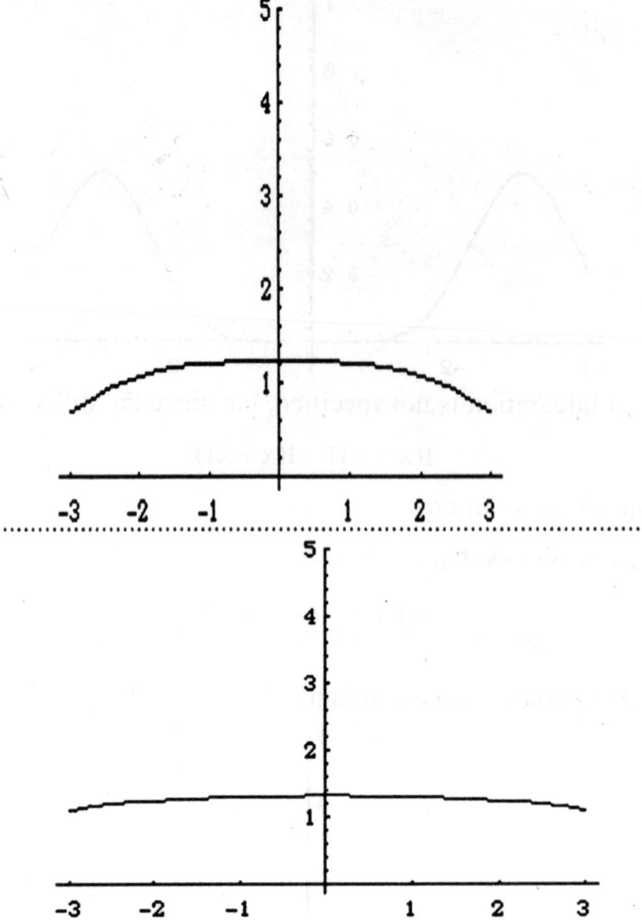

The solution with the given initial conditions keeps moving upward.

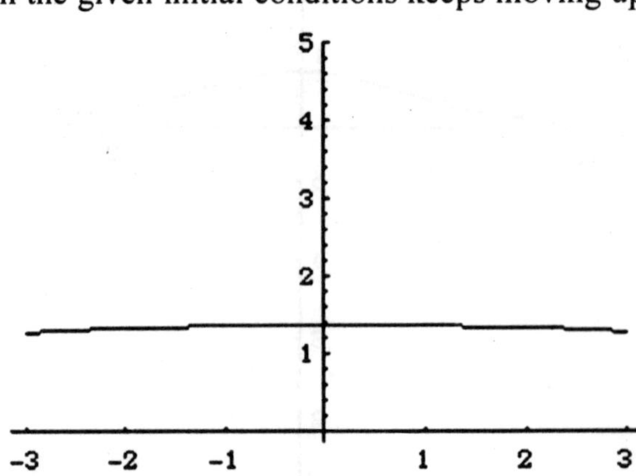

The solution with both initial conditions

u(0,x) = f(x)

ut(0,x) = g(x)

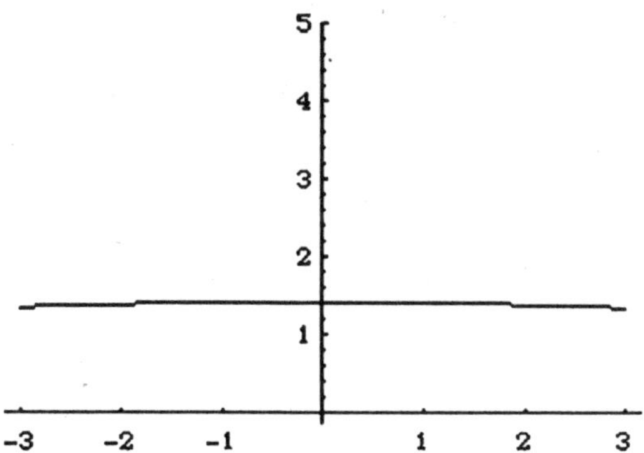

is just the sum of the previous two solutions, because of the superposition principle:

$$u(t, x) = \frac{1}{2}\left(f(x+ct)+f(x-ct)\right)+\frac{1}{2c}\int_{x-ct}^{x+ct} g(s)\, ds.$$

Example: Find the solution of the wave equation for unrestricted x, with c = 1 and initial conditions $f(x) = \sin(4\,x)$, $g(x) = x^2 - 1/3$

Solution. Using software to calculate with formula, the answer is:

$$u(t, x) = \frac{2}{3}\left(t^3 + 3tx^2 - t\right)$$

Although the formula for the solution is fairly simple, the wave forms it describes can be complicated. For example, at unit time steps beginning at t = 0, this solution looks as follows:

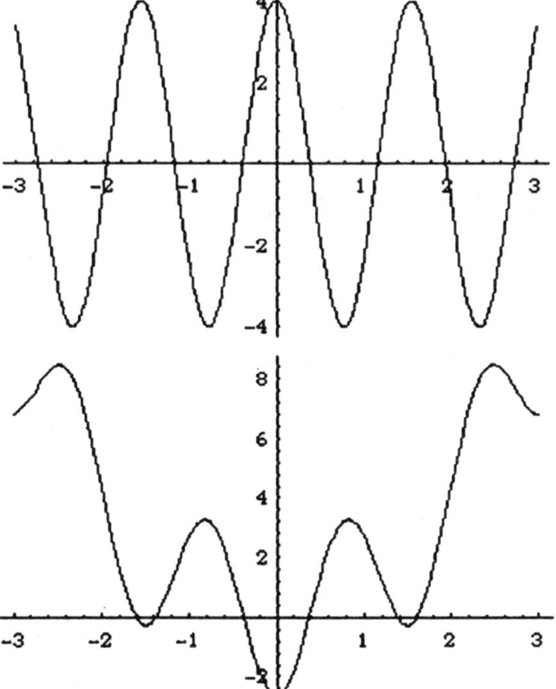

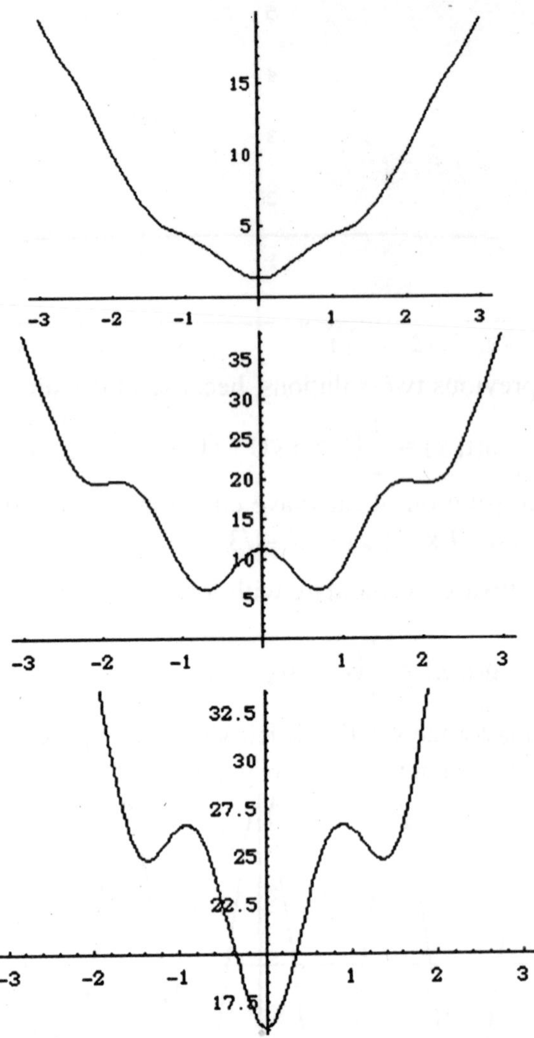

The method of d'Alembert is great if there is no boundary condition, but you may think it fails to be of use when there are boundary conditions. Not so!

Here is the trick. Suppose that we have (say) DBC at 0, and we watch the vibrating string through a window where we can see only what is going on at $x > 0$. We can't tell whether the string is actually clamped down at 0, but the height of the string just happens to be 0 at 0 (at all times). The way we can arrange this is to exploit symmetries. When we formalize this idea in a later chapter and a somewhat different context, it will be called the *method of images*

If f(x) and g(x) are defined on $-\infty < x < \infty$ and are even about the point 0 (i.e., $f(-x) = f(x)$ and $g(-x) = g(x)$), then the solution by d'Alembert's method will also be even, since

$$u(t, x) = \frac{1}{2}(f(-x+ct)+f(-x-ct))+\frac{1}{2c}\int_{x-ct}^{x+ct} g(s)\,ds$$
$$= \frac{1}{2}(f(x-ct)+f(x+ct))+\frac{1}{2c}\int_{x-ct}^{x+ct} g(r)\,dr$$
$$= u(t,x)$$

The substitution in the middle line is r = – s.

It works the same way if f and g are even about some point other than 0. Likewise if f and g are odd, then so is the solution.

Example: Solve the wave equation (with c = 1) for $0 < x < \infty$ with the following initial and boundary conditions. Use the superposition principle and the uniqueness theorem for this problem to simplify your analysis.

(a) $u(0,x) = 0$, $u_t(0,x) = 0$, $u(t,0) = t^2$.

(b) $u(0,x) = \sin(x)$, $(v(0,x) = 0$, $u(t,0) = t^2$.

(c) $u(0,x) = 0$, $u_t(0,x) = 0$, $u_x(t,0) = t \exp(-t)$.

Solutions.

(a) $u(0,x) = 0$, $u_t(0,x) = 0$, $u(t,0) = t^2$.

The uniqueness theorem for this problem tells us that it has only one solution. The point of uniqueness theorems is often lost on students, but in fact they can be extremely practical: If you can solve a different problem from the one which was posed, but it happens to satisfy all the conditions, then it must be the solution of the original problem, too.

In this case, we know how to solve the wave equation if there is no boundary and we are given initial conditions for all x. D'Alembert discovered that the general solution in that case is a superposition of a wave moving to the right and a wave moving to the left. The shapes of the moving waves can be arbitrary until they are fixed by the boundary conditions. With this in mind, imagine a string with $-\infty < x < \infty$ carrying with a rightward moving wave, which happens to be of zero amplitude when t = 0 and x > 0. In other words, $u(t,x) = \psi(x-t)$, where $\psi(x-0) = \psi(x) = 0$ when x > 0. This satisfies the initial conditions, so it must be the boundary condition at x = 0 which determines the unknown function y: Putting x = 0, we see that

$$(0-t) = t^2 \text{ for all } t > 0$$

We now know $\psi(x) =$ for all values of x, because given any negative x , we can substitute x = – t and use this last observation to find that $(x) = (-x)^2 = x^2$. In the traveling wave, we replace x by x – t, so the solution is:

$$u(t,x) = \psi(x-t) = (x-t)^2 \text{ when } 0 < x <= t, \text{ and otherwise } 0.$$

$$u(0,x) = \sin(x),\ u_t(0,x) = 0,\ u(t,0) = t^2.$$

Let us think about the superposition principle here to simplify things, rather than beginning the whole analysis at the beginning. A linear combination of solutions of the wave equation is again a solution of the wave equation, but the boundary condition is non-homogeneous. This means that we can't just add two solutions such that $u(t,0) = t^2$; if we did the boundary condition satisfied by the sum would be $2\,t^2$, not t^2. If, however, we added a function satisfying our boundary condition to one having a zero Dirichlet condition, the sum would equal t^2 at the boundary. This suggests the following strategy:

1 Solve problem a), with $u(0,x) = 0$, which we did above.

2 Solve a simpler problem, the wave equation with

$$u(0,x) = \sin(x),\ u_t(0,x) = 0,\ u(t,0) = 0.$$

3. Add the two solutions.

Since the sum solves the wave equation and the initial and boundary conditions, it is the one and only correct answer to problem b).

Here is how to solve b'). If we extend x to $-\infty$ and $u(0,x) = \sin(x)$ there, then our initial condition would be odd, and we might expect that by symmetry, $u(t,0)= 0$ when $t > 0$. The solution of this problem, by d'Alembert's method, is:

$$(1/2)\ (\sin(x + t) + \sin\ (x - t)),$$

and when $x = 0$, this is $(1/2)\ (\sin(t) + \sin\ (-t)\ )$, which is indeed zero, since the sine is an odd function. It may also help to notice that by the sine-sum formula,

$$(1/2)\ (\sin(x + t) + \sin(x - t)) = \sin(x)\cos(t)$$

The solution to problem b) can be written:

$$u(t,x) = (x - t)^2 + \sin(x)\cos(t),\ \text{when } 0 < x <= t,$$

and $$\sin(x)\cos(t),\ \text{otherwise.}$$

$$u(0,x) = 0,\ u_t(0,x) = 0,\ u_x(t,0) = t\exp(-t).$$

We can solve this problem just as for problem a), by imagining that the solution is a traveling wave coming from the left of 0, but which hasn't disturbed the string at $x > 0$ before it arrives at $t = 0$. Thus $u(t,x) = (x - t)$, where $(x- 0) = (x) = 0$ for $x > 0$.

The difference is the boundary condition. This time we know that

$$u_t = - y'(x-t),\ \text{and at time 0, we have}$$

$$- \psi\,'(0 - t) = t\exp(-t).$$

Perhaps it is best here to change variables so that $x = - t$. Then:

$$- \psi'(x) = - x\exp(x),$$

so, by integrating, we find that $(x) = (x-1)\exp(x) + C$.

The condition that the solution will be 0 when x = 0 fixes the conmstant, so we get

$$1 + (-1 - t + x)\,\exp(x-t)$$

$$\text{for } x - t\ 0 \text{ (and 0 for } x - t > 0).$$

If we are given initial conditions f and g only between 0 and L, we are free to imagine that they are defined for x > L as well as for negative x, in any way we wish. If we extend the definition of f and g so as to be odd functions about the two points 0 and L, and imagine solving the problem of an infinite string with d'Alembert's method, then the boundary conditions will continue to be satisfied when t > 0. That has to be the same as our solution for the finite problem, because of uniqueness.

Example: Suppose that g(x) = 0, but f(x) = x(1 – x), 0 < x < 1. Extend these functions to all of x and thereby solve the initial-boundary-value problem with these initial conditions and zero Dirichlet boundary conditions at 0 and 1.

Solution. We need to extend f and g so as to be odd around both 0 and 1. Obviously, we extend g(x) as 0 for all x, but f(x) is a little trickier For x from –1 to 0, we change f(x) to – f(– x).

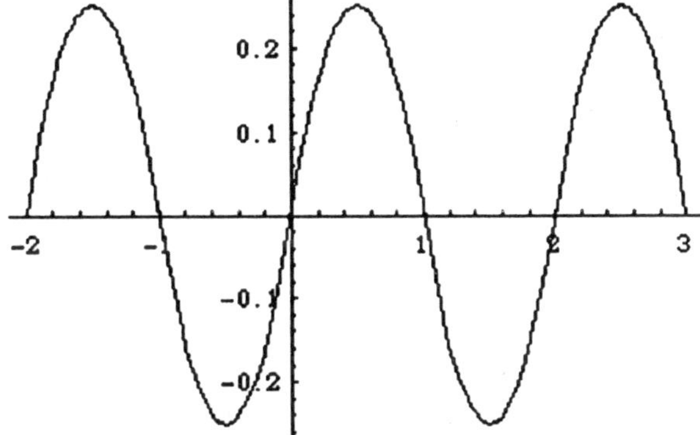

It looks much like a sine function, but numerically it isn't. We could go on like this all day, or we could use modular arithmetic as represented by the commands floor and frac in Maple or the commands Floor and FractionalPart in Mathematica. (Some programming languages use int and frac.)

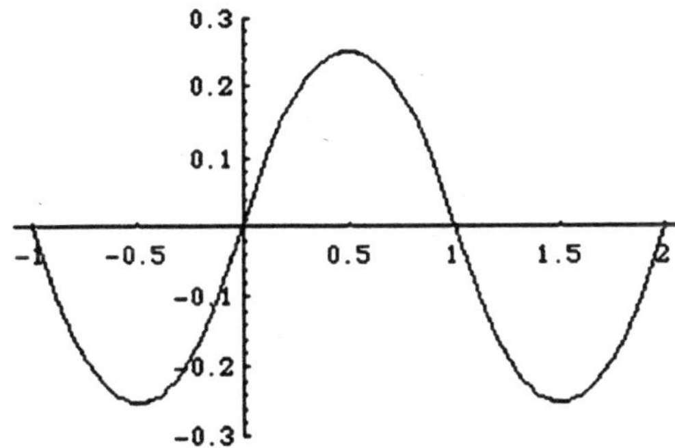

As shown in the Maple worksheet or the Mathematica notebook for this chapter, you can see how to piece a function together so that it is periodic with, say, period 2, as shown here: With the software, we can then solve the IBVP with formula:

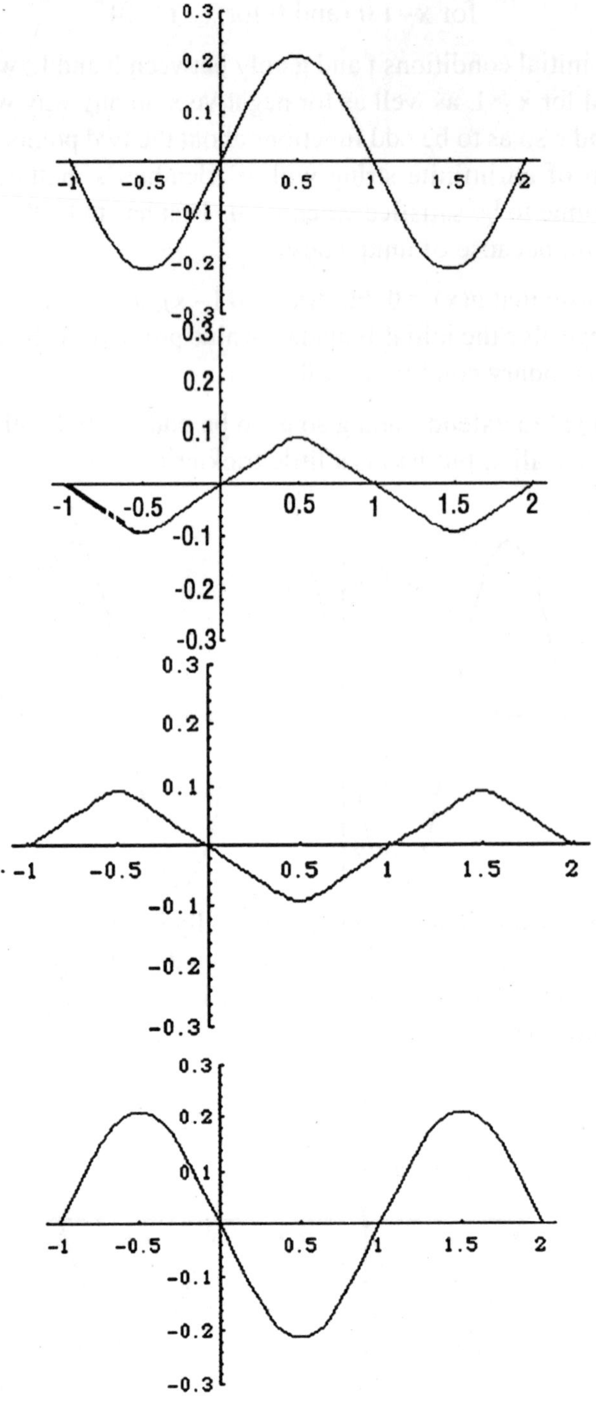

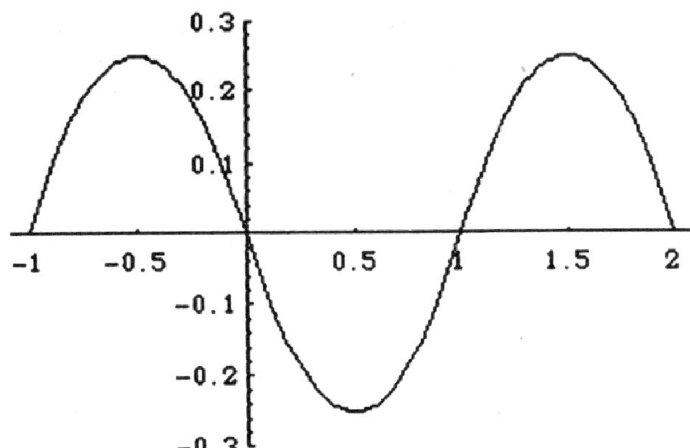

With software it is possible to animate this and see the standing wave motion. Notice that the wave does not simply jiggle up and down; its shape is changing, too.

The wave equation is the prototype for a class of partial differential equations, the *hyperbolic* equations. Other ones exhibit wave motion, although the speed may vary and dispersion and dissipation may occur. (In more ordinary language, the waves may spread and die down.) Another feature of the hyperbolic equations is that they are quite tolerant of irregularities. The general solution makes no reference to the differentiability of and , and indeed, even if and are discontinuous, that formula has a reasonable interpretation as a superposition of traveling waves. Such an expression is referred to as a *weak* or *generalized* solution of the wave equation, and is important in applications because it describes shock waves, sonic booms, etc.

A weak solution can be defined formally by integration by parts. Suppose that a function is put into an inner product with another function 5 t,x) which

1. Is differentiable at all orders, and
2. Equals 0 for $|x|$ sufficiently large or for $|t|$ sufficiently large.

The standard inner product here is defined as in chapter II, but by integrating over all x and t, $-\infty < x,t < \infty$.

I will refer to the region on which $5 \neq 0$, together with the boundary of this region, as its *support*. If u(t,x) is differentiable twice with respect to t and x (and the derivatives are continuous functions), and A denotes the wave operator defined above, then integration by parts shows that

$$< A\ 5,u> = <5\ , A\ u> = 0$$

For all such 5. The functions 5 are referred to as *test functions*, and you can think of a way of testing whether u is a solution of the wave equation without differentiating it directly.

Definition. A function u which satisfies for all test functions is a *weak solution* of the wave equation.

The method of d'Alembert works in a modified form for other hyperbolic, and even parabolic, partial differential equations, and is more generally known as the method of characteristics. Instead of simple, traveling waves, the solutions consist of waves which travel along special curves, and which can have variable speeds and can change their shapes. If the equations are not linear, the waves will not satisfy the superposition principle, but the method of characteristics is less closely tied to linearity than is the method of separation of variables.

THE MATHEMATICS OF HOT RODS

In many ways the simplest of our three model partial differential equations is the heat equation, also known as the diffusion equation,

$$u_t = k \nabla^2 u$$

or, specializing for now to one dimension,

$$u_t = k\, u_{kk}$$

The heat equation is the simplest model of a *parabolic* partial differential equation. A physical derivation of the heat equation is given in an appendix; the constant

$$k = \sigma/\rho\, \kappa_s,$$

where σ is the thermal conductivity, ρ is the density, and κ_s is the specific heat capacity of a homogeneous substance. Ordinary substances have values of k ranging from about 5 to 9000 cm^2/hr.

We shall solve the heat equation with the method of separation of variables, as we did for the wave equation. The solutions will be quite similar. We begin with the one-space-dimensional heat equation, which can be written with the help of a heat operator

$$H_k\,[f]: = \left(\frac{1}{k}\frac{\partial}{\partial t} - \frac{\partial^2}{\partial x^2}\right) f(t,x),$$

The heat equation is the equation for the kernel of this operator. This equation describes the temperature u(t,x) in a long, thin rod along the x-axis, if the rod is wrapped along its length with a perfect insulator. The heat equation is a linear, homogeneous equation, so, just as for the wave equation, the superposition principle holds; if we can find some solutions $u_1,\ldots, u_k$, then any linear combination of these solutions is again a solution.

With the method of separation of variables, we try to find solutions of a special form,

$$u = T(t)\, X(x).$$

As before, boundary and initial conditions will be important, and arise from the physical situation we wish to study. The end points of the rod may be in thermal contact with ice water, which will ensure that we have homogeneous Dirichlet boundary conditions (DBC)

$$u(t,a) = u(t,b) = 0.$$

(Actually, stirred wet crushed ice would guarantee these boundary conditions in the laboratory better than ice water.)

The fact that these boundary conditions are homogeneous (something equals 0) is going to be useful in our analysis, so you may wonder if something goes wrong if we use a different temperature scale or hold the ends at a different fixed temperature T different from 0. Fortunately, we only need to redefine u by a constant to have these same boundary conditions. Since $H_k[T] = 0$, if

$$v(t,x) := u(t,x) - T,$$

then

$$H_k[v] = H_k[u] = 0.$$

By this simple change, the problem with some other fixed temperature at the ends is converted into out standard problem with homogeneous boundary conditions of the Dirichlet type.

Neumann boundary conditions (NBC),

$$u_x(t,a) = u_x(t,b) = 0,$$

Correspond physically to the situation when the ends of the rod are also insulated, since the heat flow through either end of the rod is proportional to the temperature gradient at that end. If there is no heat flow, the gradient must be 0 (recall the derivation).

We begin by considering the case of Dirichlet boundary conditions, and suppose that the rod has length 1, with a = 0 and b = 1. We shall solve this problem with the method of separation of variables, by looking for special solutions analogous to normal modes for the wave equation. Ignoring the initial conditions for the time being, we begin by seeking product solutions, $u(t,x) = T(t)\ X(x)$ to the heat equation (HE), which also satisfy the homogeneous boundary conditions.

The efficient way to proceed is to divide the wave equation by the product solution:

$$\frac{1}{T(t)X(x)} H_k[t(t)X(x)] = \frac{T'}{kT} - \frac{X''}{X}.$$

If u solves the heat equation, this expression must be zero. Thus

$$T'/(k\ T) = X''/X.$$

Since the left side is independent of x and the right side is independent of t, both sides must in fact a constant. We don't know its value yet, so let us call it $-\mu$. (It will turn out to be negative.)

Thus

$$-X''(x) = \mu\ X(x)$$

We know that the eigenfunctions and eigenvalues are:

$$X(x): = \sin(n \pi x),$$

$$u = (n \pi)^2.$$

The other part of the partial differential equation which has been separated differs from what we got with the wave equation:

$$T'(t) = -\mu_n k T(t)$$

This is a simple first-order equation, the solutions of which are exponential functions,

$$T[t] = C \exp(-\mu_n k t).$$

The eigenvalue problem fixed the possibilities for μ_n, leaving us with normal modes of the form

$$c_n \exp(-n^2 \mu^2 t) \sin(n \pi x), n = 1,...$$

With DBC at a and b, b – a = L, we would get:

$$c_n \exp(-n^2 \pi^2 k t) \sin(n \pi (x-a)/L), n = 1,...;$$

the easiest way to see this is to change variables to x' = (x-a)/L. In passing, we note that we could have set k = 1 throughout the derivation of the normal modes, and then changed variables to t' = k t at the end.

Since the heat equation is linear, we can make linear combinations of the solutions we have found and thereby produce new solutions. The most general solution to the heat equation that we get in thisway is:

$$\mathrm{Sum}_{\{n=1\}}\{c_n \exp(-n^2 \pi^2 t) \sin(n \pi x)\}.$$

At this stage it may not be completely clear that this is the absolutely most general solution, but the Fourier sine series will guarantee that this can indeed describe the solution to the heat equation with an arbitrary square-integrable initial condition.

The initial conditions appropriate to the heat equation are somewhat different from what we had for the wave equation, again because the time variable enters differently. The heat equation is first-order in time, so we need only one initial condition, the initial temperature distribution in the rod,

$$u(t,x) = f(x).$$

There is no need to look at u_t at t = 0. Physically, this would be inappropriate, because our experience teaches us that the temperature changes only in response to the initial temperature, not to some unmeasurable effect of the rate of change of the initial temperature. If this were not the case, cookbooks would be nearly impossible.

Fortunately, the initial conditions to bake a cake are, typically, batter at room temperature, and if we have identical batter in identical pans at room temperature, we need to bake the cake for identical lengths of time. No other physical initial condition is necessary.

Model Problem: Find the future temperature within a rod of length 1, k = 1, insulated along its length, and with ends in thermal contact with ice water, if at time 0, the rod is at a uniform temperature u(0,x) = 50.

Solution: At time t = 0, the general solution becomes a Fourier sine series,

$$\sum_{n=1} c_n \sin(n\pi x),$$

So the coefficients c_n are determined as the Fourier *sine* series coefficients of the initial function, f(x) = 50. We know the formula for the Fourier sine coefficients, so we simply calculate the integral to get:

$$c_n = \frac{100}{n\pi}(1-(-1)^n).$$

This is 0 when n is even and otherwise 200/(n π). Our solution to the heat equation is thus:

$$u(t, x) = \sum_{n\,odd} (200/n\pi)\exp(-n^2\pi^2 t)\sin(n\pi x).$$

Suppose now that the rod is insulated not only along its length but also on its ends. Then u(t,0) and u(t,L) are not specified, but instead we have:

$$\partial u[t, 0]/\partial x = 0, \quad \partial u[t,L]/\partial x = 0$$

Equation: Set up the general solution of the insulated rod with Neumann BC., with the simplifications k = L = 1.

Solution.

This is very similar to the analysis, and the eigenfunctions are the basis functions of the Fourier *cosine* series,

$$\{1, \cos(n\pi x)\}.$$

The normal modes are

$$c_n \exp(-n^2 \pi^2 t) \cos(n \pi x),\ n = 0, 1,...$$

and the general solution is:

$$\sum_{n=0} \exp(-n^2\pi^2 t)\cos(n\pi x),$$

Notice that after a long time, all of the terms of this series will become negligible except the one corresponding to n=0. Remembering that c_0 is nothing other than the average temperature, we see that if the rod is entirely insulated, then after a long time the temperature has settled down to equilibrium at the one temperature which is consistent with conservation of energy.

Another way to see the same effect is as follows. (I shall do the calculation in three dimensions, since it is not much harder than in one dimension.) In the derivation of the heat equation, we reason that the rate of change of heat content is

$$\frac{dQ}{dt} = \iint_{\pi} \sigma n \nabla u(t,x) d^2x.$$

If the object is insulated, then ndot (Grad u) = 0. (This higher-dimensional Neumann condition is what simplifies to u x = 0 in one space dimension.) It follows that the total heat content

$$Q = \iiint_{\pi} \rho \kappa_5 u(t,x) d^3x$$

Remains constant. For a homogeneous material, this is just a multiple of the average temperature.

Something more subtle happens in the case of DBC, where energy is lost (or gained) through the ends of the rod. In that case, the general solution ensures that not only will the temperature settle down to an equilibrium u(t,x) – > 0 as $x \to \infty$, but it usually will do so with a particular temperature profile. Assuming that c_1 is not 0, we find that

$$\exp(\pi^2 t)\ u(t,x) = c_1 + r(t,x),$$

where the remainder terms in

$$r(t,x) = \text{Sum}_{\{n=2\}}\ c_n \exp(-(n^2-1)\pi^2 t) \sin(n \pi x)$$

all vanish rapidly as t – > infinity. The temperature profile of the rod rapidly approaches that of a simple sine function, times a time-varying factor tending to 0.

Example: Suppose now that we have non-homogeneous boundary conditions, such as what we would get if the end of the rod at x = 0 is held at temperature 100 and the end at x = 1 at temperature 0. Find the general solution to the heat equation with k =1.

Solution: We can no longer convert this problem to the usual one by the simple trick. A better idea is to think about what should happen at equilibrium in this problem. At equilibrium, $u_t = 0$, so the temperature will depend on space alone according to

$$u(eq)_{xx} = 0.$$

In this one-dimensional situation, this means that $u(eq)(x) = c_1\ x + c_2$ – the general solution to (8.6). With the given boundary conditions, we can easily solve for the constants, and find that

$$u(eq)(x) = 100 - 100\ x.$$

Since the heat equation is linear we can adjust the solution to obtain our familiar boundary conditions by subtracting the equilibrium, letting

$$v(t,x) := u(t,x) - u_{eq}(x).$$

It is easy to see that

$$H_k[u - u_{eq}] = 0,$$

and the general solution is thus

$$u(t,x) = 100 - 100\,x + \text{Sum}_{\{n=1\}}\, c_n \exp(-\,n^2\,\pi^2\,t)\,\sin(n\,\pi\,x)$$

The properties of a solution to the heat equation are quite different from those of the wave equation. Instead of wave motion and a tolerance for discontinuity, we will find

1. Equilibration - after a long time, the solution becomes independent of time.
2. Intolerance for extremes and discontinuity. Even if the solution starts off as quite a wild function f(x), it will instantaneously become a smooth function, which can be differentiated arbitrarily often.

We investigated equilibration above, but have not yet seen how the heat equation enforces smoothness. One aspect of this phenomenon can be seen from the general solution. We learned in chapters II-III that a function can behave rather terribly and still be the sum of a Fourier series, where the squares of the coefficients must tend to 0 as $n \to \infty$, owing to the Parseval formula.

Because of this, suppose that all we know about the coefficients is that they are bounded by some tremendous number B. Now, if we could say that we are safe in neglecting the terms in a Fourier series beyond some finite point, $n > N$, then we would be left with a finite sum of sines or cosines, and that would clearly be differentiable as often as we would like. We also learned in chapter V that we can safely differentiate Fourier series, provided that the differentiated series makes sense.

With this in mind, let's differentiate our general solution, say, 900 times with respect to x. It is not hard to see from (8.4) that we get

$$\sum_{n=1} c_n \exp(-n^2\pi^2 t)(n\pi)^{900} \sin(n\pi x)$$

where all we know about the c_n is that they are bounded by B. (We would have some minuses or cosines if we differentiated a number of times not divisible by 4.) If $t = 0$, we have amplified all the coefficients, but if $t > 0$, then for large n, the exponential goes to zero faster than any power of n, and the series converges uniformly. This happens no matter how small t is, and shows that the 900th x-derivative of u exists as a continuous function (a theorem from advanced calculus states that a uniform limit of continuous functions is continuous).

A careful analysis suggested by this argument proves that even if the initial condition is extremely irregular, a solution of the heat equation instantly becomes smooth. A precise definition of what we mean by "smooth" is as follows.

Definition: A function is *smooth*, or C ∞, if it can be differentiated arbitrarily often.

Theorem: Let u solve the heat equation on an interval $a < x < b$, and $t > 0$, with initial condition $u(0,x) = f(x)$ square integrable and either Dirichlet boundary conditions or Neumann boundary conditions. Then for all $t > 0$, $u(t,x)$ is smooth.

Later we shall consider the heat equation on higher dimensional regions Ω. The same theorem holds for all x in the interior of Ω. Also, other standard kinds of boundary conditions can be allowed in this theorem.

Another important property of the heat equation is the *maximum principle.*

Theorem: Let u solve the heat equation. If we consider any finite space-time region, such as a rectangle $a \leq x \leq b$, $c \leq t \leq d$, then the greatest value of the temperature occurs on the boundary of the region, in this case when $x = a$ or b or when $t = c$ or d. In other words, a hot spot cannot be spontaneously generated in the interior of the material.

The same is true of the minimum, a potential cold spot - the minimum always occurs on the boundary.

As with the smoothing theorem, the maximum principle also applies to higher-dimensional regions.

To prove the theorem, let us first consider the heat equation with a *heat sink*, which absorbs energy. The equation describing this would be

$$u_t = u_{xx} - Q(t,x),$$

Where $Q \geq 0$. For now, suppose that Q is strictly positive, so $Q \geq \varepsilon > 0$ throughout the rod. Suppose that there is a hot spot, i.e., a local maximum, at (t_0,x_0), in the interior of the space-time region. By basic calculus, the space-time gradient of u must be 0 there (we know that the solution is differentiable), that is, $u_t(t_0,x_0) = u_x(t_0,x_0) = 0$. If the point is really a maximum, then $u_{xx}(t_0,x_0) \leq 0$. But this is a contradiction, since

$$u_{xx} = u_t + Q \geq \varepsilon > 0.$$

This shows that there can be no local maximum in the interior of the region, so the only possibility is for the maximum to be on the boundary.

Now suppose that there is no sink, so we have a solution of the heat equation with no sinks or sources,

$$u_t = u_{xx},$$

and that there is a point (t_0,x_0) where the temperature exceeds that on the boundary of a finite region. To be specific, suppose that we replace u with the function

$$v(t,x) := u(t,x) + \delta \exp(x),$$

Where δ is a small but positive quantity, so small that the maximum value of v is also attained in the interior - this is possible by the intermediate value theorem of calculus, which applies to continuous functions. What equation does v solve?

$$v_t = u_t,$$

while

$$v_{xx} = u_{xx} + \delta \exp(x) = u_t + \delta \exp(x) = v_t + \delta \exp(x).$$

Solving for v_t, we have

$$v_t = v_{xx} - Q,$$

where $Q = \delta \exp(x)$ is strictly positive. We already considered this case, and found that it cannot happen. The transformation shows that it cannot happen for the basic heat equation, either.

The statement about the minimum follows because -u solves the heat equation whenever u does.

Chapter 27

Applications of Differential Equation

In this chapter we will see applications of Differential Equation. Now that we know how to compute the derivatives of many common functions, we can give a few examples of why the derivative is such a useful tool. In this chapter, we will look at four different applications of the derivative. The first application, is to use the derivative to find the velocity and acceleration of a particle moving in a straight line. When we are given a function $f(t)$ describing the position of a particle at time t, the velocity of the particle at time t is the derivative $f'(t)$ and the acceleration is the second derivate $f''(t)$.

The second application is the analysis of graphs of functions . We can use the derivative to find critical points and inflection points on graphs, from which a reasonably good sketch of a function can be constructed.

The second applications is related to the third, optimization of functions . For example, one may encounter a function in the business world that gives the total profit of producing a certain number of goods. It would then be natural to try to maximize such a function.

The fourth and final application concerns related rates . Suppose water is flowing into a gigantic ice cream cone at a fixed rate (for some strange reason). Through a clever application of differentiation, it is possible to determine how quickly the water level will be rising when it reaches any particular height in the cone.

ANALYSIS OF GRAPHS

Derivatives can be used to gather information about the graph of a function. Since the derivative represents the rate of change of a function, to determine when a function is increasing, we simply check where its derivative is positive. Similarly, to find when a function is decreasing, we check where its derivative is negative.

The points where the derivative is equal to 0 are called critical points. At these points, the function is instantaneously constant and its graph has horizontal tangent line. For a function representing the motion of an object, these are the points where the object is momentarily at rest.

THE FIRST DERIVATIVE TEST

A local minimum (resp. local maximum) of a function f is a point $(x_0, f(x_0))$ on the graph of f such that $f(x_0)$d"$f(x)$ (resp. $f(x_0)$e"$f(x)$) for all x in some interval containing x_0. Such a point is called a global minimum (resp. global maximum) of a function f if the appropriate inequality holds for all points in the domain. In particular, any global maximum (minimum) is also a local maximum (minimum).

It is intuitively clear that the tangent line to the graph of a function at a local minimum or maximum must be horizontal, so the derivative at the point is 0, and the point is a critical point. Therefore, in order to find the local minima/maxima of a function, we simply have to find all its critical points and then check each one to see whether it is a local minimum, a local maximum, or neither. If the function has a global minimum or maximum, it will be the least (resp. greatest) of the local minima (resp. maxima), or the value of the function on an endpoint of its domain (if any such points exist).

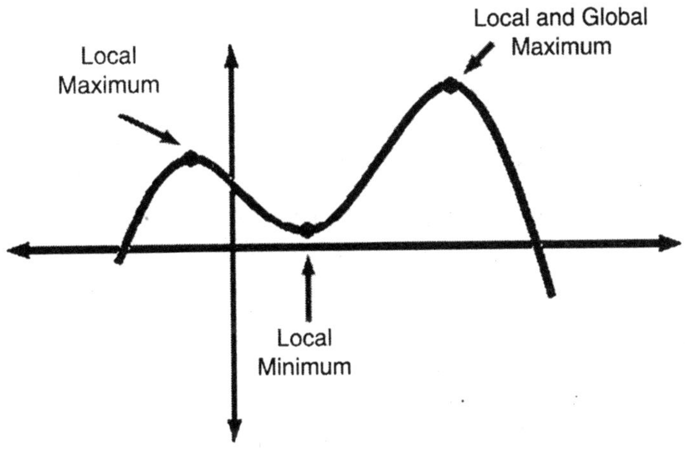

Fig. Examples of Global and Local Extrema

Clearly, the behavior near a local maximum is that the function increases, levels off, and begins decreasing. Therefore, a critical point is a local maximum if the derivative is positive just to the left of it, and negative just to the right. Similarly, a critical point is a local minimum if the derivative is negative just to the left and positive to the right. These criteria are collectively called the first derivative test for maxima and minima.

There may be critical points of a function that are neither local maxima or minima, where the derivative attains the value zero without crossing from positive to negative. For instance, the function $f(x) = x^3$ has a critical point at 0 which is of this type. The derivative $f'(x) = 3x^2$ is zero here, but everywhere else f' is positive. This function and its derivative are sketched below.

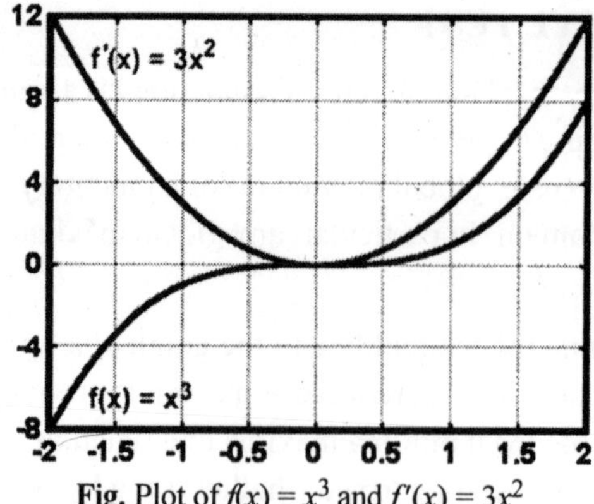

Fig. Plot of $f(x) = x^3$ and $f'(x) = 3x^2$

THE SECOND DERIVATIVE TEST

Once we have found the critical points, one way to determine if they are local minima or maxima is to apply the first derivative test. Another way uses the second derivative of f. Suppose x_0 is a critical point of the function $f(x)$, that is, $f'(x_0) = 0$. We have the following three cases:

- $f''(x_0) > 0$ implies x_0 is a local minimum
- $f''(x_0) < 0$ implies x_0 is a local maximum
- $f''(x_0) = 0$ is inconclusive

The first two of these options derive from the observation that $f''(x_0)$ is the rate of change of $f'(x)$ at x_0, which will be positive if the derivative crosses zero from negative to positive, and negative is the derivative crosses zero from positive to negative. This is called the second derivative test for maxima and minima. The third, inconclusive case is considered below. The first and second derivative tests employ essentially the same logic, examining what happens to the derivative $f'(x)$ near a critical point x_0. The first derivative test says that maxima and minima correspond to f' crossing zero from one direction or the other, which is indicated by the sign of f' near x_0. The second derivative test is just the observation that the same information is encoded in the slope of the tangent line to $f'(x)$ at x_0.

CONCAVITY AND INFLECTION POINTS

A function $f(x)$ is called concave up at x_0 if $f''(x_0) > 0$, and concave down if $f''(x_0) < 0$. Graphically, this represents which way the graph of f is "turning" near x_0. A function that is concave up at x_0 lies above its tangent line in a small interval around x_0 (touching but not crossing at x_0). Similarly, a function that is concave down at x_0 lies below its tangent line near x_0. The remaining case is a point x_0 where $f''(x_0) = 0$, which is called an inflection point. At such a point the function f holds closer to its tangent line than elsewhere, since the second derivative represents the rate at which the function turns away from the tangent line. Put another way, a function usually has the same value and derivative as its tangent line at the point of tangency; at an inflection point, the second derivatives of the function and its tangent line also agree. Of course, the second derivative of the tangent line function is always zero, so this statement is just that $f''(x_0) = 0$.

Inflection points are the critical points of the first derivative $f'(x)$. At an inflection point, a function may change from being concave up to concave down (or the other way around), or momentarily "straighten out" while having the same concavity to either side. These three cases correspond, respectively, to the inflection point x_0 being a local maximum or local minimum of $f'(x)$, or neither.

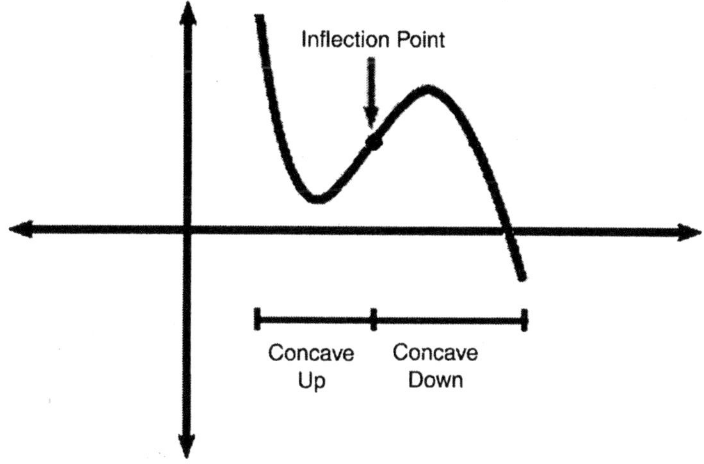

Fig. Example of Concavity and Inflection Points

Example: Find the co-ordinates of the point of contact of the tangent line to the curve $y = x \log x$, which is inclined at an angle of 450 with the x-axis

Solution: Let $P(x, y)$ be the point of contact

The curve is $y = x \log x$

$$\therefore \quad \frac{dy}{dx} = x\left[\frac{1}{x}\right] + \log x[1]$$

$$\therefore \quad \left(\frac{dy}{dx}\right)_p = 1 + \log x$$

$\therefore$ Slope of the tangent at P = 1 + log x

Since this tangent has inclination of 45° with x-axis, its slope is

$\therefore \quad \tan 45° = 1$

$\therefore \quad 1 + \log x = 1$

$\therefore\ \log x = 0 \Rightarrow x = 1$

But P lies on the curve

$$y = x \log x$$

$\therefore\ y = 1 \log 1 = 0$

$\therefore\ P$ has co-ordinates = (1,0)

Example: Find the angle between the two curves $2y^2 = x^3$ and $y^2 = 32x$ at their point of intersection in the Ist quadrant

Solution: $2y^2 = x^3$ and

$$y^2 = 32x$$

Solving equations, we get

$$2\,(32x) = x^3$$

$\therefore\ \ x^3 - 64x = 0$

$\therefore\ x(x^2 - 64) = 0$

$\therefore x = 0$ and $x = \pm 8$

As the point of intersection is in the I[st] quadrant it must be +ve and $\neq 0$

$\therefore\qquad x = 8$

Putting in $\ y^2 = 32x$

we get $\ \ y^2 = 32(8)$

$\therefore\qquad y = 256$

$\therefore\qquad y = \pm 8$

Accepting y = 16, we have the point of intersection is P = (8,16) which lies in the Ist quadrant

Now Differentiating equations w. r. to x, we get

$$4y\frac{dy}{dx} = 3x^2$$

$$\therefore\quad \frac{dy}{dx} = \frac{3x^2}{4y}$$

$$\therefore\quad \left(\frac{dy}{dx}\right)_p = \frac{3(8)^2}{4\times 16} = 3$$

$\therefore$ slope of tangent PT_1 $(m_1) = 3$

$$\text{Again } 2y\frac{dy}{dx} = 32$$

$$\therefore \quad \left(\frac{dy}{dx}\right)_p = \frac{16}{16} = m_1$$

$$= 1$$

$\therefore$ Slope of tangent PT2 (m_2) = 1

Let θ be the acute angle between curves (1) and (2) at *P*.

$$\text{Then } \tan\theta = \left|\frac{m_1 - m_2}{1 + m_1 \cdot m_2}\right|$$

$$= \left|\frac{3-1}{1+3(1)}\right|$$

$$= \frac{2}{4}$$

$$= \frac{1}{2}$$

$$\therefore \quad \theta = \tan^{-1}(1/2)$$

Example: Prove that the curves $y^2 = 16x$ and $2x^2 + y^2 = 4$ cut each other orthogonally.

Solution: Let P (x_1, y_1) be a point of intersection of the curves

$$y^2 = 16x \text{ and}$$

$$2x^2 + y^2 = 4$$

$$\therefore \quad y_1^2 = 16x_1 \text{ and}$$

$$2x_1^2 + y_1^2 = 4$$

Let m_1 and m_2 be the two slopes of the two tangents to the curve at P respectively.

By Differentiating w. r. to x, the equation, we get

$$2y\frac{dy}{dx} = 16$$

$$\therefore \quad \frac{dy}{dx} = \frac{8}{y}$$

$$\therefore \quad m_1 = \left(\frac{dy}{dx}\right)_{p(x_1, y_1)}$$

$$= \frac{8}{y_1}$$

By Differentiating w. r. to x, the equation, we get

$$4x + 2y\frac{dy}{dx} = 0$$

$$\therefore \quad \frac{dy}{dx} = -\frac{2x}{y}$$

$$\therefore \quad m_2 = \left(\frac{dy}{dx}\right)_{p(x_1,y_1)}$$

$$= \frac{-2\times 1}{y_1} \quad ...(6)$$

from equations we have

$$m_1 \times m_2 = \frac{-16\times 1}{y_1^2}$$

$$= \frac{-16\times 1}{16\times 1}$$

$$= -1 \text{ by } (3)$$

Thus the curves intersect orthogonally at P

Interpretation of the Sign of Derivative

We will now cover the meaning of the sign of the derivative and see what inference can be drawn from it and also study its applications to some practical problems from onwards.

We will assume that the functions that we consider are continuous and differentiable at all points and all intervals in questions unless otherwise stated. Further the numbers, denoted by h or Δ, will always be taken as positive.

Locally Increasing or Decreasing Functions

1. Definition of a function increasing at a point:

   If there exists a number, such that $f(c - h) < f(c) < f(c) + h)$ for all h satisfying $0 < h < \Delta$ then the function 'f' is said to increasing at $x = c$

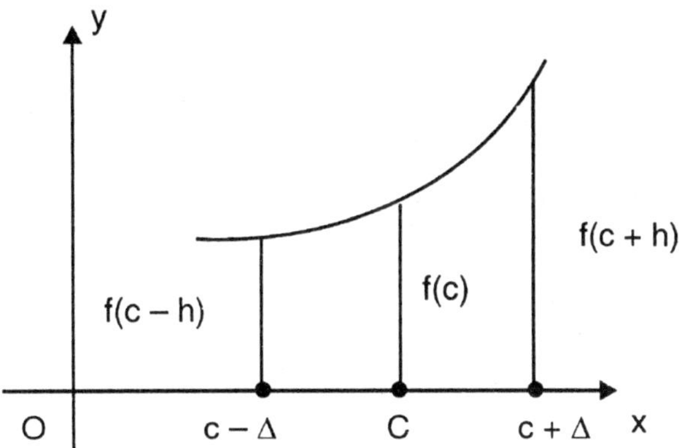

2. Definition of a function decreasing at a point:

   If there exists a number Δ, such that $f(c-h) > f(c) > f(c+h)$ for all h satisfying $0 < h < \Delta$ then the function 'f' is said to be decreasing at $x = c$

   The careful study of these definitions, clearly states that the function $y = f(x)$ is increasing at $x = c$ if $\left(\frac{dy}{dx}\right)_{x=c} > 0$ i.e. '$f'(c) > 0$ and $y = f(x)$ is decreasing at x = c if $\left(\frac{dy}{dx}\right)_{x=c} < 0$ i.e. '$f'(c) < 0$. Thus if a function is increasing at a point, then the derivative is positive at that point and if a function is decreasing at a point, then the derivative is negative at that point. Also it is clear that a function $f(x)$ is increasing in an interval (a, b) if $f'(x) > 0$, at every $x \in (a, b)$. Similarly a function $f(x)$ is decreasing in an interval (a, b) if $f'(x) < 0$, at every $x \in (a, b)$.

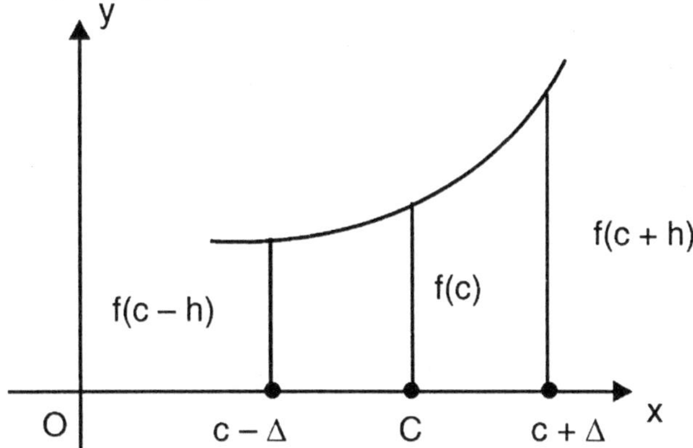

Critical Points

Points on the graph of a function at which the derivative is zero or does not exist at all are called critical points i.e. The point $(x, f(x))$ is a critical point of $f(x)$, if x is in the domain of 'f' and either $f'(x) = 0$ or DNE. Geometrically speaking, the tangent at such a point to the curve representing the function $y = f(x)$ is either horizontal, vertical or DNE.

Turning Points

There are two types of turning points. They are separately called maximum points and minimum points. At such points the curves turns. At such points the tangent to the curve is parallel to x - axis and so, at these points $\frac{dy}{dx} = 0$

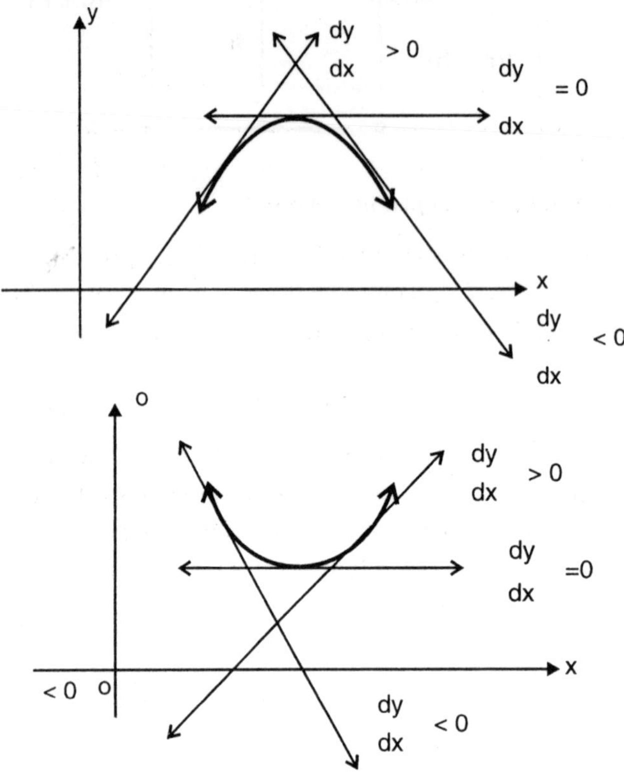

Extreme Value Theorem

In a certain interval, it is possible to determine the maximum and minimum values of the function when we know the critical points of the function. The extreme value theorem guarantees both the maximum and minimum value of a function under certain conditions.

It is stated as:

If a function $f(x)$ is continuous on a closed interval $[a, b]$, then $f(x)$ has both maximum and minimum value on $[a, b]$

Procedure:

I. First establish the continuity of $f(x)$ on $[a, b]$

II. Determine all critical points of $f(x)$ in $[a, b]$

III. Evaluate the function $f(x)$ at these critical points and at the extremities of $[a, b]$

IV. The largest value is the maximum value and the smallest value is the minimum value of $f(x)$ on $[a, b]$

Example:

1. If $f'(x) > 0$ for all x in the interval I, then 'f' is an increasing in I and
2. If $f'(x) < 0$ for all x in I then 'f' is decreasing function in I. Is the converse of these two rules true?

Justify your answer with a suitable example.

Solution: Consider the curve $y = x^3$ as shown in the figure.

x	–4	–3	–2	--1	0	1	2	3	4
$y = x^3$	–64	–27	–8	–1	0	1	8	27	64

From the graph the curve is increasing at $x = 0$. But $f'(0) = 0$ i.e. $f'(0)$ is not positive Hence the converse of (1) is not true. In a similar way by considering $y = x^3$ at $x = 0$, we can show that the curve is decreasing at $x = 0$ but $f'(0) = 0$. Hence the curve decreasing at $x = 0$ but $f'(0)$ is not negative i.e. converse of (II) is also true.

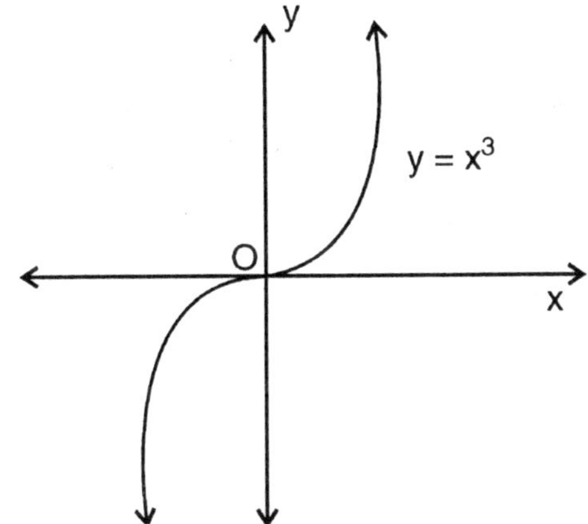

Thus $f'(c) > 0$ is only a sufficient condition for $f(x)$ to be increasing at x = c and $f'(c) < 0$ is only a sufficient condition for $f(x)$ to be decreasing at $x = c$ but it is not a necessary condition.

Example: Find the values of x for which the function $f(x) = x^3 - 12x + 5$ is (I) decreasing and (II) increasing.

Solution: $f(x) = x^3 - 12x + 5$

$\therefore \quad f'(x) = 3x^2 - 12$

$= 3(x^2 - 4)$

Now 'f' is decreasing if $f'(x) < 0$

i.e. $3\ (x^2 - 4) < 0$

i.e. $(x^2 - 4) < 0$

i.e. $x^2 < 4$

i.e. $-2 < x < 2$

$\therefore f(x)$ decreases in $(-2, 2)$

Similarly 'f' increases if $f'(x) > 0$

i.e. $3(x^2 - 4) > 0$

i.e. $x^2 > 0$

i.e. $x \leftarrow 2$ and $x > 2$

THE MEAN-VALUE THEOREM

For the theoretical purpose, the mean-value theorem is one of the most useful tools. The special and easy case of mean value theorem is the 'Rolle's' theorem, which will serve as a lemma for our main results.

Rolle's Theorem

If 'f' is continuous on $[a, b]$ and differentiable in (a, b) and if, further $f(b) = f(a)$, then their exists at least one point $x = c$ in (a, b) such that $f'(c) = 0$

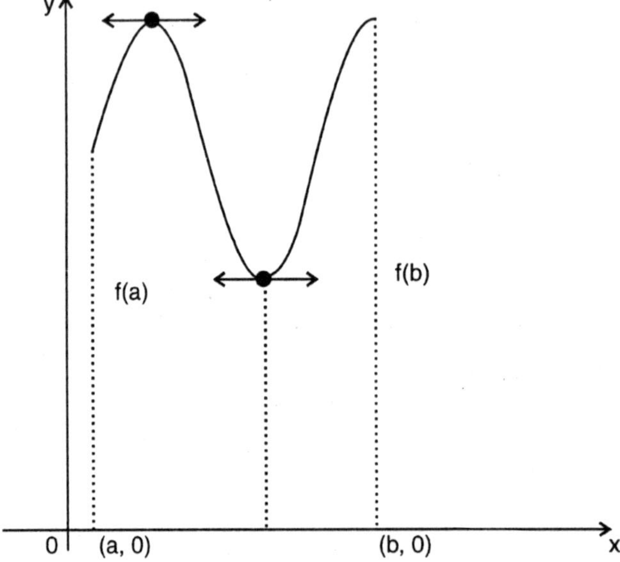

In this theorem, $y = f(x)$ is continuous on $[a, b]$ and differentiable in (a, b). Also $f(b) = f(a)$. Hence the curve $y = f(x)$ can be drawn as in the adjoining figure. The result $f'(c) = 0$ means the tangent at $x = c$ is parallel to x - axis. Hence $y = f(x)$ is continuous on $[a, b]$, differentiable in

(a, b) and $f(b) = f(a)$, then there exists at least one point on this curve at which the tangent is parallel to x - axis.

Lagrange's Mean Value Theorem

It is the generalization of the Rolle's theorem. Let $f(x)$ be any continuous function on $[a, b]$ and differentiable in (a, b), then there exists at least one value of $x = c \in (a, b)$ such that

$$\frac{f(b) - f(a)}{b - a} = f'(c)$$

i.e. $$f(b) - f(a) = (b - a) f'(c)$$

For, consider a function $\phi(x) = f(x) + Ax$

Where A is constant to be determined such that $\phi(a) = \phi(b)$

$\therefore$ $$f(a) + Aa = f(b) + Ab$$

$\therefore$ $$A = -\frac{f(b) - f(a)}{b - a}$$

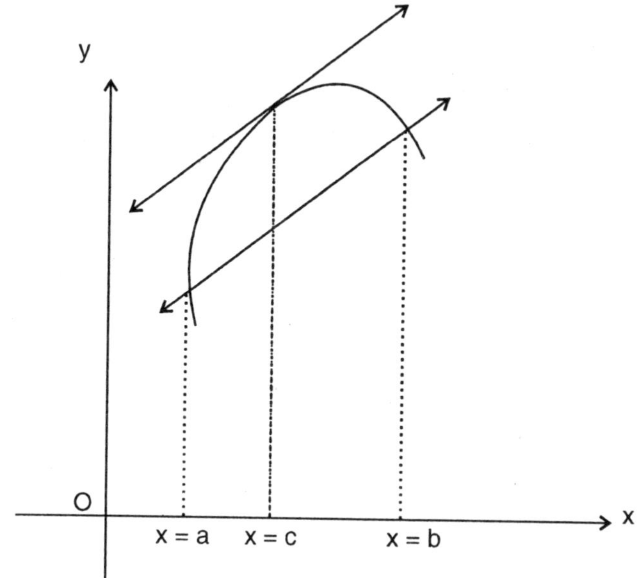

Now $\phi(x)$ satisfies all conditions of the Rolle's theorem. Hence there exists at least one $x = c$ in (a, b) such that $\phi'(c) = 0$

$\therefore$ $$f'(c) + A = 0$$

$\therefore$ $$A = -f'(c), \text{ we get}$$

$\therefore$ $$-f'(c) = \frac{f(b) - f(a)}{b - a} \text{ or } f(b) = (b - a) f'(c)$$

Geometrically

If the graph of $y=f(x)$ is continuous, be drawn between $x=a$ and $x=b$ then $A\,(a, f(a))$ and $B\,(b, f(b))$.

Slope of the chord $AB = \dfrac{f(b)-f(a)}{b-a}$ by L.M.V., this is equal to the slope of the tangent to $y=f(x)$ at $x=$ c i.e. $f'(c)$.

Hence the theorem states that there exists at least one tangent at $x=c$ which is parallel to the chord AB.

First Derivative Test For Local Extrema

If the derivative of a function changes its sign while passing through a critical point along a given curve (i.e. around the critical point), the function possesses a Local (relative) Extrema at that point.

1. If the derivative changes its sign from positive (increasing function) to negative (decreasing function), the function has a Local maxima at that critical point
2. If the derivative changes its sign from negative (decreasing function) to positive (increasing function), the function has a Local minima at that critical point.

Technique

1. Find $f'(x)$
2. Find roots of $f'(x) = 0$

   i.e. critical points of $f(x)$
3. Let $x = c$ be a critical point of $f(x)$ find $f'(c-h)$ and $f'(c+h)$, $h > 0$ (however small)

   a. If $f'(c-h)$, $h > 0$ and $f'(c+h) < 0$

      Then 'f' has a Local maximum at $x = c$

      i.e. $f(c) =$ Local maximum

   b. If $f'(c+h) < 0$ and $f'(c+h) > 0$

      Then 'f' has a Local minimum at $x = c$

      i.e. f (c) = Local minimum
4. Use the same procedure for the other roots.

SECOND DERIVATIVE TEST FOR LOCAL EXTREMA

Maximum Points

Consider the point P. The gradient at P is zero for the curve $y = f(x)$. The gradient is positive for all points to the immediate left of P and negative for all points to the immediate right of P.

Thus near $P \frac{dy}{dx}$ is changing its sign from positive, through zero to negative values, therefore at $P, \frac{d}{dx}\left(\frac{dy}{dx}\right)$ i.e. $\frac{d^2y}{dx^2}$ is negative.

Therefore, for maximum point, $\frac{dy}{dx} = 0$ and $\frac{d^2y}{dx^2} < 0$

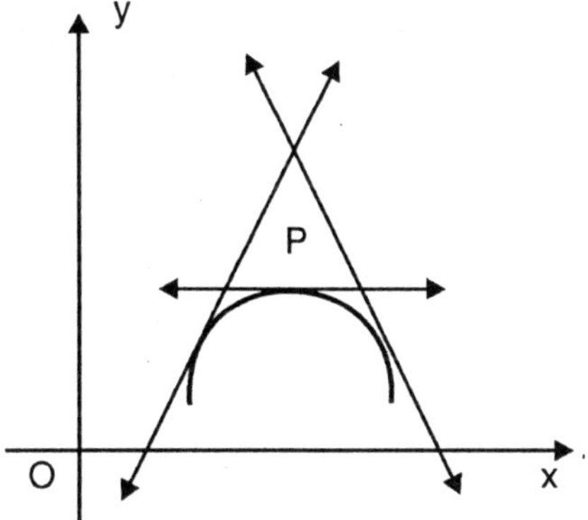

Minimum Points

Consider the point P. The gradient there is zero. For all points to the immediate right of P, the gradient is positive while for all points to the immediate left of P, the gradient is negative. Thus near $P, \frac{dy}{dx}$ changes from negative, through zero to positive values.

Therefore, at $P \frac{d}{dx}\left(\frac{dy}{dx}\right) = \frac{d^2y}{dx^2}$ is positive.

Therefore,for a minimum point $\frac{dy}{dx} = 0$ and $\frac{d^2y}{dx^2} > 0$.

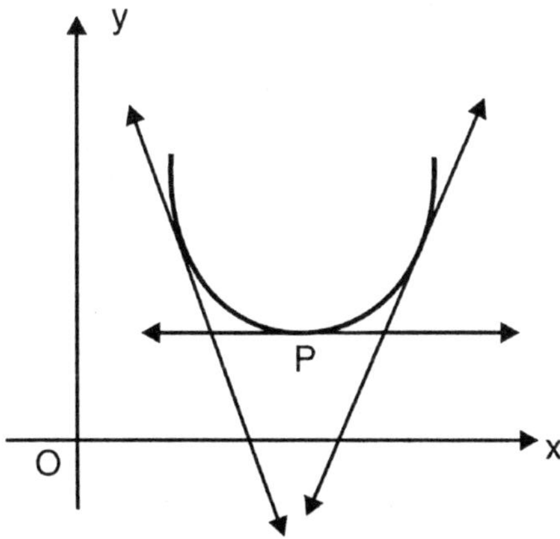

Technique

1. Find $f'(x)$ and $f''(x)$.
2. Find the roots (i.e. critical points) of $f'(x) = 0$.

   Let $x = a$, $x = b$, $x = c$ etc be these critical points of $f(x)$.
3. If $f''(a) < 0$, then $f(x)$ has a local maximum at $x = a$ and if $f''(a) > 0$, then $f(x)$ has a local minimum at $x = a$. But if $f''(a) = 0$. Then the test fails.
4. Use the same procedure for all critical points of $f(x)$ as it has been given in (3).

Drawbacks:

(I)

1. $f'(x) = 0$ and $f''(x) = 0$
2. $f'(x) = 0$ and $f''(x)$ DNE
3. $f'(x)$ DNE

   Under any of these conditions, the first derivative test would have to be used instead the second derivative test.

II. In case of some functions, the second derivatives are very difficult and tedious to calculate. Thus relent back to the first derivative test to find the Local extrema.

Stationary Points

1. The condition that $f'(x)$ i.e. $= 0$ at $x = c$, for $f(x)$ should be maximum or minimum is necessary but it is not sufficient.

   Consider $f(x) = x^3$

   Now $\dfrac{dy}{dx} = 3x^2$

   $\dfrac{dy}{dx} = 0$ at $x = 0$

   Also, $\dfrac{d^2y}{dx^2} = 6x$ then at $x = 0, \dfrac{d^2y}{dx^2} = 0$

   $y = x^3$

   Hence $\dfrac{d^2y}{dx^2}$ at $x = 0$ is neither negative nor positive

   hence at $x = c$ is neither a maximum nor a minimum.

2. The maxima and minima of $f(x)$ are also known as 'turning values' of f(x) since $\dfrac{dy}{dx}$ changes its sign from positive to negative or from negative to positive. They are also known as "extrema or extreme values."

3. The values of the function for $\frac{dy}{dx} = 0$ are known as *critical values* or *stationary values* since $f(x)$ is neither increasing nor decreasing at these values (points). It is a value or point at which the gradient $\frac{dy}{dx}$ is zero.

Note that a stationary point is not necessarily a turning point. It is clear from the above example given. At (0, 0) = 0 and the x – axis is the tangent there. However, (0, 0) $\frac{dy}{dx} = 0$ is not a turning point because $f(x)$ continues to increase when passing through it. The origin is stationary since the curve $y = f(x)$ is momentarily stationary there.

Concavity and Points of Inflection

The second derivative of a function may be also be used to determine the general shape of $y = f(x)$ on a selected interval.

A point of *inflection* on a curve is a point where the curve changes its direction of curvature. You will notice that at P the curve changes from being concave upward to

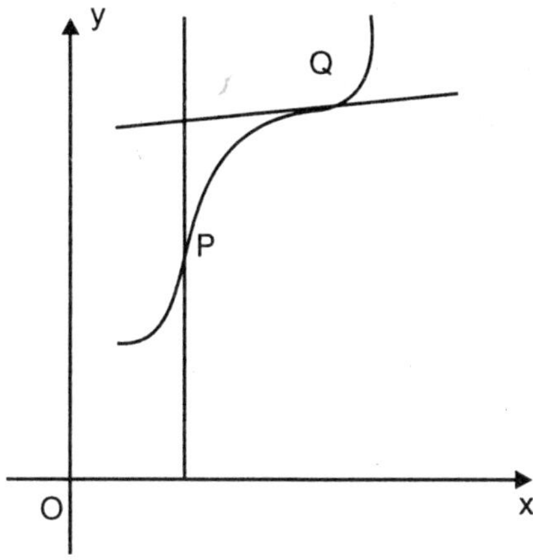

being concave downwards. At Q, it changes from concave downwards to being concave upwards. Since $f''(x)$ i.e. $\frac{d^2y}{dx^2}$ is the rate of change of the gradient $\left(\frac{dy}{dx}\right)$, a graph is concave upwards where $f''(x) > 0$ and concave downwards where $f''(x) < 0$.

Now at a point of inflection, the gradient stops increasing and begins decreasing, or it stops decreasing and begins increasing. Thus a point of inflection is one at which $\frac{dy}{dx}$ is either a maximum or a minimum. Thus the condition for a point of inflection is that $\frac{d^2y}{dx^2} = 0$ at the point and it changes sign in passing through it i.e. $\frac{d^2y}{dx^2} = 0$ and $\frac{d^3y}{dx^3} \neq 0$. Note that, the tangent to a curve at a point of inflection crosses the curve, or more accurately, it cuts the curve at three coincident points.

Rate Measure (Distance, Velocity and Acceleration)

If $y = s\ (t)$ represents the position function, then $y = s'\ (t)$ represents the instantaneous $\lim_{\Delta t \to 0} \frac{\Delta s}{\Delta t} = \frac{ds}{dt} = v$ velocity i.e. and $a = v'(t) = s''$ (t) represents the instantaneous acceleration at time t i.e. $a = \lim_{\Delta t \to 0} \frac{\Delta v}{\Delta t} = \frac{dv}{dt} or \frac{d^2 s}{dt^2}$

Note that $a = \frac{dv}{dt} = \frac{dv}{ds} \cdot \frac{ds}{dt} = v \cdot \frac{dv}{ds}$

1. If $v > 0$ (i.e. positive velocity) indicates that the position is increasing with increasing time while $v < 0$ (i.e. negative velocity) indicates the decreasing position as time decreases.
2. If $v = 0$ implies that distance 's' remains constant on the given interval of time.
3. If $a > 0$ implies that velocity is increasing with respect to time and $a < 0$ implies that the velocity is decreasing with respect to time.
4. If v = constant on an interval of time, then $a = 0$ on that interval.

Related Rates

We have seen that $\frac{ds}{dt}$ is the derivative of $s\ (t)$ w. r. to t and can be interpreted as the velocity. This physical interpretation can be extended further. If (x, y) are the co-ordinates of a moving particle then both x and y are functions of 't'.

$\frac{dx}{dt}$, then, is the rate of change of x w. r. to 't' and $\frac{dy}{dt}$ is the rate of change of y w. r. to 't' $\frac{dx}{dt}$ and $\frac{dy}{dt}$ since both x and y are related by a function $y = f(x)$, $\frac{dx}{dt}$ and $\frac{dy}{dt}$ are also related. If one of is known the other can be found using the relation between these rates.

Problem of this type are known as problem of related rates.

Technique

$y = f(x)$ where both x and y are dependent on time 't'

Hence

$$\frac{dy}{dt} = \frac{d}{dx} f(x) \cdot \frac{dx}{dt}$$

i.e.

$$\frac{dy}{dx} = f'(x) \frac{dx}{dt}$$

If one of $\frac{dx}{dt}$ and $\frac{dy}{dt}$ is known the other can be found out using the above relation.

Differentials: Approximation and Errors

Recall the definition of the derivative of a function $y = f(x)$

$$f'(x) = \lim_{\Delta x \to 0} \frac{f(x+\Delta x) - f(x)}{\Delta x}$$

This represents the slope of tangent to the curve $y = f(x)$ at some point $(x, f(x))$

Now if Δx is very small ($\Delta x \neq 0$), then the slope of tangent is approximately the same as the slope of the secant through $(x, f(x))$

$$\text{i.e. } f'(x) \approx \frac{f(x+\Delta x) - f(x)}{\Delta x}$$

$$\therefore \qquad f'(x)\Delta x \approx f(x+\Delta x) - f(x)$$

Now the differential of the independent variable x is written as dx and it is the same as Δx

$$\therefore \qquad dx = \Delta x,\ \Delta x \neq 0$$

Hence, $f'(x).\ dx \approx f(x + \Delta x) - f(x)$... (1)

Similarly $dy = \Delta y$ for the variable y, $\Delta y \neq 0$.

we also have that $\Delta y = f(x + \Delta x) - f(x)$

$$\therefore \qquad dy = f(x + Dx) - f(x)...(2)$$

Therefore from relations (1) and (2) we have $dy = f'(x)\ dx = \Delta y$

Thus, when the change in x ($\Delta x = dx$) is relatively very small, then the differential of y (dy) is approximately equal to the exact change in y (Δy) i.e. The smaller the change in x, the closer dy to Δy. This fact enables us to approximate the functional value close to $f(x)$

An alternate definition for approximation is as

If a and $a+h$ belongs to the domain of a differentiable function 'f', then the approximate value of $f(a+h)$ will be $f(a+h) \approx f(a) + hf'(a)$, $f'(a) \neq 0$. By using this, approximate problems can be solved easily.

DIFFERENTIATION OF FUNCTIONS OF SEVERAL VARIABLES

We shall mainly be concerned with differentiation and integration of functions of more than one variable. We describe

1. *How* each process can be done;
2. *Why* it is interesting, in terms of applications; and

3. How to interpret the process geometrically.

Functions of Several Variables

Last year, you did a significant amount of work studying functions, typically written as $y = f(x)$, which represented the variation that occurred in some (dependent) variable y, as another (independent) variable, x, changed. For example you might have been interested in the height y after a given time x, or the area y, enclosed by a rectangle with sides x and 10 - x. Once the function was known, the usual rules of calculus could be applied, and results such as the time when the particle hits the ground, or the maximum possible area of the rectangle, could be calculated. In the earlier part of the course, we have extended this work by taking a more rigorous approach to a lot of the same ideas.

We are going to do the same thing now for functions of several variables. For example the height y of a particle may depend on the position x and the time t, so we have $y = f(x, t)$; the volume V of a cylinder depends on the radius r of the base and its height h, and indeed, as you know, $V = r^2h$; or the pressure of a gas may depend on its volume V and temperature T, so $P = P(V, T)$. Note the trick I have just used; it is often convenient to use P both for the (defendant) variable, and for the function itself: we don't always need separate symbols as in the $y = f(x)$ example.

When studying the real world, it is unusual to have functions which depend solely on a single variable. Of course the single variable situation is a little simpler to study, which is why we started with it last year. And just as last year, we shall usually have a "standard" function name; instead of $y = f(x)$, we often work with $z = f(x, y)$, since most of the extra complications occur when we have two, rather than one (independent) variable, and we don't need to consider more general cases like $w = f(x, y, z)$, or even $y = f(x_1, x_2,..., x_n)$.

Graphing functions of Several Variables

One way we tried to understand the function $y = f(x)$ was by drawing its graph. We then used such a graph to pick out points such as the local minimum at $x = 3/2$, and to see how we could get the same result using calculus.

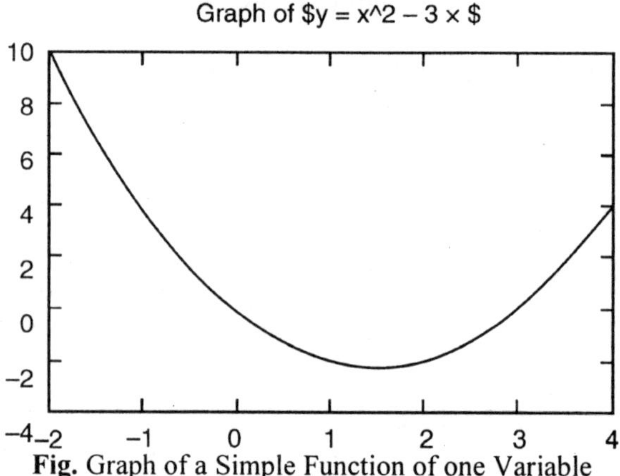

Fig. Graph of a Simple Function of one Variable

Working with two or more independent variables is more complicated, but the ideas are familiar. To plot $z = f(x, y)$ we think of z as the height of the function f at the point (x, y), and then try to sketch the resulting surface in three dimensions. So we represent a function as a surface rather than a curve.

Example Sketch the surface given by $z = 2 - x/2 - 2y/3$

Solution: We know the surface will be a plane, because z is a linear function of x and y. Thus it is enough to plot three points that the plane passes through.

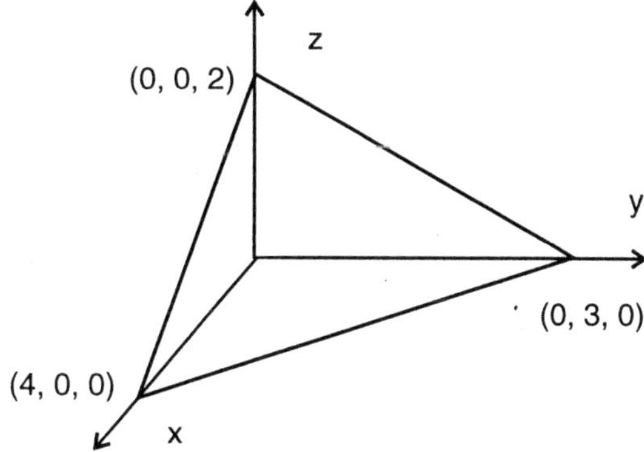

Fig. Sketching a Function of two Variables

Of course it is easy to sketch something as simple as a plane. There are graphical difficulties when dealing with more complicated functions, which make sketching and visualisation rather harder than for functions of one variable. And if there are three or more independent variables, there is really no good way of visualising the behaviour of the function directly. But for just two independent variables, there are some tricks.

Example: Sketch the surface given by $z = x^2 - y^2$.

Solution: We can represent the surface directly by drawing it as shown.

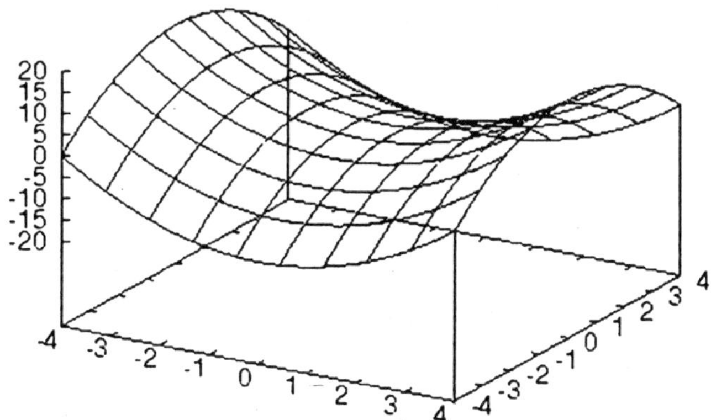

Fig. Surface Plot of $z = x^2 - y^2$.

Such a representation is easy to create using suitable software and Fig. shows the resulting surface. We now describe how to looking at similar examples without such a program. One approach is to draw a contour map of the surface, and then use the usual tricks to visualise the seurface.

For the surface $z = x^2 - y^2$, the points where $z = 0$ lie on $x^2 = y^2$, so form the lines $y = x$ and $y = - x$. We can continue in this way, and look at the points where $z = 1$; so $x^2 - y^2 = 1$. This is one of the hyperbolae indeed, fixing z at different values shows the contours (lines of constant height or z value) are all the same shape, but with different constants. We this get the alternative representation as a contour map.

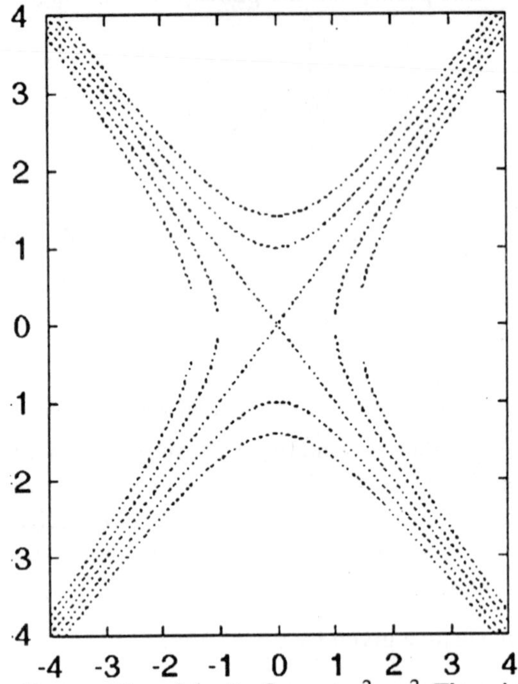

Fig. Contour Plot of the Surface $z = x^2 - y^2$. The missing points near the x-axis are an Artifact of the Plotting Program.

A final way to confirm that you have the right view of the surface is to section it in different planes.

So far we have looked at the intersection of the planes $z = k$ with the surface $z = x^2 - y^2$ for different values of the constant k. If instead we fix x, at the value a, then $z = a^2 - y^2$. Each of these curves is a parabola with its vertex upwards, at the point $y = 0$, $z = a^2$.

Continuity

As you might expect, we say that a function f of two variables is continuous at (x_0, y_0) if

$$\lim_{x \to x_0 \to \infty} f(x, y) \quad = f(x_0, y_0).$$

The only complication comes when we realise that there are many different ways if which $x \to x_0$ and $y \to y_0$. We illustrate with a simple example.

Example: Investigate the continuity of $f(x, y) = \dfrac{2xy}{x^2 + y^2}$ at the point (0, 0).

Solution: Consider first the case when $x \to 0$ along the x-axis, so that throughout the process, $y = 0$. We have

$$f(x, 0) = \frac{2x \cdot 0}{x^2 + y^2} = 0 \to 0 \quad \text{as } x \to 0.$$

Next consider the case when $x \to 0$ and $y \to 0$ on the line $y = x$, so we are looking at the special case when $x = y$. We have

$$f(x, x) = \frac{2x^2}{x^2 + x^2} = 1 \to 1 \quad \text{as } x \to 0.$$

Of course f is only continuous if it has the same limit however $x \to 0$ and $y \to 0$, and we have now seen that it doesn't; so f is not continuous at (0, 0).

Although we won't go into it, the usual "putting together" theorems show that f is continuous everywhere else.

Partial Differentiation

The usual rules for differentiation apply when dealing with several variables, but we now require to treat the variables one at a time, keeping the others constant. It is for this reason that a new symbol for differentiation is introduced. Consider the function

$$f(x, y) = \frac{2y}{y + \cos x}$$

We can consider y fixed, and so treat it as a constant, to get a partial derivative

$$\frac{\partial f}{\partial x} = \frac{2y \sin x}{(2y + \cos x)^2}$$

where we have differentiated with respect to x as usual. Or we can treat x as a constant, and differentiate with respect to y, to get

$$\frac{\partial f}{\partial y} = \frac{(2y \cos x) \cdot 2 - 2y \cdot 2}{(2y + \cos x)^2} = \frac{2 \cos x}{(2y + \cos x)^2}$$

Although a partial derivative is itself a function of several variables, we often want to evaluate it at some fixed point, such as (x_0, y_0). We thus often write the partial derivative as

$$\frac{\partial f}{\partial y}(x_0, y_0).$$

There are a number of different notations in use to try to help understanding in different situations. All of the following mean the same thing:

$$\frac{\partial f}{\partial y}(x_0, y_0), \qquad f_1(x_0, y_0), \qquad f_x(x_0, y_0) \text{ and } D_1F(x_0, y_0).$$

Note also that there is a simple definition of the derivative in terms of a Newton quotient:-

$$\frac{\partial f}{\partial y} = \lim_{\delta x \to \infty} \frac{f(x_0 + \delta x, y_0) - f(x_0, y_0)}{\delta x}$$

provided of course that the limit exists.

Example Let $z = \sin(x/y)$. Compute $x\frac{\partial z}{\partial x} + y\frac{\partial z}{\partial y}$.

Solution: Treating first y and then x as constants, we have

$$\frac{\partial z}{\partial x} = \frac{1}{y} \cos\left(\frac{x}{y}\right) \text{ and } \frac{\partial z}{\partial y} = \frac{-x}{y^2} \cos\left(\frac{x}{y}\right).$$

thus

$$x\frac{\partial z}{\partial x} + y\frac{\partial z}{\partial y} = \frac{x}{y} \cos\left(\frac{x}{y}\right) - \frac{x}{y} \cos\left(\frac{x}{y}\right) = 0.$$

Theorem Assume that f and all its partial derivatives f_x and f_y are continuous, and that $x = x(t)$ and $y = y(t)$ are themselves differentiable functions of t. Let

$$F(t) = f(x(t), y(t)).$$

Then F is differentiable and

$$\frac{dF}{dt} = \frac{\partial f}{\partial x}\frac{dx}{dt} + \frac{\partial f}{\partial y}\frac{dy}{dt}.$$

Proof: Write $x = x(t)$, $x_0 = x(t_0)$ etc. Then we calculate the Newton quotient for F.

$$\begin{aligned} F(t) - F(t_0) &= f(x, y) - f(x_0, y_0) \\ &= f(x, y) - f(x_0, y) + f(x_0, y) - f(x_0, y_0) \\ &= \frac{\partial f}{\partial x}(\xi, y)(x - x_0) + \frac{\partial f}{\partial x}(x_0, \eta)(y - y_0) \end{aligned}$$

Here we have used the Mean Value Theorem to write

$$f(x, y) - f(x_0, y) = \frac{\partial f}{\partial x}(\xi, y)(x - x_0)$$

for some point ξ between x and x_0, and have argued similarly for the other part. Note that ξ, pronounced "Xi" is the Greek letter "x"; in the same way η, pronounced "Eta" is the Greek letter "y". Thus

$$\frac{F(t) - F(t_0)}{t - t_0} = \frac{\partial f}{\partial x}(\xi, y)\frac{x - x_0}{t - t_0} + \frac{\partial f}{\partial y}(x_0, \frac{y - y_0}{t - t_0})$$

Now let tt_0, and note that in this case, xx_0 and yy_0; and since ξ and η are trapped between x and x_0, and y and y_0 respectively, then also $\xi \to x_0$ and $\eta \to y_0$. The result then follows from the continuity of the partial derivatives.

Example: Let $f(x, y) = xy$, and let $x = \cos t$, $y = \sin t$.Compute $\frac{\partial f}{\partial t}$ when $t = \pi/2$.

Solution: From the chain rule,

$$\frac{\partial f}{\partial t} = \frac{\partial f}{\partial x}\frac{dx}{dt} + \frac{\partial f}{\partial y}\frac{dy}{dt} = -y(t)\sin t + x(t)\cos t = -1.\sin(p/2) = -1.$$

The chain rule easily extends to the several variable case; only the notation is complicated. We state a typical example

***Proposition*:** Let $x = x(u, v)$, $y = y(u, v)$ and $z = z(u, v)$, and let f be a function defined on a subset $U \in R^3$, and suppose that all the partial derivatives of f are continuous. Write

$$F(u, v) = f(x(u, v), y(u, v), z(u, v)).$$

Then

$$\frac{\partial F}{\partial u} = \frac{\partial f}{\partial x}\frac{dx}{du} + \frac{\partial f}{\partial y}\frac{dy}{du} + \frac{\partial f}{\partial z}\frac{\partial z}{\partial u}$$

and

$$\frac{\partial F}{\partial v} = \frac{\partial f}{\partial x}\frac{dx}{dv} + \frac{\partial f}{\partial y}\frac{dy}{dv} + \frac{\partial f}{\partial z}\frac{\partial z}{\partial v}$$

The introduction of the domain of f above, simply to show it was a function of three variables is clumsy. We often do it more quickly by saying

Let $f(x, y, z)$ have continuous partial derivatives

This has the advantage that you are reminded of the names of the variables on which f acts, although strictly speaking, these names are not bound to the corresponding places. This is an example where we adopt the notation which is very common in engineering maths. But note the confusion if you ever want to talk about the value $f(y, z, x)$, perhaps to define a new function $g(x, y, z)$.

Example: Assume that $f(u, v, w)$ has continuous partial derivatives, and that

$$u = x - y,\ v = y - z\ w = z - x$$

Let

$$F(x, y, z) = f(u(x, y, z), v(x, y, z), w(x, y, z))$$

Show that

$$\frac{\partial F}{\partial x} + \frac{\partial F}{\partial x} + \frac{\partial F}{\partial z} = 0$$

Solution: We apply the chain rule, noting first that from the change of variable formulae, we have

$$\frac{\partial u}{\partial x} = 1,$$

$$\frac{\partial u}{\partial y} = -1,$$

$$\frac{\partial u}{\partial z} = 0,$$

$$\frac{\partial v}{\partial x} = 0, \frac{\partial w}{\partial x} = -1,$$

$$\frac{\partial v}{\partial y} = 1, \frac{\partial w}{\partial y} = 0,$$

$$\frac{\partial v}{\partial z} = -1, \frac{\partial w}{\partial z} = 1,$$

Then

$$\frac{\partial F}{\partial x} = \frac{\partial f}{\partial u}.1 + \frac{\partial f}{\partial v}.0 + \frac{\partial f}{\partial w}.-1 = \frac{\partial f}{\partial u} - \frac{\partial f}{\partial w}$$

$$\frac{\partial F}{\partial y} = \frac{\partial f}{\partial u}.-1 + \frac{\partial f}{\partial v}.1 + \frac{\partial f}{\partial w}.0 = -\frac{\partial f}{\partial u} + \frac{\partial f}{\partial v}$$

$$\frac{\partial F}{\partial z} = \frac{\partial f}{\partial u}.0 + \frac{\partial f}{\partial v}.-1 + \frac{\partial f}{\partial w}.1 = -\frac{\partial f}{\partial v} + \frac{\partial f}{\partial w}$$

Adding then gives the result claimed.

Higher Derivatives

Note that a partial derivative is itself a function of two variables, and so further partial derivatives can be calculated. We write

$$\frac{\partial}{\partial x}\frac{\partial f}{\partial x} = \frac{\partial^2 f}{\partial x^2}, \frac{\partial}{\partial x}\frac{\partial f}{\partial y} = \frac{\partial^2 f}{\partial x \partial y}, \frac{\partial}{\partial y}\frac{\partial f}{\partial x} = \frac{\partial^2 f}{\partial y \partial x}, \frac{\partial}{\partial y}\frac{\partial f}{\partial y} = \frac{\partial^2 f}{\partial y^2}.$$

This notation generalises to more than two variables, and to more than two derivatives in the way you would expect.

There is a complication that does not occur when dealing with functions of a single variable; there are *four* derivatives of second order, as follows:

$$\frac{\partial^2 f}{\partial x^2}, \frac{\partial^2 f}{\partial x \partial y}, \frac{\partial^2 f}{\partial y \partial x} \text{ and } \frac{\partial^2 f}{\partial y^2}$$

Fortunately, when f has mild restrictions, the order in which the differentiation is done doesn't matter.

***Proposition*:** Assume that all second order derivatives of f exist and are continuous. Then the mixed second order partial derivatives of f are equal. i.e.

$$\frac{\partial^2 f}{\partial x \partial y} = \frac{\partial^2 f}{\partial y \partial x}.$$

Example: Suppose that $f(x, y)$ is written in terms of u and v where $x = u + \log v$ and $y = u - \log v$. Show that, with the usual convention,

$$\frac{\partial^2 f}{\partial u^2} = \frac{\partial^2 f}{\partial x^2} + 2\frac{\partial^2 f}{\partial x \partial y} + \frac{\partial^2 f}{\partial y^2}$$

and

$$v^2 \frac{\partial^2 f}{\partial v^2} = \frac{\partial f}{\partial y} - \frac{\partial f}{\partial x} + \frac{\partial^2 f}{\partial x^2} + 2\frac{\partial^2 f}{\partial x \partial y} + \frac{\partial^2 f}{\partial y^2}$$

You may assume that all second order derivatives of f exist and are continuous.

Solution: Using the chain rule, we have

$$\frac{\partial f}{\partial u} = \frac{\partial f}{\partial x}\frac{\partial x}{\partial u} + \frac{\partial f}{\partial y}\frac{\partial y}{\partial u} = \frac{\partial f}{\partial x} + \frac{\partial f}{\partial y}$$

and

$$\frac{\partial f}{\partial v} = \frac{\partial f}{\partial x}\frac{\partial x}{\partial v} + \frac{\partial f}{\partial y}\frac{\partial y}{\partial v} = \frac{1}{v}\frac{\partial f}{\partial x} - \frac{1}{v}\frac{\partial f}{\partial y}$$

Thus using both these and their operator form, we have

$$\frac{\partial^2 f}{\partial u^2} = \frac{\partial}{\partial u}\left(\frac{\partial f}{\partial x} + \frac{\partial f}{\partial y}\right) = \frac{\partial}{\partial x} + \frac{\partial}{\partial y}\left(\frac{\partial f}{\partial x} + \frac{\partial f}{\partial y}\right)$$

$$= \frac{\partial^2 f}{\partial x^2} + \frac{\partial^2 f}{\partial x \partial y} + \frac{\partial^2 f}{\partial y \partial y} + \frac{\partial^2 f}{\partial y^2},$$

While differentiating with respect to $v, *we have*

$$\frac{\partial^2 f}{\partial v^2} = \frac{\partial}{\partial v}\left(\frac{1}{v}\frac{\partial f}{\partial x} - \frac{1}{v}\frac{\partial f}{\partial y}\right)$$

$$= \frac{1}{v^2}\frac{\partial f}{\partial x} + \frac{1}{v}\frac{\partial}{\partial v}\left(\frac{\partial f}{\partial x}\right) + \frac{1}{v^2}\frac{\partial f}{\partial y} - \frac{1}{v}\frac{\partial}{\partial v}\left(\frac{\partial f}{\partial y}\right)$$

$$= -\frac{1}{v^2}\frac{\partial f}{\partial x} + \frac{1}{v}\left(\frac{1}{v}\frac{\partial^2 f}{\partial x^2} - \frac{1}{v}\frac{\partial^2 f}{\partial x \partial y}\right) + \frac{1}{v^2}\frac{\partial f}{\partial y} - \frac{1}{v}\left(\frac{1}{v}\frac{\partial^2 f}{\partial x \partial y} - \frac{1}{v}\frac{\partial^2 f}{\partial y^2}\right)$$

$$= \frac{1}{v^2}\left(\frac{\partial f}{\partial y} - \frac{\partial f}{\partial x} + \frac{\partial^2 f}{\partial x^2} - 2\frac{\partial^2 f}{\partial x \partial y} + \frac{\partial^2 f}{\partial y^2}\right)$$

SOLVING EQUATIONS BY SUBSTITUTION

One of the main interests in partial differentiation is because it enables us to write down how we expect the natural world to behave.

We move away from 1-variable results as soon as we have properties which depend on e.g. at least one space variable, together with time. We illustrate with just one example, designed to whet the appetite for the whole subject of mathematical physics.

Assume the displacement of a length of string at time t from its rest position is described by the function $f(x, t)$. The laws of physics describe how the string behaves when released in this position, and allowed to move under the elastic forces of the string; the function f satisfies the *wave equation*

$$\frac{\partial^2 f}{\partial t^2} = c^2 \frac{\partial^2 f}{\partial x^2}.$$

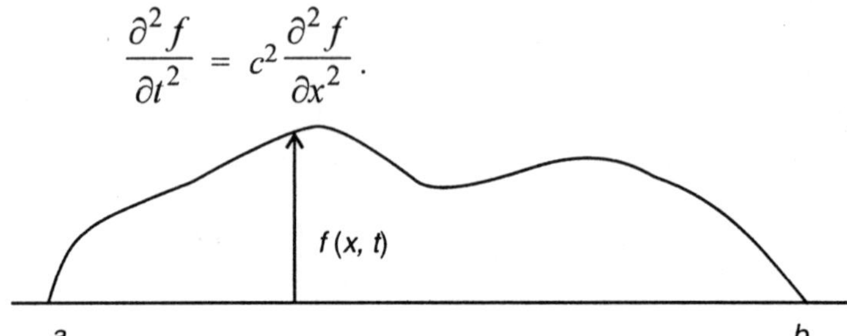

Fig. A String Displaced from the Equilibrium Position

Example: Solve the equation $\frac{\partial^2 F}{\partial u \partial v} = 0.$

Solution: Such a function is easy to integrate, because the two variables appear independently. So $\frac{\partial F}{\partial v} = g_1(v)$, where g_1 is an arbitrary (differentiable) function. since when differentiated with respect to u we are given that $\frac{\partial^2 F}{\partial u \partial v} = 0$. Thus we can integrate with respect to v to get

$$F(u, v) = \int g_1(v)dv + h(u) = g(v) + h(u),$$

where h is also an arbitrary (differentiable) function.

Example: Rewrite the wave equation using co-ordinates $u = x - ct$ and $v = x + ct$.

Solution: Write $f(x, t) = F(u, v)$ and now in principle confuse F with f, so we can tell them apart only by the names of their arguments. In practice we use different symbols to help the learning process; but note that in a practical case, all the F's that appear below, would normally be written as f's By the chain rule

$$\frac{\partial}{\partial x} = \left(\frac{\partial}{\partial u}\right).1 + \left(\frac{\partial}{\partial v}\right).1$$

and

$$\frac{\partial}{\partial t} = \left(\frac{\partial}{\partial u}\right).(-c) + \left(\frac{\partial}{\partial v}\right).c.$$

differentiating again, and using the operator form of the chain rule as well,

$$\frac{\partial^2 f}{\partial t^2} = \left(c\frac{\partial}{\partial v} - c\frac{\partial}{\partial u}\right)\left(c\frac{\partial F}{\partial v} - c\frac{\partial F}{\partial u}\right)$$

$$= c^2\frac{\partial^2 F}{\partial v^2} - c^2\frac{\partial^2 F}{\partial u \partial v} - c^2\frac{\partial^2 F}{\partial u \partial u} + c^2\frac{\partial^2 F}{\partial u^2}$$

$$= c^2\left(\frac{\partial^2 F}{\partial v^2} + \frac{\partial^2 F}{\partial u^2}\right) - 2c^2\frac{\partial^2 F}{\partial u \partial v}$$

and similarly

$$\frac{\partial^2 f}{\partial x^2} = \left(\frac{\partial^2 F}{\partial v^2} + \frac{\partial^2 F}{\partial u^2}\right) + 2\frac{\partial^2 F}{\partial u \partial v}$$

Substituting in the wave equation, we thus get

$$4c^2\frac{\partial^2 F}{\partial u \partial v} = 0,$$

an equation which we have already solved. Thus solutions to the wave equation are of the form $f(u) + g(v)$ for any (suitably differentiable) functions f and g. For example we may have $\sin(x - ct)$. Note that this is not just any function; it must be constant when $x = ct$.

Exercise: Let $F(x, t) = \log(2x + 2ct)$ for $x > -ct$, where c is a fixed constant. Show that

$$\frac{\partial^2 F}{\partial t^2} - c^2\frac{\partial^2 F}{\partial t^2} = 0.$$

Tangent Planes

Consider the surface $F(x, y, z) = c$, perhaps as $z = f(x, y)$, and suppose that f and F have continuous partial derivatives. Suppose now we have a smooth curve on the surface, say $\phi(t) = (x(t), y(t), z(t))$. Then since the curve lies in the surface, we have

$$F(x(t), y(t), z(t)) = c,$$

and so, applying the chain rule, we have

$$\frac{\partial F}{dt} = \frac{\partial F}{dx}\frac{dx}{dt} + \frac{\partial F}{dy}\frac{dy}{dt} + \frac{\partial F}{dz}\frac{dz}{dt} = 0$$

or, writing this in terms of vectors, we have

$$\nabla F.\mathrm{v}(t) = \left(\frac{\partial F}{\partial x}, \frac{\partial F}{\partial z}, \frac{\partial F}{\partial z}\right).\left(\frac{dx}{dt}, \frac{dy}{dt}, \frac{dz}{dt}\right) = 0$$

Since the RH vector is the velocity of a point on the curve, which lies on the surface, we see that the left hand vector must be the normal to the curve.

Note that we have defined the gradient vector ∇F associated with the function F by

$$\nabla F = \left(\frac{\partial F}{\partial x}, \frac{\partial F}{\partial z}, \frac{\partial F}{\partial z}\right)$$

Theorem: The tangent to the surface $F(x, y, z) = c$ at the point (x_0, y_0, z_0) is given by

$$\frac{dF}{\partial x}(x - x_0) + \frac{dF}{\partial y}(y - y_0) + \frac{dF}{\partial z}(z - z_0) = 0$$

***Proof*:** This is a simple example of the use of vector geometry. Given that (x_0, y_0, z_0) lies on the surface, and so in the tangent, then for any other point (x, y, z) in the tangent plane, the vector $(x - x_0, y - y_0, z - z_0)$ must lie in the tangent plane, and so must be normal to the normal to the curve (i.e. to ∇F). Thus $(x - x_0, y - y_0, z - z_0)$ and ∇F are perpendicular, and that requirement is the equation which gives the tangent plane. □

***Example*:** Find the equation of the tangent plane to the surface $F(x, y, z) = x^2 + y^2 + z - 9 = 0$ at the point $P = (1, 2, 4)$.

Solution: We have $\nabla F|_{(1, 2, 4)} = (2, 4, 1)$, and the equation of the tangent plane is

$2(x - 1) + 4(y - 2) + (z - 4) = 0.$

Linearisation and Differentials

We obtained a geometrical view of the function $f(x, y)$ by considering the surface $z = f(x, y)$, or $F(x, y, z) = z - f(x, y) = 0$. Note that the tangent plane to this surface at the point $(x_0, y_0, f(x_0, y_0))$ lies close to the surface itself. Just as in one variable, we used the tangent line to approximate the graph of a function, so we shall use the tangent plane to approximate the surface defined by a function of two variables.

The equation of our tangent plane is

$$\frac{\partial F}{\partial x}(x - x_0) + \frac{\partial F}{\partial y}(y - y_0) + \frac{\partial F}{\partial z}(z - f(x_0, y_0)) = 0,$$

and writing the derivatives in terms of f, we have

$$\frac{\partial f}{\partial x}(x_0, y_0)(x - x_0) + \frac{\partial f}{\partial y}(x_0, y_0)(y - y_0) + (-1)(z - f(x_0, y_0)) = 0,$$

or, writing in terms of z, the height of the tangent plane above the ground plane,

$$z = f(x_0, y_0) + \frac{\partial f}{\partial x}(x_0, y_0)(x - x_0) + \frac{\partial f}{\partial y}(x_0, y_0)(y - y_0).$$

Our assumption that the tangent plane lies close to the surface is that $z\,f(x, y)$, or that

$$f(x, y) \approx f(x_0, y_0) + \frac{\partial f}{\partial x}(x_0, y_0)(x - x_0) + \frac{\partial f}{\partial y}(x_0, y_0)(y - y_0).$$

We call the right hand side the linear approximation to f at (x_0, y_0).

We can rewrite this with $h = x - x_0$ and $k = y - y_0$, to get

$$f(x, y) - f(x_0, y_0) \approx h\left.\frac{\partial f}{\partial x}\right|(x_0, y_0) + k\left.\frac{\partial f}{\partial y}\right|(x_0, y_0),$$

$$\text{or } df \approx h\left.\frac{\partial f}{\partial x}\right|_{(x_0, y_0)} + k\left.\frac{\partial f}{\partial y}\right|_{(x_0, y_0)}$$

This has applications; we can use it to see how a function changes when its independent variables are subjected to small changes.

Example: A cylindrical oil tank is 25 m high and has a radius of 5 m. How sensitive is the volume of the tank to small variations in the radius and height.

Solution: Let V be the volume of the cylindrical tank of height h and radius r. Then $V = r^2H$, and so

$$dV = \left.\frac{\partial f}{\partial r}\right|_{(x_0, h_0)} dr + \left.\frac{\partial f}{\partial h}\right|_{(x_0, h_0)} dh,$$

$$= 250\ \pi dr + 25\ \pi dh.$$

Thus the volume is 10 times as sensitive to errors in measuring r as it is to measuring h.

Example: A cone is measured. The radius has a measurement error of 3%, and the height an error of 2%. What is the error in measuring the volume?

Solution: The volume V of a cone is given by $V = \pi r^2 h/3$, where r is the radius of the cone, and h is the height. Thus

$$dV = \frac{2}{3}\pi\, rh\, dr + \frac{1}{3}\pi\, r^2\, dh$$

$$= 2.(0.03).\frac{\pi r^2 h}{3} + \frac{\pi r^2 h}{3}.(0.02) = V(0.06 + 0.02)$$

Thus there is an 8% error in measuring the volume.

Exercise: The volume of a cylindrical oil tank is to be calculated from measured values of r and h. What is the percentage error in the volume, if r is measured with an accuracy of 2%, and h measured with an accuracy of 0.5%.

Implicit Functions of Three Variables

Finally in this section we discuss another application of the chain rule. Assume we have variables x, y and z, related by the equation $F(x, y, z) = 0$. Then assuming that $F_z \neq 0$, there is a version of the implicit function theorem which means we can *in principle*, write $z = z(x, y)$, so we can "solve for z". Doing this gives

$$F(x, y, z(x, y)) = 0.$$

Now differentiate both sides partially with respect to x. We get

$$0 = \frac{\partial}{\partial x} F(x, y, z(x, y)) = \frac{\partial F}{\partial x}\frac{\partial x}{\partial x} + \frac{\partial F}{\partial y}\frac{\partial y}{\partial x} + \frac{\partial F}{\partial z}\frac{\partial z}{\partial x},$$

and so since $\frac{\partial}{\partial x} \neq 0$,

$$0 = \frac{\partial F}{\partial x} + \frac{\partial F}{\partial z}\frac{\partial z}{\partial x} \quad \text{and} \quad \frac{\partial z}{\partial x} = -\frac{\frac{\partial F}{\partial x}}{\frac{\partial F}{\partial z}}$$

Now assume in the same way that $F_x \neq 0$ and $F_y \neq 0$, so we can get two more relations like this, with $x = x(y, z)$ and $y = y(z, x)$. Then form three such equations,

$$\frac{\partial z}{\partial x}\cdot\frac{\partial x}{\partial y}\cdot\frac{\partial y}{\partial z} = -\left(\frac{\frac{\partial F}{\partial x}}{\frac{\partial F}{\partial z}}\right)\left(\frac{\frac{\partial F}{\partial y}}{\frac{\partial F}{\partial x}}\right)\left(\frac{\frac{\partial F}{\partial z}}{\frac{\partial F}{\partial y}}\right) = -1!$$

Maxima and Minima

As in one variable calculations, one use for derivatives in several variables is in calculating maxima and minima. Again as for one variable, we shall rely on the theorem that if f is continuous on a closed bounded subset of R^2, then it has a global maximum and a global minimum. And again as before, we note that these must occur *either* at a local maximum or minimum, or else on the boundary of the region. Of course in R, the boundary of the region usually consisted of a pair of end points, while in R^2, the situation is more complicated. However, the principle remains the same. And we can test for local maxima and minima in the same way as for one variable.

Definition: Say that $f(x, y)$ has a critical point at (a, b) if and only if

$$\frac{\partial f}{\partial x}(a, b) = \frac{\partial f}{\partial y}(a, b) = 0.$$

It is clear by comparison with the single variable result, that a necessary condition that f have a local extremum at (a, b) is that it have a critical point there, although that is not a sufficient condition. We refer to this as the first derivative test.

We can get more information by looking at the second derivative. Recall that we gave a number of different notations for partial derivatives, and in what follows we use f_x rather than the more cumbersome $\frac{\partial f}{\partial x}$ etc. This idea extends to higher derivatives; we shall use

f_{xx} instead of $\frac{\partial^2 f}{\partial x^2}$, and f_{xy} instead of $\frac{\partial^2 f}{\partial x \partial y}$ etc.

Theorem: (Second Derivative Test) Assume that (a, b) is a critical point for f. Then

1. If, at (a, b), we have $f_{xx} < 0$ and $f_{xx}f_{yy} - f^2_{xy} > 0$, then f has a local maximum at (a, b).
2. If, at (a, b), we have $f_{xx} > 0$ and $f_{xx}f_{yy} - f^2_{xy} > 0$, then f has a local minimum at (a, b).
3. If, at (a, b), we have $f_{xx}f_{yy} - f^2_{xy} < 0$, then f has a saddle point at (a, b).

The test is inconclusive at (a, b) if $f_{xx}f_{yy} - f^2_{xy} = 0$, and the investigation has to be continued some other way.

Note that the discriminant is easily remembered as

$$\Delta = \begin{vmatrix} f_{xx} & f_{xy} \\ f_{yx} & f_{yy} \end{vmatrix} = f_{xx}f_{yy} - f^2_{xy}$$

A number of very simple examples can help to remember this. After all, the result of the test should work on things where we can do the calculation anyway!

Example: Show that $f(x, y) = x^2 + y^2$ has a minimum at $(0, 0)$.

Of course we know it has a global minimum there, but here goes with the test:

Solution: We have $f_x = 2x$; $f_y = 2y$, so $f_x = f_y$ precisely when $x = y = 0$, and this is the only critical point. We have $f_{xx} = f_{yy} = 2$; $f_{xy} = 0$, so $\Delta = f_{xx}f_{yy} - f^2_{xy} = 4 > 0$ and there is a local minimum at $(0, 0)$.

Exercise: Let $f(x, y) = xy$. Show there is a unique critical point, which is a saddle point

Proof: We give an indication of how the theorem can be derived — or if necessary how it can be remembered. We start with the two dimensional version of Taylor's theorem, see section 5.6. We have

$$f(a+h, b+k) \sim f(a, b) + h\,\frac{\partial f}{\partial x}\,(a, b) + k\frac{\partial f}{\partial y}\,(a, b)$$

$$\frac{1}{2}\left(h^2\frac{\partial^2 f}{\partial x^2} + 2kh\frac{\partial^2 f}{\partial x \partial y} + k^2\frac{\partial^2 f}{\partial y^2}\right)$$

where we have actually taken an expansion to second order and assumed the corresponding remainder is small.

We are looking at a critical point, so for any pair (h, k), we have $h\frac{\partial f}{\partial x}(a, b) + k\frac{\partial f}{\partial y}(a, b) = 0$ and everything hinges on the behaviour of the second order terms. It is thus enough to study the behaviour of the quadratic $Ah^2 + 2Bhk + Ck^2$, where we have written

$$A = \frac{\partial^2 f}{\partial x^2},$$

$$B = \frac{\partial^2 f}{\partial x \partial y},$$

$$\text{and } C = \frac{\partial^2 f}{\partial y^2}.$$

Assuming that $A \neq 0$ we can write

$$Ah^2 + 2Bhk + Ck^2 = A\left(h + \frac{Bk}{A}\right)^2 + \left(C - \frac{B^2}{A}\right)k^2$$

$$= A\left(h + \frac{Bk}{A}\right)^2 + \left(\frac{\Delta}{A}\right)k^2$$

where we write $\Delta = CA - B^2$ for the discriminant. We have thus expressed the quadratic as the sum of two squares. It is thus clear that

1. If $A < 0$ and $D > 0$ we have a local maximum;
2. If $A > 0$ and $D > 0$ we have a local minimum; and
3. If $D < 0$ then the coefficients of the two squared terms have opposite signs, so by going out in two different directions, the quadratic may be made either to increase or to decrease.

Note also that we could have completed the square in the same way, but starting from the k term, rather than the h term; so the result could just as easily be stated in terms of C instead of A.

Example: Let $f(x, y) = 2x^3 - 6x^2 - 3y^2 - 6xy$. Find and classify the critical points of f. By considering $f(x, 0)$, or otherwise, show that f does not achieve a global maximum.

Solution: We have $f_x = 6x^2 - 12x - 6y$ and $f_y = -6y - 6x$. Thus critical points occur when $y = -x$ and $x^2 - x = 0$, and so at (0, 0) and (1, –1). Differentiating again, $f_{xx} = 12x - 12$, $f_{yy} = -6$ and $f_{xy} = -6$. Thus the discriminant is $\Delta = -6.(12x - 12) - 36$. When $x = 0$, $\Delta = 36 > 0$ and since $f_{xx} = -12$, we have a local maximum at (0, 0). When $x = 1$, $\Delta = -36 < 0$, so there is a saddle at (1, – 1). To see there is no global maximum, note that $f(x, 0) = 2x^3(1 - 3/x) \to \infty$ as $x \to \infty$, since $x^3 \to \infty$ as $x \to \infty$.

Example: An open-topped rectangular tank is to be constructed so that the sum of the height and the perimeter of the base is 30 metres. Find the dimensions which maximise the surface area of the tank. What is the maximum value of the surface area? [You may assume that the maximum exists, and that the corresponding dimensions of the tank are strictly positive.

Solution: Let the dimensions of the box be as shown.

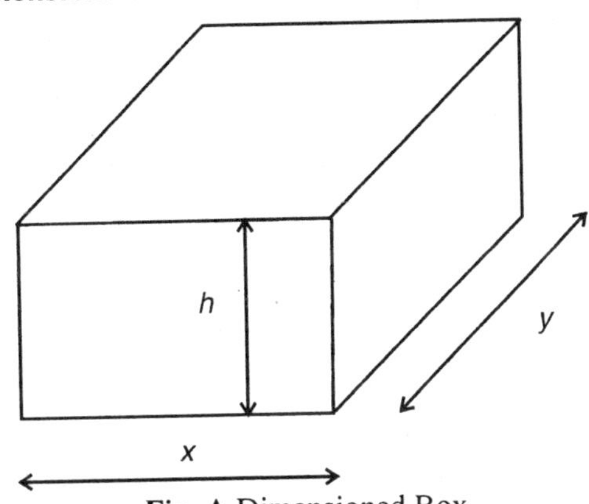

Fig. A Dimensioned Box

Let the area of the surface of the material be S. Then

$$S = 2xh + 2yh + xy,$$

and since, from our restriction on the base and height,

$$30 = 2(x + y) + h,$$

we have

$$h = 30 - 2(x + y).$$

Substituting, we have

$$S = 2(x + y)(30 - 2(x + y)) + xy = 60(x + y) - 4(x + y)^2 + xy,$$

and for physical reasons, S is defined for $x \geq 0, y \geq 0$ and $x + y \leq 15$.

A global maximum (which we are given exists) can only occur on the boundary of the domain of definition of S, or at a critical point, when $\frac{\partial S}{\partial x} = \frac{\partial S}{\partial y} = 0$. On the boundary of the domain of definition of S, we have $x = 0$ or $y = 0$ or $x + y = 15$, in which case $h = 0$. We are given that we may ignore these cases. Now

$$S = -4x^2 - 4y^2 - 7xy + 60x + 60y, \quad \text{so}$$

$$\frac{\partial S}{\partial x} = -8x - 7y + 60 = 0,$$

$$\frac{\partial S}{\partial y} = -8y - 7x + 60 = 0.$$

Subtracting gives $x = y$ and so $15x = 60$, or $x = y = 4$. Thus $h = 14$ and the surface area is $S = 16(-4 - 4 - 7 + 15 + 15) = 240$ square metres. Since we are given that a maximum exists, this must be it. [If both sides of the surface are counted, the area is doubled, but the critical proportions are still the same.

Sometimes a function necessarily has an absolute maximum and absolute minimum — in the following case because we have a continuous function defined on a closed bounded subset of ^{2}, and so the analogue of 4.35 holds. In this case exactly as in the one variable case, we need only search the boundary (using ad–hoc methods, which in fact reduce to 1-variable methods) and the critical points in the interior, using our ability to find local maxima.

Example: Find the absolute maximum and minimum values of $f(x, y) = 2 + 2x + 2y - x^2 - y^2$ on the triangular plate in the first quadrant bounded by the lines $x = 0$, $y = 0$ and $y = 9 - x$.

Solution: We know there is a global maximum, because the function is continuous on a closed bounded subset of R^2. Thus the absolute max will occur either in the interior, at a critical point, or on the boundary. If $y = 0$, investigate $f(x, 0) = 2 + 2x - x^2$, while if $x = 0$, investigate $f(0, y) = 2 + 2y - y^2$. If $y = 9 - x$, investigate

$$f(x, 9 - x) = 2 + 2x + 2(9 - x) - x^2 - (9 - x)^2$$

for an absolute maximum. In fact extreme may occur when $(x, y) = (0, 1)$ or $(1, 0)$ or $(0, 0)$ or $(9, 0)$ or $(0, 9)$, or $(9/2, 9/2)$. At these points, f takes the values -41/2, 2, 3, - 61.

Next we seek critical points in the interior of the plate,

$$f_x = 2 - 2x = 0 \quad \text{and} \quad f_y = 2 - 2y = 0.$$

so $(x, y) = (1, 1)$ and $f(1, 1) = 4$, so this must be the global maximum. Can check also using the second derivative test, that it is a *local* maximum.

Maxima and Minima

Now that we have attained a certain mastery of the problem of differentiating the elementary functions and the functions compounded from them, we are in a position to make a variety of applications. We shall consider here the simplest of these applications - the theory of maxima and minima of a function - in conjunction with a geometrical discussion of the second derivative – and then, in the next section, we shall again take up the general theory.

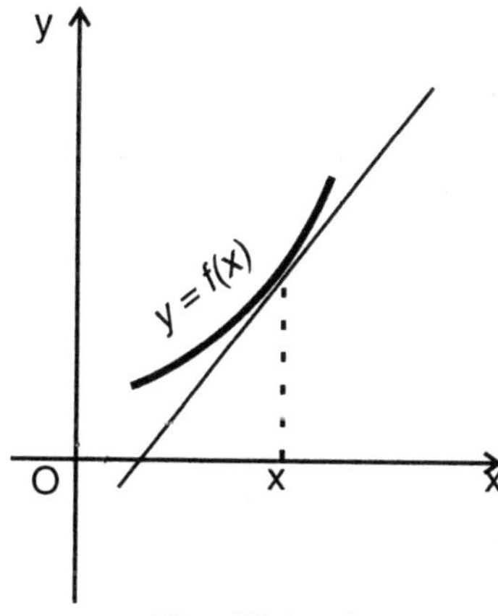

Fig. $f''(x) > 0$

Convexity or Concavity of Curves

By definition, the derivative $df(x)/dx$ of a function $f(x)$ gives the slope of the curve $y=f(x)$. This slope itself can be represented by a curve $y=df(x)/dx$ - the derived curve of the given curve. The slope of this last curve will be given by $df'(x)/dx = df(x)/dx = f''(x)$ - the second derivative of $f(x)$ – and so on. If the second derivative $f''(x)$ is positive at a point x - so that owing to continuity (which we here assume) it is positive in a certain neighbourhood of the point x - then the derivative $f'(x)$ must increase as it passes this point in the direction of increasing values of x. Hence, the curve $y = f(x)$ turns its convex side towards the direction of decreasing values of y. The opposite is true if $f'(x)$ is negative. Hence, in the first case, the curve in the neighbourhood of the point lies above the tangent, in the second case below the tangent.

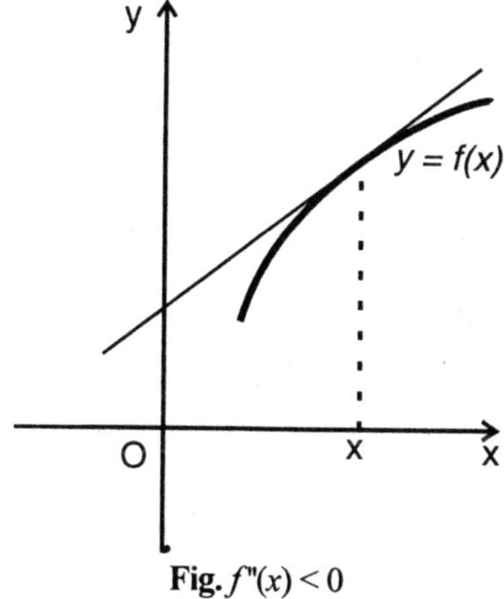

Fig. $f''(x) < 0$

Special consideration is required only in the case of points where $f''(x) = 0$. As a rule, on passing through such a point, the second derivative $f''(x)$ will change its sign. Such a point will then be a point of transition between the two cases indicated above; that is, the tangent will on one side be above the curve and on the other side below it, so that, besides touching the curve, it will also cross it. This is a point of inflection of the curve, and the corresponding tangent is called an inflectional tangent.

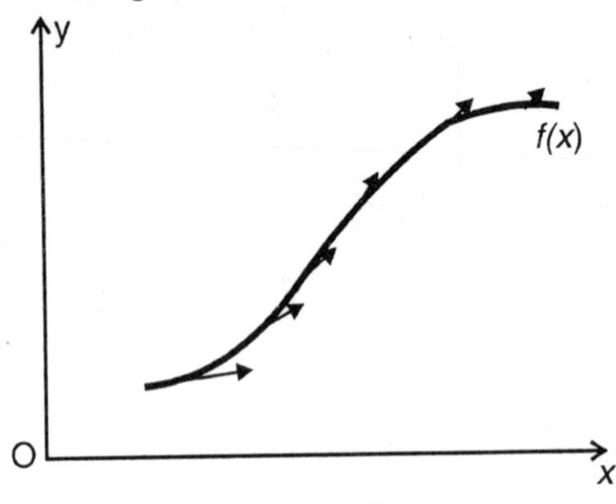

Fig. Point of Inflection

The simplest example is given by the function $y = a$, the cubical parabola, for which the x-axis itself is an inflectional tangent at the point $x = 0$. Another example is given by the function $f(x) = \sin x$, for which

$$f'(x) = d(\sin x)/dx = \cos x \text{ and } f''(x) = d(\sin x)/dx = -\sin x$$

Consequently, $f'(0) = 1$ and $f''(0) = 0$; since the sign of $f''(x)$ changes at $x = 0$, the sine curve has at the origin an inflectional tangent, inclined at an angle of 45 to the x-axis.

However, it must be noted that points can exist where $f''(x) = 0$, although the tangent does not cut the curve, but remains entirely on one side of it.

For example, the curve $y = x^4$ lies entirely above the x-axis, although the second derivative $f''(x)$ vanishes for $x = 0$.

Maxima and Minima

We say that a continuous function or a curve $f(x)$ has a maximum (minimum) at a point ξ, if in at least some neighbourhood of the point $x = \xi$ the values of the function $f(x)$ for $x \neq \xi$ are all less than $f(\xi)$ (greater than $f(\xi)$).We mean by a neighbourhood of a point an interval $\alpha < x < \beta$ which contains the point ξ in its interior.

Geometrically speaking, such maxima and minima are the wave-crests and wave-troughs of the curve, respectively. The value of the maximum at the point P_5 may very well be less than the value of the minimum at another point P_2; thus, on account of the restriction to a certain neighbourhood, the concept of maximum and minimum is always relative to some extent.

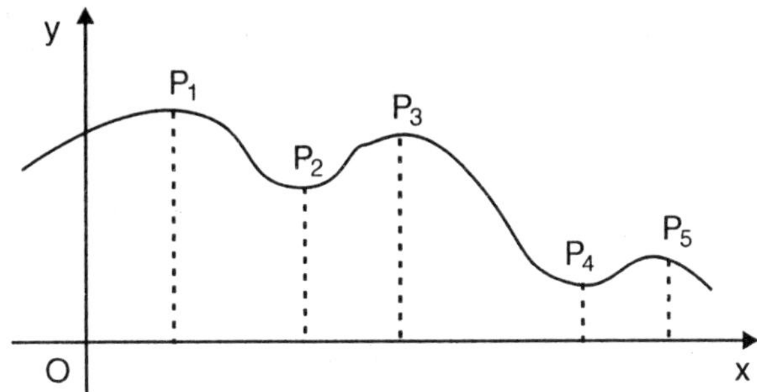

Fig. Maxima and Minima

If we wish to focus on the actual greatest or least value of a function, we must employ special means for deciding how this value is to be selected from among the maxima or minima.

Our objective at present is to find the (relative) maxima or minima, or, to use a word that covers both maxima and minima, the relative extreme values (extrema) of a given function or curve. This problem, which is *very* frequently encountered in geometry, mechanics, and physics and which occurs in many other applications, was in the Seventeenth Century one of the principal incentives for the development of the differential and integral calculus.

The expressions turning value, turning point, are also used. On the other hand, the terms stationary value, stationary point, include inflections as well as maxima and minima.

We see at once that, if a function is assumed to be differentiable, the tangent to its curve at an extreme value ξ must be horizontal. Hence, the condition

$$f'(x) = 0$$

is a necessary condition for an extreme value; by solving this equation for the unknown ξ, we obtain the points at which an extreme value may possibly occur. Our condition, however, is by no means a sufficient condition for an extreme value; there may be points at which the derivative vanishes, i.e., at which the tangent is horizontal, although the curve has there neither a maximum nor a minimum. This occurs, if at the given point the curve has a horizontal inflectional tangent cutting it, as in the above example of the function $y = x^3$ at the point $x = 0$.

However, if we have found a point at which $f'(x)$ vanishes, we may immediately conclude that the function has a maximum at that point, if $f''(\xi) < 0$, a minimum, if $f''(\xi) > 0$. In fact, in the first case, the curve in the neighbourhood of this point lies completely below the tangent, in the second case, completely above the tangent.

Instead of basing the deduction of our necessary condition on intuition, we could, of course, have given an easy proof by purely analytical methods (cf. the exactly analogous considerations for Rolle's theorem. If the function $f(x)$ has a maximum at the point ξ, then, for all sufficiently small values of h other than 0, the expression $f(\xi) - f(\xi + h)$ must be positive, whence the quotient $[f(\xi+h) - f(\xi)]/h$ will be positive or negative, according to whether h is negative or positive. Thus, if h tends to zero through negative values, the limit of this quotient cannot be negative, while if h tends to zero through positive values, the limit cannot be positive. However, since we have assumed that the derivative exists, these two limits must be equal to each other and, in fact, to $f'(\xi)$, which therefore can only have the value zero; we must have $f'(\xi) = 0$. A similar proof holds for the case of a minimum.

We can also formulate and prove analytically conditions, which are necessary and sufficient for the occurrence of a *maximum* or a minimum, without involving the second derivative. We assume that the function $f(x)$ is continuous and has a continuous derivative $f'(x)$ which vanishes only at a finite number of points.

Then $f(x)$ has a maximum or a minimum at the point $x = \xi$, if and only if the derivative $f'(x)$ changes sign on passing through this point; in particular, the function has a minimum, if the derivative is negative to the left of ξ and positive to the right of it, while, in the opposite case, it has a maximum.

We prove this by using the mean value theorem. First, we observe that there exist to the left and right of ξ intervals $\xi_1 < x < \xi$ and $\xi < x < \xi_2$ (extending to the nearest points at which $f'(x) = 0$) at the each of which $f'(x)$ has only one sign. If the signs of $f'(x)$ in these two intervals differ, then $f(\xi+h) - f(\xi) = hf'(\xi+\theta h)$ has the same sign for all numerically small values of h, whether h is positive or negative, so that $f(\xi)$ is an extreme value. If $f(x)$ has the same sign in both intervals, then $hf'(\xi+\theta h)$ changes sign with h so that $f(\xi + \theta h)$ is greater than $f(\xi)$ on one side and less than $f(\xi)$ on the other side, whence there is no extreme value. Our theorem has thus been proved.

At the same time, we see that the value $f(\xi)$ is a greatest or least value of the function in every interval, containing the point ξ, in which the only change of sign of $f'(x)$ occurs at ξ itself.

The mean value theorem, on which this proof is based, can still be used even if $f(x)$ is not differentiable at an end-point of the interval to which it is applied, provided that $f(x)$ is differentiable at all the other points of the interval; for example, the above proof still holds if $f'(x)$ does not exist at $x = \xi$. This leads us to the following more general result: If the function $f(x)$ is continuous in an interval containing the point ξ and everywhere in this interval, with the possible exception of ξ itself, has a derivative $f'(x)$ which vanishes at not more than a finite number of points, then $f(x)$ has an extreme value at the point $x = \xi$ if, and only if, the point ξ separates two intervals in which $f'(x)$ has different signs. For example, the function $y = |x|$ has a minimum at $x = 0$, since $y' > 0$ for $x > 0$ and $y' < 0$ for $x < 0$. The function $y = \sqrt[3]{x^2}$ likewise has a minimum at the point x=0, even though its derivative $2x^{-1/}$ $2/3$ is infinite there.

In addition, we make the following remark on the general theory of maxima and minima: The finding of maxima and minima is not directly equivalent to the finding of the greatest and least values of a function in a closed interval. In the case of a monotonic function, these greatest and least values will be assumed at the ends of the interval and are therefore not maxima and minima in our sense; in fact, this latter concept refers to a complete neighbourhood of the location in question. For example, the function $f(x) = x_0 \}x\} 1$ in the interval assumes its greatest value at the point $x = 1$, and its least value at $x = 0$, and a corresponding statement holds for every monotonic function. The function y=artan x with the derivative l/(l + x) is monotonic for $-\}\}x\}\}$ and has in that open interval neither a maximum nor a minimum nor a greatest or a least value.

If, after finding the zeros of $f'(x)$, we wish to make sure that we have thereby found the points at which the function has its extreme values, we can often employ the criterion:

A point ξ, at which $f'(x)$ vanishes, yields the least or greatest value of the function $f(x)$ in an entire interval, if throughout that interval $f''(x) > 0$ or $f''(x) < 0$, respectively.

In fact, if both, ξ and $\xi + h$, belong to the interval, by the mean value theorem

Hencc, at the point $x = \xi + h$, the derivative $f'(x)$ has the same sign as h or the opposite sign, according to whether $f''(x) > 0$ $f''(x) < 0$; the statement then follows from the remark following the theorem above.

Examples of Maxima and Minima

1. Of all rectangles of given area, find that with the least perimeter.

Let a be the area of the rectangle and x the length of one side (we must consider here x as ranging over the interval $0 < x < \}$), then the length of the other side is aú□/x and half the perimeter is given by

$$f(x) = x + \frac{a^2}{x},$$

We have

$$f'(x) = 1 - \frac{a^2}{x^2},\ f''(x) = \frac{2a^2}{x^3}.$$

The equation f '$(\xi) = 0$ has the single positive root $\xi = a$. For this value, $f''(x)$ is positive (as it is for any positive value of x), whence it yields the required least value and we obtain the very plausible result that of all rectangles of given area the square has the smallest perimeter.

2. Of all triangles with given base and area, find that with the least perimeter.

In order to solve this problem, we place the x-axis along the given base AB and the central point of AB as the origin. If C is the vertex of the triangle, h its altitude (which is fixed) and (x,h) are the co-ordinates of the vertex, then the sum of the two sides AC and BC of the triangle, which are to be determined, will be given by

$$f(x) = \sqrt{\{(x+a)^2 + h^2\}} + \sqrt{\{(x-a)^2 + h^2\}},$$

where $2a$ is the length of the base, whence

$$f'(x) = \frac{x+a}{\sqrt{\{(x+a)^2+h^2\}}} + \frac{x-a}{\sqrt{\{(x-a)^2+h^2\}}},$$

$$f''(x) = \frac{-(x+a)^2}{\sqrt{\{(x+a)^2+h^2\}^2}} + \frac{1}{\sqrt{\{(x+a)^2+h^2\}}} + \frac{-(x-a)^2}{\sqrt{\{(x-a)^2+h^2\}^2}} + \frac{1}{\sqrt{\{(x-a)^2+h^2\}}}$$

$$= \frac{h^2}{\sqrt{\{(x+a)^2+h^2\}^3}} + \frac{h^2}{\sqrt{\{(x-a)^2+h^2\}^3}}$$

We see at once (1) that $f'(0)$ vanishes, (2) that $f''(x)$ is always positive, whence at $x = 0$ there is a least value. In fact, since $f''(x) > 0$, the first derivative $f'(x)$ always increases and therefore cannot be equal to zero at any other point, so that the point $x = 0$ must really yield the least value of $f(x)$. Hence, this least value is given by the isosceles triangle. Similarly, we find that of all triangles with given perimeter and given base the isosceles triangle has the greatest area.

3. Find a point on a given straight line such that the sum of its distances from two given fixed points is a minimum.

Let there be given a straight line and two fixed points A and B on the same side of the line. We wish to find a point P on the straight line such that the distance $PA + PB$ has the least possible value.

We take the x-axis as the given line and use the notation of Fig.. Then the distance in question is given by

$$f(x) = \sqrt{(x^2+h^2)} + \sqrt{\{(x-a)^2+h_1^2\}},$$

and we obtain

$$f'(x) = \frac{x}{\sqrt{(x^2+h^2)}} + \frac{x-a}{\sqrt{(x-a)^2+h_1^2}},$$

$$f''(x) = \frac{-x^2}{\sqrt{(x^2+h^2)^3}} + \frac{1}{\sqrt{(x^2-h^2)}} + \frac{-(x-a)^2}{\sqrt{\{(x-a)^2+h_1^2\}^3}} + \frac{1}{\sqrt{\{(x-a)^2+h_1^2\}}}$$

$$= \frac{h^2}{\sqrt{(x^2+h^2)^3}} + \frac{h_1^2}{\sqrt{\{(x-a)^2+h_1^2\}^3}}$$

Hence. the equation $f'(\xi) = 0$ yields

$$\frac{\xi}{\sqrt{(\xi^2+h^2)}} = \frac{a-\xi}{\sqrt{\{(\xi-a)^2+h_1^2\}}} \text{ or } \cos\alpha = \cos\alpha$$

which means that the two lines PA and PB must form equal angles with the given line. The positive sign of $f''(x)$ shows us that we really have a least value.

The solution of this problem is closely linked to the optical law of reflection. By an important principle of optics known as Fermat's principle of least time, the path of a light ray is determined by the property that the time taken by the light to go from a point A to at point B under known conditions must be the least possible. If the condition is imposed that a ray of light shall on its way from A to B pass through some point on a given straight line (say, on a mirror), we see that the shortest time will be taken along the ray for which the angle of incidence is equal to the angle of reflection.

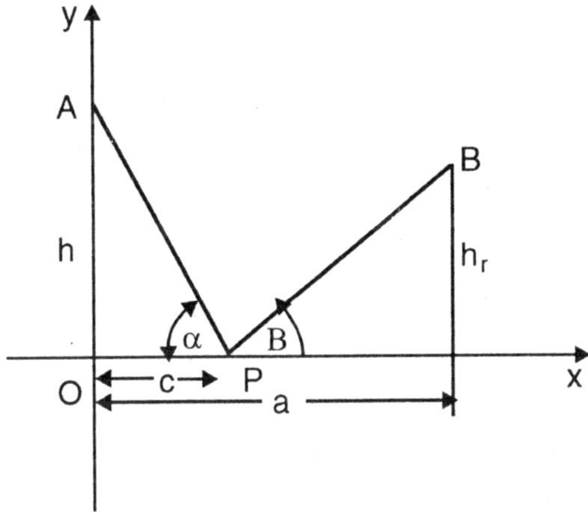

Fig. Law of Reflection

The Law of Refraction

Let there be given two points A and B on opposite sides of the x-axis. Which path from A to B corresponds to the shortest possible time, if the velocity on one side of the x-axis is c_1 and on the other side c_2?

It is clear that the shortest path must lie along two portions of straight lines meeting each other at a point P on the x-axis. Using the notation of Fig., we obtain the two expressions

$$\sqrt{(h^2+x^2)} \text{ and } \sqrt{\left(h_1^2+(a-x)^2\right)}$$

for the lengths PA, PB, respectively, and we find the time of passage along this path by dividing the lengths of the two segments by the corresponding velocities and adding the results. This yields for the time taken

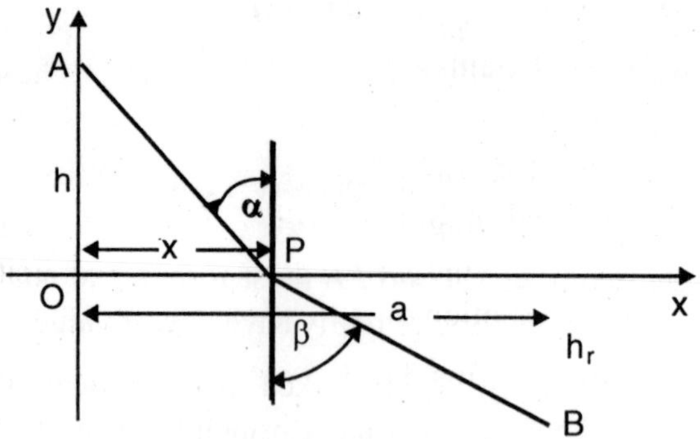

Fig. Law of Refraction

$$f(x) = \frac{1}{c_1}\sqrt{(h^2+x^2)} + \frac{1}{c_2}\sqrt{\left\{h_1^2+(a-x)^2\right\}}$$

By differentiation, we obtain

$$f'(x) = \frac{1}{c_1}\frac{x}{\sqrt{(h^2+x^2)}} - \frac{1}{c_2}\frac{a-x}{\sqrt{\left\{h_1^2+(a-x)^2\right\}}},$$

$$f''(x) = \frac{1}{c_1}\frac{h^2}{\sqrt{(h^2+x^2)^3}} - \frac{1}{c_2}\frac{h_1^2}{\sqrt{\left\{h_1^2+(a-x)^2\right\}^3}},$$

As we readily see from the figure, the equation $f'(x) = 0$, i.e.,

$$\frac{1}{c_1}\frac{x}{\sqrt{h^2+x^2}} = \frac{1}{c_2}\frac{a-x}{\left\{h_1^2+(a-x)^2\right\}},$$

is equivalent to the condition

$$\frac{1}{c_1}\sin\alpha = \frac{1}{c_2}\sin\beta, \text{or} \frac{\sin\alpha}{\sin\beta} = \frac{c_1}{c_2}$$

We leave it to the reader to prove that there is only one point which satisfies this condition and that this point actually yields the required least value. The physical meaning of our example is again given by the optical principle of least time. A ray of light travelling between two points describes the path of shortest time.

If c_1 and c_2 q are the velocities of light on either side of the boundary of two optical media, the path of the light will be that given by our result, which accordingly yields Snell's law of refraction.

MOTION, VELOCITY AND ACCELERATION

We have developed the essential theory for defining and analyzing the mathematical function. The derivative of a function was defined as the instantaneous rate of change of a function at any given point or moment. Geometrically the derivative was simply the slope of the tangent to the graph at a particular point.

As a student you are probably still a little confused by the concept of instantaneous rate of change. To strengthen your understanding of the derivative, let us scientifically analyze a physical phenomena and see how exactly the derivative is defined for the situation. The situation we will study is that of the dynamic motion of a body.

Motion is characterized by a changing distance between a reference point and the object itself. Distance is measured in the dimension of length with units of meters, miles, or feet. We know how to measure distance but how do we measure a changing distance? Whenever a dimension changes with respect to itself it constitutes an action that defines time. Thus a changing distance has to be measured relative to a change in time.

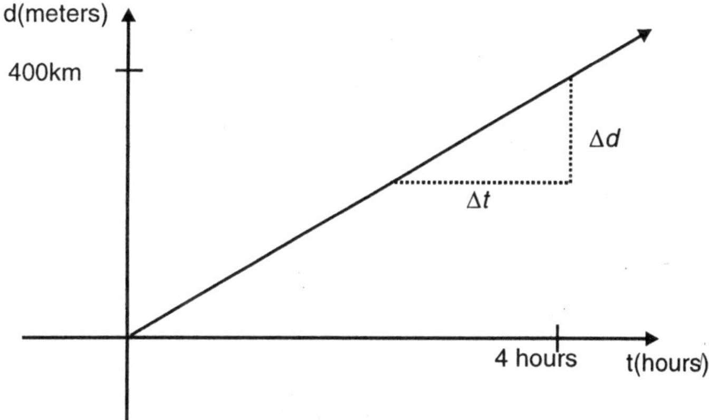

We can define velocity to be a measure of how fast an object in motion moves. Thus velocity is the rate at which the distance is changing relative to time. A simpler definition is that velocity is the distance covered per unit time. If an object moves 100 m in one second, then its velocity for that interval is $\frac{100m}{1s} = 100m/s$. The units of velocity are meters per second or change in distance per change in unit time. Consider a for a sports car moving at a constant velocity of 100 *km/hr*. If we plot graph distance covered from a reference point, we get.

The graph is a linearly increasing straight line, where the distance covered increases directly with time.

The steepness or rate of change of this graph is by definition the change in distance over the change in time $\frac{\Delta s}{\Delta t}$ where is an abbreviation for distance. Since motion is characterized by a change in position relative to time, then velocity is only defined over that definite time interval. If an object moves from a to b, the velocity of the object over this interval is this interval distance covered, $s_b - s_a$ divided by the time taken to cover this distance, $t_b - t_a$.

Thus velocity is the rate of change of distance covered with respect to time.

$$\frac{s_b - s_a}{t_b - t_a} = \frac{\Delta s}{\Delta t} = \text{velocity}$$

For example, the car above takes one hour to go from a station 400 kilometers down the highway to another station 500 kilometers down the same highway. The velocity of the car is then $\frac{\Delta d}{\Delta t}\frac{500-400}{1}$ =100km/hr

Since the graph of the car's distance covered with respect to time is a straight line, it has a constant rate of change, such that $\frac{\Delta d}{\Delta t} = v$, where v is a constant velocity. This tells us that the velocity over the entire journey is a constant or 100km/hr. The graph of this velocity function is just a straight line, or is the same at any time t.

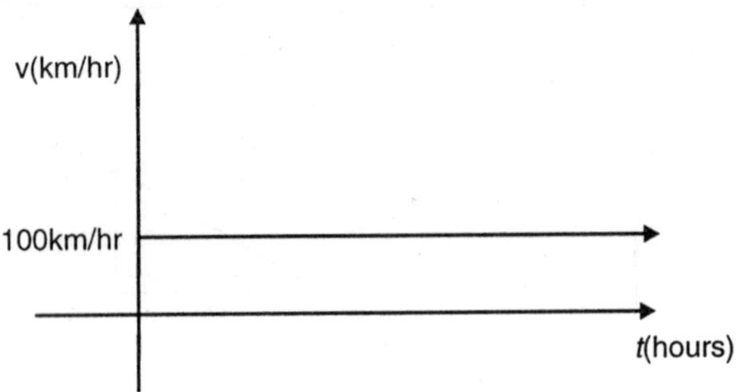

This confirms logic since throughout the journey the speedometer reads a constant 100 km./hr. Therefore, distance covered is directly related to the elapsed time, in each hour it will cover a 100 more kilometers.

Now let us define acceleration. Acceleration is the rate of change of velocity with respect to time. Acceleration is to velocity as what velocity is to distance. The concept of acceleration refers to a changing velocity per unit time. The units of acceleration are $\frac{m/s}{s} = \frac{m}{s^2}$

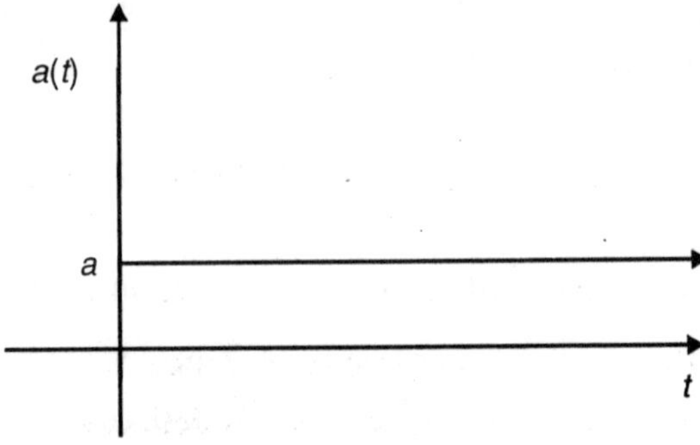

For example if the acceleration of a Ferrari is 4 km/hr per second, all this means is that each second the velocity increases or changes by 4 km/hr. If the velocity at $t = 3$ seconds is 48

km/hr, then the velocity at $t = 6$ seconds will be:

$$48\text{km/hr} + \left(4\frac{km}{hr}\Big/\text{sec}\right).(3\text{sec}) = 60\text{km/hr}$$

Lets see how this applies to a car moving with constant acceleration. Since the acceleration is constant, the graph is a horizontal line, where the acceleration of the car is at any time t is the same.

The graph of velocity as a function of time, $v(t)$ will increase directly with time for a car moving with constant acceleration.

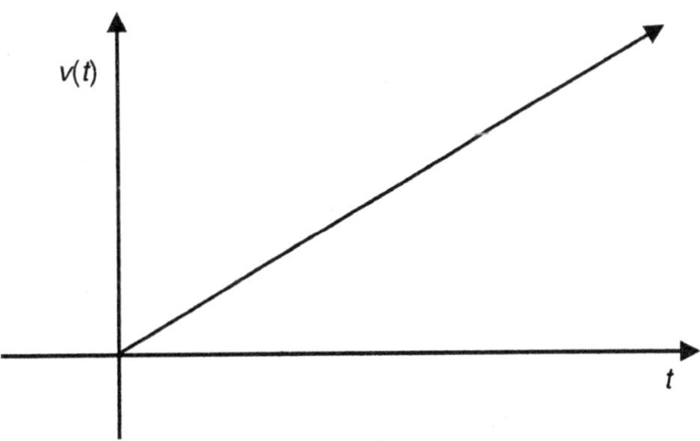

Acceleration is therefore the rate of change of velocity with respect to time or $\frac{\Delta v}{\Delta t}$. An acceleration of 4 m/s/s means that every second the velocity increases by 4 m/s.

The concept of acceleration and velocity are fairly obvious to understand when dealing with constant accelerations and velocities. We now need to define a more precise way of explaining velocities as the derivative of the position or distance function with respect to time and acceleration as the derivative of the velocity function with respect to time.

INSTANTANEOUS VELOCITY AND ACCELERATION

When an object's distance changes with time, its velocity is the rate at which the distance is changing with respect to time, while its acceleration is the rate at which the velocity is changing with respect to time. As our time interval goes to zero, the velocity and acceleration of an object take on instantaneous values at a certain moment. These instantaneous rate of changes represent the derivatives with respect to time.

To understand how the derivative relates to a moving object, consider a Porche that accelerates *from rest* at a constant rate of 15 km/hr/s from a starting point d=0. It continues at this acceleration until its velocity is 160 km/hr after which it stops accelerating and maintains its velocity.

If we freeze the moment when 4 seconds have past then its speedometer will read a velocity of exactly 60 km/hr *at that instant only*. However at $t = 4.1$ seconds the velocity will be slightly higher since the car is accelerating. This is why we use Calculus to analyze how these accelerations give rise to a changing velocity that results in a changing distance that is covered.

The graph of its distance from the starting point as a function of time is:

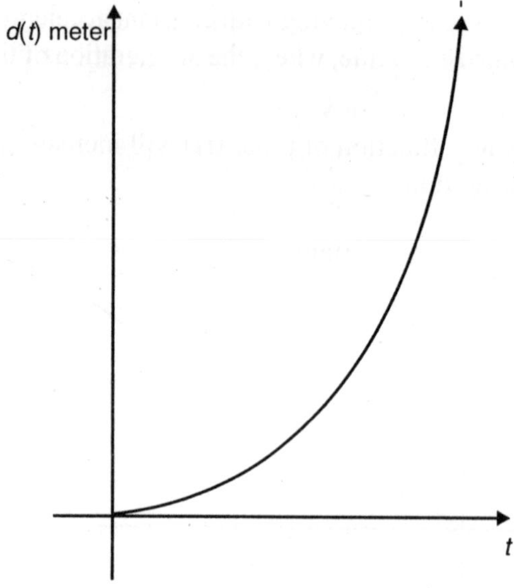

From the graph we can see that at t = 4 seconds the car has covered 2 kilometers. Since the speedometer at that moment reads 60 km/hr, then we can say that at t = 4 seconds, its velocity is a constant 60 km./hr or $\frac{\Delta d}{\Delta t}$ =182 km/hr t = 4.1 seconds; however, the velocity has changed due to the car's and the speedometer now reads, 61 km/hr..

We can assume that from t = 4.0 seconds to t = 4.1 seconds the velocity of the car is a constant 60 km/hr. By definition velocity is the distance covered divide by the time taken to cover the distance or $\frac{\Delta d}{\Delta t}$ Thus, the distance covered by the car in this small time interval, divided by the time,.1 seconds, will give us 60 km/hr.

Since the velocity of the car is increasing, due to its constant rate of acceleration, the velocity of the Porche at any instant, t, will be whatever the speedometer reads at that moment.

If we were given the relationship for distance covered as a function of time t, then velocity of the car at any time t can be found by calculating the distance covered $d(t+\Delta t)-d(t)$ over a time interval, Δt :

$$v(t) = \frac{d(t+\Delta t)-d(t)}{\Delta t}$$

Since the car is accelerating, its velocity is not constant over the interval Δt . We can assume the velocity is constant over an infinitely small time interval, Δt

Therefore we have to take the limit as Δt goes to zero to find the instantaneous rate of change of distance with respect to time. The instantaneous rate of change of distance will correspond exactly to what the speedometer reads at time t.

$$v(t) = \lim_{M \to 0} \frac{d(t+\Delta t)-d(t)}{\Delta t}$$

From the definition of the derivative:

$$d'(t) = \lim_{\Delta t \to 0} \frac{d(t+\Delta t)-d(t)}{\Delta t}$$

This leads to the extremely important result:

$$v(t) = d'(t) = \frac{dd}{dt}$$

The velocity at any time t is the instantaneous rate of change of the distance function at a time t. By definition the derivative is the instantaneous rate of change of a function over an infinitely small interval. thus the derivative of the distance function, $d'(t)$ with respect to time is the velocity function for the object $v(t)$

We now need to derive an expression for acceleration as function of time. In the same way that velocity is the rate of change of distance with respect to time, acceleration is the rate of change of velocity with respect to time.

$$a(t) = \frac{v(t+\Delta t)-v(t)}{\Delta t}$$

To find the instantaneous acceleration at any time t, we need to take the limit as Δt goes to zero. Without taking the limit, Δt is a discrete value such that the calculated acceleration is the average acceleration is for that interval. By taking the limit as $\Delta t \to 0$, we are assuming the acceleration is constant over that time interval.

$$a(t) = \lim_{\Delta x \to 0} \frac{v(t+\Delta t)-v(t)}{\Delta t}$$

$$a(t) = v'(t)$$

This proves that acceleration is the derivative of the velocity function with respect to time. Since velocity is the first derivative of the distance function with respect to time, the acceleration function is the second derivative of the distance function. In other words the acceleration function is obtained by differentiating the distance function twice. Our results can be summarized as follows:

$$v(t) = d'(t)$$
$$a(t) = v'(t) = d''(t)$$
$$a(t) = d''(t)$$